Barron's Review Course Series

Let's Review:

Earth Science

Dr. Graig

Edward J. Denecke, Jr.
Staff Development Specialist
Science Technical Assistance Center
Whitestone, New York

All inquiries should be addressed to:
Barron's Educational Series, Inc.
250 Wireless Boulevard
Hauppauge, NY 11788

Library of Congress Catalog Card No. 94-39685

International Standard Book No. 0-8120-1568-1

Library of Congress Cataloging-in-Publication Data
Denecke, Edward J.
     Let's review : earth science / Edward J. Denecke, Jr.
        p.   cm. — (Barron's review course series)
     Includes index.
     ISBN 0-8120-1568-1
     1. Earth sciences—Outlines, syllabi, etc. 2. Earth sciences—
Study and teaching (Secondary)—New York (State) I. Title. II. Series.
QE41.D46   1995
550—dc20   94-39685
                                           CIP

PRINTED IN THE UNITED STATES OF AMERICA

678   8800   15 14 13 12 11

Table Of Contents

UNIT FOUR – Weathering, Erosion, and Deposition 156

Optional/
Extended
TOPIC C –
Oceanography/Coastal Processes 211

Optional/
Extended
TOPIC D –
Glacial Geology 238

UNIT FIVE – Earth's History 267

Preface

About this Book

Earth Science is laced with the challenge and excitement of new theories, new discoveries, and new problems to be solved. It is a science in which sweeping new theories are being tested and applied to puzzling new observations. It is a science in which revolutionary advances in our knowledge of the Earth and the other planets of our solar system are being made almost daily. I hope that studying Earth Science will fill you with wonder and delight at the intricacies of our planet Earth.

Let's Review: Earth Science is designed to be used as a review text for the New York State course in Regents Earth Science on the secondary level. Because the material in this book provides such comprehensive coverage of topics in Earth Science, it can be used as a review text to supplement virtually any secondary course in Earth Science taught in the United States, using any major textbook.

The organization of this book generally follows the **Program Modification Syllabus** by the Earth Science Program Resource Innovation Team (E.S.P.R.I.T.), a dedicated group of Regents Earth Science mentors brought together by the New York State Education Department. The book is divided into nine core units and six optional/extended topics. The optional/extended topics follow the core unit from which they logically extend.

In this book, I have tried to make Earth Science as understandable to the student as possible by incorporating the following features:

- Each unit or topic begins with a brief overview (Key Ideas) and a summary of what should be learned (Key Objectives).
- Important terms are printed in *italic type* where they are defined in the text.
- Explanations of concepts and understandings are detailed, yet simply and clearly stated.
- The illustrations are designed to clarify concepts that students typically have difficulty understanding.
- Many of the illustrations are similar to those used in Regents Examination questions in order to familiarize students with the type of diagrams they will be asked to interpret on a Regents Examination.
- Each unit and topic ends with a wide range of Review Questions, including free-response type questions.

- A complete Glossary and Index make it easy for the student to find the definition of a specific term or the place in the book where the specific topic is covered.
- Three full-length Regents Examinations provide students with the opportunity to test their knowledge of Earth Science and practice their question-answering skills before taking the Regents Examination.

I wish to thank my wife Gerry for her infinite patience and my children Meredith, Abigail, and Benjamin for their loving support during the preparation of this manuscript.

UNIT ONE _____

Dimensions of the Earth

KEY IDEAS The shape and size of the Earth can be readily determined by combining Earth-based measurements with simple observations of the sky. Observations show that the Earth is a very slightly oblate sphere with an equatorial diameter of 12,757 kilometers.

The Earth has three basic components that differ in composition, density and phase of matter: the atmosphere, hydrosphere, and lithosphere.

Locating and mapping positions on the Earth's surface are essential to a wide range of human activities, from engineering to urban planning to national defense. Topographic maps represent the three-dimensional shape of the Earth's surface in two dimensions. A field is a region of space with a measurable quantity at every point. Field maps can be drawn to represent any quantity that varies in a region of space.

KEY OBJECTIVES
Upon completion of this unit, you should be able to:

- Describe the Earth's shape and explain how it can be determined by simple observations.
- Explain how the size of the Earth can be calculated from Earth-based measurements and simple observations of the sky.
- Compare and contrast the compositions, densities, and phases of matter of the lithosphere, hydrosphere, and atmosphere.
- Use the latitude and longitude coordinate system to locate points on the Earth's surface.
- Construct field maps and calculate gradient within a field.
- Read and interpret a topographic map.

A. HOW CAN THE EARTH'S SHAPE BE DETERMINED?

A-1. Nature of the Earth's Shape and Surface

The Earth is very nearly a perfect sphere. A perfect sphere has exactly the same diameter when measured in any direction. The Earth's actual measurements deviate slightly from this ideal. The polar diameter, or diameter measured from the North Pole through the center of the Earth to the South Pole, is 12,714 kilometers. The equatorial diameter, or diameter measured from a point at the Equator through the center of the Earth to the Equator, is 12,756 kilometers. Thus, the Earth "bulges" very slightly at the Equator and is also slightly "flattened" at the poles, resulting in a shape called an *oblate (flattened) spheroid.*

The Earth is so close to being a perfect sphere that its oblateness cannot be detected by the eye. If you were to view the earth from space, it would appear perfectly round, and any cross section of the Earth looks like a perfect circle. To give you an idea of how close to being perfectly round the Earth is, let's suppose we made a scale model of the Earth—a globe. If we used a scale of 1 centimeter = 1,000 kilometers, we would end up with a globe that had a polar diameter of 12.714 centimeters and an equatorial diameter of 12.756 centimeters—a little bigger than a softball. The difference in diameters would be 0.042 centimeter, or less than 0.5 millimeter! You would need a micrometer to measure the difference in diameters.

Look carefully at the circle in Figure 1.1. Its polar diameter is 6.357 centimeters and its equatorial diameter is 6.378 centimeters—both measurements exactly half the size of those for the globe described above. Can you tell that the circle is not perfectly round?

The Earth's oblateness is the result of forces produced by the Earth's rotation on its axis. Just as a loose skirt will swirl outward if the person wearing it spins around, the Earth "swirls" outward when it rotates. However, since the Earth is much stiffer than a skirt, the distance it moves outward is much less.

In addition to being round, the Earth is very smooth. In comparison to its diameter, the irregulari-

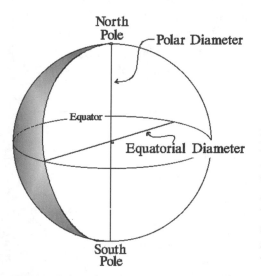

Figure 1.1 Schematic Diagram of Earth, Showing Polar and Equatorial Diameters.

ties of the Earth's surface (mountains, valleys, and ocean basins) are relatively small. Mount Everest protrudes 8,848 meters, or about 8.8 kilometers, above sea level. That height is about 7/10,000 of the Earth's diameter (8.8 km/12,756 km). If we were to draw Mount Everest to scale on the circle in Figure 1.1, it would stick out of the surface less than 1/20 of a millimeter. Mark off a millimeter on a blank piece of paper, and then try to make dots small enough so that 20 will fit between the lines. If you put one of those dots on the outside of the circle, it will stick out of the circle as much as Mount Everest rises above the Earth. As you can see, the dot barely affects the smoothness of the circle.

Now consider that most of the protrusions and indentations on the Earth's surface are much smaller than Mount Everest, and you can see why the Earth appears so smooth. The only reason mountains and valleys seem so high and so deep to us humans is that we are so tiny in comparison to the Earth. If you were to draw a human to scale on our model globe, you would need a powerful microscope to see him or her.

A-2. Evidence of the Earth's Shape

The true shape of the Earth is a question that has captured the minds and imaginations of humans for thousands of years. Records dating back to the sixth century B.C. indicate that Pythagoras theorized that the universe was a giant sphere and that all the objects in it—the Earth, Sun, Moon, and planets—were also spheres. This giant sphere had the Earth at its center, with the Sun, Moon, and planets moving in concentric circles around the Earth. In the fourth century B.C., a follower of Pythagoras named Aristarchus of Samos supposed that "the fixed stars and the Sun are immovable, but that the Earth is carried around the Sun in a circle." Aristarchus was correct, but his radical idea was not accepted and it died with him.

The concept of a spherical Earth gained some acceptance in the third century B.C. when Aristotle, the teacher of Alexander the Great, concluded that the Earth is a sphere. Aristotle based his conclusion upon simple observations of the sky and on Earth-based measurements. This use of observation to support his ideas, rather than merely stating them as opinions, places Aristotle among the first scientists. First, Aristotle observed that the shadow of the Earth during a lunar eclipse is definitely round. Although this could happen also if the Earth were a cone or a cylinder, additional evidence indicated a sphere. Travelers reported that, as they went farther north or south, some stars appeared lower and lower in the sky. Furthermore, Aristotle noted that ships disappeared bow-first over the horizon, no matter in which direction they sailed. These observations could occur only if persons or objects were constantly moving along a curved surface that changed the angle of view.

Today, a number of other observations also provide evidence of the Earth's shape.

Photographs Taken from Space

The most direct evidence is provided by photographs of the Earth taken from space. These photographs can be analyzed, and precise measurements of the Earth's image in the photographs can be made. These measurements indicate that the Earth's polar and equatorial diameters are indeed slightly different and confirm that its shape is an oblate spheroid.

Observations of the Altitude of Polaris (the North Star)

Observations of Polaris, a distant star, provide information that can be interpreted, using simple geometry, to provide evidence of the Earth's spherical shape and its oblateness. Polaris is located almost exactly over the North Pole and almost in line with the Earth's axis of rotation. Polaris is very far from the Earth, literally trillions of times the Earth's diameter away. To observers at the farthest edges of the Earth, the direction to Polaris varies by only a minute fraction of a degree (see Figure 1.2).

The direction to Polaris from all points on earth is not measurably different.

Figure 1.2 Direction to Polaris.

The *altitude* of Polaris is the angle between the star and the horizon, with the observer at the vertex, as shown in Figure 1.3. The *latitude* of an observer is the angle of the observer north or south of the Equator with the vertex at the Earth's center, as shown in Figure 1.3(b).

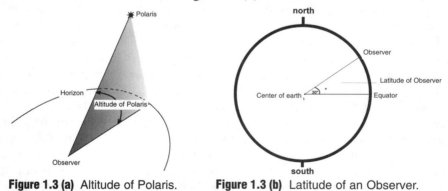

Figure 1.3 (a) Altitude of Polaris.   **Figure 1.3 (b)** Latitude of an Observer.

If the Earth were a flat disc, the altitude of Polaris would always be almost exactly the same at any location and at all times (see Figure 1.4).

4

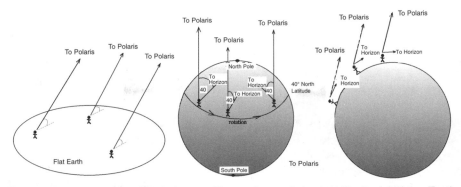

Figure 1.4 The Altitude of Polaris on a Flat and on a Spherical Earth. (a) If the Earth were a flat disc, the altitude of Polaris would always be almost exactly the same at any location and at all times. (b) To a fixed observer on a spherical earth, the altitude of Polaris does not appear to change as the earth rotates. (c) However, if the observer travels north or south, the altitude does change, a result that would be expected if the earth were spherical.

It is true that, to a fixed observer, the altitude of Polaris does not appear to change as the Earth rotates (Figure 1.4b). If the observer travels north or south however, the altitude does change, a result that would be expected if the Earth were spherical (see Figure 1.4c).

If the Earth were a perfect sphere, then the altitude of Polaris would be the same as an observer's latitude and the distance traveled to change the altitude of Polaris by 1° would always be the same (see Figure 1.5).

What is observed, however, is that the altitude of Polaris is not always exactly equal to the observer's latitude, and the distance traveled to change the altitude of Polaris by 1° also varies. This is evidence that the earth is an oblate spheroid (see Figure 1.6).

Measurements of gravity

Gravity is the force of attraction that exists among all objects. The gravity between any two objects is proportional to the masses of the two objects and inversely proportional to the square of the distance between the centers of the objects. In simple terms, the more massive the objects, and the closer the cen-

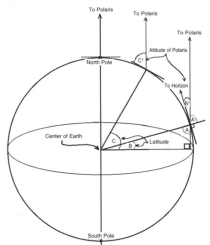

Figure 1.5 Relationship of Altitude to Latitude of Observer. By simple geometry it can be shown that the altitude of Polaris is the same as an observer's latitude. Since the sum of the angles of a triangle is equal to 180°, $\angle A + \angle B + 90° = 180°$. similarly, since a straight line has a measure of 180°, $\angle A' + \angle B' + 90° = 180°$. $\angle A = \angle A'$ because they are alternate interior angles. Therefore, $\angle B$ (latitude) = $\angle B'$ (altitude of Polaris).

5

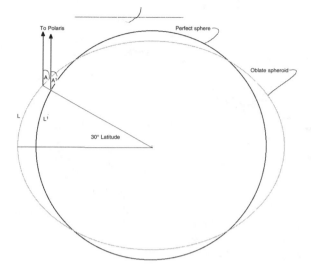

Figure 1.6 Altitude Differences at Same Latitude on a Perfect Sphere and an Oblate Spheroid. For the same latitude, the altitude of Polaris on a perfect sphere (*A'*) differs slightly from what it would be on an oblate spheroid (*A*). Similarly, for an identical change in latitude the distance covered on a perfect sphere (*L'*) would differ from the distance covered on an oblate spheroid (*L*).

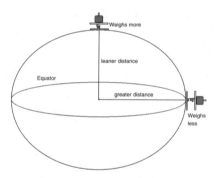

Figure 1.7 Differences in Weight of the Same Object at the Equator and near the Poles.

ters of the objects, the stronger the gravitational attraction between them.

The gravitational attraction between the Earth and an object can be measured with a spring scale and is called the object's *weight*. If the Earth were perfectly spherical, the distance between its center and any point on its surface would always be the same. Therefore, if we measured the weight of a specific object at several points on the surface of the Earth, the weight should always be the same.

If the Earth is an oblate spheroid, however, the distance between the center of an object and the center of the Earth will be greater at the Equator than at the poles. If the distance is greater, the gravitational attraction and the weight are less. Therefore, if we measured the weight of a specific object at the Equator and again near the poles, it should weigh less at the Equator and more near the poles. Even taking into account the outward force on an object at the Equator due to the Earth's rotation, this difference in weight is exactly what is observed, as shown in Figure 1.7, and is further evidence of the Earth's oblateness.

B. HOW CAN THE EARTH'S SIZE BE DETERMINED?

Modern measuring instruments based on lasers allow us to measure the Earth with great precision. However, the Earth's size was estimated quite accurately more than 2,000 years ago. The principle is a simple one: If the Earth is a sphere, then its circumference is a circle, and all of the mathematical relationships among the various parts of a circle pertain to the earth.

These relationships make it possible to determine the Earth's circumference without measuring the entire circumference directly. If the circumference is a circle, it consists of 360°, and a 1° angle will intersect 1/360 of the circumference. Therefore, if two observers at two locations on a north-south line simultaneously determine the altitude of Polaris to differ by 1°, the Earth's circumference must be 360 times the distance between these locations.

In 235 B.C., Eratosthenes used this technique but substituted the Sun for a star. Eratosthenes lived in Alexandria and had reliable reports that on June 21, the summer solstice, the Sun was directly overhead at midday in the city of Syene. Eratosthenes believed that Alexandria was due north of Syene, and he knew that a camel caravan traveling 100 stades a day typically took 50 days to get to Syene. This meant that Syene was 50 × 100 or 5,000 stades from Alexandria. The final piece of information he needed was the angle of the Sun at midday on June 21 in Alexandria. Using a gnomon (a vertical column), he determined the angle to be 7°12′, or about 1/50 of 360°. Therefore the distance between Syene and Alexandria was about 1/50 of the Earth's circumference—5,000 stades × 50 = 250,000 stades. The problem is that there is some dispute as to exactly how long a stade is. The likely figure is about 0.1 mile, making Eratosthenes's estimate of the Earth's circumference to be 25,000 miles (40,230 km), compared to the actual figure of 24,862 miles (40, 010 km)—a remarkable achievement for his time (see Figure 1.8).

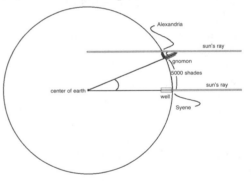

Figure 1.8 Eratosthenes's Method of Determining Earth's Circumference.

Once the Earth's circumference is known, many of its other dimensions can be calculated:

C = circumference, D = diameter, r = radius, A = surface area, V = volume, and π(pi) = 3.1462,

where

$C = \pi D$, $D = 2r$, $A = 4\pi r^2$, $V = {}^{4}/_{3}\pi r^3$

C. HOW ARE THE SEVERAL SPHERES OF THE EARTH SYSTEM RELATED?

Looking back at the Earth from Apollo 8 as it orbited the Moon, astronaut William Anders said, "We saw the beautiful orb of our planet coming up over this relatively stark, inhospitable lunar horizon and it brought back to me that... it was really the Earth that was most important to us." The photographs of the Earth taken by the Apollo 8 astronauts from their vantage point near the Moon show a hauntingly beautiful sphere, with deep blue seas and green lands veiled in swirling white clouds suspended against the black backdrop of space. These images reveal much about the structure of the Earth and its components.

The Earth does not consist of a single material, but has distinct layers like an onion. Unlike an onion, however, each layer has a different composition. The Apollo 8 pictures show three of these layers quite clearly: a solid surface, mostly covered by a layer of water, all surrounded by a gaseous envelope. All three components are spherical in shape and are called, respectively, the *lithosphere, hydrosphere,* and *atmosphere.*

C-1. The Lithosphere

Anywhere we examine the Earth's surface we find solid rock or pieces of solid rock. Therefore, we think of the entire Earth as a solid, but that impression is not correct. Consider a volcanic eruption. Clearly, liquid rock is emerging from beneath the surface of the Earth, so at least part of what lies beneath the surface is not solid.

Extensive investigation has revealed that the Earth's interior has a layered structure (this topic will be discussed in detail in Unit 3) and that the outermost layer of solid, brittle rock covers a more plastic interior. The lithosphere is Earth's solid outer layer of soil and rock, extending from the surface to a depth of about 100 kilometers. It is made up mainly of compounds of about eight to ten abundant elements, such as silicon, oxygen, calcium, potassium, aluminum, iron, magnesium, and sodium.

C-2. The Hydrosphere

The hydrosphere consists of a thin layer of water that rests upon the lithosphere. It includes all the Earth's oceans, lakes, streams, underground water and ice. A tiny fraction exists in the atmosphere as water vapor. The hydrosphere covers about 71 percent of the lithosphere's surface to an average depth of about 3.8 kilometers, which is very thin compared to the diameter of the Earth. If you dipped a basketball in water, the water wetting its surface would be deeper in many places than the hydrosphere is on the Earth. Since the hydrosphere is water, it consists of the elements hydrogen and oxygen.

The hydrosphere plays a crucial role in many geologic processes. Water is the chief transporter of loose rock and the main shaper of the Earth's surface. The oceans are a vast reservoir for most of the soluble materials on Earth and act as heat absorbers, keeping temperatures from fluctuating too drastically. Water is also essential to living things, not only as drinking water, but as the very stuff of which their cells are composed.

C-3. The Atmosphere

The atmosphere is a thin shell bound to the Earth by gravity. It consists mostly of gases but also contains water, ice, dust, and other particles. This mixture of gases and other substances is called air (see Figure 1.9).

Dry air is composed mainly of nitrogen and oxygen, but contains traces of other gases as well (see Table 1.1). The water vapor content of air may range from 0 percent over deserts and polar ice caps to as much as 4 percent in a tropical jungle.

As you travel upward through the atmosphere, the temperature changes. The atmosphere is divided into four distinct layers according to temperature: the *troposphere*, *stratosphere*, *mesosphere*, and *thermosphere*. Figure 1.10 shows the changes in temperature that occur in the atmosphere. Notice that, near the ground, the temperature decreases with altitude. At about 11 kilometers it stops getting colder and starts to become warmer. This change signals that you have crossed the boundary from one layer to another—in this case from troposphere to stratosphere. The boundary is called the *tropopause* because, at the top of the troposphere, the temperature stops decreasing, or pauses. You can see from the temperature graph in Figure 1.10 that such a change happens twice more, at the *stratopause* and *mesopause*.

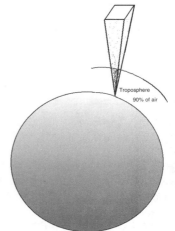

Figure 1.9 A Sectional Slice through Earth's Atmosphere. Most of the gas molecules in the atmosphere are clustered near the Earth's surface forming the troposphere. Only the most energetic molecules can overcome gravity and reach higher levels in the atmosphere.

TABLE 1.1 COMPOSITION OF DRY AIR BY VOLUME

Gas	Percent by Volume in Dry Air
Nitrogen	78
Oxygen	21
Argon	Almost 1
Neon	Trace
Helium	Trace
Krypton	Trace
Xenon	Trace
Hydrogen	Trace
Ozone	Trace
Carbon dioxide	Trace
Nitrous oxide	Trace
Methane	Trace

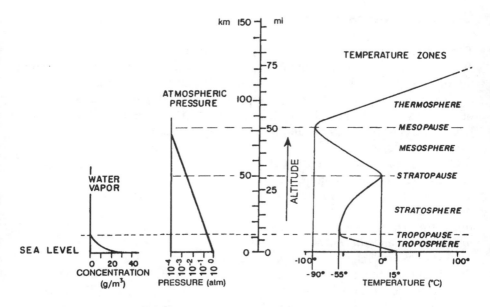

Figure 1.10 Selected Properties of Earth's Atmosphere
Source: The State Education Department, *Earth Science Reference Tables,* 1994 ed. (Albany, New York, The University of the State of New York).

These changes in temperature are the result of differences in the composition of the different layers. The upper stratosphere contains a gas called *ozone*. Ozone absorbs ultraviolet light from the Sun, and this warms the stratosphere. Once past the ozone, the temperature decreases again in the mesosphere. In the thermosphere, gases and charged particles absorb ultraviolet light and other wavelengths of energy and turn them into heat energy; once again, temperatures rise. In fact, the amount of energy absorbed by this outermost layer is so great that temperatures exceed the boiling point of water. Some gas molecules absorb so much energy that they lose or gain electrons and become ions, that is, charged particles. These ions in the thermosphere form a layer called the *ionosphere*, which can reflect radio waves.

As you can see from the atmospheric pressure graph in Figure 1.10, most of the air in the atmosphere and all of the water vapor are confined to the troposphere. Air exerts pressure in proportion to the amount of air that is present. At the tropopause, air exerts only a tenth of the pressure that it does at sea level. This means that there is only about a tenth as much air! At the stratopause there is only about a thousandth, and by the mesopause less than one ten-thousandth of the air that exists at sea level. The atmosphere extends for hundreds of kilometers above the lithosphere and hydrosphere, but most of the air is confined to the 8–10 kilometers closest to the surface of these layers.

C-4. The Positions of the Lithosphere, Hydrosphere, and Atmosphere

The relative positions of the lithosphere, hydrosphere, and atmosphere can be explained in terms of density differences. Rock is a solid with an average density in the range of 2.8–3.0 grams per cubic centimeter. Water is a liquid with a density of about 1 gram per cubic centimeter. Air is a mixture of gases with a density of 1.3×10^{-3} grams per cubic centimeter. In a mixture of such materials, it is expected that the materials will separate into layers with the densest at the bottom and the least dense on top.

D. HOW CAN SURFACE FEATURES OF THE EARTH BE MAPPED?

A *map* is a model of the Earth's surface. Although models are different from the real thing, they are tools for learning about the things they represent. A map is meant to communicate a sense of place, of where one point is in relation to another. A map can be anything from a quick sketch showing a friend how to get to the park from school to an elaborate scale model of the Earth complete with mountain ranges and ocean basins. The nature of a map depends on the purpose for which it was created.

D-1. Coordinate Systems

Geometric shapes and relationships can be described in terms of symbols and numbers. For example, the position of any point on a map or any other flat surface can be described by two numbers, or *coordinates*. The coordinates may be two distances from a point, an angle and a distance from a point, two angles from a point, and so on (see Figure 1-11). The particular method used to locate a point by means of coordinates is called a coordinate system. Often a coordinate system is represented by a series of intersecting lines, or a grid, on which any point can be described by the two lines that intersect there. The system commonly used to locate points on the Earth's surface is the latitude-longitude system. This is not the only system that has ever been devised, but it is the most widely used at present.

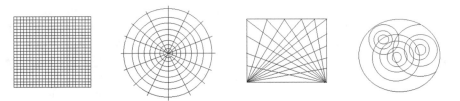

Figure 1.11 Various Coordinate Systems

The Latitude-Longitude Coordinate System

The latitude-longitude system consists of two sets of lines that cross each other. *Latitude* lines run in an east-west direction; *longitude* lines, in a north-south direction. This type of system and the words *latitude* and *longitude* were already being used to describe such lines at the time of the Roman scholar Claudius Ptolemy in A.D. 150. To draw these lines on a globe or map it is necessary to have starting points, or points of reference.

One point of reference used in the latitude-longitude system is the Equator, which an east-west line midway between the North and South poles. Since the Earth is a sphere, lines that run east-west, such as the Equator, actually form circles. The circles formed by lines of latitude are called *parallels*, because if you drew a series of east-west lines they would all be parallel to one another. The location parallels are described by their angular distances north or south of the equator. Latitude is 0° at the Equator and 90° at the poles. Notice, though, that the farther the latitude from the Equator, the smaller the circle. The east-west line at 60° north of the equator is only half as long as the Equator (see Figure 1.12).

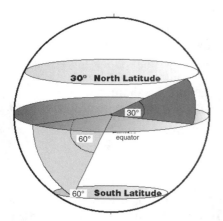

A second point of reference used in the latitude-longitude system is the *prime meridian*. North-south lines, or lines of longitude, are commonly called meridians of longitude, or simply *meridians*. The zero meridian, or starting point for measuring the angular distance east or west of any other meridian, is the prime meridian. Each meridian goes only halfway around the Earth—from pole to pole—and has a twin on the other side of the Earth.. The pairs of meridians form circles, called *great circles*, that cut the Earth into two equal halves, or hemispheres. Only one parallel, the Equator, is a great circle.

Figure 1.12 Cutaway drawing of the Earth, Demonstrating How Latitude Is Determined. It shows that latitude is a measure of the angle between the plane of the Equator and lines projected from the center of the Earth.

Unlike the Equator, which is the only line halfway between the poles, the prime meridian is an arbitrarily chosen line and could be almost anywhere. Over the years, on different maps, it *has* been located in different places. A map maker usually began numbering the lines of longitude at whichever meridian passed through the site of his national observatory. As recently as 1881, 14 different prime meridians were being used on topographic survey maps. In 1884, however, the United States hosted an international conference in Washington whose purpose was to agree on a prime meridian. This conference agreed that the meridian that runs through the Royal Observatory in

Greenwich, England, would be the prime meridian. Longitude would be measured in degrees east or west of this reference line up to 180°. East longitude would be plus, and west longitude would be minus. This choice was made mainly because the prime meridian's twin, the 180° meridian, runs through the Pacific Ocean and touches almost no habitable land. Thus, it is a good choice for an international date line, a line marking the transition from one date to the next.

Measuring Latitude and Longitude

Both the latitude and the longitude of a location can be determined by making simple observations. As described earlier, an observer in the Northern

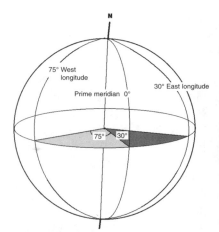

Figure 1.13 Cutaway Drawing of the Earth, Showing How Longitude Is Determined. Longitude is seen to be a measure of the angle between the planes of two meridian semicircles, one of which is the prime meridian.

Hemisphere can determine his or her latitude by measuring the altitude of Polaris (see Figure 1.5). Observers in the Southern Hemisphere use the positions of other stars and make corrections for their deviation from the Earth's axis of rotation.

Longitude is measured by observing time and the position of the Sun. Local noon is the time at which the Sun is at its highest altitude of the day. At that moment, a line drawn from the Sun to the center of the Earth will pass through that location's meridian. A few moments later, the Earth will have rotated from east to west and another meridian will align with the Sun and experience local noon. Thus, different longitudes will experience local noon at different times.

Since the Earth makes one complete 360° rotation every 24 hours, we can calculate the relationship between time and degrees rotated. For example, if the Earth takes 24 hours to rotate 360°, in 1 hour it will rotate 360/24, or 15°. Now suppose you were at a location and it was exactly local noon. One hour later the earth would have rotated 15°, and a location 15° of longitude away would be experiencing local noon. Two hours later, the earth would have rotated 30°, and a location 30° of longitude away would experience local noon. By comparing the difference in time between the local noons at any two locations, you can easily calculate their difference in longitude; every hour of difference equals 15° of longitude. But this tells us only a location's longitude relative to another point, not its actual longitude.

Suppose, however, that one of the times you knew was the time at which local noon occurred in *Greenwich*! Then you would know the difference in the longitude of your location from *zero* longitude—and that would be your

13

actual longitude. So, in order to measure longitude you need to know two times—local noon and the time of local noon in Greenwich. You already know that each hour of difference in the time of local noon equals 15° of longitude. Since the Earth rotates from west to east, locations east of Greenwich will experience local noon earlier than Greenwich, while those west of Greenwich will experience local noon later than Greenwich. For example, if Old Forge, New York, experiences local noon about 5 hours later than Greenwich, England, its longitude is 5 × 15°, or 75°. Also, since local noon in Old Forge occurs after local noon in Greenwich, the 75° is west longitude.

To measure longitude, then, you need a clock, or radio broadcast, that tells you what time it is in Greenwich when you experience local noon. You can then look up the time of local noon in Greenwich on that day in an astronomical table and, using the difference in times, calculate longitude. The need for such clocks resulted in the development of ever more precise timekeeping devices, culminating with today's atomic clocks.

D-2 Field Maps

A *field* is a region of space that has a measurable quantity at every point. Some examples of field quantities are temperature, pressure, magnetism, gravity, and elevation. A *field map* can be used to represent any quantity that varies in a region of space. One way to represent field quantities on a two-dimensional field map is to use isolines. *Isolines* connect points of equal field values. For example, a temperature field map would contain lines connecting points of equal temperature, or isotherms (see Figure 1.14).

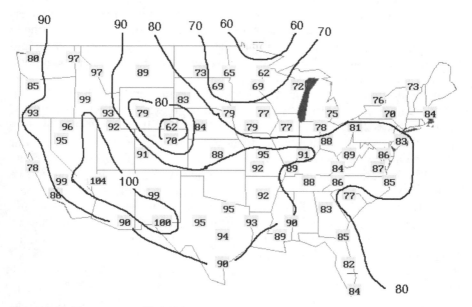

Figure 1.14 Temperature Field Map

14

Within a field, the field value changes as you move from place to place. The rate at which the field value changes is called its *gradient*. Gradient can be calculated as follows:

$$\text{Gradient} = \frac{\text{Amount of change in field value}}{\text{Distance through which change occurs}}$$

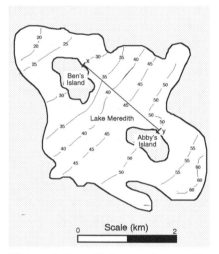

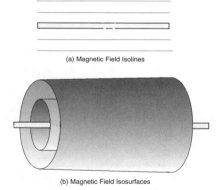

Figure 1.15 Pollution Levels in Meredith Lake.

Figure 1.16 Magnetic Field Isolines and Isosurfaces Around a Wire Carrying Electric Current.

For example, Figure 1.15 shows the pollutant levels that were measured in a large lake. Isolines connect points of equal pollutant concentration, measured in parts per billion. Abby's Island and Ben's Island are 2 kilometers apart, and the pollutant concentration between them changes from 30 to 50 parts per billion. The rate of change, or gradient, is 20 parts per billion per 2 kilometers or 10 parts per billion per kilometer.

Field maps can also be represented in three dimensions by using isosurfaces. Isolines show field values, for example, air pressure at ground level, in a particular two-dimensional plane. An *isosurface* is a three-dimensional surface on which every point has the same field value. Isosurfaces provide insights into the nature of a field that may not be evident in a two-dimensional map (see Figure 1.16).

Fields seldom remain unchanged over time. A field map shows a field at a particular point in time—the time at which its values were measured. For example, the temperature field over the United States changes drastically from July to December. The Earth's magnetic field has changed position many times since the Earth formed. Therefore, field maps need to be updated periodically in order to represent current conditions. On weather maps, which represent rapidly changing fields, values are updated as often as every hour.

D-3. Topographic Maps

A *topographic map* is a scale model of a part of the Earth's surface that shows its three-dimensional shape in two dimensions. A topographic map is actually a type of field map in which the field value is elevation above sea level of points on the Earth's surface. The isolines that connect points of equal elevation are called *contour lines* because they represent the shape, or contours, of the Earth's surface. A contour line shows the shape that would be formed if the land surface were sliced by a horizontal plane at a particular elevation above sea level (see Figure 1.17). Successive lines are separated by an unvarying vertical distance called the *contour interval*. Contour lines around an enclosed depression (see Figure 1.18) are shown with hatchure marks pointing into the depression in order to distinguish them from small hills.

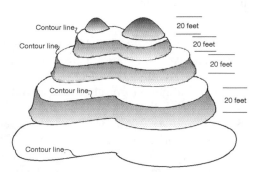

Figure 1.17 Mountain Sliced by Horizontal Plane, Showing Contour Intervals.

Map Symbols

A topographic map provides a view of the ground as seen from vertically above. Surface features are represented by map symbols. The four most common are blue color for bodies of water, black and red for human-made structures, and brown for contour lines and other relief symbols. An extensive list of map symbols is given in Figure 1.19.

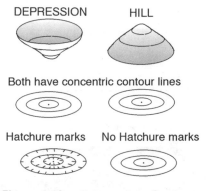

Figure 1.18 Depression Contours Versus Regular Contours.

Map Scale

A map scale is the ratio between the distance shown on a map and the actual distance on the ground. In this ratio, such as 1:100,000, both numbers have the same units, but these units can be anything. For example, the ratio 1:100,000 may mean that 1 centimeter on the map equals 100,000 centimeters on the ground, or that 1 inch on the map equals 100,000 inches on the ground. On most topographic maps, the map scale is also represented by a graphic scale such as the one shown in Figure 1.20. The graphic scale can be used as a ruler to measure distances on the map.

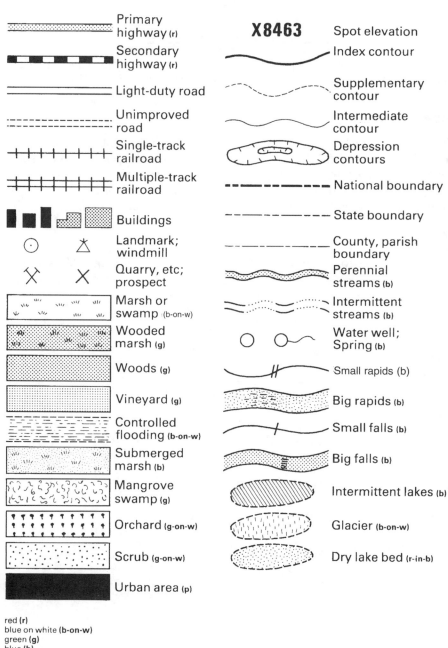

Primary highway (r)	
Secondary highway (r)	
Light-duty road	
Unimproved road	
Single-track railroad	
Multiple-track railroad	
Buildings	
Landmark; windmill	
Quarry, etc; prospect	
Marsh or swamp (b-on-w)	
Wooded marsh (g)	
Woods (g)	
Vineyard (g)	
Controlled flooding (b-on-w)	
Submerged marsh (b)	
Mangrove swamp (g)	
Orchard (g-on-w)	
Scrub (g-on-w)	
Urban area (p)	

X8463 Spot elevation

Index contour

Supplementary contour

Intermediate contour

Depression contours

National boundary

State boundary

County, parish boundary

Perennial streams (b)

Intermittent streams (b)

Water well; Spring (b)

Small rapids (b)

Big rapids (b)

Small falls (b)

Big falls (b)

Intermittent lakes (b)

Glacier (b-on-w)

Dry lake bed (r-in-b)

red (r)
blue on white (b-on-w)
green (g)
blue (b)
green on white (g-on-w)
pink (p)
red in blue (r-in-b)

Figure 1.19 Selected Topographic Map Symbols.
From: Earth Science on FIle. © Facts on File, Inc., 1988.

Figure 1.20 Legend on a Topographic Map. Included are the name of the map quadrangle, the contour interval, the map scale stated as a ratio, and a graphic map scale.

Map Direction

The convention used in most topographic maps is that the top of the map is north. However, since north is determined with a compass, and geographic north differs from magnetic north, most topographic maps include arrows showing both. The map in Figure 1.21 shows magnetic north with an arrow labeled "MN" and geographic north with an arrow labeled "GN".

Map Profiles

It is often useful to construct a profile from a topographic map. A profile is what a cross section of the land would look like between two points. Figure

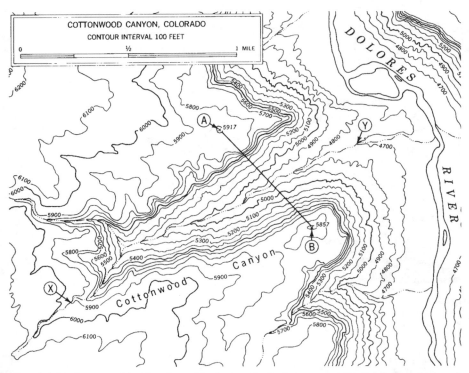

Figure 1.21 Topographic Map from *Exploring Earth Science*, by Walter A. Thurber and Robert E. Kilburn, Allyn and Bacon, 1965. Used with permission of Prentice-Hall.

1.22 shows how a profile can be constructed. First, the two points between which the profile is to be drawn are chosen, and a line is drawn to connect them. Next, a piece of paper is placed along the line, and the edge of the paper is marked wherever it intersects a contour line (Figure 1.22a). This paper is then placed against a piece of graph paper on which a vertical scale has been marked to match the contour lines intersected. At each point where a contour line crosses the edge of the paper, a line of the appropriate height is drawn on the graph paper. Finally, the endpoints of the lines are connected in a smooth line to form the finished profile (Figure 1.22b).

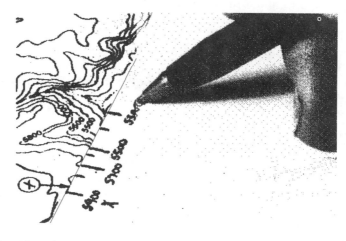

a. Place the edge of a sheet of scrap paper along the line from *X* to *Y* and mark off the places where contour lines cross the edge of the paper. Label the elevation of the contour lines on the scrap paper.

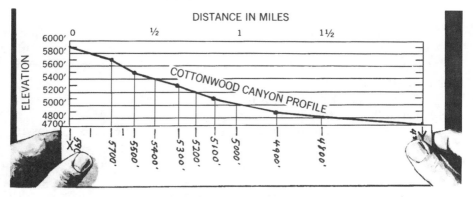

b. Place the scrap paper along graph paper which has a vertical scale marked on it and draw lines at the appropriate height. Smoothly connect the tops of the lines to form the profile.

Figure 1.22 Steps in Constructing a Map Profile from *Exploring Earth Science*, by Walter A. Thurber and Robert E. Kilburn, Allyn and Bacon, 1965. Used with permission of Prentice-Hall.

Conventions Used in Topographic Maps

When topographic maps are drawn or read, the following conventions apply:

- All points on a contour line have the same elevation.
- Every fifth line, called an *index line*, is generally printed in boldface type and labeled with its elevation.
- All contour lines are closed, but they may run off the map.
- Two contour lines of different elevations may not cross each other.
- Contour lines may merge at a cliff or waterfall.
- The spacing of contour lines indicates the nature of the slope. The closer the contour lines are spaced, the steeper is the slope. Even spacing indicates a uniform slope.
- Where contour lines cross a stream, they always form a V whose apex points up the valley.
- Where contour lines cross a ridge, they often form a V whose apex points down the valley.

REVIEW QUESTIONS FOR UNIT 1

1. Which diagram most accurately shows the cross-sectional shape of the earth?

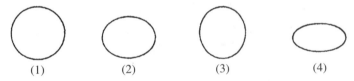

      (1)   (2)   (3)   (4)

2. The north-south distance between the Earth's Equator (0°) and North Pole (90°N.) is 10.002 kilometers. The distance between 0° and 10° N. is 1,106 kilometers. Which statement is best supported by this information?
   (1) The shape of the Earth is not perfectly spherical.
   (2) The lines of longitude are not parallel.
   (3) The north-south distance for every 10° of latitude is a constant value.
   (4) The Earth's equatorial radius and polar radius are equal.

3. An observer watching a sailing ship at sea notes that the ship appears to be "sinking" as it moves away. Which statement best explains the observation?
   (1) The surface of the ocean has depressions.
   (2) The Earth has a curved surface.
   (3) The Earth is rotating.
   (4) The Earth is revolving.

4. At sea level, which location would be farthest from the center of the Earth?
   (1) the North Pole
   (2) 23½° South latitude
   (3) 45° North latitude
   (4) the Equator

5. The best evidence of the Earth's nearly spherical shape is obtained through
   (1) telescopic observations of other planets
   (2) photographs of the Earth from an orbiting satellite
   (3) observations of the Sun's altitude made during the day
   (4) observations of the Moon made during lunar eclipses

6. According to the *Earth Science Reference Tables,* the Earth's radius is approximately
   (1) 637 km
   (2) 6,370 km
   (3) 63,700 km
   (4) 637,000 km

7. The circumference of the Earth is about 4.0×10^4 kilometers. This value is equal to
   (1) 400 km
   (2) 4,000 km
   (3) 40,000 km
   (4) 400,000 km

8. According to the data below, what is the exact shape of the Earth?

   Actual Dimensions of the Earth

Equatorial Radius	6,378 km
Polar Radius	6,357 km
Equatorial Circumference	40,076 km
Polar Circumference	40,008 km

   (1) slightly flattened at both the Equator and the poles
   (2) slightly bulging at both the Equator and the poles
   (3) slightly flattened at the Equator and slightly bulging at the poles
   (4) slightly flattened at the poles and slightly bulging at the Equator

9. In which group are the spheres of the Earth listed in order of increasing density?
   (1) atmosphere, hydrosphere, lithosphere
   (2) hydrosphere, lithosphere, atmosphere
   (3) lithosphere, hydrosphere, atmosphere
   (4) lithosphere, atmosphere, hydrosphere

10. According to the *Earth Science References Tables*, the element oxygen is found in the Earth's
(1) troposphere, only
(2) hydrosphere, only
(3) troposphere and hydrosphere, only
(4) troposphere, hydrosphere, and crust

11. Which profile when drawn to true scale most accurately shows the relationship between the ocean width of 1,500 km and a maximum depth of 8 km?

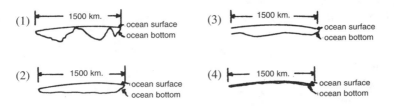

Note that question 12 has only three choices.

12. The hydrosphere is mostly
(1) solid rock
(2) liquid water
(3) gaseous air

13. Which statement correctly describes the Earth's crust?
(1) It is of uniform thickness.
(2) It is thicker under the poles than under the Equator.
(3) It is thinnest under the center of continents.
(4) It is thinnest under the oceans.

14. According to the *Earth Science Reference Tables,* in which zone of the Earth's interior is the melting point of the rock inferred to be lower than the actual temperature of the rock?
(1) outer core
(2) inner core
(3) crust
(4) mantle

15. According to the *Earth Science Reference Tables,* in which group are the zones of the Earth's interior correctly arranged in order of increasing average density?
(1) crust, mantle, outer core, inner core
(2) crust, mantle, inner core, outer core
(3) inner core, outer core, mantle, crust
(4) outer core, inner core, mantle, crust

22

16. According to the *Earth Science Reference Tables*, the temperature at the center of the Earth is estimated to be
    (1) 1,000 K
    (2) 2,800 K
    (3) 5,000 K
    (4) 6,000 K

17. The Earth's core is believed to be composed primarily of
    (1) oxygen and silicon
    (2) aluminum and silicon
    (3) iron and nickel
    (4) carbon and iron

18. According to the *Earth Science Reference Tables*, what is the approximate percentage by volume of oxygen in the crust of the Earth?
    (1) 20%
    (2) 30%
    (3) 70%
    (4) 90%

19. According to the *Earth Science Reference Tables*, approximately how far below the Earth's surface is the interface between the mantle and the outer core?
    (1) 5 to 30 km
    (2) 700 to 900 km
    (3) 2,900 to 3,000 km
    (4) 5,000 to 5,200 km

20. Which statement most accurately describes the Earth's atmosphere?
    (1) The atmosphere is layered, with each layer possessing distinct characteristics.
    (2) The atmosphere is a shell of gases surrounding most of the Earth.
    (3) The atmosphere's altitude is less than the depth of the ocean.
    (4) The atmosphere is more dense than the hydrosphere but less dense than the lithosphere.

21. According to the *Earth Science Reference Tables*, nearly all the water vapor in the atmosphere is found within the
    (1) mesosphere
    (2) thermosphere
    (3) troposphere
    (4) stratosphere

22. Which layer of the atmosphere is found immediately above the tropopause?
    (1) ionosphere
    (2) troposphere
    (3) hydrosphere
    (4) stratosphere

23. According to the *Earth Science Reference Tables,* the greatest atmospheric pressure occurs in the
(1) troposphere
(2) stratosphere
(3) mesosphere
(4) thermosphere

24. The lithosphere has an average thickness of about
(1) 100 km
(2) 200 km
(3) 300 km
(4) 400 km

25. According to the *Earth Science Reference Tables,* the *rate* of temperature increase below the Earth's surface is greatest between depths of
(1) 250 and 500 km
(2) 1,500 and 2,500 km
(3) 2,500 and 3,500 km
(4) 3,500 and 4,000 km

26. According to data in the *Earth Science Reference Tables,* which circle graph best represents the volume of gases in the troposphere?

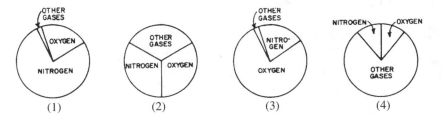

27. According to the *Earth Science Reference Tables,* which graph best represents the percent of oxygen, by volume, found in the Earth's crust (*C*), hydrosphere (*H*), and troposphere (*T*)?

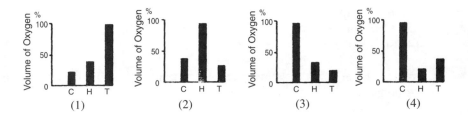

24

28. At which interface in the accompanying diagram is the moisture content of the air likely to be highest?
(1) thermosphere–mesosphere
(2) mesosphere–stratosphere
(3) stratosphere–troposphere
(4) troposphere–hydrosphere

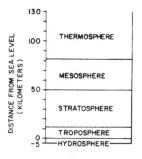

29. Two students are located 800 kilometers apart on the same meridian of longitude. They measure the altitude of the Sun at noon on the same day. They can use these measurements to calculate the
(1) Earth's circumference
(2) Earth's density
(3) Earth's mass
(4) Sun's diameter

30. According to the *Earth Science Reference Tables,* which graph best represents the relationship between altitude and pressure in the atmosphere?

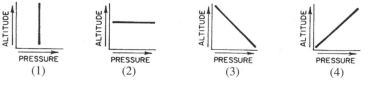

(1)   (2)   (3)   (4)

31. The diagram below shows stations *A* and *B*, which are located on a north-south line. At noon, the altitude of the Sun is 90° at station *B* and 30° at station *A*. What is the distance between stations *A* and *B*? [Earth's circumference is 40,000 km.]

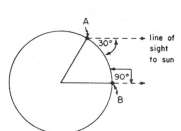

(1) 30 km   (3) 6,666 km
(2) 2,222 km   (4) 10,000 km

32. According to the geologic map of New York State in the *Earth Science Reference Tables,* Triassic bedrock is found in New York State at approximately
(1) 41° 05′ N, 74° 00′ W
(2) 44° 05′ N, 74° 00′ W
(3) 74° 00′ N, 41° 05′ W
(4) 74° 00′ N, 44° 05′ W

33. Polaris is used as a celestial reference point for the Earth's latitude system because Polaris
  (1) always rises at sunset and sets at sunrise
  (2) is located over the Earth's axis of rotation
  (3) can be seen from any place on Earth
  (4) is a very bright star

34. What is the latitude of the observer shown in the illustration?
  (1) 35° N
  (2) 55° N
  (3) 90° N
  (4) 25° N

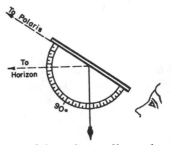

35. The diagrams below represent four systems of imaginary lines that could be used to locate positions on a planet. Which system is most similar to the latitude-longitude system used on the Earth?

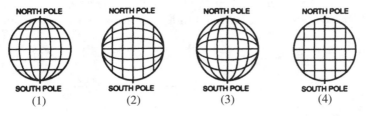

36. Based on the Generalized Bedrock Geology Map of New York State in the *Earth Science Reference Tables,* what could be the approximate location of an observer if he measured the altitude of Polaris to be 41° above the horizon?
  (1) Watertown   (3) Buffalo
  (2) Massena   (4) New York City

37. The diagram below represents a portion of the Earth's latitude and longitude system.

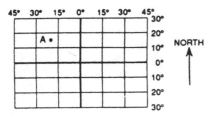

What are the approximate latitude and longitude of point A?
  (1) 15° S, 20° W   (3) 15° N, 20° W
  (2) 15° S, 20° E   (4) 15° N, 20° E

38. According to the *Earth Science Reference Tables,* what is the straight-line distance in kilometers from Elmira, New York, to Buffalo, New York?
(1) 100 km   (3) 150 km
(2) 120 km   (4) 190 km

39. On the Generalized Bedrock Geology Map of New York State, what similar pattern is found at 43° 00′ North latitude by 77° 30′ West longitude?

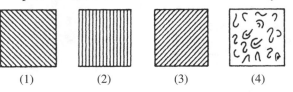

   (1)   (2)   (3)   (4)

40. According to the *Earth Science Reference Tables,* most of the surface bedrock in New York State south of latitude 43° N and west of longitude 75° W was formed during which period?
(1) Silurian   (3) Cambrian
(2) Devonian   (4) Ordovician

41. According to the *Earth Science Reference Tables,* what type of landscape region is located at 44° 30′ N and 74° 30′ W?
(1) plateau   (3) coastal lowland
(2) plain   (4) mountainous area

42. An observer in New York State measures the altitude of Polaris to be 44°. According to the *Earth Science Reference Tables,* the location of the observer is nearest to
(1) Watertown   (3) Buffalo
(2) Elmira   (4) Kingston

43. In the accompanying diagram, the thermometer held 2 meters above the floor shows a temperature of 30°C. The thermometer on the floor shows a temperature of 24°C.

What is the temperature gradient between the two thermometers?
(1) 6 C°/m
(2) 2 C°/m
(3) 3 C°/m
(4) 4 C°/m

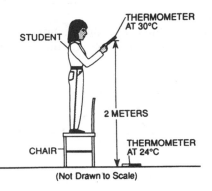

27

44. The diagram below shows the isothermal pattern obtained at a height of 1 meter above the floor in a classroom during a temperature study in November.

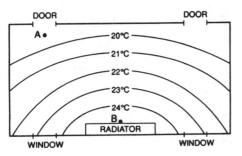

Which conclusion is best supported by this diagram?
(1) A heat source is located at *A*.
(2) A heat sink is located at *B*.
(3) Room temperature at the floor and at the ceiling can be determined.
(4) The isotherms indicate the temperatures at only one level in the room.

45. A topographic map is a two-dimensional model that uses contour lines to represent points of equal
(1) barometric pressure
(2) temperature gradient
(3) elevation above sea level
(4) magnetic force

46. The contour lines on the map below represent a hill.

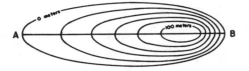

Which hill shape shown below best represents a profile drawn along line *AB* of the contour map?

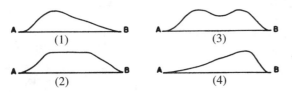

47. The temperature data (in °C) shown on the map below were taken at the same elevation and time in a room. Letters *A* through *D* represent the corners of the room.

A _____ B

26 •	24 •	22 •	21•	20•
28 •	25•	22 •	20•	19 •
27 •	27•	22 •	19•	18 •
24 •	25•	24 •	18•	17 •
22 •	22 •	20 •	17•	16 •
18•	18 •	17 •	16•	15 •

C _____ D

Which diagram best represents the temperature field in the room?

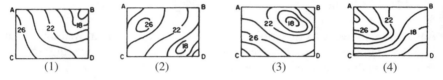

    (1)   (2)   (3)   (4)

48. A stream has a source at an elevation of 1,000. meters. It ends in a lake that has an elevation of 300. meters. If the lake is 200. kilometers away from the source, what is the average gradient of the stream?
(1) 1.5 m/km   (3) 10. m/km
(2) 3.5 m/km   (4) 15. m/km

Base your answers to questions 49 and 50 on the diagram below, which represents a contour map of a hill.

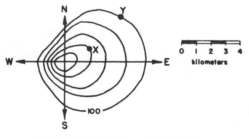

CONTOUR INTERVAL = 10 METERS

49. On which side of the hill does the land have the steepest slope?
(1) north   (3) east
(2) south   (4) west

50. What is the approximate gradient of the hill between points *X* and *Y*? [Refer to the *Earth Science Reference Tables*.]
(1) 1 m/km   (3) 3 m/km
(2) 10 m/km   (4) 30 m/km

29

51. The elevation of a certain area was measured for many years, and the results are recorded in the data table below.

Year	Elevation (m)
1870	102.00
1980	102.25
1910	102.50
1930	102.75
1950	103.00

If the elevation continued to increase at the same rate, what was most likely the elevation of this area in 1990?
(1) 103.25 m   (3) 103.75 m
(2) 103.50 m   (4) 104.00 m

Base your answers to questions 52 through 56 on your knowledge of earth science and the information provided by the diagram. The diagram represents a sketch drawn in a notebook by an earth science student. Lines *A, B, C, D, E,* and *F* are isolines. The points along the isolines indicate the only locations where actual measurements were made. Points *W, X, Y,* and *Z* are reference locations in the field diagram.

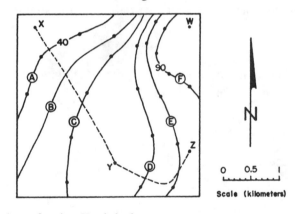

52. The value of point *Y* might be
(1) 55   (3) 75
(2) 65   (4) 95

53. What is the approximate distance from point *X* to point *Z* along the dashed line *XYZ*?
(1) 1.5 km   (3) 4.0 km
(2) 3.5 km   (4) 4.5 km

54. Between which two points is the greatest gradient?
(1) *X-W*   (3) *Y-W*
(2) *X-Y*   (4) *Z-W*

Note that question 55 has only three choices.

55. Which is the best description of the value of the measured quantities along isoline *C* on the map?
(1) The values decrease from north to south.
(2) The values increase from north to south.
(3) The values remain the same.

56. The isoline of this field diagram are for snow depth in centimeters. In which part of the diagram is the snow deepest?
(1) southeast corner
(2) northeast corner
(3) southwest corner
(4) northwest corner

Base your answers to questions 57 through 61 on your knowledge of earth science, the *Earth Science Reference Tables*, and the adjacent drawing. The drawing represents five positions of a balloon after being released from a ship. The drawings of the balloon are not to scale compared to the altitude distances, but are to scale with each other.

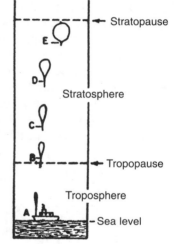

57. Which position represents the balloon when it is about 12 kilometers above sea level?
(1) *A*   (3) *D*
(2) *B*   (4) *E*

58. Why is the balloon's appearance at position *E* different from the balloon's appearance at position *A*?
(1) There is a partial vacuum inside the balloon at *A*, but not at *E*.
(2) There is more gas inside the balloon at *A* than at *E*.
(3) The outside air temperature is lower at *E* than at *A*.
(4) The outside air pressure is lower at *E* than at *A*.

59. At which position would the warmest surrounding air temperature most likely be found?
(1) *A*   (3) *C*
(2) *B*   (4) *D*

60. Which graph best represents the relative amounts of water vapor found in the atmosphere at the different balloon positions?

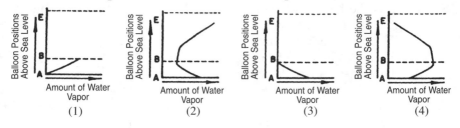

(1)   (2)   (3)   (4)

Note that question 61 has only three choices.

61. In order to make the balloon rise, the density of the gas put inside the balloon must be
(1) less than the density of the air at sea level
(2) more than the density of the air at sea level
(3) the same as the density of the air at sea level

Base your answers to questions 62 through 66 on the *Earth Science Reference Tables,* the diagram below, and your knowledge of earth science. The diagram represents three cross sections of the Earth at different locations to a depth of 50 kilometers below sea level. The measurements given with each cross section indicate the thickness and the density of the layers.

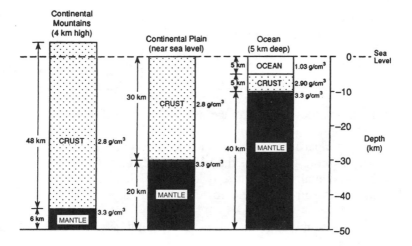

62. In which group are the layers of the Earth arranged in order of increasing average density?
(1) mantle, crust, ocean water
(2) crust, mantle, ocean water
(3) ocean water, mantle, crust
(4) ocean water, crust, mantle

63. Which material is most likely to be found 20 kilometers below sea level at the continental mountain location?
(1) basalt   (3) granite
(2) shale   (4) limestone

64. Which statement about the Earth's mantle is confirmed by the diagram?
(1) The mantle is liquid.
(2) The mantle has the same composition as the crust.
(3) The mantle is located at different depths below the Earth's surface.
(4) The mantle does not exist under continental mountains.

65. Compared with the oceanic crust, the continental crust is
(1) thinner and less dense
(2) thinner and more dense
(3) thicker and less dense
(4) thicker and more dense

66. The division of the Earth's interior into crust and mantle, as shown in the diagram, is based primarily on the study of
(1) radioactive dating   (3) volcanic eruptions
(2) seismic waves   (4) gravity measurements

Base your answers to questions 67 through 71 on the topographic map below and your knowledge of earth science.

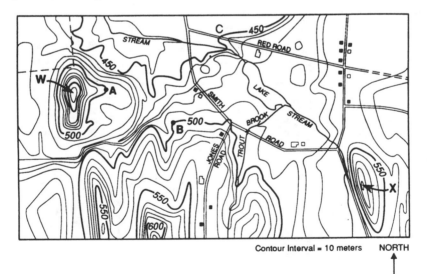

Contour Interval = 10 meters   NORTH

67. What is the elevation at the intersection of Jones Road and Smith Road?
(1) 450 m   (3) 550 m
(2) 500 m   (4) 600 m

68. What is the elevation of the highest contour line on hill *W*?
 (1) 440 m (3) 560 m
 (2) 510 m (4) 610 m

69. On which side of hill *X* is the steepest slope found?
 (1) north (3) southeast
 (2) east (4) southwest

70. In which general direction is Trout Brook flowing when it passes under Smith Road?
 (1) northeast (3) southeast
 (2) northwest (4) southwest

71. Which diagram best represents the profile along a straight line between points *A* and *B*?

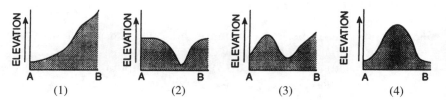

 (1) (2) (3) (4)

UNIT TWO _____

Minerals and Rocks

KEY IDEAS The Earth's crust is a rich source of materials ranging from fuels to building materials, metals, chemicals, and gems. The crust is composed mainly of rock, solid material that is a mixture of minerals.

Understanding how rocks and minerals form helps us to understand how the earth may have formed. It is important to realize that Earth's resources are limited and should be used wisely.

KEY OBJECTIVES
Upon completion of this unit, you should be able to:

- Demonstrate that most rocks are mixtures of one or more minerals.
- Define minerals as naturally occurring, inorganic elements or compounds with orderly internal structures and characteristic properties.
- Describe the characteristic properties that can be used to identify minerals.
- Explain how rocks are classified on the basis of their origins as igneous, metamorphic, or sedimentary.
- Relate the environment and the processes by which rocks and minerals form to the natural processes occurring within and upon the earth.
- Describe the rock cycle.

A. WHAT IS THE EARTH MADE OF?

If you walk outside and pick up any earth material—a rock, sand, soil, gravel, mud—you will hold minerals in your hand. Nearly all rocks are composed of one or more substances called *minerals*. When rocks are broken down into smaller pieces such as pebbles or sand, those smaller pieces are composed of the same minerals that were in the rock.

35

Nearly every single thing we use is made of minerals or contains them. Minerals are essential to the life of plants and animals. Industry is equally dependent on an abundant supply of minerals. Without minerals, none of the devices and structures that are part of our daily lives could exist.

Some minerals are rare and highly prized for their characteristics, such as gold for its conductivity, resistance to corrosion, and high luster, or diamonds for their beauty and hardness. Some minerals are raw materials from which industries manufacture the products that are the basis of a nation's wealth. Hematite, an iron ore, is needed to produce the steel from which products ranging from cars to skyscrapers are made. As such, minerals are of strategic importance, and wars have been fought to gain control of mineral resources. Most minerals, though, are common, and have little commercial value.

B. WHAT CHARACTERISTICS HELP US TO IDENTIFY MINERALS?

B-1. Characteristics of Minerals

What is a mineral? To be classified as a mineral, a substance must have certain characteristics.

Minerals are naturally occurring. A naturally occurring material is formed as result of natural processes in or on the Earth. It is not manufactured in a factory or synthesized in a laboratory. For example, most diamonds are formed naturally in the Earth and are minerals. Synthetic diamonds made in the laboratory are not minerals.

Minerals are inorganic matter. Inorganic substances are not alive, never were alive, and do not come from living things. Thus, amber (a tree resin in which insects are often found embedded) and the fossil fuels coal, petroleum, and natural gas are not true minerals. They were formed from organic substances, animal or vegetable material, that once lived on the Earth.

A chemical symbol or formula can be written for a mineral. Minerals are either elements or compounds.

Elements are substances that cannot be broken down into simpler substances by ordinary chemical means. Ninety-two different elements have been found to occur naturally on the Earth, each with distinctly different physical and chemical properties. Elements consist of particles called atoms, and all atoms of an element have the same properties. One- or two-letter symbols are used to represent the atoms of elements. For example, the symbol for the element oxygen is O and the symbol for the element silicon is Si.

Table 2.1 shows the relative abundances of different elements in the Earth's crust. Notice that only eight elements make up more than 98 percent of the Earth's crust, the two most common elements being silicon and oxygen.

TABLE 2.1 AVERAGE CHEMICAL COMPOSITION OF EARTH'S CRUST, HYDROSPHERE, AND TROPOSPHERE

Element (symbol)	Crust Percent by Mass	Percent by Volume	Hydrosphere Percent by Volume	Troposphere Percent by Volume
Oxygen (O)	46.40	94.04	33	21
Silicon (Si)	28.15	0.88		
Aluminum (Al)	8.23	0.48		
Iron (Fe)	5.63	0.49		
Sodium (Na)	2.36	1.11		
Magnesium (Mg)	2.33	0.33		
Potassium (K)	2.09	1.42		
Nitrogen (N)				78
Hydrogen (H)			66	

Source: The State Education Department, *Earth Science Reference Tables*, 1994 ed. (Albany, New York, The University of the State of New York).

Although some minerals found in the Earth's crust are pure elements, such as native copper and silver, the elements in the Earth's crust rarely exist by themselves. Most are chemically combined with other elements as compounds. *Compounds* consist of molecules, groups of atoms joined together in a definite proportion. For example, the mineral calcite is a compound of the elements calcium, carbon, and oxygen (see Figure 2.1a). Every calcite molecule contains one atom of calcium, one atom of carbon, and three atoms of oxygen. To show the chemical composition of a mineral, a formula can be written. The chemical formula for calcite is $CaCO_3$. The mineral quartz consists of molecules each containing one atom of silicon and two atoms of oxygen (Figure 2.1b). The chemical formula for quartz is SiO_2.

Every compound has distinct properties of its own. Thus, a mineral, which is either an element or a compound, has both a definite composition and distinct properties. Table 2.2 shows the chemical names and formulas of some common minerals.

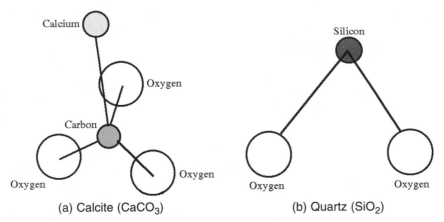

(a) Calcite ($CaCO_3$)   (b) Quartz (SiO_2)

Figure 2.1 Molecular Models of Calcite and Quartz

TABLE 2.2 SOME COMMON MINERALS

Mineral	Chemical Name	Chemical Formula
Calcite	Calcium carbonate	$CaCO_3$
Galena	Lead sulfide	PbS
Gypsum	Calcium sulfate-water	$CaSO_4 \cdot 2\,H_2O$
Olivine (fosterite)	Magnesium silicate	Mg_2SiO_4
Potassium Feldspar	Potassium aluminum silicate	$KAlSi_3O_8$
Pyrite	Iron sulfide	FeS_2
Quartz	Silicon dioxide	SiO_2

Minerals have a crystalline form. The atoms or molecules of a mineral are the same throughout that mineral. When they are joined in fixed positions as a solid, a definite pattern is formed. A solid having a definite internal structural pattern is said to have a crystalline form. If the pattern is large enough to be seen with the unaided eye, the solid is called a *crystal*. The crystal form of a mineral determines its cleavage, or the way it splits or breaks, as well as many other properties. Mica, for example, splits into thin, flat sheets (see Figure 2.2).

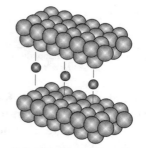

Figure 2.2 The Crystalline Pattern of Mica. Mica consists of sheets of tightly bonded silicon and oxygen atoms held together weakly by metallic ions.

B-2. Identifying Minerals

Mineralogists have examined earth materials and distinguished more than 2000 minerals. Minerals can be identified on the basis of well-defined physical and chemical properties. Several important properties commonly used in mineral identification are color, luster, streak, hardness, and cleavage. Other properties used to identify minerals include specific gravity, radioactivity, luminescence, and chemical, thermal, electrical, and magnetic properties, as well as elasticity and strength.

Since no single property can be used to identify all minerals, mineral identification is usually a process of elimination. As each property is observed, it becomes evident what a mineral is not, rather than what it is. Step by step, the possibilities are narrowed down until the identity of the mineral is determined.

Color

Color is often the first property noticed about a mineral. When observing color, it is important to use a fresh surface of a mineral since exposed surfaces are often discolored by weathering. Color alone is an unreliable property by which to identify a mineral. Although some minerals are always the same color (e.g., sulfur is yellow), others may occur in a variety of colors.

Thus quartz is found in many colors, some of which are rare (these specimens are highly prized as semiprecious gems). For example, amethyst is purple quartz, citrine is yellow quartz, and rose quartz is pink quartz. Therefore, quartz cannot be identified by color alone. Another reason why color is unreliable is that many minerals have almost the same color. For example, calcite, quartz, and halite all occur in white and transparent varieties and can look strikingly alike.

Luster

The *luster* of a mineral is the way light reflects from its surface. Luster can be metallic or nonmetallic. Nonmetallic lusters can be described as glassy, brilliant, greasy or oily, waxy, silky, pearly, or earthy.

Related to the luster of a mineral are its transparency and iridescence, or the play of colors in its interior or exterior.

Streak

Streak is the color of the powder left when a mineral is rubbed against a hard, rough surface. Unglazed porcelain, called a *streak plate,* is usually used. Although the color of a mineral may vary, the streak of a particular mineral is always the same color. For example, fluorite may range in color from green to blue, yet its streak is always white. This characteristic makes streak a useful property for identifying a mineral. It must be remembered, though, that a mineral's streak is not always the same color as the mineral. The streak of pyrite is different in color from pyrite itself.

Hardness

Hardness is a mineral's resistance to being scratched. Talc is so soft that it can be scratched by a fingernail. Diamond is so hard that no other mineral can scratch it. The hardness of minerals is usually stated in terms of Moh's scale (see Table 2.3). On this scale, ten typical minerals are arranged in order from the softest to the hardest.

TABLE 2.3 MOH'S SCALE

Mineral	Hardness	
Talc	1	SOFTEST
Gypsum	2	
Calcite	3	
Fluorite	4	
Apatite	5	
Orthoclase	6	
Quartz	7	
Topaz	8	
Corundum	9	
Diamond	10	HARDEST

To find the hardness of a mineral, you determine what minerals your sample can scratch and what minerals it cannot scratch. For example, tourmaline will scratch quartz or anything softer, but cannot scratch topaz or anything harder. The hardness of tourmaline, therefore, is between 7 and 8 (sometimes expressed as 7.5).

Cleavage, Parting and Fracture

Some minerals break in a way that helps to identify them. Cleavage (see Figure 2.3) is the tendency of a mineral to break parallel to atomic planes in its crystalline structure. Parting is the tendency to break along surfaces that follow a structural weakness caused by factors such as pressure, or along zones of different crystal types. These cleavage or parting surfaces often occur at very specific angles to one another and can be helpful in identifying a mineral.

Mineral	Cleavage	Appearance
Muscovite mica	Breaks in parallel sheets	
Pyroxene	Breaks along planes at an 88° angle to one another in a prismatic shape	
Halite	Breaks along planes at 90° to one another in a cubic shape	
Calcite	Breaks along planes at a 75° angle to one another in a rhombohedral shape	

Figure 2.3 Some Common Minerals That Display Cleavage

Some minerals have no planes of weakness, or are equally strong in all directions. They *fracture*, that is, they do not follow a particular direction when they break. Although while fracture does not occur along smooth surfaces, it can display distinctive patterns that can be very helpful in mineral identification (see Figure 2.4).

Patterns of Fracture	Description	Appearance
Conchoidal	Smooth, curved break that looks somewhat like the inside of a shell	
Fibrous or splintery	Fibers, as in asbestos Long, thin splinters	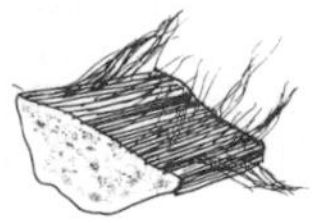
Hackly	Jagged or sharp-edged surfaces	
Uneven	Rough, irregular surfaces	

Figure 2.4 Some Common Minerals That Display Fracture

Specific Gravity

Every mineral has a certain density. If you have samples that are about the same size, you can compare the densities of two different minerals. When you hold one in each hand, the denser mineral will feel heavier than the less dense one. A more exact way to determine the relative densities of different minerals is to compare them all to one standard. *Specific gravity* is the comparison of the density of a mineral to the density of water. It tells us how much heavier than water a mineral is. For example, the density of galena is 7.5 grams per cubic centimeter; the density of water is 1 gram per cubic centimeter. Galena is 7.5 times denser than water and therefore has a specific gravity of 7.5.

Most rocks have a specific gravity of 2.5 to 3.5. Most people have a feeling for how heavy a rock of a given size should be. Sometimes, though, when you pick up a mineral and toss it in your hand, it feels heavier than you expected. This feeling of unexpected weight is termed *heft*. You can usually judge by a mineral's heft that its specific gravity is greater than 4.0.

Chemical Tests

Two chemical tests are commonly used to identify minerals in the field. The acid test uses hydrochloric acid. Some minerals bubble when a drop of

41

hydrochloric acid dropped on the mineral reacts with it. Calcite bubbles vigorously; dolomite, slowly.

The second chemical test is the taste test. Halite tastes the same as table salt, which it is, though not purified. **The taste test should be used with caution and only when directed by your teacher**.

Special Properties

Special properties that are displayed by minerals and can be used to distinguish them are summarized in Table 2.4.

TABLE 2.4 SOME SPECIAL PROPERTIES OF MINERALS

Property	Definition or Description
Magnetism	Mechanical force of attraction associated with moving electricity (magnetite)
Luminescence	Emission of light by a substance that has received energy or electromagnetic radiation of a different wavelength from an external stimulus
includes:	
phosphorescence	Light emitted after stimulus is removed (scheelite, autunite)
fluorescence	Light emitted while stimulus is applied (fluorite, willemite)
triboluminescence	Light emitted when pressure is applied (fluorite, lepidolite)
thermoluminescence	Light emitted when heated (chlorophane fluorite)
Piezoelectricity	Electricity emitted when pressure is applied (quartz)
Pyroelectricity	Electricity emitted when heated (tourmaline)
Flame color	Color of the flame when a mineral is intensely heated in a flame
Double refraction	Double image produced when an object is viewed through it (isinglas spar calcite)
Play of colors	Series of colors produced when the angle of the light shining on it is changed (labradorite, opal)
Chatoyancy and asterism	Appearance of being made of tiny parallel fibers (satin spar gypsum, cat's-eye chrysoberyl) Light scattered in a pattern that looks like a three- or six-rayed star (star rubies and sapphires)

B-3. Minerals as the Building Blocks of Rocks

Of the thousands of known minerals, about 100 are so common that they make up more than 95 percent of the Earth's crust. Not surprisingly, these minerals are generally compounds of the most abundant elements in the Earth's crust. Nearly every rock contains one or more of these so-called *rock-forming minerals* (see Table 2.5). Of the rock-forming minerals, fewer than 20 comprise most of the rocks you are likely to find.

TABLE 2.5 SOME MAJOR ROCK-FORMING MINERALS

Mineral or Mineral Group	Composed of These Elements
Olivines	Iron, magnesium, silicon, oxygen
Pyroxenes	Calcium, iron, magnesium, silicon, oxygen
Amphiboles	Sodium, calcium, aluminum, silicon, oxygen
Micas	Potassium, aluminum, iron, magnesium, silicon, oxygen, hydrogen
Chlorite	Magnesium, iron, manganese, aluminum, silicon, oxygen, hydrogen
Kaolinite	Aluminum, silicon, oxygen, hydrogen
Feldspars	Potassium, sodium, calcium, aluminum, silicon, oxygen
Quartz	Silicon, oxygen
Nepheline	Sodium, potassium, aluminum, silicon, oxygen
Hematite	Iron, oxygen
Rutile	Titanium, oxygen
Magnetite	Iron, oxygen
Spinels	Magnesium, aluminum, oxygen
Pyrites	Iron, sulfur
Gypsum	Calcium, sulfur, oxygen
Calcite	Calcium, carbon, oxygen
Dolomite	Calcium, magnesium, carbon, oxygen
Apatite	Calcium, phosphorus, oxygen, hydrogen, fluorine, chlorine
Fluorite	Calcium, fluorine

C. WHAT ARE THE THREE IMPORTANT TYPES OF ROCKS?

Rock is the naturally formed, solid material that makes up the Earth's crust. As stated earlier, most rocks are composed of one or more minerals. Both rocks and the minerals of which they are composed originate in a variety of ways. Rocks are classified on the basis of their origin into three main types: igneous, sedimentary, and metamorphic.

C-1. Igneous Rocks

Formation of Igneous Rocks

Igneous rocks form by the crystallization of a molten mixture of minerals and dissolved gases. This molten mixture is called *magma* while it is beneath the Earth's surface and *lava* when it escapes onto the surface. Most magmas can be thought of as a fiery soup of fast-moving ions, that is, electrically charged atoms or molecules. As the magma cools, these ions slow down, are attracted to one another, and settle into stable structures—molecules of mineral compounds joined in a structural array, or mineral crystals (see Figure 2.5).

The formation of mineral crystals is a process, not an instantaneous event. As magma or lava cools, the amount of time available for crystal structures to form determines how large they can become. Thus, the size of the mineral

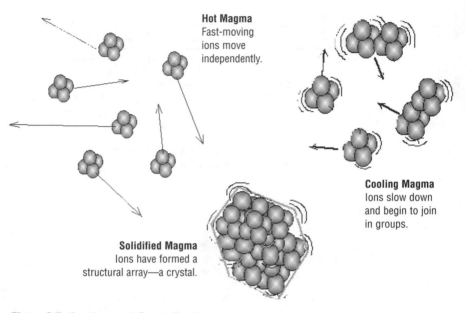

Hot Magma
Fast-moving ions move independently.

Cooling Magma
Ions slow down and begin to join in groups.

Solidified Magma
Ions have formed a structural array—a crystal.

Figure 2.5 Cooling and Crystallization

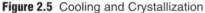

crystals in an igneous rock provides a clue to the speed at which the rock crystallized, and that in turn is a clue to its origin.

If cooling is very rapid, groups of atoms may solidify into a substance that lacks any distinctive structure—a glass. Such rapid cooling typically occurs at the Earth's surface, where molten material comes into sudden contact with air or water at temperatures that are hundreds of degrees cooler than its temperature. Volcanic glasses, such as obsidian, can be observed forming when lava flows into the ocean and rapidly cools. From this type of observational data, we can infer the origin of glass found elsewhere, far from current volcanic activity or perhaps even buried beneath layers of other rock.

If cooling occurs more slowly, mineral crystals may begin to form, but may solidify while still microscopic in size. This too, typically occurs at or near the surface of the Earth. For example, the surface of lava, which may flow out of a fissure as a glowing liquid, can be observed to quickly cool and "skin over." This thin skin of rock shields the molten lava beneath it and allows the lava to cool more slowly.

Only very slow cooling allows the formation of crystals large enough to be visible to the unaided eye. Such slow cooling occurs underground, where the cooling magma is blanketed by the surrounding rock. Rock is a poor conductor of heat, and magma surrounded by rock, like hot coffee in a Thermos bottle, loses heat very slowly. Some underground pools of magma cool only a few degrees per century! It may take thousands of years for the magma to totally crystallize. The result is large, visible crystals.

Characteristics of Igneous Rocks

Since rocks are mixtures of minerals, rocks cannot be identified in the same way as minerals. A single rock may consist of minerals of several colors, hardnesses, and densities. For example, granite may contain white quartz, pink feldspar, and black mica.

Two more general properties that are useful in identifying rocks are texture and mineral composition. *Texture* is the size, shape, and arrangement of the mineral crystals or grains in a rock. *Mineral composition* is simply the minerals that comprise the rock.

Igneous rocks are composed of randomly scattered, tightly interlocking mineral crystals. The main differences in the textures of igneous rocks involve the sizes and compositions of the mineral crystals. Most igneous rocks consist of a mixture of several of the following common minerals: quartz, feldspars, biotite, hornblende, pyroxene, and olivine.

Figure 2.6 summarizes the properties of the most common igneous rocks.

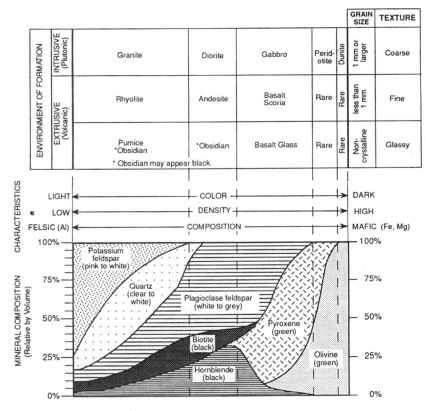

Note: The intrusive rocks can also occur as exceptionally coarse-grained rock, Pegmatite.

Figure 2.6 Scheme for Igneous Rock Identification
Source: The State Education Department, *Earth Science Reference Tables*, 1994 ed. (Albany, New York: The University of the State of New York).

C-2. Sedimentary Rocks

Sediments

All sedimentary rocks are composed of sediments. *Sediments* can be broadly defined as solid fragments of material that have been transported and then deposited by air, water, or ice. Most sediments are fragments of rock, but they can also be bits and pieces of plants and animals, or even molecules dissolved in water.

When deposited, sediments typically form loose, unconsolidated layers on the Earth's surface. These layers may then be transformed by physical and chemical changes into sedimentary rock.

Lithification

Lithification is the conversion of loose sediments into coherent, solid rock by processes such as cementation, compaction, dessication, and crystallization. This conversion may occur while the sediment is being deposited, soon after deposition, or long after deposition.

Cementation is the binding together of sediment particles by substances such as clay, carbonates, or hydrates of iron. These substances literally act as cement to hold particles together in a solid mass when they *crystallize* or otherwise accumulate between the loose sediment particles.

Compaction is the reduction in volume or thickness of a sediment layer due to the increasing weight of overlying sediments that are continually being deposited or to the pressure produced by movements of the Earth's crust. Compaction usually results in a decrease in pore space within the sediments as they become more tightly packed. This, in turn, commonly leads to *desiccation*, the drying out or drying up of water, as it is forced out of the pore spaces during compaction.

Figure 2.7 illustrates the compaction and cementation processes.

Most sedimentary rocks form as a result of the compression and cementing of sediments. In most sedimentary rocks these sediments are rock fragments that range in size from tiny particles such as clay and silt to larger particles such as sand and pebbles.

Some sedimentary rocks, however, result from the evaporation of seawater or from organic processes. Seawater has many substances dissolved in it; most are salts. When seawater evaporates, the salts are left behind. They form salt crystals, which settle out of the water in layers. Rock salt forms in this way as seawater evaporates.

Some organisms, such as corals, take substances dissolved in seawater into their bodies and form skeletons. When these organisms die, their skeletons settle to the ocean floor and can build up in layers. Fossil limestones are formed in this way.

Plant materials that build up in layers can change drastically during compaction. They become desiccated, and the substances from which they were composed break down. Since carbon is the chief component of organic mate-

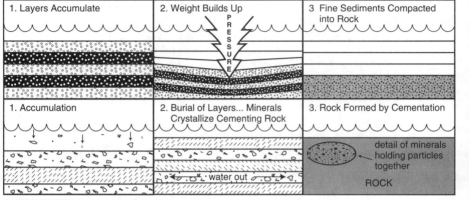

(a) COMPACTION

Figure 2.7 Compaction and Cementation.

rials, it is often the end product of the lithification of plant parts. Coal typically forms in this way.

Characteristics of Sedimentary Rocks

Several unique characteristics of sedimentary rocks are that they:

Form at or near the surface. Sedimentary rocks form at or near the surface of the Earth at normal temperatures and pressures, unlike metamorphic rocks and many igneous rocks. As a result the environment in which sedimentary rocks form is quite different from that in which most other types of rock have their origins. Therefore, the minerals contained in sediment rocks are often quite different from those found in igneous or metamorphic rocks.

Form layers. Water is the chief transporter of sediments on the Earth's surface. While being transported by water, most sediments become rounded. Sediments deposited in water settle into horizontal layers of rounded particles. This layering is preserved during lithification, so most sedimentary rocks usually contain rounded grains cemented in layers. These successive layers, known as strata, or beds, preserve evidence of past surface conditions. The individual layers can be easily distinguished because they usually differ in some way, such as color, thickness, size of particles, mineral composition, or degree of cementation.

The surfaces of layers in sedimentary rock often contain distinctive features identical to those found on modern sediments. Examples are ripple marks created by waves in shallow water, polygon-shaped cracks formed by fine sediment that was dried out by the Sun, footprints of walking animals, and the tracks and trails of crawling animals such as snails.

When sand and coarser sediments are moved along by air or water, they often form streamlined mounds such as dunes or ripples. Within these mounds, the layers of sediment are inclined in a down-wind or down-current direction. As successive layers of dunes or ripples pass over the surface, they leave behind layers of inclined sediments, called *cross-bedded sediments,* as shown in Figure 2.8.

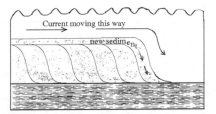

Contain fossils. Plants and animals often live in areas where sediments are being deposited. The remains of these plants and animals are often incorporated into layers of accumulating sediment and into the sedimentary rocks that form from them.

Figure 2.8 Diagram of Cross-Bedded Sediments

Observations of current sediment layers and surfaces indicate that some characteristics enable us to infer which is the top, and which the bottom, of a series of sedimentary rock layers. This, in turn, allows us to infer the relative age of the layers (bottom is oldest and top is youngest). When a mixture of sediments settles in water, the larger particles reach the bottom first, and the smallest last. The result is graded bedding, as shown in Figure 2.9. When graded bedding is preserved in a sedimentary rock, we can infer that the original top was where the smaller particles are located. Similarly, cross-bedded sediments (Figure 2.9b) allow us to infer not only top and bottom, but also the direction in which the wind or water current was moving when the beds were deposited. If a rock layer contains ripple marks, mud cracks, footprints, or trails, we can infer the top of the layer from their position. Also, when certain shells settle in water (Figure 2.9c), they end up with one side facing upward most of the time, so their position in a sedimentary rock again allows us to infer top and bottom.

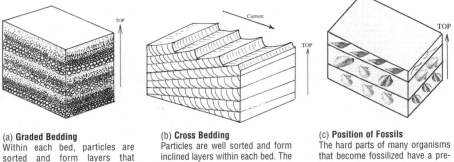

(a) **Graded Bedding**
Within each bed, particles are sorted and form layers that decrease in size from the bottom to the top of the bed.

(b) **Cross Bedding**
Particles are well sorted and form inclined layers within each bed. The layers in the bed are steeply sloping at the top and gently sloping at the bottom.

(c) **Position of Fossils**
The hard parts of many organisms that become fossilized have a preferred orientation after settling through water. For example, clam shells settle more often with the cupped side of the shell facing down.

Figure 2.9 Diagram of Graded Bedding, Cross-bedding, and Fossil Position All Showing the Top of a Series of Layers

48

C-3. Metamorphic Rocks

Formation of Metamorphic Rocks

Metamorphic rocks get their name from the Greek words *meta*, meaning "change," and *morph*, meaning "form." A metamorphic rock has literally been changed from its original form. A rock changes when it is exposed to an environment significantly different from the one in which it formed. In general, three factors cause metamorphic changes in rock: heat, pressure, and chemical activity.

Heat

Most minerals expand when heated, causing the atoms to move apart, and the bonds that hold them together to be stretched and weakened. If heated enough, all of the bonds break and the mineral melts. If this happens to a rock, a magma will form. In metamorphism, however, heating breaks *some* but not all of the bonds in the minerals of the rock. *Some* atoms break loose, migrate through the rock and join with other atoms to form new minerals—metamorphic minerals—in the rock. The result is to change both the chemical composition and the structure of the rock, that is, to change the form of the rock. Such extreme heating of rock may be caused by deep burial (temperature increases with depth) or by contact with hot materials, such as magma.

Pressure

The effect of pressure on minerals is opposite to that of heat. Pressure forces the atoms in the mineral much closer together. The resulting stress on the bonds causes some of them to break. This, in turn, enables the atoms to rearrange into a more compact structure. The result is a significant change—a denser, harder rock.

Pressure exerted on rock may be due to deep burial or to movements in the Earth's crust. Pressure due to deep burial is equal in all directions, but crustal movements may produce pressures that are greater in one direction than another, thereby deforming or flattening rocks. The effects on the particles in a rock are different in these two cases, as shown in Figure 2.10.

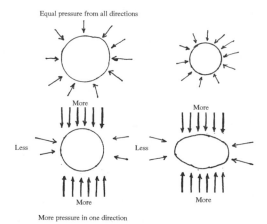

Figure 2.10 Equal Versus Directed Pressure on a Particle. Equal pressure causes the particle to decrease in size, but the shape is unchanged. Unequal pressure causes both a decrease in size and a change in shape.

49

Chemical Activity

A rock can also be changed when solutions rich in dissolved ions move through it. Such solutions are given off by cooling magmas; others come from metamorphism taking place deep underground. As these solutions move through pores in surrounding rock, they interact with minerals in the rock to form new minerals. Sometimes the water itself combines with minerals to form new ones. For example, olivine reacts with water to form talc or serpentine.

Characteristics of Metamorphic Rocks

When a rock is metamorphosed, nearly every characteristic can change, including texture and mineral composition. These changes can be so great that the metamorphic rock bears little, if any, resemblance to its original state. Therefore, to infer what a metamorphic rock once was, one needs to know what kinds of changes take place during metamorphism.

Changes in Texture

Crystalline Texture. During metamorphism the size, shape, and spacing of the grains (crystals or particles) in a rock are changed. Under both heat and pressure, the grains in a rock may partially melt and fuse together, forming larger grains (recrystallization); or, pressure may break brittle grains into smaller pieces, destroying the original texture of the rock. The heat and the pressure of metamorphism fuse any fragmental textured rock into a solid, crystalline mass. As a result, most metamorphic rocks have a crystalline texture.

Increased Density. Pressure from deep burial closes pore spaces in sedimentary rocks and openings in igneous rocks. Pressure also compresses grains so that they are smaller and closer together. The rock becomes more compact. Its mass is forced into a smaller volume, and its density increases. Thus, the igneous or sedimentary rock is changed—it becomes metamorphic rock.

Foliation and Banding. A rock is said to be foliated when its crystals are arranged in layers or bands along which the rock breaks easily. These layers may be microscopic or thick enough to be easily seen. Foliation results when mineral grains recrystallize or are flattened under pressure. Even pressure compresses the grains so that they are smaller and closer together, but retain their original shape. Uneven pressure not only compresses grains but also changes their shape. Grains that may have been randomly oriented will form parallel layers (foliation) or become aligned when deformed under pressure.

Banding occurs when minerals of different densities recrystallize under pressure and separate into layers, like a mixture of oil and water. Since light-colored minerals tend to be less dense than dark ones, the layers that form are alternating light and dark.

Figure 2.11 shows the effects of pressure and recrystallization on rocks.

Distortion of Layers. When sedimentary rocks are metamorphosed, the layers in the rock may be distorted. Softened by the heat and squeezed by the pressure, the once-flat layers become twisted and contorted. Such distortion is commonly seen in large-scale formations of metamorphic rock.

Changes in Mineral Composition. The changes described above can occur without altering the original mineral composition of the rock. When a rock is metamorphosed, however, the original minerals often disappear and new minerals form in their place. The underlying reason for this change is stability. Minerals that are stable in one environment are not stable in another. Sugar is stable at room temperature; but when heated strongly, it becomes unstable and breaks down into carbon and water. Clay minerals, such as kaolinite, provide another good example. Kaolinite forms by the weathering of feldspar. It has an open, sheetlike molecular structure that is stable at

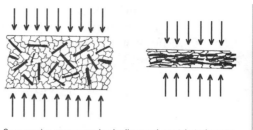

Compression causes randomly dispersed crystals to become oriented in a direction perpendicular to the pressure. The oriented crystals form thin sheets or layers called *foliation.*

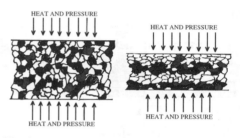

Migration of mineral cystals due to density differences produces banding—regions of light and dark colored minerals.

Figure 2.11 Random to Oriented Patterns Due to Pressure and Fusing of Grains, and Banding Caused by Density Separation

the temperatures and pressures normally found at the Earth's surface. Under the heat and pressure associated with deep burial, however, it cannot survive, and forms micas that have a more compact, sheetlike molecular structure.

DEHYDRATION (WATER GIVEN OFF)

Muscovite mica + quartz → sillimanite + potassium feldspar + water
(K, Al, Si, O, H)   (Si, O)   (Al, Si, O)   (K, Al, Si, O)   (H, O)

DECARBONATION (CARBON DIOXIDE GIVEN OFF)

Dolomite   +   quartz →   diopside   + carbon dioxide
(Ca, Mg, C, O) (Si, O)   (Ca, Mg, Si, O)   (C,O)

REPLACEMENT REACTIONS

Dolomite   +   quartz + water →   tremolite   + calcite + carbon dioxide
(Ca, Mg, C, O) (Si, O) (H, O)   (Ca, Mg, Si, O, H) (Ca, C, O)   (C, O)

COMBINING REACTIONS

Jadeite   + quartz →   albite
$(NaAlSi_2O_6)$ (SiO_2) $(NaAlSi_3O_8)$

Figure 2.12 Some Chemical Changes That Take Place During Metamorphism. Formulas that are very complex have been simplified to show only the elements they contain.

Some of the common chemical changes that take place during metamorphism (see Figure 2.12) are dehydration, decarbonation, replacement reactions involving water and carbon dioxide, and the combining of compounds to form new ones.

Classification of Metamorphic Rocks

Metamorphic rocks are classified according to mineral composition and texture, including foliation and banding, as shown in Figure 2.13.

An interesting question associated with metamorphic rocks is "When has a rock changed enough to be called metamorphic?" Rocks do not change instantaneously from one type to another; the changes occur along a continuum from little alteration to major changes. Generally, if a change can be recognized in a rock, the rock is considered metamorphic. Table 2.6 lists some common rocks and the rocks they can become when metamorphosed.

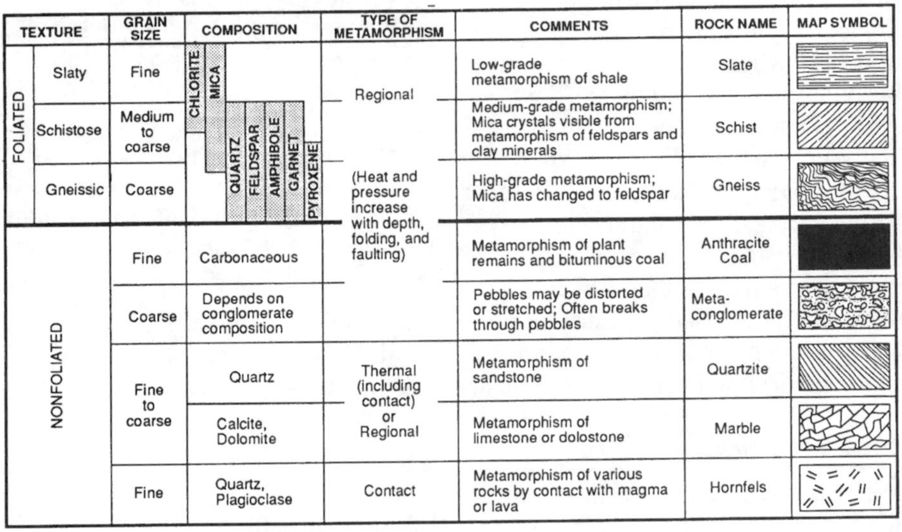

TEXTURE		GRAIN SIZE	COMPOSITION	TYPE OF METAMORPHISM	COMMENTS	ROCK NAME	MAP SYMBOL
FOLIATED	Slaty	Fine	CHLORITE MICA / QUARTZ FELDSPAR AMPHIBOLE GARNET PYROXENE	Regional	Low-grade metamorphism of shale	Slate	
	Schistose	Medium to coarse			Medium-grade metamorphism; Mica crystals visible from metamorphism of feldspars and clay minerals	Schist	
	Gneissic	Coarse		(Heat and pressure increase with depth, folding, and faulting)	High-grade metamorphism; Mica has changed to feldspar	Gneiss	
NONFOLIATED		Fine	Carbonaceous		Metamorphism of plant remains and bituminous coal	Anthracite Coal	
		Coarse	Depends on conglomerate composition		Pebbles may be distorted or stretched; Often breaks through pebbles	Meta-conglomerate	
		Fine to coarse	Quartz	Thermal (including contact) or Regional	Metamorphism of sandstone	Quartzite	
			Calcite, Dolomite		Metamorphism of limestone or dolostone	Marble	
		Fine	Quartz, Plagioclase	Contact	Metamorphism of various rocks by contact with magma or lava	Hornfels	

Figure 2.13 Scheme for Metamorphic Rock Identification.
Source: The State Education Department, *Earth Science Reference Tables*, 1994 ed. (Albany, New York: The University of the State of New York).

TABLE 2.6 DERIVATION OF COMMON METAMORPHIC ROCKS

Parent Rock	Metamorphic Rock
Shale	Compaction and reorientation → slate
	Slate → micas and chlorite form → phyllite
	Phyllite → grains recrystallize and get coarser → schist
	Schist → layering occurs → gneiss
Shale	Contact with magma → granofels
Sandstone	Quartzite
Limestone or dolostone	Marble
Basalt	Hydration reactions → greenschist
	Greenschist → dehydration → amphibolite
	Amphibolite → layering occurs → gneiss

D. WHAT IS THE ROCK CYCLE?

Where do rocks come from? You might say igneous rock comes from magma, sedimentary rock comes from sediments, and metamorphic rock comes from other rocks that have changed. But consider that magma is melted rock, sediments are broken rock, and there had to be an already-existing rock to be metamorphosed. Basically, then, rocks come from other rocks.

Rocks are constantly changing from one type to another in a never-ending cycle called the *rock cycle*, shown in Figure 2.14. The outside circle shows the different forms in which rock matter can exist: magma, igneous rock, sediment, and so on. Arrows leading from one form to another are labeled with the change that is taking place. Arrows inside the circle show some alternative paths in the rock cycle. For example, an igneous rock does not always break down into sediments. If buried, it may be metamorphosed; therefore, there is an arrow inside the circle leading from igneous rock to metamorphic rock.

Let's follow a rock through one possible cycle, beginning with magma. A magma solidifies and forms an igneous rock. The igneous rock is uplifted and exposed at the surface, where weathering breaks it down into sediments. Rain washes the sediments into a stream. The stream carries them to the ocean, where they are deposited and buried. Compaction and cementation produce a sedimentary rock. The sedimentary rock is exposed to heat and pressure, perhaps as two of the plates in the Earth's crust collide, and a metamorphic rock is formed. If the metamorphic rock is forced deep enough into the crust, it melts and forms magma.

Notice that we have come full circle back to where we started. This is the rock cycle. It is, however, only one of many possible paths. Except for the

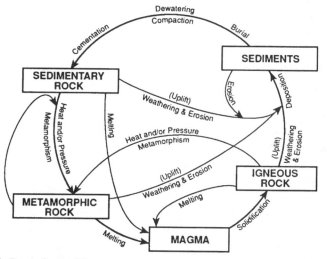

Figure 2.14 Rock Cycle Diagram.
Source: The State Education Department, *Earth Science Reference Tables*, 1994 ed. (Albany, New York: The University of the State of New York).

53

stray meteorite that reaches the surface, the Earth is essentially a closed system. Rock doesn't enter or leave, but is constantly recycling from one form to another.

The rock cycle diagram has no beginning and no end, but of course the rock cycle really did start somewhere. There is much evidence that the Earth was originally totally molten. No solid rock existed, only magma. Therefore, it is thought that the original rocks formed as this magma cooled, and the rock cycle began with igneous rocks.

REVIEW QUESTIONS FOR UNIT 2

1. All rocks contain
   (1) minerals
   (2) intergrown crystals
   (3) sediments
   (4) fossils

2. Which rock would most likely be monominerallic (contains only one mineral)?
   (1) granite
   (2) rhyolite
   (3) basalt
   (4) rock salt

3. Which two elements are most abundant by weight in the Earth's crust?
   (1) oxygen and aluminum
   (2) oxygen and silicon
   (3) iron and aluminum
   (4) iron and silicon

4. Which element comprises most of the Earth's crust both by weight and by volume?
   (1) nitrogen
   (2) hydrogen
   (3) oxygen
   (4) silicon

5. The physical properties of a mineral are due chiefly to the
   (1) hardness and cleavage of the mineral
   (2) number of atoms present
   (3) number of oxygen-silicon tetrahedra
   (4) internal arrangement of its atoms

6. The hardness of a mineral such as quartz is due to
   (1) the internal arrangement of its atoms
   (2) large amounts of impurities
   (3) its characteristic luster and color
   (4) its formation in certain rock types

7. Which property is most useful in mineral identification?
   (1) hardness
   (2) color
   (3) size
   (4) texture

8. Certain minerals usually break along flat surfaces, while other minerals break unevenly. This characteristic is due to the
   (1) luster of the mineral
   (2) age of the mineral
   (3) internal arrangement of the mineral's atoms
   (4) force with which the mineral is broken

9. Which property of a mineral most probably determines the extent to which the mineral will resist being mechanically eroded?
   (1) density
   (2) streak
   (3) luster
   (4) hardness

10. Rocks are classified on the basis of
    (1) how they were formed
    (2) the shape of the sample
    (3) their age in millions of years
    (4) the mass of the sample

11. Which properties of a rock provide the most information about the environment in which the rock formed?
    (1) texture, composition, and structure
    (2) mass, composition, and color
    (3) density, volume, and mass
    (4) color, texture, and volume

12. Rocks are classified as igneous, metamorphic, or sedimentary according to
    (1) chemical composition
    (2) density
    (3) grain size
    (4) origin

13. Most igneous rocks form by which processes?
    (1) melting and solidification
    (2) heat and pressure
    (3) erosion and deposition
    (4) compaction and cementation

14. The diagrams below represent magnifications of rocks. Which is most likely a diagram of a *nonsedimentary* rock?

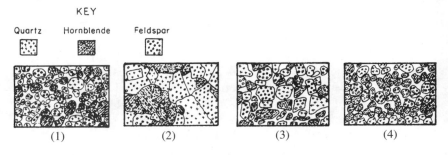

15. According to the *Earth Science Reference Tables,* which property would be most useful for identifying igneous rocks?
(1) kind of cement
(2) mineral composition
(3) number of minerals present
(4) types of fossils present

Base your answers to questions 16 and 17 on your knowledge of earth science and the diagrams below, which represent cross sections of four rock samples. Each cross section illustrates the sediments, minerals, or structural appearance of the rock sample.

16. Which rock sample is most likely monomineralic?
(1) A   (3) C
(2) B   (4) D

Note that question 17 has only three choices.

17. Which rock sample is most likely a nonsedimentary rock?
(1) A   (3) C
(2) B

18. Large crystal grains in an igneous rock indicate that the rock was formed
(1) near the surface
(2) under low pressure
(3) at a high temperature
(4) over a long period of time

19. Extremely small crystal grains in an igneous rock are an indication that the crystals formed
(1) under high pressure
(2) over a short period of time
(3) from an iron-rich magma
(4) deep below the surface of the Earth

20. Which igneous rock crystallized quickly near the surface of the Earth, is light in color, and contains quartz and plagioclase feldspars?
(1) granite   (3) basalt
(2) gabbro   (4) rhyolite

21. Which would most likely occur during the formation of igneous rock?
(1) compression and cementation of sediments
(2) recystallization of unmelted material
(3) solidification of molten materials
(4) evaporation and precipitation of sediments

22. The diagrams below represent four rock samples. Which rock took the longest time to solidify from magma deep within the Earth?

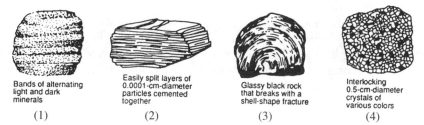

Bands of alternating light and dark minerals
(1)

Easily split layers of 0.0001-cm-diameter particles cemented together
(2)

Glassy black rock that breaks with a shell-shape fracture
(3)

Interlocking 0.5-cm-diameter crystals of various colors
(4)

23. Which process would form a sedimentary rock?
(1) cooling of molten magma within the Earth's crust
(2) recrystallization of unmelted material within the Earth's crust
(3) cooling of a lava flow on the Earth's surface
(4) precipitation of minerals as the seawater evaporates

24. According to the *Earth Science Reference Tables,* a sedimentary rock formed as the result of compaction and cementation of uniformly sized fine sediments with a diameter of 0.1 centimeter is called
(1) conglomerate
(2) sandstone
(3) shale
(4) slate

25. Which sedimentary rock is most likely of organic origin?
(1) limestone   (3) shale
(2) halite   (4) conglomerate

26. According to the *Earth Science Reference Tables,* which sedimentary rock could form as a result of evaporation?
(1) conglomerate   (3) shale
(2) sandstone   (4) limestone

27. According to the *Earth Science Reference Tables,* which type of sedimentary rock contains the greatest range of particle sizes?
(1) conglomerate   (3) shale
(2) sandstone   (4) siltstone

28. According to the rock cycle diagram in the *Earth Science Reference Tables,* which type(s) of rock can be the source of deposited sediments?
(1) igneous and metamorphic rocks, only
(2) metamorphic and sedimentary rocks, only
(3) sedimentary rocks, only
(4) igneous, metamorphic, and sedimentary rocks

29. Where are the Earth's sedimentary rocks generally found?
(1) in regions of recent volcanic activity
(2) deep within the Earth's crust
(3) along the midocean ridges
(4) as a thin layer covering much of the continents

30. According to the *Earth Science Reference Tables,* which sedimentary rock is composed of fragmented skeletons and shells of sea organisms compacted and cemented together?
(1) shale   (3) sandstone
(2) limestone   (4) gypsum

31. The adjacent diagram represents a conglomerate rock. Some of the rock particles are labeled.

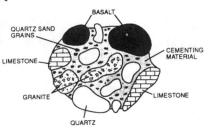

Which conclusion is best made about the rock particles?
(1) They are the same age.
(2) They originated from a larger mass of igneous rock.
(3) They all contain the same minerals.
(4) They have different origins.

32. Which property best describes a rock that has formed from sediments?
(1) crystalline structure
(2) distorted structure
(3) banding or zoning of minerals
(4) fragmental particles arranged in layers

33. According to the *Earth Science Reference Tables,* compaction and cementation of pebble-size particles would form the sedimentary rock known as
(1) shale
(2) conglomerate
(3) sandstone
(4) siltstone

34. According to the *Earth Science Reference Tables,* sedimentary rocks formed by compaction and cementation of land-derived sediments are classified on the basis of
(1) composition
(2) type of cement
(3) particle size
(4) rate of formation

35. The information below is a classification of six common rocks, based on how they were formed. In which group would conglomerate rock be placed in this classification?

Group A	Group B	Group C
basalt	sandstone	marble
granite	shale	gneiss

(1) Group A
(2) Group B
(3) Group C

36. Which process(es) would result in the formation of a nonsedimentary rock?
(1) compression and cementation
(2) solidification of molten material
(3) evaporation and precipitation
(4) biologic processes

37. The adjacent diagram represents a cross section of a coarse-grained nonsedimentary rock.

According to the *Earth Science Reference Tables,* this rock is most likely to be
(1) basalt
(2) rhyolite
(3) gabbro
(4) granite

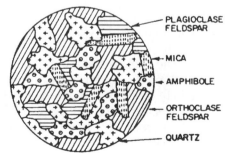

PLAGIOCLASE FELDSPAR

MICA

AMPHIBOLE

ORTHOCLASE FELDSPAR

QUARTZ

38. Which properties are most often used to distinguish metamorphic rocks from other kinds of rocks?
(1) organic composition and density
(2) banding and distortion of structure
(3) mineral color and hardness
(4) layering and range of particle sizes

39. The recrystallization of unmelted material under high temperature and pressure results in
(1) metamorphic rock   (3) igneous rock
(2) sedimentary rock   (4) volcanic rock

40. Metamorphic rocks result from the
(1) erosion of rocks
(2) recrystallization of rocks
(3) cooling and solidification of molten magma
(4) compression and cementation of soil particles

41. Which rock is most likely a nonsedimentary rock?
(1) a rock showing mud cracks
(2) a rock containing dinosaur bones
(3) a rock consisting of layers of rounded sand grains
(4) a rock composed of distorted light-colored and dark-colored mineral bands

42. What is the main difference between metamorphic rocks and most other rocks?
(1) Many metamorphic rocks contain only one mineral.
(2) Many metamorphic rocks have an organic composition.
(3) Many metamorphic rocks exhibit banding and distortion of structure.
(4) Many metamorphic rocks contain a high amount of oxygen-silicon tetrahedra.

43. The metamorphism of sandstone rock will cause the rock
(1) to be melted   (3) to become more dense
(2) to contain more fossils   (4) to occupy a greater volume

44. Which characteristics would indicate that a rock has undergone meta-morphic change?
(1) The rock shows signs of being heavily weathered and forms the floor of a large valley.
(2) The rock is composed of intergrown mineral crystals and shows signs of deformed fossils and structure.
(3) The rock becomes less porous when exposed at the surface and is finely layered.
(4) The rock contains a mixture of different-sized, rounded grains or both felsic and mafic silicate minerals.

45. Which statement is supported by information in the rock cycle diagram in the *Earth Science Reference Tables?*
(1) Metamorphic rock results directly from melting and crystallization.
(2) Sedimentary rock can be formed only from igneous rock.
(3) Igneous rock always results from melting and solidification.
(4) All sediments turn directly into sedimentary rock

46. Which drill core represents rocks formed in an environment of high pressure and temperature?

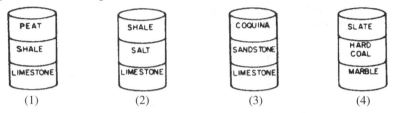

| (1) | (2) | (3) | (4) |

Base your answers to questions 47 through 51 on your knowledge of earth science, the *Earth Science Reference Tables,* and the diagrams of five rock samples (including metamorphic gneiss) shown below.

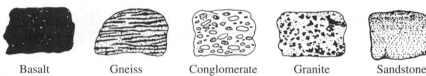

| Basalt | Gneiss | Conglomerate | Granite | Sandstone |

47. Which rock is composed of sediments that have a range of sizes and that originate from different rock types?
(1) basalt   (3) conglomerate
(2) gneiss   (4) granite

48. Which rock shows banding that formed as a result of the recrystallization of unmelted material under high temperature and pressure?
(1) gneiss   (3) granite
(2) conglomerate   (4) sandstone

49. The granite most likely was formed by the processes of
(1) erosion and deposition
(2) compaction and cementation
(3) heating and metamorphism
(4) melting and solidification

50. The adjacent diagram represents the percentage by volume of each mineral found in a sample of basalt. Which mineral is represented by the letter *X* in the diagram?
(1) orthoclase feldspar   (3) quartz
(2) plagioclase feldspar   (4) mica

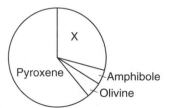

51. Which rock was formed by the compression and cementation of sediment with particle sizes ranging from 0.08 to 0.1 centimeter?
(1) basalt
(3) granite
(2) conglomerate
(4) sandstone

Base your answers to questions 52 through 56 on your knowledge of earth science, the *Earth Science Reference Tables,* and the two graphs below. Graph I represents the percentages of sedimentary and nonsedimentary rock that make up the Earth's crust by volume. Graph II represents, of those rocks that are exposed at the surface (outcrops), the percentages that are sedimentary rocks and nonsedimentary rocks.

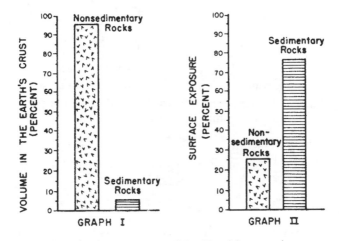

52. Approximately what percentage of the Earth's crust is composed of sedimentary rock?
(1) 5%
(3) 75%
(2) 25%
(4) 95%

53. Which is the most abundant element present in the rocks shown in Graph I?
(1) nitrogen
(3) silicon
(2) oxygen
(4) hydrogen

54. All of the rocks represented in Graph I must contain
(1) fossils
(2) intergrown crystals
(3) sediments
(4) minerals

55. Most sedimentary rock has been formed by which two processes?
(1) uplifting and melting
(2) extrusion and intrusion
(3) compaction and cementation
(4) faulting and folding

56. Which statement is best supported by the data shown in the graphs?
  (1) The crust of the Earth is composed mostly of sedimentary rocks.
  (2) Rock outcrops on the Earth's surface are chiefly of the nonsedimentary type.
  (3) Most nonsedimentary rocks are composed of the melted remains of sedimentary rocks.
  (4) Most sedimentary rock is found at or near the surface of the Earth.

Base your answers to questions 57 through 61 on your knowledge of earth science, the *Earth Science Reference Tables,* and the table below. The table provides data about the textures and mineral compositions of four different igneous rock samples having the same volume

Rock	Texture	Orthoclase feldspar	Quartz	Plagioclase feldspar	Mica	Hornblende	Pyroxene
A	coarse	62%	20%	7%	7%	4%	0%
B	coarse	24%	40%	19%	10%	7%	0%
C	fine	6%	16%	41%	14%	23%	0%
D	fine	0%	0%	50%	0%	6%	44%

57. Which igneous rock sample contains the most quartz by volume?
  (1) *A*   (3) *C*
  (2) *B*   (4) *D*

58. According to the rock cycle diagram in the *Earth Science Reference Tables,* all four rock samples have undergone
  (1) compaction and sedimentation
  (2) volcanic eruption
  (3) solidification from a molten state
  (4) deposition and burial

59. Which two igneous rocks most likely formed closest to the surface of the Earth?
  (1) *A* and *B*   (3) *B* and *C*
  (2) *C* and *D*   (4) *A* and *D*

60. Which rock sample is probably basalt?
  (1) *A*   (3) *C*
  (2) *B*   (4) *D*

61. Which rock sample has the greatest density and also contains the most magnesium?
  (1) *A*   (3) *C*
  (2) *B*   (4) *D*

TOPIC A _____

Rocks, Minerals, and Other Natural Resources

A. HOW DO MINERALS DIFFER FROM EACH OTHER?

Minerals are grouped according to their chemical compositions. Major mineral groups include the silicates, sulfides, oxides, carbonates, and sulfates. Minor mineral groups include the nitrates, borates, chromates, phosphates, halides, hydroxides, and native elements.

A-1. Major Mineral Groups

Silicates

If we look at the chemical composition of the Earth's crust (see Table A.1), we see that oxygen is the most abundant element, by both mass and volume, and silicon is second, by mass. It is not surprising, then, that minerals composed mainly of silicon and oxygen are abundant and widespread. In all silicates, one ion (charged atom) of silicon is always joined to four ions of oxygen in the shape of a tiny tetrahedron (*plural, tetrahedra*) as shown in Figure A.1. The bonds in this structure, which we will call the silicon tetrahedron, are very strong.

Silicon tetrahedra can join with positive ions of common elements in a wide variety of structures. Consider the next most abundant elements in the Earth's crust after silicon and oxygen; aluminum, iron, calcium, sodium, magnesium, and potassium are all metals that form positive ions. When the positive ions of these elements join with negative silicon tetrahedra, a

64

tremendous number of different silicates can be formed. Rings, chains, sheets, and frameworks are some of the different structures that silicon tetrahedra can form by joining with positive ions and sharing oxygen atoms.

TABLE A.1 AVERAGE CHEMICAL COMPOSITION OF THE EARTH'S CRUST

Element (Symbol)	Crust	
	Percent by Mass	Percent by Volume
Oxygen (O)	46.40	94.04
Silicon (Si)	28.15	0.88
Aluminum (Al)	8.23	0.48
Iron (Fe)	5.63	0.49
Calcium (Ca)	4.15	1.18
Sodium (Na)	2.36	1.11
Magnesium (Mg)	2.33	0.33
Potassium (K)	2.09	1.42

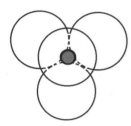
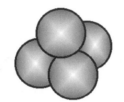

An overhead view, as though looking through the large oxygen ions at the smaller silicon atom between them

The natural position of the silicon tetrahedron. Notice that the silicon ion is nestled among the oxygen ions and is hidden from view

An exploded view, showing the positions of the silicon ion and oxygen ions

Figure A.1 Three Views of a Silicon Tetrahedron. The silicon ion has a +4 charge (Si^{+4}), and each oxygen ion has a -2 charge (O^{-2}), so the net charge of the silicon tetrahedron is -4 $(SiO_4)^{-4}$.

In the complex structures shown in Figure A.2, many other elements and groups of elements, such as extra oxygen, water, aluminum, and other metal ions, and hydroxide ions, promote electrical and mechanical balance so that these structures are stable

The physical properties of minerals depend on the arrangement and bonding of their atoms, as is very clearly seen in the structures formed by silicon tetrahedra. Look at a typical sheet silicate, mica, in Figure A.2. The tetrahedra all share oxygens and are tightly bound together. The metal ions between the sheets, however, are isolated from each other and are less tightly bound. They form a plane of weakness in the structure—a weakness that manifests itself in the tendency of mica to break into thin sheets. Quartz, on the other hand, is a framework. All of the tetrahedra are bound tightly to each other, and no place in its structure is weaker than any other place. Therefore, quartz does not break apart easily; its structure resists outside forces, so it is hard and strong. When it does break, there are no planes of weakness, and the break is random and uneven.

How Do Minerals Differ from Each Other?

Independent Tetrahedra
e.g., olivine, zircon, garnet

Ring Silicates
e.g., tourmaline, beryl

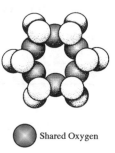

Shared Oxygen

Single-Chain Silicates
pyroxenes: e.g., enstatite, ferrosilite, augite,
diopside, jadeite, spodumene

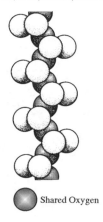

Shared Oxygen

Double-Chain Silicates
amphiboles: e.g., hornblendes,
anthophyllite, tremolite, glaucophane

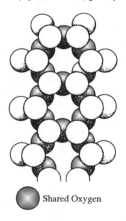

Shared Oxygen

Sheet Silicates
micas: e.g., muscovite, biotite, lepidolite
also, talc, chlorite, serpentine

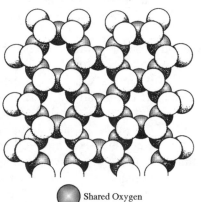

Shared Oxygen

Framework Silicates
e.g., feldspars, quartz, nephaline, sodalites

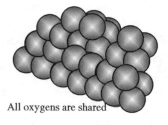

All oxygens are shared

Figure A.2 Some of the Structures That Form with Silicon Tetrahedra

This relationship between the arrangement and bonding of atoms and the physical properties of minerals extends to all of the mineral compounds. Knowing the structure of a mineral helps us to understand why that mineral displays certain physical properties

Silicate minerals are so abundant and widespread that almost every rock contains one or more. Although there are more groups of nonsilicate minerals than there are of silicates, nonsilicates make up only a very small part of the Earth's crust, and very few are important rock-forming minerals. A brief overview of the nonsilicate minerals follows.

Sulfides

Sulfides are compounds in which one or more ions of sulfur are combined with a metallic ion. Sulfide minerals usually display a metallic luster, and many are important ores of metals. Some examples of sulfides are: galena (PbS), pyrite (FeS), marcasite (FeS_2), cinnabar (HgS), sphalerite (ZnS), and stibnite (Sb_2S_3).

Oxides

Oxides are compounds in which oxygen is joined with ions of other elements, usually metals. Common oxides include hematite (Fe_2O_3), magnetite ($FeO\ Fe_2O_3$), corundum (Al_2O_3), ilmenite ($FeTiO_5$), chromite ($FeCr_2O_4$), and spinel ($MgAl_2O_4$). Magnetite is the only mineral attracted to a magnet. Iron oxides such as magnetite and hematite are mined to make iron and steel. Corundum is extremely hard and is widely used as an abrasive. When it forms with impurities, corundum produces red rubies and blue sapphires.

Carbonates and Sulfates

In oxides, oxygen forms a single ion and acts alone. In other minerals that contain oxygen, the oxygen ions join with other ions to form a group called a polyatomic ion. Two very common polyatomic ions are the carbonate ion $(CO_3)^{-2}$, and the sulfate ion, $(SO_4)^{-2}$. Polyatomic ions are electrically charged and can combine with ions of the opposite charge to form mineral compounds.

When carbon dioxide dissolves in water, it often forms carbonate ions. These react with other ions in the water to form carbonate compounds that can precipitate out of the water or be left behind when the water evaporates. Some organisms take carbon dioxide from the environment and form, in their bodies as skeletons, compounds containing carbonate ions. In this way, carbonate minerals have accumulated into huge sedimentary rock formations and in some cases have been metamorphosed. Calcite (calcium carbonate) is the chief mineral in limestone and marble. Dolomite (calcium magnesium carbonate) is the chief mineral in dolomite and dolostone

Gypsum (hydrated calcium sulfate) is probably the best known sulfate mineral. It is mined for use in making wallboard, plaster, cement, fertilizer, and paint and as a filler in paper. Barite ($BaSO_4$) and celestite ($SrSO_4$) are important sources of barium and strontium, respectively, which are used in medicines and drilling.

A-2. Minor Mineral Groups

There are also other, less abundant ions that form minerals: nitrates (nitrogen and oxygen), borates (boron and oxygen), chromates (chromium and oxygen), phosphates (phosphorus and oxygen), and hydroxides (hydrogen and oxygen).

Halides are compounds of metallic ions with halogen elements such as chlorine, fluorine, and bromine. The only major rock-forming mineral in this group is halite (NaCl), a sedimentary mineral that forms when seawater evaporates.

Native elements are found as pure elements in the crust. They include gold, silver, copper, and sulfur. These were among the first minerals to be mined by humans.

B. WHAT CAN BE LEARNED FROM A STUDY OF IGNEOUS ROCKS?

All igneous rocks are formed by the cooling of magma or lava. Igneous rocks are divided into two major types, intrusive and extrusive, based on whether they formed from magma or lava. Magma and lava solidify into rock under very different conditions and thus produce rocks with different characteristics. Except for the volcanic glasses, all igneous rocks consist of tightly interlocking mineral crystals (see Figure A.3). The texture of these crystals is a key to how a particular rock was formed.

Figure A-3 Tightly Interlocking Crystals of an Igneous Rock.

B-1. Crystal Textures of Igneous Rocks

Intrusive Rocks

Magma is molten rock inside the Earth. Magma can move around by pushing its way into cracks or crevices in surrounding rock or by intruding into them. If the magma remains trapped underground and cools enough to solidify, an igneous rock will form. Igneous rock formed by the cooling of magma is called *intrusive*, or *plutonic*, rock.

Magma generally cools slowly underground because it is blanketed by the surrounding rocks. Rocks are poor conductors of heat and do not allow the heat in the magma to escape very rapidly. When magma cools slowly, ions in the magma have time to align themselves in orderly structures called crystals. The slower the cooling, the more ions are able to align and the bigger the crystal. As a result, intrusive rocks generally have crystals large enough to be seen with the unaided eye. While you may not think a 1-millimeter crystal is large when you look at it, remember that millions upon millions of ions had to move into alignment with each other to reach that size. Rocks with large, visible crystals are said to have a coarse texture.

Granite, gabbro, and pegmatite are typical intrusive rocks.

Extrusive Rocks

When magma reaches the surface and pours out of the Earth, it is called *lava*. Lava is usually pushed out, or extruded, from inside the Earth in volcanoes or through large cracks in the Earth's crust. Therefore, igneous rock formed from lava is called *extrusive*, or *volcanic*, rock.

Lava cools and solidifies into rock rapidly. When lava is extruded from a volcano or cracks in the crust, it is instantly exposed to an environment that may be 1,000°C colder than where the lava was just moments before. Since lava cools and solidifies rapidly, the ions in it have less time to align themselves into crystals, and the crystals that form are generally too small to be seen by the unaided eye. A magnifier or microscope may be needed to see the crystals in an extrusive rock. Rocks with crystals that can be seen only with magnification have a fine texture.

In some cases, magma cools so rapidly that no crystalline structures at all can form and the ions literally become frozen in place. When a rock of this type is examined with a microscope, no crystals can be seen. The physical characteristics of these unusual rocks are the same as those of glass, so they are called *volcanic glasses*. Their texture, in which no mineral crystals or grains are visible, is termed glassy. Obsidian is a common volcanic glass.

In some cases, as lava is being extruded, gases dissolved in the lava escape, and a lava froth forms similar to the froth that overflows from a champagne bottle after the cork is popped. When this froth cools and solidifies, a rock filled with holes is formed. Pumice and scoria are examples of this type of rock.

When lava is very forcibly extruded, as in an explosive volcanic eruption, a whole array of unusual rock materials form. These range from the thin, glassy strands, called *Pele's hair,* produced when very liquid lava is sprayed out, to large, streamlined globs, called *volcanic bombs,* that solidify while hurtling through the air. Such shapes are clues to the way in which the rock formed.

Basalt and rhyolite are typical extrusive rocks. The differences in the grain sizes of typical intrusive and extrusive rocks are shown in Figure A.4.

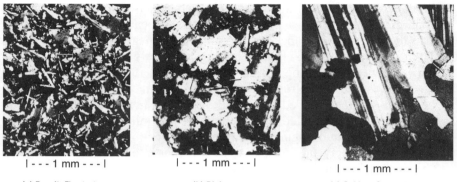

| - - - 1 mm - - - | | - - - 1 mm - - - | | - - - 1 mm - - - |

(a) Basalt: Fine texture   **(b) Diabase**   **(c) Gabbro:** Coarse texture

Figure A-4 Comparison of the Grain Sizes in a Continuum from Extrusive (a) and Intrusive (b&c) Rocks. Basalt (a) forms at the surface. Diabase (b) may form as an intrusion near the surface as in a dike or sill. Gabbro (c) cools very slowly deep underground.

Physical Geology, 2nd Edition, by Richard Foster Flint, Brian J. Skinner, John Wiley and Sons, New York 1974, 1977.

Porphyrys

In most igneous rocks the crystal grains are all roughly the same size, all large or all small. One type, however, has an unusual texture. In porphyrys, coarse mineral grains are embedded in a mass of fine mineral grains (see Figure A.5). The isolated, large crystals are called *phenocrysts*; the surrounding fine-grained material is the *groundmass*. Porphyrys are thought to form in two stages: (1) a magma begins to solidify slowly at depth, and (2) the magma rises rapidly to the surface, where it finishes solidifying. The crystals formed at depth are large because the rate of cooling is slow. Nearer the surface, cooling is more rapid and the crystals formed are tiny.

Figure A.5 Porphyritic Texture—Large Phenocrysts of Plagioclase Feldspar in a Basalt Matrix

Physical Geology, 2nd Edition, by Richard Foster Flint, Brian J. Skinner, John Wiley and Sons, New York 1974, 1977.

B-2. Mineral Composition of Igneous Rocks

The mineral composition of an igneous rock is the product of the magma from which the rock solidifies. All common igneous rocks are mixtures of one or more of the following minerals: quartz, potassium feldspar, plagioclase feldspar, biotite mica, hornblende, pyroxene and olivine. All are silicates, but they vary in composition, color, density, and other properties.

These silicate minerals are divided into two main groups: the *felsic* group, consisting of quartz and the feldspars, and the *mafic* group, consisting of silicates rich in magnesium and iron, such as olivine, pyroxene, horneblende, and biotite. The word *felsic* comes from *fel*dspar and *si*lica (the name of the compound SiO_2, of which quartz is composed). *Mafic* comes from *ma*gnesium and the symbol *Fe* for iron. Felsic minerals are light in color and low in density compared to the darker colored, denser mafic minerals. Felsic rocks are common in the continents, while mafic rocks are more frequently found in the ocean basins.

B-3. The Scheme for Igneous Rock Identification

There are more than 1,500 different igneous rocks, and the system for classifying them is very complex. Petrologists are trying to simplify the system. Figure A.6 can be used to make a very basic classification of an igneous rock. The scheme, which is based on texture, color, density, mineral composition, and other characteristics, has two main parts: an upper half, in which some of the most common igneous rocks are arranged by characteristics, and a lower half, which lists the percent mineral compositions of these rocks.

In the upper half, rocks like granite and rhyolite, which are composed of light-colored minerals, are on the left. Rocks like basalt and gabbro, which are composed of dark-colored minerals, are on the right. Since the dark-colored minerals tend to be denser than the light-colored minerals, the rocks on the right are denser than the rocks on the left. Coarse-grained, intrusive rocks, such as granite and gabbro, are on the top; fine-grained extrusive rocks, such as rhyolite and basalt, in the middle; glassy extrusive rocks, such as obsidian and pumice, on the bottom.

In the lower half, the relative percents of the minerals of which the rock is composed are shown. As you can see, a particular type of rock can have a mineral composition that varies within a certain range. Look at the two different granites. Granite A consists of about 50 percent potassium feldspar, 30 percent quartz, 10 percent plagioclase feldspar, and 5 percent each of biotite and horneblende. Granite B consists of only 10 percent potassium feldspar, 35 percent quartz, 25 percent plagioclase feldspar, and 15 percent each of biotite and horneblende.

The following is a brief description of the rocks shown on this chart.

GRANITE consists mainly of quartz and feldspars. It is light colored and coarse grained and is thought to have been formed by very slow cooling deep within the earth. Granite is the most abundant of all the igneous rocks. Most of the crust beneath the continents is made of granite. Granite is commonly used as a building stone and in making monuments.

RHYOLITE contains the same minerals as granite, but it is fine grained and is much rarer than granite. It is formed in lava flows and can be found in volcanic regions.

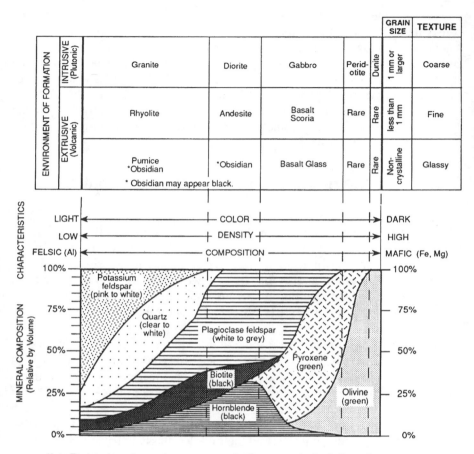

Figure A.6 Scheme for Igneous Rock Identification
Source: The State Education Department, *Earth Science Reference Tables*, 1994 ed. (Albany, New York: The University of the State of New York).

DIORITE is coarse grained and contains about equal amounts of light- and dark-colored minerals, so that it has a "salt and pepper" appearance.

ANDESITE is named for the Andes mountains, where these rocks were first studied. Andesites are fine grained and usually gray or green in color. They have the same composition as diorites.

GABBRO, a dark, coarse-grained rock, is a common intrusive rock in New England and other mountain areas. It is sometimes used as a building stone.

BASALT, the most abundant extrusive rock, is dark colored and fine grained. Most lava flows form basalt, and the ocean is mostly basalt. Sometimes basalt is crushed and used in road and railroad beds.

OBSIDIAN, sometimes called volcanic glass, is dark colored and has a glassy texture. It forms when lava cools very rapidly. Obsidian is very hard and brittle, and it breaks along shell-like depressions, forming sharp edges. Primitive peoples used it to make knives.

PUMICE and SCORIA hardened while gases were escaping from lava. The cooling rate was so fast that gases were trapped in the rock. As a result, pumice and scoria are full of holes! They look like rocky sponges. Pumice is light colored, scoria, dark. Because of the holes these rocks are so light they float on water.

C. WHAT CAN BE LEARNED FROM A STUDY OF SEDIMENTARY ROCKS?

All sedimentary rocks are composed of sediments that have been lithified.

C-1. Types of Sediment Particles

There are three major types of sediment particles: clastic, organic, and chemical.

Clastic sediments are rock or mineral fragments formed by the breakdown of rock due to weathering. The vast majority of sediments are of this type. These particles are grouped and named by size according to the Wentworth scale (Table A.2).

TABLE A.2 THE WENTWORTH SCALE

Particle Size Range (diameter, cm)	Particle Name	Sediment Composed of That Particle
Greater than 25.6	Boulder	Boulder gravel
6.4–25.6	Cobble	Cobble gravel
0.4–6.4	Pebble	Pebble gravel
0.2–0.4	Granule	Granule gravel
0.006–0.2	Sand	Sand
0.0004–0.006	Silt	Silt
Less than 0.0004	Clay	Clay

Organic sediments are particles produced by the life activities of plants or animals. The most abundant organic sediment consists of the shells of aquatic animals. These may range from the large calcium carbonate shells of clams or snails to the microscopic silica skeletons of amoebas and diatoms. There are even some algae, which absorb calcium carbonate from seawater and incorporate it into their skeletons. When these organisms die, their shells and skeletons settle to the bottom and accumulate in layers of mud.

In warm seas, coral reefs form when huge populations of little animals called corals secrete calcium carbonate. Wave action smashes the reef into pieces, along with the shells of other animals living on the reef, producing a coarse sand or granule-sized sediment.

Organic sediments also include the woody tissue of plants (tree trunks, branches, twigs, and leaves), as well as spores and pollen. *Peat* is an organic sediment formed from plant material that accumulates in swamps, where stagnant water conditions prevent the oxidation of the plant material. Over time, buried peat may be transformed into coal.

The fatty or waxy tissues of land plants and the soft tissues of aquatic plants and animals do not form particles. However, this soft organic matter is usually mixed in with sediment particles and is considered to be the raw material from which petroleum forms underground.

Fossils in sedimentary rocks provide evidence of the environment in which they formed. The fossil of an organism that is clearly a fish, together with fossilized clam shells, is evidence of an aquatic environment. Pollen from citrus or palm trees in a rock is evidence of a warm terrestrial environment.

Chemical sediments consist of particles (crystals) that have crystallized out of solutions at or near the Earth's surface. All water on the Earth's surface has some material dissolved in it. Seawater, for example, contains over 80 elements and hundreds of different compounds. A number of conditions can cause some of the substances dissolved in water to crystallize.

Let's use a simple analogy. Think of a solution as a container in which a substance is being held. A glass of juice is a good example. The solvent is like the glass, and the solute is like the juice. If there is more juice than can fit into the cup, the juice will overflow and spill to the ground. Now think of seawater. Water is the solvent, or glass; salts are the solute, or juice. Overflow is like crystallization; spilling to the ground is like precipitation.

In our example, there are two things that can change—the size of the cup and the amount of juice. Having more solvent is like having a bigger cup. Similarly, if more water is added, seawater can hold more salt. On the other hand, removing water is like having a smaller cup. If water is removed, less salt can be held.

Evaporation removes water from seawater. When the water in a solution evaporates, the solution becomes more and more concentrated. When it reaches saturation, the substance dissolved in the solution begins to crystallize out. Sediments made of particles that crystallized as a result of evaporation are called *evaporites*. Halite crystals are a common chemical sediment that forms along the edges of bodies of water undergoing extensive evaporation.

If you have ever tried to dissolve sugar in iced tea, you know that less sugar dissolves in cold tea than hot tea. The temperature of a solution affects the amount of material that can be dissolved in it. As the temperature of a solution falls, its ability to hold dissolved materials decreases. Cooling a solution is like making the cup in our example smaller without removing any of the juice. The result is overflow. For example, when the mineral-laden water of a hot spring cools, calcite or even opal may precipitate from the water.

Particles may also crystallize out of a solution if more solute is added than the solution can hold. If you add juice to an already full glass, the glass will overflow. For example, aquatic plants and animals can increase the amount of carbon dioxide (CO_2) dissolved in the water around them and thereby cause calcium carbonate to crystallize out and sink to the bottom, or precipitate. And, if a chemical reaction in a solution produces a substance that doesn't dissolve in water, it will also precipitate.

C-2. The Scheme for Sedimentary Rock Identification

Sedimentary rocks are classified as *fragmental, organic,* or *chemical,* depending on the sediments from which they lithified. As with igneous rocks, texture and mineral composition are the main properties used to classify and name sedimentary rocks, as shown in Figure A.7. In sedimentary rocks, the mineral composition of the sediment particles tells us where the particles came from, and the texture provides clues about the processes that deposited the sediments.

In Figure A.7, sedimentary rocks are divided into two main groups. In the upper part of the chart are "Inorganic Land-Derived Sedimentary Rocks," that is, rocks formed from clastic sediments. In the lower part are "Chemically and/or Organically Formed Sedimentary Rocks," that is rocks formed from chemical or organic sediments.

Clastic sedimentary rocks are classified on the basis of grain size. Nonclastic rocks are identified primarily by composition and texture.

INORGANIC LAND-DERIVED SEDIMENTARY ROCKS					
TEXTURE	GRAIN SIZE	COMPOSITION	COMMENTS	ROCK NAME	MAP SYMBOL
Clastic (fragmental)	Mixed, silt to boulders (larger than 0.001 cm)	Mostly quartz, feldspar, and clay minerals; May contain fragments of other rocks and minerals	Rounded fragments	Conglomerate	
			Angular fragments	Breccia	
	Sand (0.006 to 0.2 cm)		Fine to coarse	Sandstone	
	Silt (0.0004 to 0.006 cm)		Very fine grain	Siltstone	
	Clay (less than 0.0006 cm)		Compact; may split easily	Shale	

CHEMICALLY AND/OR ORGANICALLY FORMED SEDIMENTARY ROCKS					
TEXTURE	GRAIN SIZE	COMPOSITION	COMMENTS	ROCK NAME	MAP SYMBOL
Nonclastic	Coarse to fine	Calcite	Crystals from chemical precipitates and evaporites	Chemical Limestone	
	Varied	Halite		Rock Salt	
	Varied	Gypsum		Rock Gypsum	
	Varied	Dolomite		Dolostone	
	Microscopic to coarse	Calcite	Cemented shells, shell fragments, and skeletal remains	Fossil Limestone	
	Varied	Carbon	Black and nonporous	Bituminous Coal	

Figure A.7 Scheme for Sedimentary Rock Identification.
Source: The State Education Department, *Earth Science Reference Tables*, 1994 ed. (Albany, New York: The University of the State of New York).

D. WHAT CAN BE LEARNED FROM A STUDY OF METAMORPHIC ROCKS?

D-1. Metamorphism

Igneous and sedimentary rocks can be changed to different kinds of rock by the influence of temperature, pressure, and the circulation of chemically active fluids. The changes that occur in solid rock under these conditions are known as *metamorphism*, from the Greek *meta,* meaning "change," and *morph,* meaning "form." The new rock that forms is called metamorphic rock. Metamorphic rocks are distinctively different from igneous or sedimentary rocks in appearance and method of formation. They commonly contain twisted, contorted shapes and layers, as well as exotic minerals found nowhere else. They form under conditions midway between those that produce sedimentary rocks and those that produce igneous rocks.

Metamorphism occurs while rocks are still in the solid state. Changes produced by weathering, although they occur while rocks are solid, are not considered metamorphism. Neither are igneous processes such as the melting of rocks in high-temperature zones. True metamorphism occurs beneath the surface at high temperatures and pressures. Under the conditions that produce metamorphism, ordinarily rigid, brittle rocks behave like soft modeling clay. The rocks can be bent and twisted without breaking. Rocks that originally had a clastic texture can recrystallize so that the metamorphic rock has a crystalline texture. Conversely, rocks that were originally crystalline can be crushed and broken into pieces and then relithified to form a clastic texture. The preexisting rock from which a metamorphic rock forms is called the *parent rock*. By investigating the conditions that produce metamorphic changes, it is possible to infer the parent rock from the mineral composition and structure of most metamorphic rocks.

Metamorphic rocks occur on a continuum from little alteration in the parent rock to major changes. Rarely are there sharp boundaries between parent rocks and metamorphic rocks; instead, gradations lead from one to the other. Limestone grades almost imperceptibly into marble. Shales grade into schists. Sandstones grade into quartzites. This characteristic has been helpful in determining the origin of certain metamorphic rocks. Metamorphic rocks are divided into groups based on the parent materials from which they were derived.

D-2. Types of Metamorphism

There are two major types of metamorphism: contact and regional. *Contact metamorphism* occurs when molten rock comes into contact with surrounding rocks. Heat given off by the cooling magma is conducted into the rock, where it causes changes in texture and mineral composition. Contact metamorphism is most intense at the contact between the magma and the surrounding rock, and decreases with distance from the magma until points are reached at which the effects of the heat are not felt. Generally, crystals are largest at the point of contact and become smaller with distance.

Regional metamorphism occurs over large areas and is generally associated with mountain building. Regional metamorphism takes its name from the large scale on which it occurs. For example, about 300 million years ago North America and Africa collided along a boundary between plates of the Earth's crust. The result was the crushing of rocks from Alabama to New York and the formation of the Appalachian Mountains. During regional metamorphism, deeply buried sedimentary and igneous rocks are squeezed both by the weight of rock layers above and by movements of the crust, are intruded by igneous rocks, and are infused with chemically active fluids.

D-3. The Scheme for Metamorphic Rock Identification

Metamorphic rocks are classified according to their textures (including foliation and banding) and compositions. In Figure A.8 you will see that metamorphic rocks are divided into two main groups, foliated and nonfoliated, based on texture. Then, within each of these groups, the rocks are classified by grain size.

If you look at the column headed "Composition," you will see that the foliates share a common composition, while each of the nonfoliates has a distinctly different composition. The majority of foliated rocks are derived from clastic sedimentary rocks rich in clay minerals, such as shale and siltstone. The "Comments" tell you what the parent rock was in each case.

TEXTURE		GRAIN SIZE	COMPOSITION	TYPE OF METAMORPHISM	COMMENTS	ROCK NAME	MAP SYMBOL
FOLIATED	Slaty	Fine	CHLORITE MICA	Regional	Low-grade metamorphism of shale	Slate	
	Schistose	Medium to coarse	QUARTZ FELDSPAR AMPHIBOLE GARNET PYROXENE		Medium-grade metamorphism; Mica crystals visible from metamorphism of feldspars and clay minerals	Schist	
	Gneissic	Coarse		(Heat and pressure increase with depth, folding, and faulting)	High-grade metamorphism; Mica has changed to feldspar	Gneiss	
NONFOLIATED		Fine	Carbonaceous		Metamorphism of plant remains and bituminous coal	Anthracite Coal	
		Coarse	Depends on conglomerate composition		Pebbles may be distorted or stretched; Often breaks through pebbles	Meta-conglomerate	
		Fine to coarse	Quartz	Thermal (including contact) or Regional	Metamorphism of sandstone	Quartzite	
			Calcite, Dolomite		Metamorphism of limestone or dolostone	Marble	
		Fine	Quartz, Plagioclase	Contact	Metamorphism of various rocks by contact with magma or lava	Hornfels	

Figure A.8 Scheme for Metamorphic Rock Identification
Source: The State Education Department, *Earth Science Reference Tables*, 1994 ed. (Albany, New York: The University of the State of New York).

Notice that slate, schist, and gneiss really form a continuum. Slate forms first as shale is slightly metamorphosed. The flakelike micas align because of compression and form surfaces along which the rock will split. If slate is further metamorphosed, chlorite changes to biotite, foliation becomes stronger, grains become coarser because of recrystallization, and a schist is formed. Still further metamorphism causes the crystals to grow even larger, alternating layers of light and dark minerals (bands) form as crystals migrate because of density differences, and a gneiss forms. The larger, interlocking crystals of a gneiss cause these rocks to break less regularly than schists, but they will often break completely across layers. In the highly metamorphic environment in which gneisses form, rock is plastic and will often be contorted into the weirdly twisted patterns seen in gneiss.

E. WHY IS THE CONSERVATION OF MINERALS AND OTHER NATURAL RESOURCES SO IMPORTANT?

E-1. Minerals

The minerals in the Earth are a treasure of almost unimaginable value. They help to fill many human needs, which, through time, have become greater. Minerals are used in making products ranging from steel to electric light bulbs. Unfortunately, minerals are a nonrenewable resource, that is, they can be used only once and then they are gone.

Imagine that every week you put a few cents of your allowance into a bank. For three years you do not take any money from the bank. Then you see a video game that you want, you open the bank, and you use the money to buy the game. In a few minutes, you have spent the money it took you three years to save. In the same way, it is easy to use up in a few years minerals that took millions of years to form. But while you will probably save more money in the future, mineral resources may not re-form within your lifetime or even a thousand lifetimes. Every time a mineral is used, that much less remains. Thus, our mineral resources will not last forever.

For this reason, people must conserve mineral resources, not just by saving these resources, but also by making sure that there is no waste in using them. Recycling and modernizing factories are parts of a good conservation policy. Another important part is research into better ways of production.

If conservation is practiced by everyone, we can make the best use of our mineral resources. Today there are recycling centers throughout the United States, and many communities have instituted mandatory recycling. Recycling centers process used aluminum cans, old bottles, scrap metals, and newspapers. These materials are then reused in the manufacture of aluminum, glass, steel, and paper products. You may have noticed on some cans a statement that they were made from recycled aluminum.

E-2. Fossil Fuels

At certain times in the Earth's history, environmental conditions have enabled plants and marine organisms to grow at a rapid rate. Over millions of years, their bodies stored energy from sunlight that was converted into chemical bonds during photosynthesis. When they completed their life cycles, their remains accumulated faster than decomposers could break them down and recycle them back into the environment. When these remains became buried by deposition, the pressure of overlying sediments and the increased temperatures at depth converted the layers of organic remains into coal beds and pools of petroleum. Since coal and petroleum are derived from organic remains and are used chiefly as fuels, they are called fossil fuels.

Fossil fuels are a vital source of energy and petrochemicals for the global economy. Transportation, urban development, industry, commerce, agriculture, and many other human activities depend on the amount and type of energy available. Most of the energy used today is obtained by burning fossil fuels. While coal was widely used in the past, recently petroleum and natural gas have become the fuels of choice because they are easier to collect than coal, have many uses in industry, and are concentrated, easily portable sources of energy for cars, trucks, airplanes, and trains. Unfortunately, the burning of fossil fuels releases into the atmosphere waste products that threaten the health of living things and have environmental risks associated with them. Current environmental conditions are not resulting in the large-scale accumulation of remains that can be converted into coal or petroleum. Therefore, for the foreseeable future, organic fossil fuels are a nonrenewable resource.

E-3. Global Implications of Resource Use and Distribution

The use and the distribution of mineral resources have global political, financial, and social implications. Highly industrialized nations use more energy and mineral resources than less developed countries. The energy and natural resources we consume contribute to our high standard of living. But this consumption has also led to a rapid depletion of the Earth's natural resources and to mounting environmental crises, ranging from marine life killed by oil-tanker spills to the breakdown of the ozone layer of the atmosphere by the chloroflourocarbons (CFC's) used in refrigerators and air conditioners.

As developing nations industrialize and their urban centers grow, they demand more energy and natural resources, thereby entering into competition with all of the other industrialized nations for these resources. Often this competition has resulted in friction between nations and even in war. If we are to live in peace and maintain our environment, we must develop ways to slow the depletion of natural resources and to use them more efficiently.

The depletion of energy and other natural resources can be slowed by decisions to use efficient technologies, to conserve, and to recycle. These decisions can be implemented at many levels ranging from the national down to the personal. Government can restrict low-priority uses of materials, such as the use of petrochemicals to manufacture purely ornamental packaging that adds to our solid-waste disposal problems. Government can also set standards of technical efficiency, for example, mileage and emission requirements for automobiles or the use of insulation in home construction. Cars that are built with more power than their function warrants waste energy. (If the speed limit is 55 mph, why build cars with the power to do 100 mph?) However, such decisions often involve trade-offs of cost and social values. For example, when other, less damaging compounds are substituted for CFC's, new compressors are required, so older units must be replaced before they have worn out. Reduced emissions depend on catalytic converters filled with toxic heavy metals. The development of wise policies for conserving and managing natural resources is one of the great challenges now facing all nations.

REVIEW QUESTIONS FOR TOPIC A

1. Which two elements listed below are most abundant by weight in the Earth's crust?
   (1) silicon and oxygen
   (2) hydrogen and iron
   (3) oxygen and magnesium
   (4) hydrogen and calcium

2. The data table below shows the composition of six common rock-forming minerals

Mineral	Composition
Mica	$KAl_3Si_3O_{10}$
Olivine	$(FeMg)_2SiO_4$
Orthoclase	$KAlSi_3O_8$
Plagioclase	$NaAlSi_3O_8$
Pyroxene	$CaMgSi_2O_6$
Quartz	SiO_2

   The data table provides evidence that
   (1) the same elements are found in all minerals
   (2) a few elements are found in many minerals
   (3) all elements are found in only a few minerals
   (4) all elements are found in all minerals

3. The diagram below represents a single silicon-oxygen tetrahedron unit. Two different minerals have these same units arranged in different patterns. How will the minerals differ?

KEY:
● SILICON ATOM
○ OXYGEN ATOM

(1) One mineral will have some physical properties different from the other.
(2) One mineral will be more radioactive than the other.
(3) One mineral will have the silicon atom outside the tetrahedron while the other will have it inside the tetrahedron.
(4) One mineral will have larger silicon atoms than the other.

4. Which object is the best model of the shape of a silicon-oxygen structure unit?

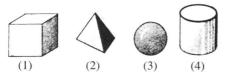

(1)   (2)   (3)   (4)

5. Which element combines with silicon to form the tetrahedral unit of structure of the silicate minerals?
(1) oxygen   (3) potassium
(2) nitrogen   (4) hydrogen

6. The most abundant minerals found in the rocks of the Earth's crust are
(1) phosphates   (3) silicates
(2) carbonates   (4) sulfates

7. The crystal characteristics of quartz shown in the accompanying diagram are the result of the
(1) internal arrangement of the elements from which quartz is formed
(2) shape of the other rock crystals in the area where the quartz was formed
(3) amount of weathering that the quartz has been exposed to
(4) age of the quartz crystal

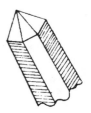

8. What causes the characteristic crystal shape and cleavage (breaking along flat surfaces) of the mineral halite as shown in the accompanying diagram?
   (1) metamorphism of the halite
   (2) the internal arrangement of the atoms in halite
   (3) the amount of erosion the halite has undergone
   (4) the shape of other minerals located where the halite formed

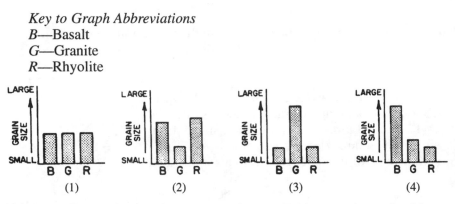

HALITE (salt)

9. Two mineral samples have different physical properties, but each contains silicate tetrahedra as its basic structural unit. Which statement about the two mineral samples must be true?
   (1) They have the same density.
   (2) They are similar in appearance.
   (3) They contain silicon and oxygen.
   (4) They are the same mineral.

10. Which characteristics would give the best evidence about the conditions under which a rock was formed?
    (1) the rock's density and size
    (2) the rock's structure and texture
    (3) the rock's minerals and color
    (4) the rock's shape and phase

11. According to the *Earth Science Reference Tables*, which graph best represents the comparison of the average grain sizes in basalt, granite, and rhyolite?

    Key to Graph Abbreviations
    B—Basalt
    G—Granite
    R—Rhyolite

12. According to the *Earth Science Reference Tables,* granite and gabbro
    (1) are both intrusive
    (2) are both extrusive
    (3) have different grain sizes
    (4) both contain potassium feldspar

13. According to the *Earth Science Reference Tables,* rhyolite is an example of a
(1) monomineralic igneous rock
(2) polymineralic igneous rock
(3) monomineralic sedimentary rock
(4) polymineralic sedimentary rock

14. Sand collected at a beach contains a mixture of pyroxene, olivine, hornblende, and plagioclase feldspar. According to the *Earth Science Reference Tables,* the rock from which this mixture of sand came is best described as
(1) dark-colored with a mafic composition
(2) dark-colored with a felsic composition
(3) light-colored with a mafic composition
(4) light-colored with a felsic composition

15. The best evidence for determining the cooling rate of an igneous rock during its solidification is provided by
(1) index fossils
(2) faults in the rock
(3) the crystal size of its minerals
(4) the disintegration of radioactive substances

16. According to the *Earth Science Reference Tables,* generally, as the percentage of felsic minerals in a rock increases, the rock's color will become
(1) darker and its density will decrease
(2) lighter and its density will increase
(3) darker and its density will increase
(4) lighter and its density will decrease

17. An igneous rock that has crystallized deep below the Earth's surface has the following approximate composition: 70 percent pyroxene, 15 percent plagioclase, and 15 percent olivine. According to the *Earth Science Reference Tables,* what is the name of this igneous rock?
(1) granite   (3) gabbro
(2) rhyolite   (4) basalt

18. A fine-grained rock has the following mineral composition: 30 percent plagioclase feldspar, 60 percent pyroxene, 5 percent olivine, and 5 percent hornblende. This rock would most likely be
(1) rhyolite   (3) gabbro
(2) granite   (4) basalt

19. Which combination of minerals could represent a rock categorized as being andesite on the Scheme for Igneous Rock Identification in the *Earth Science Reference Tables?*
(1) 10% potassium feldspar, 30% quartz, 30% plagioclase, 15% mica, 15% hornblende
(2) 40% potassium feldspar, 35% quartz, 12% plagioclase, 8% mica, 5% hornblende
(3) 80% potassium feldspar, 10% plagioclase, 5% mica, 5% hornblende
(4) 5% quartz, 55% plagioclase, 10% mica, 30% hornblende

20. According to the *Earth Science Reference Tables,* rhyolite and granite are alike in that they both are
(1) fine-grained   (3) mafic
(2) dark-colored   (4) felsic

21. According to the Scheme for Igneous Rock Identification in the *Earth Science Reference Tables,* which statement best describes the percentage of plagioclase feldspars in a sample of gabbro?
(1) The percentage of plagioclase feldspars in gabbro can vary.
(2) Gabbro always contains less plagioclase than pyroxene.
(3) Plagioclase feldspars always make up 25% of a gabbro sample.
(4) Gabbro contains no plagioclase feldspars.

22. Which characteristic of rocks tends to increase as the rocks are metamorphosed?
(1) density   (3) permeability
(2) volume   (4) number of fossils present

23. According to the *Earth Science Reference Tables,* which metamorphic rock will have visible mica crystals and a foliated texture?
(1) pumice   (3) schist
(2) shale   (4) slate

24. According to the *Earth Science Reference Tables,* metamorphic rocks form as the direct result of
(1) precipitation from evaporating water
(2) melting and solidification in magma
(3) erosion and deposition of soil particles
(4) heat and pressure causing changes in existing rock

25. Which rocks would most likely be separated by a transition zone of altered rock (metamorphic rock)?
(1) sandstone and limestone
(2) granite and limestone
(3) shale and sandstone
(4) conglomerate and siltstone

26. Which statement best explains why shark teeth found in bedrock indicate that the bedrock is sedimentary?
(1) Shark's teeth are a key component of all sedimentary rocks
(2) Shark's exist only in water and only sedimentary rocks can form in water
(3) Processes that form igneous or metamorphic rocks would destroy shark teeth
(4) Sharks have existed for millions of years, igneous and metamorphic rocks have not

27. According to the *Earth Science Reference Tables,* which sequence of events occurs in the formation of a sedimentary rock?

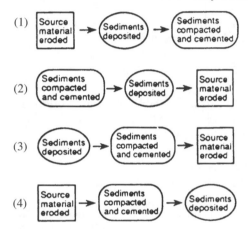

28. Rock layers showing ripple marks, cross-bedding, and fossil shells indicate that these layers were formed
(1) from solidification of molten material
(2) from deposits left by a continental ice sheet
(3) by high temperature and pressure
(4) by deposition of sediments in a shallow sea

29. Large rock salt deposits in the Syracuse area indicate that the area once had
(1) large forests
(2) a range of volcanic mountains
(3) many terrestrial animals
(4) a warm, shallow sea

30. According to the *Earth Science Reference Tables,* which sedimentary rock would be formed by the compaction and cementation of particles 1.5 centimeters in diameter?
(1) shale   (3) conglomerate
(2) sandstone   (4) siltstone

85

31. Which rock is composed of materials that show the greatest variety of rock origins?
(1) a limestone composed of coral fragments cemented together by calcium carbonate
(2) a conglomerate composed of pebbles of granite, siltstone, and basalt
(3) a very fine-grained basalt with sharp edges
(4) a sandstone composed of rounded grains of quartz

32. According to the *Earth Science Reference Tables,* which rock most likely formed as a result of biologic processes?
(1) granite   (3) sandstone
(2) basalt   (4) limestone

33. According to the *Earth Science Reference Tables,* a rock that forms directly from land-derived sediments is
(1) sandstone   (3) gabbro
(2) dolostone   (4) granite

34. According to the *Earth Science Reference Tables,* limestone, gypsum, and salt are rocks formed by the processes of
(1) melting and solidification
(2) evaporation and precipitation
(3) erosion and deposition
(4) weathering and metamorphism

35. According to the *Earth Science Reference Tables,* which characteristic determines whether a rock is classified as a shale, a siltstone, a sandstone, or a conglomerate?
(1) the absolute age of the sediments within the rock
(2) the mineral composition of the sediments within the rock
(3) the particle size of the sediments within the rock
(4) the density of the sediments within the rock

36. Some sedimentary rocks are composed of rock fragments that had different origins. Which statement best explains why this could occur?
(1) Fossils are often found in sedimentary rock
(2) Sedimentary rocks form from the weathered products of any type of rock
(3) When molten lava solidifies to form sedimentary rock, it often contains foreign particles
(4) Under high heat and pressure, recrystallization results in the formation of many minerals

37. Which rock is formed when rock fragments are deposited and cemented together?
(1) dolostone   (3) rhyolite
(2) sandstone   (4) gabbro

38. According to the *Earth Science Reference Tables,* some sedimentary rocks form as the direct result of the
(1) solidification of molten magma
(2) recrystallization of material
(3) melting of minerals
(4) cementation of rock fragments

39. When dilute hydrochloric acid is placed on the sedimentary rock limestone and the nonsedimentary rock marble, a bubbling reaction occurs with both. What does this indicate?
(1) The minerals of these two rocks have similar chemical compositions
(2) The molecular structures of these two rocks have been changed by heat and pressure
(3) The physical properties of these two rocks are identical
(4) The two rocks originated at the same location

40. According to the *Earth Science Reference Tables,* which is a sedimentary rock that forms as a result of precipitation from seawater?
(1) conglomerate   (3) basalt
(2) gypsum   (4) shale

Base your answers to questions 41–43 on the diagram below, the *Earth Science Reference Tables,* and your knowledge of earth science. The diagram shows the elements found in four minerals.

	O	Si	Al	Fe	Ca	Na	C
Quartz	▨	▨					
Feldspar	▨	▨	▨		▨	▨	
Olivine	▨	▨		▨	▨		
Diamond							▨

▨ = element present

41. Which mineral contains the greatest variety of elements?
(1) quartz   (3) olivine
(2) feldspar   (4) diamond

42. Which of the four minerals shown is *not* a silicate mineral?
(1) quartz   (3) olivine
(2) feldspar   (4) diamond

43. Which of the following elements is the most abundant in Earth's crust?
(1) iron   (3) oxygen
(2) silicon   (4) calcium

Base your answers to questions 44 and 45 on the chart below. The chart shows information about selected mineral and energy resources.

Group	Mineral Resource	Uses
Native Elements	Gold	coins, jewelry, investment, electrical conductors, dental fillings
	Copper	electrical wiring, plumbing, coins
	Graphite	lubricants, pencil "lead"
Mineral Compounds	Hematite (ore of iron)	construction, motor vehicles, machinery parts
	Halite	food additive, melting of ice, water softeners, chemicals
	Garnet	abrasives (sandpaper), jewelry
	Feldspar	porcelain, glass, ceramics
Fuels	Coal	heating, electric generation plants, plastics and other synthetic chemicals
	Petroleum	automobile fuel, lubricants, plastics and other synthetic chemicals, medicines

44. Which natural resource would last longer if people used public transportation and set thermostats to lower indoor temperatures in the winter?
(1) petroleum
(2) copper
(3) feldspar
(4) graphite

45. What is the primary source of all of the materials listed in the chart above?
(1) recycled and discarded waste materials
(2) deposits within the Earth's crust
(3) substances extracted from ocean water
(4) meteorites from outer space

Base your answers to questions 46 through 50 on your knowledge of earth science, the *Earth Science Reference Tables*, and the table of minerals below. The table shows the physical properties of nine minerals.

Mineral	Color	Luster	Streak	Hardness	Density (g/ml)	Chemical Composition
biotite mica	black	glassy	white	soft	2.8	$K(Mg,Fe_3)$ $(AlSi_3O_{10})$ (OH_2)
diamond	varies	glassy	colorless	hard	3.5	C
galena	gray	metallic	gray-black	soft	7.5	PbS
graphite	black	dull	black	soft	2.3	C
kaolinite	white	earthy	white	soft	2.6	$Al_4(Si_4O_{10})$ $(OH)_8$
magnetite	black	metallic	black	hard	5.2	Fe_3O_4
olivine	green	glassy	white	hard	3.4	$(Fe,Mg)_2$ SiO_4
pyrite	brass-yellow	metallic	greenish-black	hard	5.0	FeS_2
quartz	varies	glassy	colorless	hard	2.7	SiO_2

Chemical Symbols

Al—aluminum	Pb—lead	C—carbon
Si—silicon	Fe—iron	K—potassium
H—hydrogen	S—sulfur	Mg—magnesium
O—oxygen		

Definitions

Luster: the way a mineral's surface reflects light
Streak: the color of a powdered form of the mineral
Hardness: the resistance of a mineral to being scratched
(soft—easily scratched; hard—not easily scratched)

46. Which mineral has a different color in its powdered form than in its original form?
    (1) pyrite   (3) kaolinite
    (2) graphite   (4) magnetite

47. Which mineral contains iron, has a metallic luster, is hard, and has the same color and streak?
    (1) biotite mica   (3) kaolinite
    (2) galena   (4) magnetite

48. Which mineral is commonly found in granite?
(1) quartz   (3) magnetite
(2) olivine   (4) galena

49. Why do diamond and graphite have different physical properties, even though they are both composed entirely of the element carbon?
(1) Only diamond contains radioactive carbon
(2) Only graphite consists of organic material
(3) The minerals have different arrangements of carbon atoms
(4) The minerals have undergone different amounts of weathering

50 Which mineral would most likely be weathered most after being placed in a container and shaken for 10 minutes?
(1) pyrite   (3) magnetite
(2) quartz   (4) kaolinite

Base your answers to questions 51 through 55 on the Scheme for Igneous Rock Identification in the *Earth Science Reference Tables,* the diagram below, and your knowledge of earth science. The diagram shows a top view of the bedrock geology of a portion of the Earth's surface. Two faults (F_1 and F_2) and three periods of igneous activity have occurred in this area.

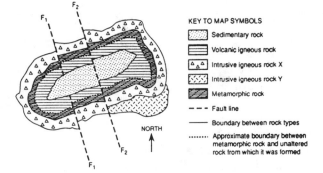

51. Fossil remains would most likely be found in the rock represented by which symbol?

(1)   (2)   (3)   (4)

52. The best evidence that rock Y formed after rock X would be finding rock Y
(1) present as broken pieces within rock X
(2) intruded into rock X
(3) cut by a fault that also cuts rock X
(4) weathered more than rock X

53. When did faults F_1 and F_2 most likely occur?
  (1) before the metamorphic rock was formed
  (2) before the intrusion of igneous rock X
  (3) at two different times
  (4) after the sedimentary rock was deposited

54. The metamorphic rock was most likely which rock originally?
  (1) the sedimentary rock
  (2) the volcanic igneous rock
  (3) intrusive igneous rock X
  (4) intrusive igneous rock Y

55. The diagram below shows a sample of igneous rock Y.

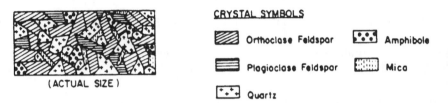

(ACTUAL SIZE)

CRYSTAL SYMBOLS

▨ Orthoclase Feldspar   ▦ Amphibole

▤ Plagioclase Feldspar   ▦ Mica

▦ Quartz

Based on its mineral composition and crystal size, what is igneous rock Y?
  (1) rhyolite   (3) conglomerate
  (2) basalt   (4) granite

The answers to questions 56 through 60 must be based on information provided in the table below.

				Mineral					
Property	Biotite	Corundum	Galena	Kaolinite	Limonite	Magnetite	Microcline	Quartz	Sphalerite
color	black	varies	gray	white	brown	black	varies	varies	varies
luster	glassy	glassy	metallic	earthy	sub-metallic	metallic	glassy	glassy	resinous
streak	white		gray-black	white	rust	black	white		yellow-white
hardness	2.5	9.0	2.5	2.0	5.0	6.0	6.5	7.0	3.5
specific gravity	2.8	4.0	7.5	2.6	4.0	5.2	2.5	2.65	4.0
cleavage	1 plane perfect	rhombo-hedral	cubic, perfect	none	none	imperfect	2 planes at right angles	none	6 planes
chemical composition	$K(Mg,Fe)_3$ $(AlSi_4O_{10})$ $(OH)_2$	Al_2O_3	PbS	Al_4 (Si_4O_{10}) $(OH)_8$	$FeO(OH)$ $\cdot nH_2O$	Fe_3O_4	$KAlSi_3O_8$	SiO_2	ZnS

Chemical Symbols

Al—aluminum	O—oxygen	H—hydrogen	K—potassium	Fe—iron
Si—silicon	Pb—lead	S—sulfur	Mg—magnesium	Zn—zinc

56. Which mineral in the table would scratch all of the others listed?
  (1) biotite   (3) kaolinite
  (2) corundum   (4) quartz

57. A geologist grinds one of these minerals to a gray-black powder; he then analyzes it and discovers that it contains sulfur. What is the mineral?
    (1) biotite   (3) magnetite
    (2) galena   (4) sphalerite

58. A given volume of kaolinite weights one-half as much as an equal volume of which mineral?
    (1) corundum   (3) magnetite
    (2) galena   (4) quartz

59. Which is the potassium-containing mineral with two cleavage planes?
    (1) biotite   (3) magnetite
    (2) galena   (4) microcline

60. Which is an iron-bearing mineral that is chemically combined with water?
    (1) corundum   (3) limonite
    (2) kaolinite   (4) magnetite

Base your answers to questions 61 through 65 on the *Earth Science Reference Tables* and the information and graph below.

Several drill-core samples of sandstone and siltstone were collected from various depths below the Earth's surface. The porosities of the sandstone and siltstone were determined by analyzing the drill-core samples. The graph shows the relationship between the porosity of the rock samples and the depth below the Earth's surface.

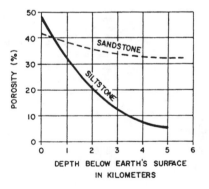

DEPTH BELOW EARTH'S SURFACE
IN KILOMETERS

61. According to the *Earth Science Reference Tables,* both of these sedimentary rocks are formed by the
    (1) chemical replacement of skeletons and shell fragments
    (2) burial and recrystallization of clay particles
    (3) compaction and cementation of rock particles
    (4) precipitation of soluble minerals from evaporating water

62. From which layer of the Earth were the drill-core samples most likely taken?
(1) crust (3) outer core
(2) mantle (4) inner core

63. As the distance below the Earth's surface increases from 1 to 4 kilometers, the porosity of the sandstone
(1) decreases, only
(2) increases, only
(3) decreases, then increases
(4) remains the same

64. Collecting drill-core rock samples and measuring their porosities to determine the data shown on the graph are examples of
(1) predicting the results of laboratory tests
(2) extending the sense by the use of instruments
(3) altering the properties of similar rock material
(4) calculating the percent of deviation

65. At approximately what depth are the porosities equal for the sandstone and siltstone tested in this investigation?
(1) 0 km (3) 3 km
(2) 0.5 km (4) 5 km

UNIT THREE _____

The Dynamic Crust

KEY IDEAS The Earth is a dynamic planet. Its surface is constantly being changed by weathering and erosion. Its interior also changes, causing earthquakes, the eruption of volcanoes, the uplift of mountains, and the opening of ocean basins. Earth's internal changes involve deformation of the rocks in the crust and the mantle.

Our understanding of the structure of the Earth's interior is based on analysis of seismic waves that have passed through the Earth. Our ideas about the Earth change as new discoveries are made.

KEY OBJECTIVES
Upon completion of this unit, you should be able to:

- Explain how earthquake waves can be used to create and refine a model of the Earth's interior.
- Compare and contrast the composition, density, and phase of matter of: the inner core, outer core, mantle, and crust.
- Identify regions of high earthquake and volcanic activity and relate them to the structure of the Earth's crust.
- Describe evidence of crustal movements.
- Relate specific forms of crustal activity to the various types of plate boundaries.

A. WHY ARE EARTHQUAKES AND VOLCANOES IMPORTANT TO US?

A-1. Earthquakes

An *earthquake* is a sudden trembling of the ground. Seismologists, scientists who study earthquakes, estimate that over 1 million quakes occur each year—about 1 every second! During most earthquakes, the shaking is so slight it is barely noticeable to the senses. During minor earthquakes the ground feels a bit like a railway station when the train rumbles in. Major earthquakes cause violent shaking and lurching of the ground. The ground may split open, buildings may topple, pipes and power lines may be broken, and roadways may collapse. A destructive earthquake is a catastrophe, something to be prepared for and guarded against.

Causes of Earthquakes

The major cause of earthquakes is *faulting*, the sudden movement of rock along planes of weakness, called faults, in the Earth's crust. However, some earthquakes are caused by volcanic eruptions, and a few by explosions set off by humans. Nuclear testing is usually detected by the earthquakes it causes.

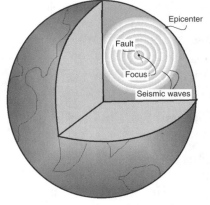

Figure 3.1 Fault, Focus, Epicenter, and Emanating Seismic Waves.

The elastic rebound theory explains the mechanism by which faulting causes earthquakes. At some places in the crust, immense forces push or pull on the rock. Under this stress, rock can bend elastically, much as a thin wooden ruler can bend. However, if the stress increases, a point is reached where the rock can bend no farther without breaking; it is at its elastic limit. If stressed beyond this point, the rock snaps and the two broken edges whip back, or rebound. Great masses of rock suddenly scrape past each other, and the shock of this wrenching action jars the crust and sets an earthquake in motion.

The point where the rock breaks is called the *focus* of the earthquake. The focus may be just beneath the surface or hundreds of kilometers down. The earthquake is first felt at the *epicenter*, a point on the Earth's surface directly above the focus (see Figure 3.1).

Earthquake Waves

When faulting occurs, vibrations called *seismic waves* spread out in all directions from the focus. As these seismic waves move through the crust, they cause the rock to shake and tremble. When they reach the Earth's surface, they are transmitted to loose rock, soil, water, buildings, or any other objects resting on the surface, so that they too shake and tremble.

The vibrations produced by faulting are called seismic *waves* because they behave like waves. Seismic waves refract, or bend, when they travel from one material into another. Also, they vary in speed according to the density of the material through which they travel; in general, the denser the medium, the greater the speed at which the wave moves. And just as some, but not all, wave types will travel through liquids, some seismic waves travel through liquids and others do not. Together, these characteristics of seismic waves enable seismologists to make inferences about the nature of the Earth's interior (see Figure 3.2) based on how seismic waves behave as they travel through the Earth.

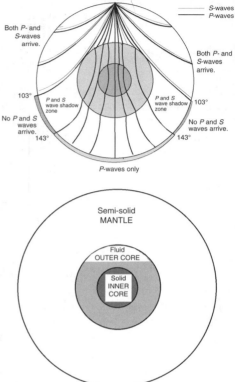

Figure 3.2 Regions of the Earth's Interior Inferred from Seismic Wave Behavior

Measuring Earthquakes

As seismic waves travel farther from the focus, they lose energy and their effects lessen. At a distance of 100 kilometers from the focus, the energy of an earthquake is only about 1/10,000 of what it was at 1 kilometer from the focus. The Richter and the Mercalli scales are used to measure the strengths of earthquakes. The Richter scale (see Figure 3.3) is based on the energy released by the earthquake and is determined by direct measurements of the motions of the crust using seismic instruments. The Mercalli scale (see Table 3.1) is based on descriptions of earthquake damage and is concerned more with the impact of an earthquake on structures made by humans and on human activities.

Effects of Earthquakes

The effects of earthquakes depend mainly on the strength of the seismic waves and the nature of the material through which they pass. Earthquake effects include ground shaking and failure, surface faulting, changes in wells, and geysers and tsunamis.

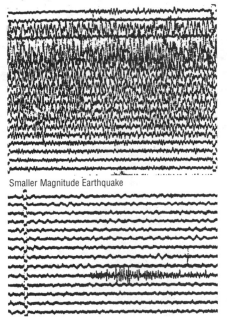

Larger Magnitude Earthquake

Smaller Magnitude Earthquake

Figure 3.3 The Richter Magnitude Scale. This is based on the motion of bedrock as recorded by seismographs.

Ground shaking affects solid rock the least, and loose, water-soaked ground the most. The same shock that causes solid bedrock to sway slightly may cause intense jolting in nearby wet clay or landfill. Tapping a bowl of Jell-O is a good analogy. The solid bowl vibrates but hardly moves, while the Jell-O jiggles and quivers noticeably. In ground failure, strong vibrations cause loose or water-soaked ground to break up and settle, forming cracks and fissures. On hillsides these cracks can trigger landslides, slumping, and mudflows.

Surface faulting occurs when movement along a fault causes the surface to be lifted up, lowered, or shifted sideways. The result is displacement of surface features and structures.

As seismic waves travel through water-saturated rock or soil, water is compressed and forced through pore spaces. Like a hand squeezing and releasing a soaked sponge in a bucket, seismic waves force water in and out of rock and soil surrounding wells, causing their water levels to fluctuate. In rock fissures a similar effect is produced on geysers.

Tsunamis are immense sea waves caused by earthquakes beneath the ocean floor or by undersea landslides. Tsunamis may be only a few meters high, but they are very long and travel much more rapidly than ordinary ocean waves. Tsunamis have been clocked moving faster than 500 kilometers per hour with wavelengths as great as 200 kilometers. When tsunamis reach shallow water, they are slowed by friction with the ocean bottom. In bays and narrow channels, however, their high speed and long wavelength may be funneled into huge breaking waves more than 20 meters high. The force exerted by such an enormous, fast moving mass of water can do extensive damage.

Earthquakes are geologic hazards to humans, but loss of lives and property can be minimized by anticipating the hazards that earthquakes pose and preparing to meet them.

TABLE 3.1 THE MODIFIED MERCALLI SCALE

Intensity Value	Description of Effects
I	Not felt.
II	Felt by persons at rest; felt on upper floors because of sway.
III	Felt indoors. Hanging objects swing. Feels like a passing truck.
IV	Hanging objects swing. Feels like heavy truck passing, or a jolt is felt. Standing cars rock. Windows and dishes rattle. Glasses clink. In the upper range of IV, wooden frames and walls creak.
V	Felt outdoors and direction can be estimated. Sleepers are awakened. Liquids are disturbed, some spill. Small, unstable objects upset. Doors swing open and close. Shutters and pictures move.
VI	Felt by all. Persons walk unsteadily, and many are frightened and run indoors. Windows, dishes, and glassware are broken. Furniture is moved or overturned. Pictures fall from walls. Weak plaster cracks. Trees and bushes visibly shaken or heard to rustle.
VII	Difficult to stand. Noticed by drivers of cars. Hanging objects quiver. Furniture is broken. Weak chimneys break at roof line. Fall of plaster, loose bricks, masonry. Waves on ponds and water muddied. Small slides and cave-ins of sand and gravel banks.
VIII	Steering of moving cars is affected. Partial collapse of masonry structures. Chimneys and smokestacks twist and fall. Frame houses move on foundation if not bolted down. Branches broken from trees. Wet ground and steep slopes crack.
IX	General panic. Weak masonry destroyed, stronger masonry cracks, is seriously damaged. Frame structures not bolted shift off foundations. Frames cracked. Reservoirs seriously damaged. Underground pipes broken. Conspicuous cracks in ground.
X	Masonry and frame structures destroyed along with their foundations. Some bridges destroyed. Serious damage to reservoirs, dikes, dams, and embankments. Large landslides. Water thrown out of lakes, canals, rivers, etc. Rails bent slightly.
XI	Rails bent greatly. Underground pipelines completely destroyed and out of service.
XII	Damage nearly total. Large rock masses displaced. Objects thrown into the air. Lines of sight and level are distorted.

Modified from: H.O. Wood and F. Neumann, "Modified Mercalli Intensity Scale of 1931," *Bulletin of the Seismological Society of America,* Vol. 21, No. 4, pp. 277–288.

A-2. Volcanoes

Origin of Magma

There is much evidence that the Earth's interior is very hot. Two examples of this evidence are the yellow-hot molten rock that flows from the Earth during a volcanic eruption and the hot water and steam that emerge from the ground in hot springs and geysers. Also, measurements made in mines and drill holes in the Earth's crust show that, for every kilometer of depth beneath the sur-

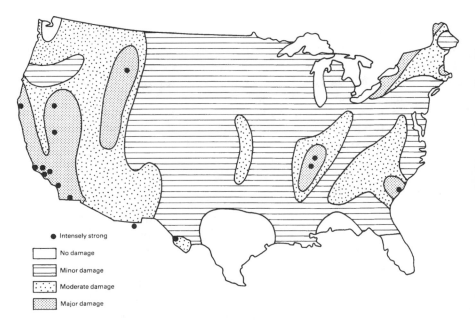

Figure 3.3a Seismic Risk Map of the United States.
From: Earth Science on FIle. © Facts on File, Inc., 1988.

● Intensely strong

▢ No damage

▤ Minor damage

▨ Moderate damage

▩ Major damage

face, the temperature rises about 30°C. From this we can infer that, at the core, temperatures may reach 5,000°C or more. Possible sources of this internal heat include the decay of radioactive elements within the Earth, residual heat from the formation of the Earth, and internal pressure and friction.

Although temperatures within the Earth are high enough to melt rock, the enormous pressures deep within the Earth prevent the rock from turning into a liquid. Melting occurs only in places where heat is concentrated or pressure is reduced. Molten rock inside the Earth is called *magma*. Most magma forms in the upper mantle or lower crust at depths of 40–60 kilometers. The process by which magma rises into the crust and is extruded is called *volcanism*.

Formation of Volcanoes

A *volcano* is both the opening in the crust through which magma erupts and the mountain built by the erupted material. Volcanoes form where cracks in the Earth's crust lead to a magma

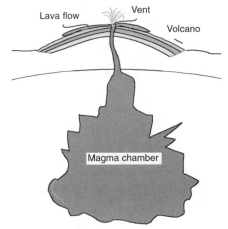

Figure 3.4 A Cross Section of a Volcano. Magma rises from chambers in the upper mantle until it exits the crust through an opening called the vent, after which it is called lava, and accumulates in a mound called a volcano.

99

chamber. Since liquid magma is less dense than the surrounding solid rocks and is under pressure, it rises toward the surface through these cracks. As magma rises, dissolved gases in it expand and are released, giving the magma an upward boost. The closer to the surface, the less confining pressure there is to overcome and the faster the magma and gases move until they exit the crust through an opening called the vent. Depending on its viscosity, magma may either pour out quietly onto the surface as a flood of molten rock, or spurt out explosively, showering the surrounding terrain with solid rock and globs of molten rock. Once magma emerges from the surface, it is called *lava* (see Figure 3.4.)

Volcanic Structures

Erupted materials spread out in all directions around the vent, or central opening of a volcano. With each new eruption, more material piles up around the vent and a cone-shaped mound forms. The shape of the cone depends on the viscosity of the magma.

Fluid lava tends to flow quietly through the vent, spreading out in wide, thin sheets. The result is a flat, wide cone called a *shield volcano* (for its resemblance to a shield when viewed from overhead).

Thick, pasty lava containing a lot of dissolved gas does not flow readily through the vent, which can easily become clogged or blocked completely. As a result gases cannot escape and pressure builds up until a violent explosion takes place. During an explosive eruption, rock and soil surrounding the

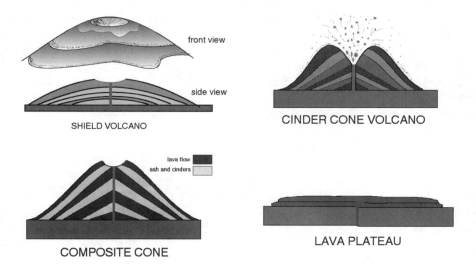

Figure 3.5 Typical Volcanic Structures. Broad, gently sloping shield. Volcanoes form from successive lava flows. Steep-sided cinder cones form from ash and cinders. Composite cones of intermediate slope form when periods of lava flows are interspersed with eruptions of ash and cinder. Very fluid lava that erupts along a fissure spreads out to form a lava plateau.

vent are thrown upward, along with chunks and droplets of lava blown out of the vent. The lava cools as it hurtles through the air and is generally solid by the time it reaches the ground. Collectively, all of the fragments of solidified lava ejected during a volcanic eruption are called *tephra*. This "rain" of solid particles produces a steep, narrow cone called a *cinder cone volcano* (for the particles' resemblance to cinders).

Alternating explosive and quiet eruptions give rise to large, symmetrical cones, called *composite cones,* made of alternating layers of solidified lava and volcanic rock particles. In some cases, magma emerges through long, open cracks, or fissures. Lava pouring out of fissures spreads out in wide, thin sheets; but instead of forming a cone, as it would around a central vent, it forms a sheet, called a *lava plateau,* that blankets the surrounding land.

Figure 3.5 shows the various types of volcanic structures.

Intrusive Activities and Structures

Some magma moves around within the Earth but never reaches the surface. Underground flows of magma, or *intrusions,* move into and through the countless underground cracks in the rocks of the Earth's crust. There, intrusions may solidify into rock structures called *plutons* (see Figure 3.6). Plutons are named for their shape and include dikes, sills, laccoliths, batholiths, and stocks. *Dikes* are flat, slablike structures that cut across layers of rock like a wall. *Sills* have a similar shape, but lie parallel to layers of rock (as a window sill is parallel to the ground). A *laccolith* (the name means "lake-rock") is a large structure that has a flat bottom and arched top and looks like an inverted lake. Laccoliths form when magma is forced between rock layers faster than it can spread out, causing it to push the overlying layers of rock upward to form a dome mountain. Large, irregularly shaped plutons are called *stocks* if they cover less than 75 square kilometers and *batholiths* if they cover more.

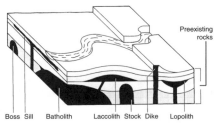

Boss Sill Batholith Laccolith Stock Dike Lopolith

Figure 3.6 Types of Plutons

A-3. Zones of Earthquake and Volcanic Activity

If the locations of earthquakes and volcanic eruptions are plotted on a map of the Earth, an interesting pattern emerges, as shown in Figure 3.7. Rather than being randomly spread over the earth, volcanoes and earthquakes occur in distinct, narrow zones. These zones include the mid-ocean ridges, the rim of the Pacific Ocean (called the *Ring of Fire* for its many active volcanoes), and the Mediterranean Belt. The locations of volcanoes and earthquake epicenters reveal the locations of boundaries between the plates of the Earth's crust.

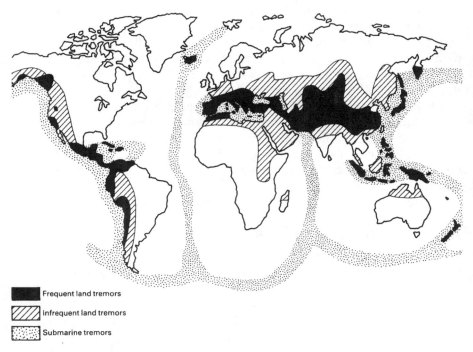

Frequent land tremors

Infrequent land tremors

Submarine tremors

Figure 3.7 Earthquake and Volcano Distribution

B. WHAT EVIDENCE SUGGESTS THAT THE LITHOSPHERE IS DYNAMIC?

The lithosphere is almost constantly in motion. This motion may be abrupt, as in an earthquake, or so gradual that it is imperceptible to the senses. These movements of the lithosphere exert forces on rocks, causing them to be deformed. Evidence that the lithosphere is moving ranges from direct observation of motion to inferences based on the displacement of structures and the deformation of rock.

B-1. Evidence of Crustal Movements

Earthquakes are unmistakable evidence of crustal movement. During an earthquake, movement of the crust occurs along faults. Every one of the nearly 1 million earthquakes that occur annually involves movement of the Earth's crust.

Volcanic eruptions also involve movements of the crust. As magma moves beneath and then out of a volcano, the crust around the volcano rises and sinks measurably, often causing the brittle crust to crack and form fissures. These breaks are direct evidence of movement.

Displaced structures are also evidence of crustal movement. When the crust moves, structures built on its surface will move along with it.

Bench marks are permanent metal tablets set in the ground that give the exact locations and precisely determined elevations of specific points. Bench marks are used as references in geologic surveys and tidal observations. Measurements of bench-mark elevations for more than 100 years reveal that in large areas of the United States the ground is slowly moving upward or downward (see Figure 3.8). Crustal motion can also be detected and measured using satellite laser technology.

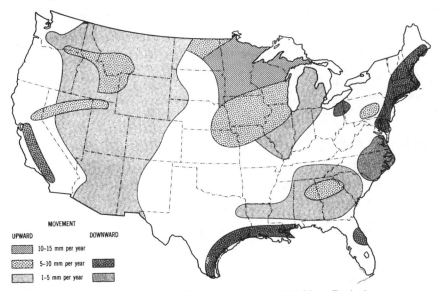

MOVEMENT
UPWARD   DOWNWARD
▨ 10–15 mm per year
▨ 5–10 mm per year ▨
▨ 1–5 mm per year   ▨

Figure 3.8 Changes in Bench-Mark Elevations over a 100-Year Period.
Physical Geology, 2nd Edition, by Richard Foster Flint, Brian J. Skinner, John Wiley and Sons, New York 1974, 1977.

Tilted or folded rock layers are also evidence of crustal movement. Sedimentary strata, lava flows, and tephra are originally deposited in horizontal layers. Therefore, where such materials are observed to be steeply tilted, or bent and folded, it can be inferred that they have been moved from their original positions.

Sedimentary rock layers at high elevations are further evidence of crustal movement. Most layers of sediments are deposited at or below sea level, and may be buried deeper before changing into sedimentary rock. Nevertheless, in many places layers of sedimentary rock are found high atop mountains or plateaus. Some are located as much as 2 kilometers above current sea level, far higher than any known body of water, past or present. From this fact it can be inferred that these rock layers have been uplifted.

Fossils provide striking evidence of crustal movement. Fossils of marine life such as corals and clams found at mountain tops in the Himalayas are evidence of uplift. Conversely, fossils of terrestrial or shallow-water organisms beneath the deep ocean bottom indicate sinking, subsidence of the crust.

Thick layers of shallow water sediments, as much as 15 kilometers thick in some places, are interesting evidence of crustal movement. The average depth of the ocean is 3.8 kilometers with several isolated points having depths of about 10 kilometers. How can 15 kilometers of sediment that is typically deposited in shallow water accumulate in a body of water without filling it completely? One probable inference is that the crust beneath the sediment is sinking while the sediment is accumulating, so that sediments layers build up while the water depth remains fairly constant.

B-2. Causes and Effects of Crustal Movements

Crustal movement is caused by unbalanced forces acting on the crust. Some of the forces acting on the Earth's crust are gravity (including gravitational attraction by the Sun and Moon), forces produced by the Earth's rotation, the expansion and contraction of rock material due to heating and cooling, and forces produced by density currents in the Earth's mantle.

Stress and Strain

The many forces acting on rock are called *stress*. Stress can act on rock in several different ways. A rock is under *uniform stress* when the stress in all directions is equal. *Tension* stresses act in opposite directions, pulling rock apart or stretching it. *Compression* stresses act toward each other, pushing or squeezing rock together. *Shear* stresses may act toward or away from each other, but they do so along different lines of action, causing rock to twist or tear.

When stress acts upon a rock, the rock *strains*, that is, it changes in size or shape or both (see Figure 3.9). Uniform stress causes rock to change in size but not in shape. Tension, compression, and shear cause a change in shape and may also cause a change in size.

When rock is stressed, it goes through a series of changes (see Figure 3.10). The first step is *elastic deformation,* in which the rock will strain, but the change will not be permanent. If the stress is removed, the rock will return to its original shape and size. The next step is *ductile deformation,* which begins

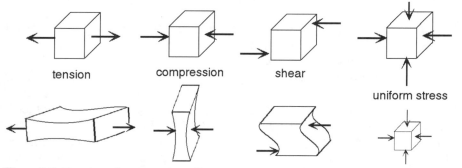

tension compression shear uniform stress

Figure 3.9 Tension, Compression, Shear, and Uniform Stress and Their Effects on Rock.

Elastic Deformation—force removed, object returns to original size and shape

Ductile Deformation—force removed, object remains deformed

Fracture—rock breaks

Figure 3.10 Elastic Deformation, Ductile Deformation, and Fracture.

when stress reaches a point called the elastic limit. At the elastic limit, the stress exceeds the strength of the rock's internal bonding and permanent changes occur. Even if the stress is removed, the deformation is permanent and the rock no longer returns to its original shape or size. Ductile behavior is similar to what happens when you squeeze or stretch modeling clay without breaking it. The final step is *fracture*, in which the stress actually causes the rock to break. In general, high temperatures and pressures favor ductile behavior and make fracture less likely to occur.

Folds

Ductile deformation of layered rock forms bends or warps called *folds* (see Figure 3.11). Folding is due to compression stresses. An upward, arched fold is an *anticline*; a downward, valleylike fold is a *syncline*. The sides of a fold are called its *limbs*. Folds range from simple *monoclines*, in which only one limb is bent, to *recumbent* folds that are bent back on themselves almost horizontally.

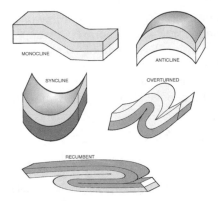

Figure 3.11 Various Types of Folds

Joints and Faults

Joints and faults (see Figure 3.12) are the fractures that form when rock is stressed to the point where it breaks. A *joint* is a fracture along which there has been no movement of the rock. A *fault* is a fracture along which the rock has moved and one side of which is displaced relative to the other side. Faulting is always associated with earthquakes. A fault is classified by the angle of the fracture and the direction of relative movement along it.

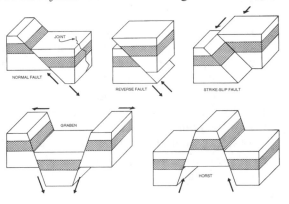

Figure 3.12 Various Types of Faults

C. WHAT IS PLATE TECTONICS?

Plate tectonics is the only hypothesis ever proposed that explains all of the Earth's features, as well as earthquakes, volcanic activity, and crustal movements. As stated earlier, analysis of earthquake waves led to the discovery that the Earth's interior consists of shells that have different properties. On the basis of density differences, the Earth can be divided into the inner and outer core, the mantle, and crust. If, instead of density, we consider the *rigidity* of rock, another pattern emerges (see Figure 3.13) (which encompasses the crust and upper mantle). To a depth of 100 kilome-

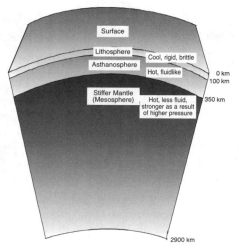

Figure 3.13 Lithosphere and Asthenosphere.

ters the rock is strong and rigid, forming a shell called the *lithosphere*. The region of the upper mantle between 100 and 350 kilometers in depth is weak and behaves like a viscous fluid; this shell is called the *asthenosphere*.

C-1. Two Key Ideas

In simple terms, the plate tectonics hypothesis consists of two key ideas. First, the rigid lithosphere consists of great slabs called plates; and these plates move sideways around the Earth, sliding on the fluidlike asthenosphere. Second, as the plates move laterally around the Earth, their edges interact in one of three ways: they spread apart, they collide, or they slide past each other.

C-2. Plate Boundaries

Divergent boundaries are places where adjacent plates are moving apart. Divergent boundaries create tension stresses, causing rock to fracture. These fractures result in earthquakes and opens rifts through which magma can rise to the surface and solidify to form new lithosphere. The mid-ocean rifts are divergent boundaries, and the mid-ocean ridges are volcanic mountains formed by magma emerging from the rift.

Convergent boundaries are places where adjacent plates are moving toward each other and colliding. When plate edges collide, compression and shear stresses cause rocks to fold, fracture, and move along faults. Plates consisting of ocean and continental crust interact differently along divergent boundaries.

Since ocean crust is denser than continental crust, ocean crust tends to plunge under continental crust when plates collide, forming a *subduction zone.* In subduction zones, the plunging plate is forced into the mantle and melts. If two plates carrying continental crust collide, they both crumple and fold. The Andes and Himalaya mountains formed along convergent boundaries.

Transform boundaries are places where plates move past each other in strike-slip motions. Shear stresses cause rock to fracture, forming numerous faults. The plates on either side of transform fault margins smash and rub against each other like the sides of two ships that came too close together and are grinding past each other. The rock along these faults is intensely shattered, and the sliding motion of the plates causes many earthquakes. If the plates separate a little, some magma may "leak" through the boundary, causing small-scale volcanism. The San Andreas Fault is part of a huge transform boundary along which the Pacific plate is moving past the North American plate.

The three types of plate boundaries are diagrammed in Figure 3.14.

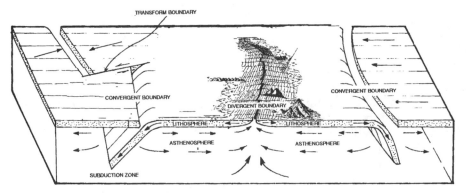

Figure 3.14 Divergent, Convergent, and Transform Boundaries.

REVIEW QUESTIONS FOR UNIT 3

1. Living corals are found in warm, shallow seas. Coral fossils have been found in the sedimentary rocks of Alaska. These findings suggest that
   (1) Alaska once had a tropical marine environment
   (2) Alaska's cold climate fossilized the corals
   (3) corals usually develop in cold climates
   (4) ocean currents carried the corals to Alaska

2. A line of former beaches along a coast, all 50 meters above sea level, is evidence of
   (1) present erosion
   (2) the present melting of polar icecaps
   (3) land uplift
   (4) a decrease in the deposition of marine fossils

3. Fossils of organisms that lived in shallow water can be found in horizontal sedimentary rock layers at great ocean depts. This fact is generally interpreted by most Earth scientists as evidence that
   (1) the cold water deep in the ocean kills shallow-water organisms
   (2) sunlight once penetrated to the deepest parts of the ocean
   (3) organisms that live in deep water evolved from species that once lived in shallow water
   (4) sections of the Earth's crust have changed their elevations relative to sea level

4. The best evidence of crustal movement would be provided by
   (1) dinosaur tracks found in the surface bedrock
   (2) marine fossils found on a mountaintop
   (3) weathered bedrock found at the bottom of a cliff
   (4) ripple marks found in sandy sediment

5. The most frequent cause of major earthquakes is
   (1) faulting
   (2) folding
   (3) landslides
   (4) submarine currents

6. The immediate result of a sudden slippage of rocks within the Earth's crust will be
   (1) isostasy
   (2) erosion
   (3) an earthquake
   (4) the formation of convection currents

7. Where does most present-day faulting of rock occur?
   (1) in regions of glacial activity
   (2) in the interior areas of continents
   (3) at locations with many lakes
   (4) at interfaces between moving parts of the crust

8. Where have earthquakes occurred most frequently during the last 100 years?
   (1) in the polar regions
   (2) in the interior of continental areas
   (3) along the Pacific Ocean coastlines
   (4) along the Atlantic Ocean coastlines

9. Two different cities experience earthquakes with a magnitude of 5.5 on the Richter scale. However, on the Modified Mercalli Intensity Scale, the earthquake was rated V in one city and VII in the other city. The best explanation for the difference is that
   (1) one city is nearer the Equator, the other is near a pole
   (2) the earthquake is one city occurred at 8 P.M. and the other occurred at 4 P.M.
   (3) one city is built on bedrock while the other is built on loose sediments
   (4) one city is in a drier climate zone than the other

10. Recent volcanic activity in different parts of the world supports the inference that volcanoes are located mainly in
    (1) the centers of landscape regions
    (2) the central regions of the continents
    (3) zones of crustal activity
    (4) zones in late stages of erosion

11. A large belt of mountain ranges and volcanoes surrounds the Pacific Ocean. Which events are most closely associated with these mountains and volcanoes?
    (1) hurricanes   (3) tornadoes
    (2) sandstorms   (4) earthquakes

12. Two geologic surveys of the same are, made 50 years apart, showed that the area had been uplifted 5 centimeters during the interval. If the rate of uplift remains constant, how many years will it take for this area to be uplifted a total of 70 centimeters?
    (1) 250   (3) 500
    (2) 350   (4) 700

13. Which diagram indicates the *least* amount of crustal activity?

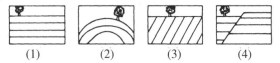

    (1)   (2)   (3)   (4)

14. The diagrams below show cross sections of exposed bedrock. Which cross section shows the *least* evidence of crustal movement?

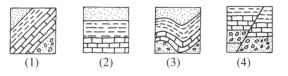

    (1)   (2)   (3)   (4)

Note that question 15 has only three choices.

15. As the depth within the Earth's crust increases, the amount of sedimentary rock, compared to the amount of nonsedimentary rock, will generally
(1) decrease   (3) remain the same
(2) increase

16. The source of energy for the high temperatures found deep within the Earth is
(1) tidal friction
(2) incoming solar radiation
(3) decay of radioactive materials
(4) meteorite bombardment of the Earth

17. According to the *Earth Science Reference Tables,* as the depth within the Earth's interior increases, the
(1) density, temperature, and pressure increase
(2) density, temperature, and pressure decrease
(3) density and temperature increase, but pressure decreases
(4) density increases, but temperature and pressure decrease

18. Which observation provides the strongest evidence for the inference that convection cells exist within the Earth's mantle?
(1) Sea level has varied in the past.
(2) Marine fossils are found at elevations high above seal level.
(3) Displaced rock strata are usually accompanied by earthquakes and volcanoes.
(4) Heat-flow readings vary at different locations in the Earth's crust.

19. Which Earth movement is best represented by the drawing at the right?
(1) folding   (3) jointing
(2) faulting   (4) volcanism

20. How was the valley shown in the diagram below most likely formed?

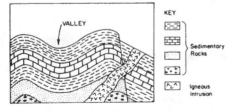

(1) by the deposition of sediments
(2) by the extrusion of igneous material
(3) by the faulting of rock layers
(4) by the folding of rock layers

21. A sandstone layer is found tilted at an angle of 75° from the horizontal. What probably caused this 75° tilt?
   (1) The sediments that formed this sandstone layer were originally deposited at a 75° tilt.
   (2) This sandstone layer has changed position because of crustal movement.
   (3) The sandstone layer has recrystallized because of contact metamorphism.
   (4) Nearly all sandstone layers are formed from wind-deposited sands.

22. Which Earth process most likely formed the depression now occupied by the lake shown in the diagram below?
   (1) glaciation
   (2) climate changes
   (3) erosion
   (4) faulting

23. The diagram below represents a geologic cross section of a portion of the Earth's crust. Which geologic event that is shown in the diagram would suggest past crustal movement at that location?

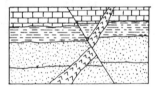

KEY:
[ΛΛΛ] Igneous Intrusion
[▦] Sedimentary Rocks

   (1) deposition of sediments
   (2) intrusion of molten material (magma)
   (3) folding of rock layers
   (4) faulting of rock layers

24. Which cross-sectional diagram below best represents a landscape region that resulted from faulting?

(1)   (2)   (3)   (4)

25. The diagram below represents a cross section of a portion of the Earth's crust.

Which past activity in this region is suggested by the shape of these sedimentary rock layers?
(1) widespread volcanic activity
(2) horizontal sorting
(3) glacial deposition
(4) crustal movements

26. Folded sedimentary rock layers are usually caused by
(1) deposition of sediments in folded layers
(2) differences in sediment density during deposition
(3) a rise in sea level after deposition
(4) crustal movement occurring after deposition

Base your answers to questions 27 through 31 on your knowledge of earth science, the *Earth Science Reference Tables*, and the map below. The map shows the epicenters and intensities of recent earthquakes within New York State. The state has been subdivided into four regions (*A, B, C, D*). In the key, VIII represents the most intense earthquakes and IV represents the least intense.

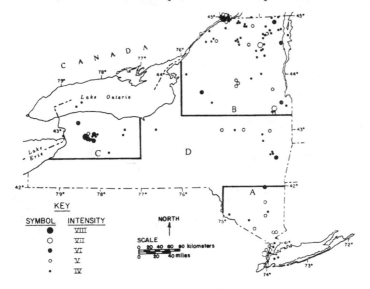

27. Which levels of earthquake intensity have occurred most frequently in New York State?
(1) V and VII
(2) IV and VIII
(3) VI and VII
(4) IV and V

112

28. Which is the best approximation of the length of the northern boundary line of zone *A*?
(1) 80 km   (3) 120 km
(2) 100 km   (4) 155 km

29. Which city is most likely to experience an earthquake at some future time?
(1) Massena   (3) Jamestown
(2) Elmira   (4) Utica

30. Which landscape region occupies most of earthquake region *B*?
(1) Erie-Ontario Lowlands
(2) Adirondack Highlands
(3) Appalachian Uplands
(4) New England Highlands

31. What type of rocks are found surrounding the epicenters of region *C*?
(1) intensely metamorphosed
(2) slightly metamorphosed
(3) igneous
(4) sedimentary

Base your answers to questions 32 through 36 on your knowledge of earth science, the *Earth Science Reference Tables,* and the map and information below.

One eruption of Mt. St. Helens in Washington State resulted in the movement of volcanic ash across the northwestern portion of the United States. The map shows the movement of the ash along paths at three different altitudes above sea level: path *A* at 1.5 km, path *B* at 3.0 km, and path *C* at 5.5 km. The lines across each path indicate the time interval between the eruption and the position of the leading edge of the ash. Points *X* and *Y* are places on the Earth's surface.

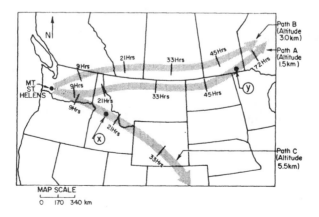

113

32. The data for paths *A*, *B*, and *C* were collected in which region of the atmosphere?
(1) troposphere   (3) mesosphere
(2) stratosphere   (4) thermosphere

Note that question 33 has only three choices.

33. How would the amount of insolation reaching point ×24 hours after the eruption compare with the amount of insolation this point usually receives?
(1) The amount of insolation would be less.
(2) The amount of insolation would be greater.
(3) The amount of insolation would be the same.

34. Approximately how many hours did it take for the ash front of path *A* to arrive at point *Y*?
(1) 39   (3) 59
(2) 45   (4) 72

35. Why did the general direction of the ash front path *C not* follow the general direction of the ash fronts along path *A* and path *B*?
(1) The eruption threw material toward the southeast.
(2) Northwest winds existed at the 5.5-kilometer level.
(3) Mountain ranges southeast of the volcano are 1.5 kilometers high.
(4) Only the ash along path *C* was affected by the Earth's rotation.

36. Approximately how long would it take for an earthquake *P*-wave caused by the eruption to travel from Mt. St. Helens to point *Y*? (Use the map scale.)
(1) 7 minutes   (3) 11 minutes
(2) 2 minutes   (4) 4 minutes

37. The lines on the accompanying map represent faults and fractures in the bedrock of New York State.

At which location on this map are earthquakes likely to occur?
(1) New York City
(2) Buffalo
(3) Kingston
(4) Mt. Marcy

OPTIONAL/EXTENDED
TOPIC B _____

Earthquakes and the Earth's Interior

A. HOW CAN WE LEARN ABOUT EARTHQUAKES?

The major cause of earthquakes is faulting. Faulting is an event that occurs beneath the Earth's surface, hidden from view and difficult to predict. Therefore, the study of earthquakes, or *seismology*, is largely a study of the effects of this event on surrounding matter.

When faulting occurs, rock fractures and the rock on either side of the fracture moves abruptly. As the fractured rock moves, it pushes against the rock surrounding it, causing an elastic deformation (discussed in Unit 3). As the rock rebounds from this deformation, it again pushes against rock surrounding it, causing another deformation. The energy of this back-and-forth motion, or vibration, is transmitted as a series of deformations spreading out through the rock in all directions.

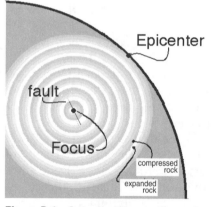

Figure B.1 Seismic Waves Spreading Out from the Focus of an Earthquake.

The motion of a change in a medium is called *wave motion*. The deformations of rock that spread out from the focus of an earthquake are therefore called *earthquake waves*, or *seismic waves*. Waves transport energy from one place to another through the motion of a change in the medium. Seismic waves transport energy from the focus through the surrounding rock to the Earth's surface (see Figure B.1).

A-1. Understanding Waves

Wave Behavior

A pulse in a rope is a simple analogy that will help you visualize what is going on when seismic waves travel through the Earth. It will also help you to understand some of the inferences that can be drawn from the behavior of seismic waves.

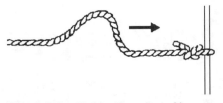

Figure B.2 Pulse Traveling Along a Rope.

If you give one end of a stretched rope a quick shake, you will see that a bump, or pulse, travels down the rope at some speed (see Figure B.2). If the rope is uniform and completely flexible, the pulse keeps the same shape as it moves down the rope.

As a pulse travels down a rope, its forward part is moving upward and its rear part is moving downward (see Figure B.3). The up and down motions of the string have kinetic energy. Work has to be done to produce the pulse by pulling against the tension of the string, so the deformed string has potential energy. Thus, a pulse contains both kinetic and potential energy as it travels along a rope.

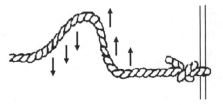

Figure B.3 Movement of Rope as Pulse Passes Through It.

The speed of the pulse depends on the properties of the rope—how dense it is and how tightly it is stretched. Pulses move slowly down a loose, heavy rope because the heavy rope has a lot of inertia and responds slowly to the forces acting on it. However, pulses move rapidly down a taut, light rope. When the rope is taut, the pulse moves faster because the tendency of the rope to straighten out is greater.

When a pulse reaches the end of the rope, it may be reflected and travel back toward its source. Depending on how the end is held, the reflected pulse may be upright, may be inverted or may disappear completely. If the end is fixed to a support (see Figure B.4), the pulse exerts a force on the support, which then exerts an equal but opposite force on the rope. This opposite force causes the rope to be

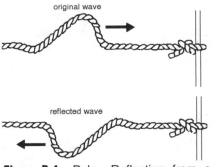

original wave

reflected wave

Figure B.4 Pulse Reflecting from a Fixed Support.

displaced in the opposite direction from the original pulse. An inverted pulse forms and travels back along the rope in the opposite direction from the original pulse. If the end is looped around a support so that it is connected but still free to move, an upright pulse is reflected. If the end is held somewhere between totally fixed and totally free, the pulse disappears completely.

Now, consider what happens if two different ropes, a heavy (denser) rope and a light (less dense) rope, are spliced together (see Figure B.5). One end is tied to a support, and the other end is shaken so that a pulse is produced. When the pulse reaches the splice, the pulse passes from the light rope to the heavy rope and is said to be transmitted. However, the transmission is not total because a reflected pulse also appears at the splice and travels back down the light rope.

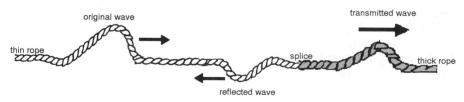

Figure B.5 Transmitted and Reflected Pulses at a Splice Between a Heavy and a Light Rope.

If one pulse follows another in a rope, the result is a periodic wave (see Figure B.6). In a periodic wave a certain waveform, or shape, is repeated at regular intervals. Three quantities are often used to describe periodic waves: wave velocity, wavelength, and frequency. *Wave velocity* is the distance a wave moves each second; *wavelength* is the distance between adjacent crests or troughs; *frequency* is the number of waves that pass a given point in a second.

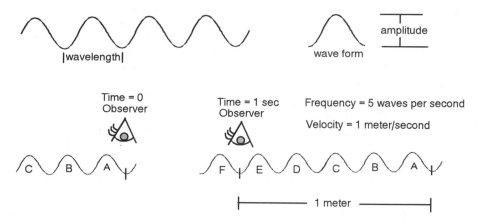

Figure B.6 Periodic Wave, Showing Waveform, Wavelength, Wave Velocity, and Frequency.

If a periodic wave travels across a boundary between a more dense and a less dense medium at an angle, it is bent, or refracted (see Figure B.7). This *refrac-*

tion occurs because, as the pulse crosses the boundary, the front of the pulse is moving at a different speed from the rear of the pulse. The effect is somewhat like that produced by an axle on which one wheel turns faster than the other—the path of the axle curves.

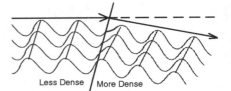

Figure B.7 Periodic Wave Crossing a Boundary Between a More Dense and a Less Dense Medium and Being Refracted.

Wave Types

Transverse waves are waves in which the particles of the medium move back and forth in a direction perpendicular to the direction of wave motion. Waves in a stretched rope are transverse waves because any point on the rope moves side to side perpendicularly to the direction in which the wave is moving, as shown in Figure B-8.

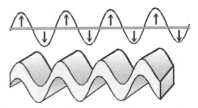

Figure B.8 Transverse Waves, Showing Perpendicular Motion.

Longitudinal waves occur when the particles of the medium move back and forth in the same direction in which the waves are traveling. A good analogy is the movement of a compression down a coil spring when the end is pushed in and out (see Figure B.9).

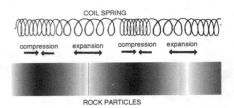

Seismic Waves

The complex motions that occur during faulting generate several types of periodic seismic waves. Some are transverse waves; others, longitudinal. Seismology took a giant step forward

Figure B.9 Longitudinal Waves in a Coil Spring and in Rock. As a longitudinal wave moves through a spring, its coils alternatively compress and expand. In a rock, the individual particles are compressed and expanded.

with the invention of the modern *seismograph*, a device that detects, measures, and records the motions of the Earth associated with seismic waves.

A seismograph's operation is based on the law of inertia: objects that are at rest will tend to remain at rest. A simple seismograph consists of a heavy, suspended weight that tends to remain at rest while the Earth moves around it. A recording device (e.g., a pen) is attached to the weight. A recording medium, such as paper, is wound around a clockwork-driven drum mounted firmly in bedrock. When a series of seismic wave pulses causes the bedrock to move back and forth, the drum moves with it, but the heavy weight and its attached pen remain motionless for a long time. As the paper moves under the motionless pen, a line is drawn on the paper that records the back-and-forth motion of the bedrock, as shown in Figure B.10a. The line recorded on paper by a seismograph is called a *seismogram*, as shown in Figure B.10b.

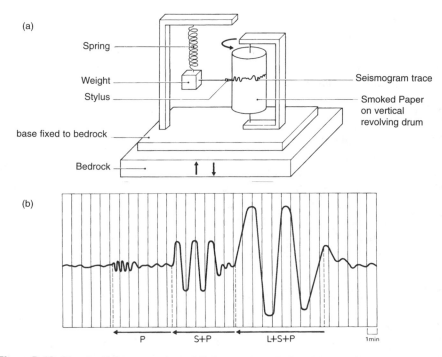

Figure B.10 Simple Seismograph and Seismogram. *P, S,* and *L* represent *P-, S-,* and *Love* waves, respectively.

From: Earth Science on FIle. © Facts on File, Inc., 1988.

Modern seismographs use a laser beam reflected from a mirror on the weight instead of a pen. The laser beam exposes a track along photographic paper that is attached to the drum. When the bedrock vibrates, the track becomes a wavy line. In most earthquake recording stations, at least three seismographs are used, one to measure vibrations in each of three dimensions: *x, y,* and *z* (typically north-south, east-west, and up-down).

A careful study of seismograms reveals four different types of seismic waves. Two types, *P*-waves and *S*-waves, transport energy through the body of the Earth and are therefore called *body waves*. Two other types, Love waves and Rayleigh waves, transport energy along the surface of the Earth and are therefore called *surface waves*. Although surface waves play a major role in the damage caused by earthquakes, it is the body waves that reveal the structure of the earth's interior and make it possible to locate an earthquake's focus and epicenter. For this reason we will focus our attention on *P*-waves and *S*-waves.

P-waves

P-waves are longitudinal waves. They are alternating pulses of compression and expansion of rock in the same direction in which the wave travels. *P*-waves, which travel at speeds of about 8 kilometers per second in the upper crust, are the first (or *primary* waves) to reach a seismograph after an earthquake occurs because they travel fastest. *P*-waves can travel through solids, liquids, and gases because all three media can be compressed and expanded. *P*-waves travel faster in rigid materials because these materials have stronger internal bonds and, like a taut rope, a greater tendency to pull back into their original volumes. Also, since denser materials tend to be more rigid, *P*-wave speed increases in denser rock.

S-waves

S-waves are transverse waves. An *S*-wave twists rock back and forth, deforming its shape in a direction perpendicular to that of wave travel. *S*-waves (or *secondary* waves) are the second waves to reach at a seismograph. They arrive after *P*-waves because they travel at a slower speed, about 4 kilometers per second in the upper crust. The speed of *S*-waves, like that of *P*-waves, increases in denser, more rigid rocks. *S*-waves can be transmitted only through solids. These waves cannot travel through liquids and gases because, when they are deformed, they do not return to their original shape.

A-2. Analyzing Seismograms

P-wave and *S*-wave recordings from a single event can be used to find the distance to the epicenter of an earthquake the and time of its origin. With recordings from three or more seismic stations, the location of the epicenter can be determined. The basis for these determinations is the difference in speed of *P*-waves and *S*-waves.

When an earthquake occurs, all four types of seismic waves start moving outward from the focus at the same time. However, since they travel at different speeds, they do not all arrive at a seismograph at the same time. As stated earlier, the *P*-waves, which travel the fastest, will arrive first, followed by the *S*-waves some time later and the surface waves last. The farther a seismograph is from the epicenter, the greater will be the difference between the arrival times of the *P*-waves and the *S*-waves.

For example, suppose *P*-waves travel 8 kilometers per second and *S*-waves travel 4 kilometers per second. The *P*-waves will arrive at a seismograph 80 kilometers away in 10 seconds. The *S*-waves will reach the same seismograph in 20 seconds. The arrival times at this first station will be 20–10 or *10 seconds* apart. The same *P*-waves will reach a seismograph 200 kilometers

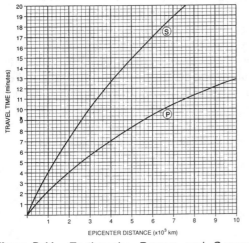

Figure B.11 Earthquake *P*-wave and *S*-wave Travel Time

away in 25 seconds and the *S*-waves will arrive in 50 seconds. The arrival times at the farther station will be 50 – 25 or *25 seconds* apart.

Determining the Distance to the Epicenter of an Earthquake

For every distance that seismic waves travel, there is a corresponding difference in arrival times. This relationship between the difference in arrival time between *P*-waves and *S*-waves can be seen clearly in Figure B.11. The travel time axis is marked off in minutes, and each minute is subdivided into three 20-second segments. The epicenter distance is marked off in thousands of kilometers ($10^3 = 1,000$), with each subdivided into five 200-kilometer segments. The lines marked *S* and *P* show the time it takes *S*-waves and *P*-waves, respectively, to travel a given distance. The vertical difference between the lines is the difference in arrival times at a given distance. Figure B.11 shows, for example, that *P*-waves travel 1,000 kilometers in 2 minutes while *S*-waves take about 4 minutes. The difference in arrival times is 2 minutes.

Using Figure B.11 and the relationship between difference in arrival times and distance, we can determine the distance between a seismograph and the epicenter of an earthquake as follows:

1. Determine the arrival times of the *P*-waves and the *S*-waves.
2. Calculate the *difference* in arrival times by subtracting the *P*-wave arrival time from the *S*-wave arrival time.
3. Using the travel time axis of the graph, mark off a distance along the edge of a piece of scrap paper equivalent to the difference in arrival times.
4. Keeping the edge of the paper vertical, find the point at which the marked-off difference in arrival times corresponds to the vertical distance between the *P*-wave line and the *S*-wave line.
5. Read the epicenter distance corresponding to that location on the graph.

Example

The seismogram in Figure B.12 is a record of an earthquake that was recorded by a seismograph. The seismogram does not tell how far away the earthquake occurred, but it does show when the *P*-waves and *S*-waves arrived.

To determine the epicenter, we proceed as follows:

1. We note that the *P*-waves arrived at 08:16:00 and the *S*-waves arrived at 08:21:00.
2. We subtract 08:21:00 – 08:16:00 = 00:05:00, or 5 min.

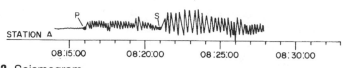

Figure B.12 Seismogram

3. Using the travel time axis in Figure B.11, we mark off 5 min on the edge of a piece of paper.
4. Keeping the paper vertical, we find the place where it just fits between the *S*-wave and *P*-wave lines.
5. We read the epicenter distance on the axis. It is 3,400 km.

These five steps are illustrated in figure B.13.

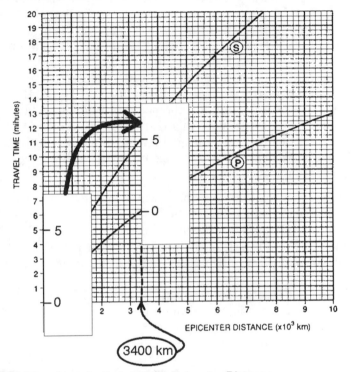

Figure B.13 Five Steps in Determining Epicenter Distance.

Determining the Origin Time of an Earthquake

A seismogram tells you when seismic waves arrive, but not when the earthquake occurred, or its *origin time*. However, once the distance to the epicenter is known, the origin time of the earthquake can be determined.

Consider a simple example. Suppose a girl lives 10 kilometers from school and walks at a speed of 5 kilometers per hour. The girl will take 2 hours to walk to school. If she arrives in school at 9:00 A.M. after walking directly from home, when did she leave home? If the girl reaches school at 9:00 A.M. and took 2 hours to get there, she must have left home at 7:00 A.M.

This same approach is used to determine the origin time of an earthquake: determine the epicenter distance, use the travel time graph (Figure B.11) to determine the *P*-wave travel time, and subtract the *P*-wave travel time from the *P*-wave arrival time to get the origin time.

Example

Now, let's apply this approach to the seismogram in Figure B.12.

1. We determined that the epicenter distance is 3,400 km.
2. Using the travel time graph in Figure B.14, we see that a *P*-wave takes 6 minutes 10 seconds to travel 3,400 km.
3. Since the *P*-waves arrived at 08:16:00 and took 00:06:10 (6 minutes 10 seconds) to get there, they must have left the epicenter at 08:16:00–00:06:10 or 08:09:50.

(Remember: When adding or subtracting time, carry 1's and 10's. There are *60* seconds in a minute and *60* minutes in an hour!)

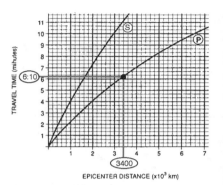

Figure B.14 The distance from the epicenter to a seismograph is 3,400 kilometers. It takes a *P*-wave 6 minutes, 10 seconds to travel from the epicenter to the seismograph.

Determining the Location of the Epicenter of an Earthquake

Knowing how far a seismograph is from the epicenter does not pinpoint the location of the epicenter. In the example we have been considering, it could be 3,400 kilometers away in any direction. All of the points that are 3,400 kilometers from the seismograph in station *A* form a circle around the seismograph with a radius of 3,400 kilometers. The epicenter is located somewhere on this circle (see Figure B-15a).

If however, the same earthquake is 2,000 kilometers from a second station, station *B*, a circle can be drawn around this station too, narrowing the possible locations down to two points, as shown in Figure B-15b. A third recording of the same earthquake, in station *C*, will eliminate one of these two points and pinpoint the location of the epicenter as shown in Figure B-15c.

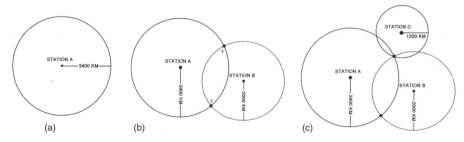

Figure B.15 Information From Three Seismograph Stations is Needed to Pinpoint the Location of the Earthquake Epicenter.

Example

The seismograms in the upper half of Figure B.16 were recorded at three seismic recording stations during the same earthquake we have been considering.

We know that the epicenter distance from station *A* is 3,400 km. The difference in *P*-wave and *S*-wave arrival times at station *B* is 7 minutes, yielding an epicenter distance of 5,400 km. The difference in *P*-wave and *S*-wave arrival times at station *C* is 9 minutes for an epicenter distance of 7,600 km. When these distances are plotted to scale on a map as shown in the lower part of Figure B.16, stations *A*, *B*, and *C* pinpoint the epicenter.

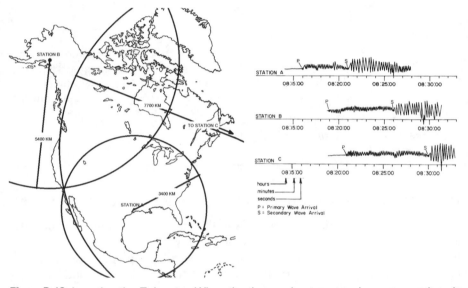

Figure B.16 Locating the Epicenter. When the three seismograms above are analyzed, the distance from each seismograph to the earthquake epicenter can be determined. Plotted on a map, only one place is the correct distance from all three stations.

B. HOW CAN WE INVESTIGATE THE STRUCTURE OF THE EARTH?

Whenever an earthquake occurs, its seismic waves are recorded by seismographs at hundreds of recording stations throughout the world. Analysis of these seismograms reveals much about the structure of the Earth's interior.

As discussed earlier, seismic waves are reflected from the surfaces of materials and refracted as they move across boundaries between substances of different densities. If the Earth had a homogeneous composition and density, seismic waves would move through it in simple, straight-line paths with reflections only when the surface was reached. They would also travel directly from the focus to all points on the Earth's surface. Furthermore, if the location of the epicenter is known, the arrival times of *P*-waves and *S*-waves at all stations should be easy to predict.

This is not the case. Predictions of arrival times based upon straight-line travel differ markedly from actual measurements. One reason is that the path of a seismic wave curves. This gradual bending is due to the gradual increase in density with depth, caused by increasing pressure. Also, there are many more reflected waves than would be expected from a homogeneous Earth. At some places on the Earth's surface *P*-waves or *S*-waves, or both, do not arrive directly from the focus. These facts have led seismologists to conclude that the Earth's interior is not homogeneous.

B-1. The Crust

In 1910, Andrija Mohorovicic discovered that two sets of *P*-waves and *S*-waves arrived at locations close to an epicenter, but only one set arrived farther away. Mohorovicic explained this phenomenon by hypothesizing a boundary between rocks of different densities that would cause the waves to refract. As shown in Figure B.17, stations close to the epicenter of an earthquake can receive both direct and refracted seismic waves. Farther away, however, waves can't reach the stations without crossing the boundary, so only refracted waves arrive.

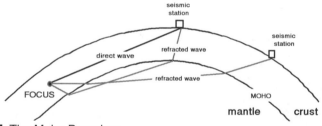

Figure B.17 The Moho Boundary

This boundary called the *Moho* in honor of Mohorovicic, marks the point where the outermost layer of the earth, the crust, ends and the mantle beneath it begins. The density of the crust varies from about 2.7 grams per cubic centimeter near the surface to about 3.1 grams per cubic centimeter near the Moho. The thickness of the crust varies, but is usually greatest beneath mountains and least beneath oceans.

B-2. The Mantle and Core

Between the epicenter and an angle of 103° from the epicenter, *P*-waves and *S*-waves arrive directly from the focus, as would be expected. Between 103° and 143°, however, they *disappear*. This is known as the *P*-wave shadow zone. Beyond 143°, in what is known as the *S*-wave shadow zone, the

P-waves reappear but the *S*-waves do not. These shadow zones, shown in Figure B.18, can be explained as follows:

A boundary at a depth of about 2,900 kilometers, between rock of markedly different densities, produces a strong refraction of the *P*-waves that cross it. Up to an angle of 103°, waves arrive directly from the epicenter. The strong refraction as soon as the boundary is encountered immediately angles

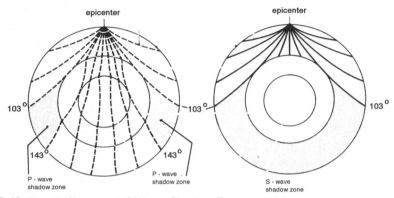

Figure B.18 The *P*-Wave and *S*-Wave Shadow Zones.

the waves away to 143°, so no waves reach the region in between. However, since, *P*-waves continue to be received past 143°, they are able to pass through the material beneath the boundary. This boundary marks the bottom of the mantle, and the layer beneath it is called the core.

The complete absence of *S*-waves beyond 103° indicates that they are unable to travel through the material beneath the mantle. Since transverse waves cannot travel through fluids, the disappearance of *S*-waves makes sense if the core is a fluid.

Careful analysis of refraction of *P*-waves that pass through the core and the presence of reflections from inside the core indicate that another boundary inside the core divides it into an inner core and an outer core. Only the outer core is thought to be a fluid.

B-3. Summary of Inferences about the Earth's Interior

Figure B.19 summarizes what has been inferred about the Earth's interior based on studies of seismic waves and other data. The top section shows a cross section of the earth with the interior layers labeled. Along the right edge, the density of each layer is shown. Notice that the mantle is divided into two regions, the asthenosphere and the stiffer mantle. Notice, too, the sharp increase in density between the mantle and the core, which explains the strong refraction of *P*-waves. Directly beneath the cross section is a graph showing pressures in *millions of atmospheres* at different depths. The dotted lines indicate boundaries between crust, mantle, and outer and inner core. On the bottom a similar graph shows actual temperatures and melting

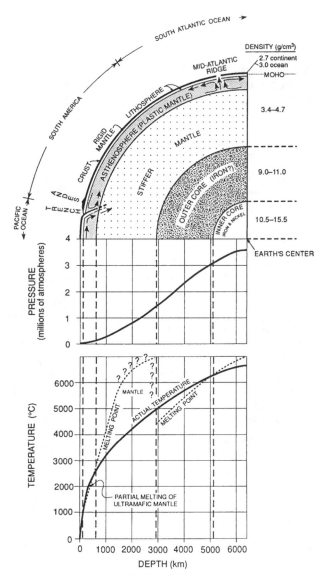

Figure B.19 Inferred Properties of the Earth's Interior.

points of rock at different depths. Notice that the actual temperature of the rocks in the outer core is higher than the melting point.

Meteorites are thought to be fragments of an Earthlike planet. Some have a composition similar to Earth rocks, but others are composed of iron-nickel alloy. An iron-nickel alloy has the same density and rigidity as are deduced for the core from seismic wave studies. Therefore, the core is believed to consist of an iron-nickel alloy. An iron-nickel core would also help explain the Earth's magnetic field.

127

C. WHAT CAUSES THE MOVEMENTS OF LITHOSPHERIC PLATES?

The development of the plate tectonic theory is an example of how a theory gains acceptance based on the evidence that supports it and its ability to explain puzzling observations. Plate tectonics makes sense out of such a wide range of phenomena that it has become a unifying principle in geology.

C-1. Earlier Ideas

Two earlier ideas—continental drift and sea-floor spreading—were based on evidence that was later incorporated into a single unifying theory—plate tectonics.

Continental Drift

Matching Shorelines
The shorelines on both sides of the Atlantic match strikingly, especially the shapes of the Atlantic coasts of Africa and South America. This match is even clearer if the edges of the continental shelves are compared, rather than the present shorelines. As a result the idea arose that the continents were once together but then drifted apart. At first the idea was rejected because it seemed ridiculous to think that anything as large as a continent could move around. In 1912, however, Alfred Wegener reintroduced the idea of continental drift, that is, that continents drift slowly across the Earth's surface, sometimes colliding and sometimes breaking into pieces. Wegener and his colleagues presented striking evidence that the world's landmasses had once been joined together, but over time had broken into the continents of today and slowly drifted to their present positions. The original landmass was dubbed Pangaea, meaning "all land" (see Figure B.20).

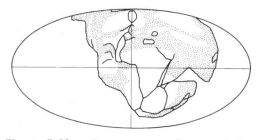

Figure B.20   Pangaea as Proposed by Wegener.
From: Earth Science on File. © Facts on File, Inc., 1988.

Continental and Ocean Crust
Statistical studies of elevations and depths indicate that there are two distinct levels for the world's surface—the continents and the ocean floors. These levels alternate and exist side by side with almost no transition between the two. This fact suggests a series of side-by-side blocks. Some blocks are continents; some are ocean floors. The thickness and composition of the ocean floors and continents differ. The ocean floors are thin and made of dense basalt, while the continents are thick and are composed less dense granite. Wegener used

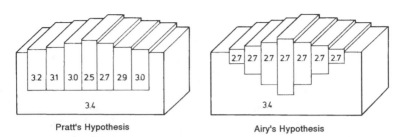

Figure B.21 In both hypotheses, the rocks of the crust float on a denser (3.4 g/cm^3) mantle. In Pratt's hypothesis, mountains are the result of less dense rock (2.5 g/cm^3) floating higher than dense rock (3.2 g/cm^3). In Airy's hypothesis, the rocks of the crust all have the same density (2.7 g/cm^3). Mountains occur where the crust is thicker, and ocean basins occur where it is thinner.
From: Earth Science on FIle. © Facts on File, Inc., 1988.

an earlier idea—isostasy—to explain the two levels of the crust. Isostasy stated that the crust was floating on hot, fluid rock in the mantle, and offered two possible explanations (see Figure B.21) for continents being higher than ocean floors: continents are higher because less dense materials float higher than more dense materials in the same fluid (Pratt's hypothesis); or continents are higher because thick objects float higher than thin objects in the same fluid (Airy's hypothesis). The existence of blocks would explain how continents could "drift"— the blocks simply moved sideways.

Correspondence of Geologic Structures
When geological structures on both sides of the Atlantic are compared, a remarkable correspondence is observed. For example, the Sierras near Buenos Aires contain a succession of beds very like those of the Cape Mountains in South Africa. The large gneiss plateau of Africa is strikingly similar to that of Brazil, and matching pockets of igneous rocks and sedimentary rocks can be found in both. The sequence of rock types—older granite, younger granite, alkaline rocks, Jurassic age volcanic rocks, and kimberlite—is also the same. The diamond fields of South Africa and of Brazil both occur in the kimberlite beds. The Falkland Islands contain rock layers almost indistinguishable from those of the African Cape yet markedly different from those of nearby Patagonia. Figure B.22, from Wegener's book *The Origin of Continents and Oceans*, shows some of the matches between rock types from South America and Africa. Similar matches of North America with Europe, Greenland with North America and Norway, Madagascar with Africa, and between other continents can be found.

Fossils and Contemporary Organisms
Fossils of certain land animals (Mesosauroidae) are found nowhere else but in southern Africa and South America. Fossils, of identical trees (*Glossopteris*) are found in Australia, India, and South America. Sixty-four percent of all reptile fossils from the Carbonaceous period are identical in Europe and in North America.

129

Contemporary organisms also show corresponding patterns of distribution. For example, *Lumbricus* earthworms are found from Japan to Spain in Eurasia, but only in the eastern United States. Freshwater perch are found in Europe and Asia, but only in the eastern United States. Similar patterns exist for pearl mussels, mud minnows, and garden snails. Common heather is found only in Europe and Newfoundland. Both European and eastern North American eels spawn in the Sargasso Sea off the North American coast. Similar patterns can be found between other, now separated continents.

Past Climates

There is also unmistakable evidence that 300 million years ago an ice sheet covered South America, southern Africa, India, and southern Australia. That ice sheet was huge, similar to the one covering Antarctica today. Glacial striations even indicate the direction of ice flow on all four continents, which is difficult to explain without continental drift. If the continents were in their present positions, the ancient ice sheet would have covered all the southern oceans and in places crossed the Equator. For this to happen, the Earth would have had to be very cold, yet we find no evidence of glaciation at that time in the Northern Hemisphere. If the continents drifted, though, then 300 million years ago they could have all been joined and located adjacent to the South Pole. If you look back at Figure B.20, you will see that this was indeed the case. Similarly, places that are today located near the poles contain deposits of coal, which could have formed only in equatorial rain forests. This fact, too can be explained by drifting continents.

All of the above evidence supports the idea that continents have moved over time. However, it does not suggest a mechanism that would move them.

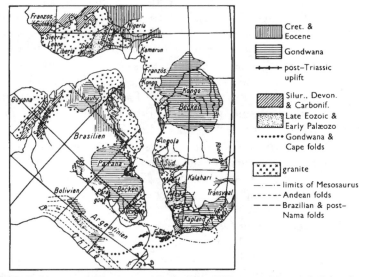

Figure B.22 Evidence of Continental Drift. Alfred Wegener's correlation of rock types and surface features along the margins of Africa and South America.

From: *The Origins of Continents and Oceans.* Alfred Wegener. By permission of Dover Publications, Inc., 1966

Ocean Floor Spreading

In the 1950s the ocean floors were explored extensively by oceanographic research vessels, and much information was gathered. On the basis of this new knowledge Professor Harry Hess of Princeton University proposed the idea that the ocean floor is spreading sideways, away from the mid-ocean ridges. The evidence supporting this idea suggests a mechanism that could move the continents apart.

Mid-ocean Ridges
New instruments made it possible to survey the ocean floor with unprecedented accuracy. The ocean basins, once thought to be flat plains, were shown to contain chains of undersea mountains that run down the center of almost every ocean in the world. The mountain chain is split by a deep rift, and the surface around it is riddled with faults. This rugged terrain suggests that movement is occurring on the ocean floors.

Young Rock
Surveys of the ocean floors showed that the rocks beneath the oceans are generally younger than the rocks of the continents. Most are only a fraction of the age of some continental rocks. In addition, the age of the rocks on the ocean bottoms increases with distance from the mid-ocean ridges, suggesting that rocks form at the ridges and then move sideways away from them.

Paleomagnetism
Basalt, the major rock of the ocean floors, is rich in iron compounds. Magnetite, pyrrhotite, and hematite are strongly affected by magnetism. When lava is extruded in the midocean rifts, crystals of minerals affected by magnetism align with the Earth's magnetic field while the lava cools. When solidified, the rock contains a record of the Earth's magnetic field locked in its crystals. Studies of ancient rocks show that the Earth's magnetic field has reversed many times in the past. The reversals seem to take place at irregular intervals ranging from 20,000 to several million years. Research ships carrying sensitive instruments that can measure small changes in magnetism have discovered that magnetism in rocks on the ocean floors reveals a striking pattern (see Figure B.23) of strips of normal and reversed magnetic fields. This pattern is identical on either side of the midocean ridges. This discovery also suggests that rocks first formed at the ridges and then moved sideways away from the ridges.

The proposed mechanism for ocean floor spreading is as follows. Lava is extruded in the rift and hardens to form new ocean floor with Earth's magnetic field "frozen" in its crystals. New lava erupts, splitting the just formed rock and pushing it aside. This new lava then hardens. As this process is repeated, new ocean floor is constantly being formed, and the floor on either side of the ridge is pushed sideways.

Trenches. While new ocean floor is constantly being formed, old ocean floor is being consumed in the trenches—deep crevices where the floor

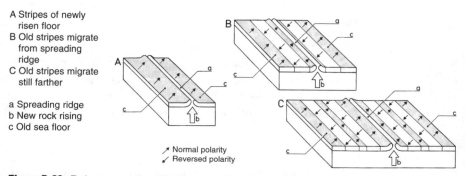

A Stripes of newly risen floor
B Old stripes migrate from spreading ridge
C Old stripes migrate still farther

a Spreading ridge
b New rock rising
c Old sea floor

↗ Normal polarity
↙ Reversed polarity

Figure B.23 Paleomagnetism Patterns on the Ocean Floor.
From: Earth Science on File. © Facts on File, Inc., 1988.

bends downward sharply. Beneath these trenches, earthquakes occur frequently and the positions of their foci show a plunging pattern. This pattern suggests that the ocean floor is descending downward in the trenches and being consumed as it melts in the mantle below (see cross section of crust in Figure B.19).

C-2. Plate Tectonics

Plate tectonics combines aspects of isostasy, continental drift, and ocean floor spreading into a single unifying theory that consists of the following ideas:

1. The Earth's crust consists of a series of rigid slabs of rock called *lithospheric plates*. Each plate is about 100 kilometers thick. The Earth's surface consists of about six major plates and several smaller pieces (see Figure B.24).

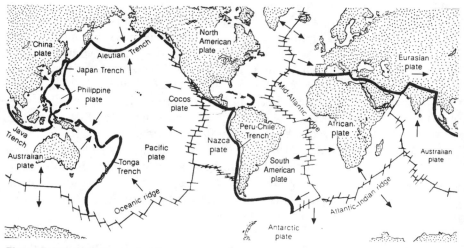

Figure B.24 Plate Tectonic Diagram.

132

2. The plates rest on a relatively fluid layer of the upper mantle and lower crust called the asthenosphere. This fluid layer lets the plates move independently of the deeper rocks in the mantle. Boundaries between plates are regions of volcanic and earthquake activity because they are where the plates collide and scrape against each other. The interiors of the plates are relatively quiet.

3. The type of crust atop each plate determines whether it carries an ocean floor, a continent, or both. The continents and ocean floors are merely passengers on the moving plates. When the plates change position, the continents and ocean floors move with them. Past positions of the plates can be inferred from evidence such as fossils, paleomagnetism, and past climates.

4. There are three types of plate boundaries: convergent, divergent, and transform. Plates collide at convergent boundaries, separate at divergent boundaries, and slide past each other at transform boundaries. When an ocean plate collides with a continental plate, the denser ocean plate is pushed under the continental plate, or *subducted*. Most mountains form at divergent and convergent boundaries.

5. In subduction zones, ocean floor plates are consumed as they are pushed into the hot mantle and melt. As the plate plunges into the mantle, it remains rigid and produces deep-focus earthquakes. Friction between the subducted plate and the adjacent plate melts rock along the interface and produces volcanic activity directly above and parallel to the trench. As a result most trenches are bordered by volcanic island arcs or volcanic mountain chains.

6. The creation of ocean floor in the mid-ocean rifts and its destruction in the trenches are in equilibrium. Therefore, the Earth remains the same size.

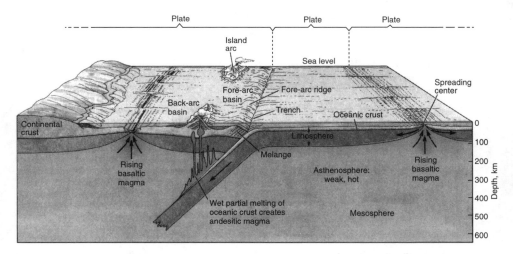

Figure B.25 Cross section of the Earth showing the structures found in plate tectonics. From *The Dynamic Earth*, 2nd Ed., by Brian J. Skinner and Stephen C. Porter, John Wiley, 1992.

7. Forces exist within the Earth that are powerful enough to move the lithospheric plates. Two hypotheses have been proposed to explain plate motion. In one view the plates are thought to be pushed by convection currents in the mantle. In the other they are thought to be pulled downward by gravity in subduction zones. Observations of heat flow support the concept of convection in the mantle. However, the scientific community is not yet unified in identifying the specific forces that propel the plates.

8. Convection in the mantle may produce "hot spots" over plumes of rising heat. As plates move over a hot spot, the base of the plate melts and a volcano forms. Volcanic chains in the center of a plate, such as the Hawaiian Islands, are thought to have formed in this way.

Figure B.25 is a simplified cross section of the Earth showing the structures described above.

D. WHAT ARE THE PROPERTIES OF THE EARTH'S CRUST?

The properties of the Earth's crust can be summarized as follows:

1. The average thickness of the continental crust is greater than that of the oceanic crust.

2. Continents tend to rise higher than ocean floors because they are composed of rocks of lower density. The theory of isostasy states that continents and ocean floors are floating in a higher density fluid rock. Observations of solid objects floating in fluids show that denser objects float lower in a fluid than less dense objects of the same size and shape, and also that objects of the same density float higher or lower depending on their thickness. Since continents tend to be thicker and less dense than ocean floors, they would be expected to float higher than adjacent ocean floors.

3. The continental crust and oceanic crust have different compositions. The ocean crust is composed mainly of basaltic rock; the continents are made mostly of rocks with a granitic composition.

REVIEW QUESTIONS FOR TOPIC B

1. Earthquakes generate compression waves (*P*-waves) and shear waves (*S*-waves). Compared to the speed of shear waves in a given earth material, the speed of compression waves is
   (1) always slower
   (2) always faster
   (3) always the same
   (4) sometimes faster and sometimes slower

Note that questions 2 and 3 have only three choices.

2. As the distance from the epicenter of an earthquake increases, the time between the arrival of the *P*-waves and *S*-waves on a seismograph will
   (1) decrease
   (2) increase
   (3) remain the same

3. Which of the following earthquake waves can travel through both solids and fluids?
   (1) *S*-waves, only
   (2) *P*-waves, only
   (3) *S*-waves and *P*-waves

4. Earthquake *S*-waves do not travel through the Earth's
   (1) crust   (3) mantle
   (2) moho   (4) core

5. The distance between an earthquake epicenter and the location of a seismograph can be calculated because
   (1) seismographs are sensitive to directions
   (2) earthquake waves decay at known rates
   (3) shear waves will not pass through liquids
   (4) shear waves and compression waves travel at different speeds

6. Which is most helpful to a seismologist when determining the distance to the epicenter of an earthquake?
   (1) time of arrival of the *P*-wave
   (2) intensity of the *P*-wave
   (3) time interval between *P*-wave and *S*-wave arrival
   (4) *S*-wave characteristics on the seismogram

7. The time that an earthquake occurs can be inferred by knowing the
   (1) arrival time of the *P*-wave
   (2) distances between seismograph stations
   (3) travel time of the *S*-waves
   (4) epicenter distance and arrival time of the *P*-waves

8. An earthquake occurred at 5:00:00 A.M. According to the *Earth Science Reference Tables,* at what time would the *P*-wave reach a seismic station 3,000 kilometers from the epicenter?
   (1) 5:01:40 A.M.
   (2) 5:04:30 A.M.
   (3) 5:05:40 A.M.
   (4) 5:10:15 A.M.

9. A *P*-wave reaches a seismograph station 2,600 kilometers from an earthquake epicenter at 12:10 P.M. According to the *Earth Science Reference Tables,* at what time did the earthquake occur?
   (1) 12:01 P.M.   (3) 12:15 P.M.
   (2) 12:05 P.M.   (4) 12:19 P.M.

10. The difference in arrival times for *P*- and *S*-waves from an earthquake is 5.0 minutes. According to the *Earth Science Reference Tables,* how far away is the epicenter of the earthquake?
   (1) 1.3×10^3 km   (3) 3.5×10^3 km
   (2) 2.6×10^3 km   (4) 8.1×10^3 km

11. A seismograph station records an earthquake with a difference of 9 minutes between the arrival times of initial *P*-waves and *S*-waves. According to the *Earth Science Reference Tables,* approximately how far from the epicenter is this station?
   (1) 2,500 km   (3) 7,700 km
   (2) 5,800 km   (4) 9,000 km

12. A seismographic station determines that its distance from the epicenter of an earthquake is 4,000 kilometers. According to the *Earth Science Reference Tables,* if the *P*-wave arrived at the station at 10:15 A.M., the time of the earthquake's origin was
   (1) 10:02 A.M.   (3) 10:10 A.M.
   (2) 10:08 A.M.   (4) 10:22 A.M.

13. If a seismograph recording station located 5,700 kilometers from an epicenter receives a *P*-wave at 4:45 P.M. at which time did the earthquake actually occur at the epicenter?
   (1) 4:24 P.M.   (3) 4:36 P.M.
   (2) 4:29 P.M.   (4) 4:56 P.M.

14. The epicenter of an earthquake is located near Massena, New York. According to the *Earth Science Reference Tables,* the greatest difference in arrival times of the *P*- and *S*-waves for this earthquake would be recorded in
   (1) Albany, New York   (3) Plattsburgh, New York
   (2) Utica, New York   (4) Binghamton, New York

15. Which graph best represents the relationship between the differences in arrival times of *P*-waves and *S*-waves for locations at varying distances from an earthquake?

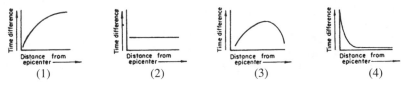

16. The circles on the accompanying
    map show the distances from three
    seismic stations, *X*, *Y*, and *Z*, to the
    epicenter of an earthquake.

    Which location is closest to the
    earthquake epicenter?
    (1) *A*
    (2) *B*
    (3) *C*
    (4) *D*

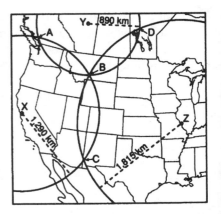

17. The seismogram below shows the arrival times of *P*- and *S*-waves from
    a single earthquake. According to the *Earth Science Reference Tables*,
    how far from the earthquake epicenter was the station that recorded this
    seismogram?

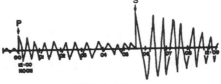

TIME (min)

(1) 1.5×10^3 km   (3) 3.0×10^3 km
(2) 2.5×10^3 km   (4) 4.0×10^3 km

Base your answers to questions 18 and 19 on your knowledge of earth science, the *Earth Science Reference Tables*, and the chart below. This chart
provides partial seismic data from an earthquake that was detected at two different seismographic stations. The earthquake occurred at 10:00:00 A.M.
eastern standard time (e.s.t.)

	P-*wave Arrival* Time (e.s.t.)	S-*wave Arrival* Time (e.s.t.)	Distance From Epicenter (km)
Triville	10:07:00 A.M.		
Endburg			9,000

18. At what time (e.s.t.) did the *S*-wave arrive at Triville?
    (1) 10:01:45 A.M.   (3) 10:07:00 A.M.
    (2) 10:04:00 A.M.   (4) 10:12:30 A.M.

19. At what time (e.s.t.) did the *P*-wave arrive at Endburg?
    (1) 10:07:00 A.M.   (3) 10:12:10 A.M.
    (2) 10:09:00 A.M.   (4) 10:22:15 A.M.

20. Four seismograph stations receive data from the same earthquake. The table below shows the differences in travel times for the *P*- and *S*-waves recorded at each station. Which station is the closest to the epicenter of the earthquake?

Station	Travel-Time Difference
A	4 min 32 sec
B	3 min 52 sec
C	3 min 10 sec
D	4 min 17 see

(1) *A*   (3) *C*
(2) *B*   (4) *D*

21. A seismic station is 2,000 kilometers from an earthquake epicenter. According to the *Earth Science Reference Tables,* how long does it take an *S*-wave to travel from the epicenter to the station?
(1) 7 min 20 sec   (3) 3 min 20 sec
(2) 5 min 10 sec   (4) 4 min 10 sec

22. Which has been used to develop a model of the composition of the Earth's core ?
(1) index fossils   (3) radioactive minerals
(2) drill cores   (4) seismic waves

23. Useful information regarding the composition of the interior of the Earth can be derived from earthquakes because earthquake waves
(1) release materials from within the Earth
(2) travel through the Earth at a constant velocity
(3) travel at different rates through different materials
(4) change radioactive decay rates of rocks

24. Which has been studied to obtain information about the Earth's inner core?
(1) seismic waves   (3) index fossils
(2) sedimentary rocks   (4) exposed bedrock

25. The theory that the outer core of the Earth is composed of liquid material is best supported by
(1) seismic studies which indicate that shear waves do not pass through the outer core
(2) seismic studies which show that compression waves can pass through the outer core
(3) density studies which show that the outer core is slightly more dense than the inner core
(4) gravity studies which indicate that gravitational strength is greatest within the core

26. Which is the most accurate statement about the interior of the Earth?
(1) The temperature within the mantle decreases with depth.
(2) The pressure within the mantle increases with depth.
(3) The continental crust is thinnest under the mountain regions.
(4) The two most common elements in the Earth's crust are iron and nickel.

27. According to the *Earth Science Reference Tables,* which zone of the Earth has the greatest density?
(1) crust (3) outer core
(2) mantle (4) inner core

Note that question 28 has only three choices.

28. A comparison of seismic graphs taken at a location in the ocean with those taken at the center of a continent would indicate that the crustal thickness under the oceans is probably
(1) less than that under the continents
(2) more than that under the continents
(3) equal to that under the continents

Base your answers to questions 29 and 30 on the diagram of the Earth below showing the observed pattern of waves recorded after an earthquake.

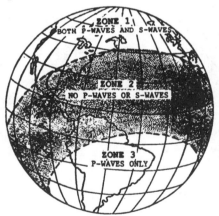

29. The lack of *S*-waves in zone 3 can best be explained by the presence within the Earth of
(1) density changes (3) a liquid outer core
(2) mantle convection cells (4) a solid inner core

30. The location of the epicenter of the earthquake that produced the observed wave pattern most likely is in the
(1) crust in zone 1 (3) crust in zone 3
(2) mantle in zone 2 (4) core of the Earth

31. Which graph best represents the relationship between depth and density for the Earth's interior? [Refer to the *Earth Science Reference Tables.*]

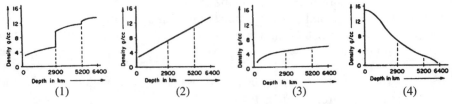

(1)   (2)   (3)   (4)

32. Theories about the composition of the Earth's core are supported by meteorites that are composed primarily of
(1) oxygen and silicon
(2) aluminum and iron
(3) aluminum and oxygen
(4) iron and nickel

Note that question 33 has only three choices.

33. According to the *Earth Science Reference Tables,* as depth within the Earth's interior increases, the density, temperature, and pressure values
(1) decrease
(2) increase
(3) remain the same

34. Which statement best supports the theory that all the continents were once a single landmass?
(1) Rocks of the ocean ridges are older than those of the adjacent sea floor.
(2) Rock and fossil correlation can be made where the continents appear to fit together.
(3) Marine fossils can be found at high elevations above sea level on all continents.
(4) Great thicknesses of shallow-water sediments are found at interior locations on some continents.

35. The theory of continental drift does *not* explain the
(1) matching of rock features on continents thousands of kilometers apart
(2) melting of glacial ice at the close of the Pleistocene Epoch
(3) apparent fitting together of many continental boundaries
(4) fossils of tropical plants in Antarctica

36. According to the *Earth Science Reference Tables,* during which geologic period were the continents all part of one landmass, with North America and South America joined to Africa?
(1) Tertiary   (3) Triassic
(2) Cretaceous   (4) Ordovician

140

37. Which statement best supports the theory of continental drift?
  (1) Basaltic rock is found to be progressively younger at increasing distances from a mid-ocean ridge.
  (2) Marine fossils are often found in deep-well drill cores.
  (3) The present continents appear to fit together as pieces of a larger landmass.
  (4) Areas of shallow-water seas tend to accumulate sediment, which gradually sinks.

38. According to the Inferred Position of Earth Landmasses information shown in the *Earth Science Reference Tables,* on what other landmass would you most likely find fossil remains of the late Paleozoic reptile called Mesosaurus shown here?

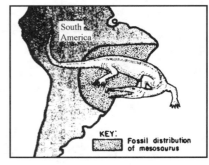

  (1) North America
  (2) Africa
  (3) Antarctica
  (4) Eurasia

39. Which is the best evidence supporting the concept of ocean floor spreading?
  (1) Earthquakes occur at greater depths beneath continents than beneath oceans.
  (2) Sandstones and limestones can be found both in North America and in Europe.
  (3) Volcanoes appear at random within the oceanic crust.
  (4) Igneous rocks along the mid-oceanic ridges are younger than those farther from the ridges.

40. Igneous materials found along oceanic ridges contain magnetic iron particles that show reversal of magnetic orientation. This is evidence that
  (1) volcanic activity has occurred constantly throughout history
  (2) the Earth's magnetic poles have exchanged their positions
  (3) igneous materials are always formed beneath oceans
  (4) the Earth's crust does not move

Note that question 41 has only three choices.

41. As the distance from the mid-oceanic ridges increases, the age of the ocean basin rock will
  (1) decrease
  (2) increase
  (3) remain the same

42. Which statement best describes the continental and oceanic crusts?
(1) The continental crust is thicker and less dense than the oceanic crust
(2) The continental crust is thicker and more dense than the oceanic crust
(3) The continental crust is thinner and less dense than the oceanic crust
(4) The continental crust is thinner and more dense than the oceanic crust.

43. In the diagram below, letters *A* and *B* represent locations near the edge of a continent.

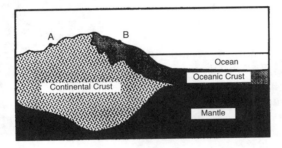

A geologist who compares nonsedimentary rock samples from locations *A* and *B* would probably find that the samples from location *A* contain
(1) more granite
(2) more basalt
(3) more fossils
(4) the same minerals and fossils

Base your answers to questions 44 and 45 on the diagram below, which represents a profile of the floor of the Atlantic Ocean between the continents of South America and Africa.

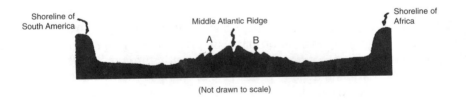

(Not drawn to scale)

44. If the rocks at *A* and *B* were compared, the results would probably show that the rocks at
(1) *A* and *B* were formed about the same time
(2) *A* were formed before those at *B*
(3) *A* are sedimentary and those at *B* are nonsedimentary
(4) *A* are nonsedimentary and those at *B* are sedimentary

45. Which evidence does *not* support the theory that Africa and South America were once part of the same large continent?
(1) correlation of rocks on opposite sides of the Atlantic Ocean
(2) correlation of fossils on opposite sides of the Atlantic Ocean
(3) correlation of coastlines on opposite sides of the Atlantic Ocean
(4) correlation of living animals on opposite sides of the Atlantic Ocean

46. Igneous rock along oceanic ridges is younger than the igneous rock farther from the ridges. This evidence supports the theory that
(1) the ocean floor is stable
(2) the ocean floor is spreading
(3) volcanoes once existed on both sides of oceanic ridges
(4) oceanic ridges are areas of subsidence

47. The adjacent map shows the position of the north magnetic pole at various times in the past.

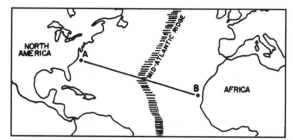

These positions have most likely been determined by using
(1) compass readings on various continents today
(2) magnetic properties of rocks formed during various geologic times
(3) seismic waves traveling through the Earth's interior
(4) fossils found in bedrock formed during various geologic times

48. The drawing below represents the ocean floor between North America and Africa.

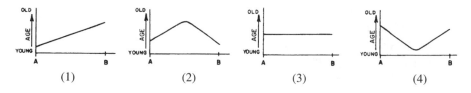

Which graph best represents the age of the bedrock in the ocean floor along line *AB*?

49. The accompanying diagram shows a cross-sectional view of the Earth's interior. The motion represented by the arrows indicates that the Earth's mantle

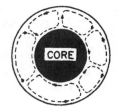

   (1) has properties of a fluid
   (2) is composed of solid metamorphic rocks
   (3) is not affected by the heat from the Earth's core
   (4) is more dense than the core

50. According to the *Earth Science Reference Tables,* during which geologic time period were the continents of North America, South America, and Africa closest together?
   (1) Tertiary
   (2) Cretaceous
   (3) Triassic
   (4) Ordovician

51. What is the primary force in the theory of ocean (sea) floor spreading?
   (1) Density differences in the mantle cause convection cells.
   (2) Planetary winds in the atmosphere blow against landmasses.
   (3) The Earth rotates on its axis.
   (4) The Earth revolves around the Sun.

52. Which is suggested by the occurrence of higher than average temperatures below the surface of the Earth in the area of the mid-Atlantic ridge?
   (1) the existence of convection cells in the mantle
   (2) the presence of heat due to an orographic effect
   (3) a high concentration of magnetism in the mantle
   (4) the existence of a thinner crust under mountains

53. Which set of Earth processes are thought to be most closely related to each other because they normally occur in the same zones?
   (1) mountain building, earthquakes, and volcanic activity
   (2) mountain building, shallow-water fossil formation, and rock weathering
   (3) volcanic activity, rock weathering, and deposition of sediments
   (4) earthquakes, shallow-water fossil formation, and shifting magnetic poles

54. The primary cause of convection currents in the Earth's mantle is believed to be the
   (1) differences in densities of earth materials
   (2) subsidence of the crust
   (3) occurrence of earthquakes
   (4) rotation of the Earth

55. Base your answer to the following question on your knowledge of earth science and the diagrams below.

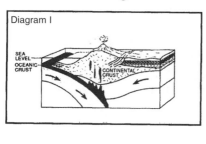

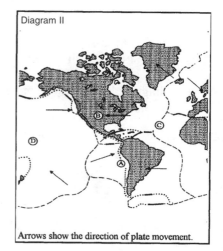

Arrows show the direction of plate movement.

At which location in Diagram II would the plate boundary shown in Diagram I most likely be found?

(1) *A*   (3) *C*
(2) *B*   (4) *D*

56. The diagram below represents a partial cross section of a model of the Earth. The arrows show inferred motions within the Earth.

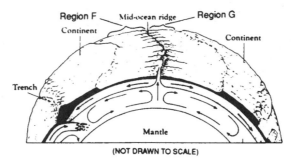

(NOT DRAWN TO SCALE)

Which property of the oceanic crust in regions *F* and *G* is a result of these inferred motions?

(1) The crystal size of the rock decreases constantly as distance from the mid-ocean ridge increases.
(2) The temperature of the basaltic rock increases as distance from the mid-ocean ridge increases.
(3) Heat-flow measurements steadily increase as distance from the mid-ocean ridge increases.
(4) The age of the igneous rock increases as distance from the mid-ocean ridge increases.

57. According to the *Earth Science Reference Tables,* which of the following locations is the site of a convergent plate boundary?
(1) the mid-Atlantic ridge
(2) the Aleutian trench
(3) the Atlantic-Indian ridge
(4) the Pacific/North American plate boundary

58. The diagram below shows one side of an oceanic ridge and a portion of the ocean floor.

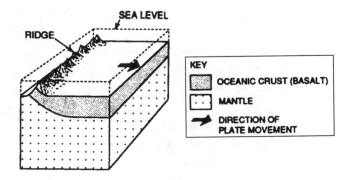

Which graph best illustrates the age of the basalt as the distance from an oceanic ridge increases?

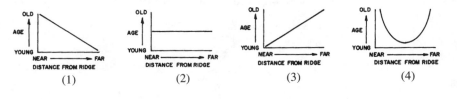

Base your answers to questions 59–62 on your knowledge of earth science, the *Earth Science Reference Tables,* and the three seismograms shown below. The seismograms were recorded at earthquake recording stations *A, B,* and *C.* The letters *P* and *S* on each seismogram indicate the arrival times of the compression (primary) and shear (secondary) seismic waves.

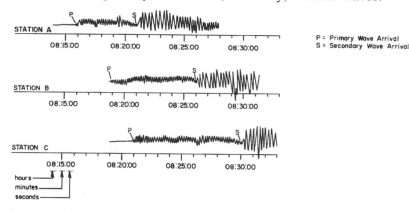

59. At what time did the *S*-wave arrive at station *A*?
(1) 08:16:00   (3) 08:27:00
(2) 08:21:00   (4) 08:30:00

60. Approximately how far from station *C* is the earthquake epicenter located?
(1) 3,500 km   (3) 5,600 km
(2) 4,300 km   (4) 6,300 km

61. A fourth station recorded the same earthquake. The *P*-wave arrived, but the *S*-wave did not arrive. The best explanation for the absence of the *S*-wave is that it
(1) was never transmitted by the earthquake
(2) stopped when it reached the liquid part of the Earth's interior
(3) stopped when it reached a solid part of the Earth's interior
(4) traveled only on the Earth's surface and did not penetrate the earth's interior

62. The radius of each circle below represents the distance from each seismograph recording station to the epicenter. Which map correctly illustrates the position of the three recording stations relative to the location of the earthquake epicenter?

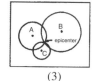

(1)   (2)   (3)   (4)

Base your answers to questions 63 through 67 on your knowledge of earth science and the diagrams below which are used to help explain the theory of continental drift.

Diagram I represents a map of a portion of the Earth showing the relative position of the mid-Atlantic ridge. *A* through *D* are locations in the ocean and points *X* and *Y* are locations on the continents.

Diagram II represents a portion of the Earth's interior with a model of probable convection cells. Locations 1 through 4 are at the Earth's surface.

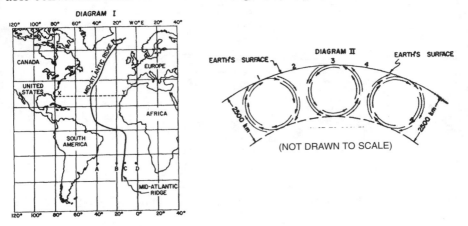

63. According to the theory of continental drift, which location in Diagram I would most likely have the best correlation of rock, mineral, and fossil evidence?
    (1) the eastern coast of the United States and the southern coast of Africa
    (2) central Canada and central Europe
    (3) the eastern coast of South America and the western coast of Africa
    (4) the central United States and central South America

64. A ship obtains a sample of igneous bedrock from each of the positions *A, B, C,* and *D.* Which rock sample would probably be the oldest?
    (1) *A*   (3) *C*
    (2) *B*   (4) *D*

65. Which locations in Diagram II would correspond to the position of the mid-Atlantic ridge in Diagram I?
    (1) location 1   (3) location 3
    (2) location 2   (4) location 4

66. Diagram II refers to convection cells which probably exist in the
    (1) crust   (3) outer core
    (2) mantle   (4) inner core

148

67. Which graph best represents the most likely pattern of heat flow along line *XY*?

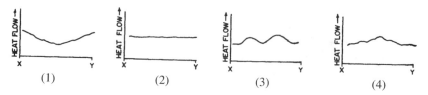

(1)   (2)   (3)   (4)

Base your answers to questions 68 through 72 on your knowledge of earth science, the *Earth Science Reference Tables,* and the map information below. The map shows the present-day relative positions of South America and Africa and the age of the rocks composing the two continents. Letters *A–H* indicate specific rock units. The apparent close correlation between rocks on the two continents provided early evidence for the theory of continental drift.

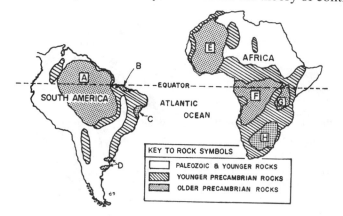

68. According the *Earth Science Reference Tables,* when did Africa and South America completely separate and move away from each other into two distinct landmasses as shown above?
(1) before the Cambrian Period
(2) during the Carboniferous Period
(3) during the Triassic Period
(4) after the Cretaceous Period

69. Which Precambrian rock unit in South America is most probably a former section of rock unit *F* in present-day Africa?
(1) unit *A*   (3) unit *C*
(2) unit *B*   (4) unit *D*

70. What present-day Atlantic Ocean feature is part of the evidence that these two continents are still drifting apart?
(1) a magnetic pole   (3) a deep ocean trench
(2) land sediments   (4) a mid-ocean ridge

149

71. Which model best indicates the overall direction of movement of the African landmass relative to the position of South America during this time of continental drift?

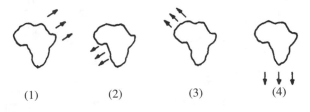

(1)　　　　　　(2)　　　　　　(3)　　　　　　(4)

Note that question 72 has only three choices.

72. Compared to the age of the rocks composing these continental landmasses, the age of the oceanic crust between them is mostly
(1) younger
(2) older
(3) approximately the same

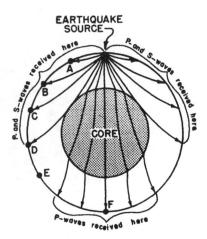

Base your answers to questions 73 through 77 on your knowledge of earth science, the *Earth Science Reference Tables,* and the adjacent diagram. The diagram represents a cross section of the Earth showing the paths of earthquake waves from a single earthquake source. Seismograph stations are located at points *A* through *F,* and they are all located in the same time zone.

73. At which station is the difference in time between the arrival of *P*- and *S*-waves the greatest?
(1) *A*　　　　　　　(3) *C*
(2) *B*　　　　　　　(4) *D*

74. Station *E* did *not* receive any *P*-waves or *S*-waves from this earthquake because the *P*-waves and *S*-waves
(1) cancel each other out
(2) are bent, causing shadow zones
(3) are changed to sound energy
(4) are converted to heat energy

75. What explanation do scientists give for the reason that station *F* did *not* receive *S*-waves?
  (1) The Earth's inner core is so dense that *S*-waves cannot pass through.
  (2) The Earth's outer core is liquid, which does not allow *S*-waves to pass.
  (3) *S*-waves do not have enough energy to pass completely through the Earth.
  (4) *S*-waves become absorbed by the Earth's crust.

76. Seismograph station *D* is 7,700 kilometers from the epicenter. If the *P*-wave arrived at this station at 2:15 P.M., at approximately what time did the earthquake occur?
  (1) 1:56 P.M.   (3) 2:04 P.M.
  (2) 2:00 P.M.   (4) 2:08 P.M.

77. Seismograph station *B* recorded the arrival of *P*-waves at 2:10 P.M. and the arrival of *S*-waves at 2:15 P.M. Approximately how far is station *B* from the earthquake epicenter?
  (1) 1,500 km   (3) 3,500 km
  (2) 2,500 km   (4) 4,500 km

Base your answers to questions 78 through 82 on your knowledge of earth science, the *Earth Science Reference Tables,* and the diagram below. The diagram represents the age of the basaltic ocean crust in the Atlantic Ocean between the United States and Africa. Line *AB* is drawn for reference purposes only.

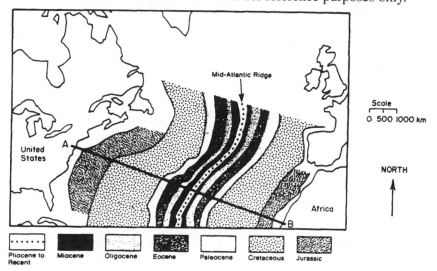

78. Which statement is best supported by the diagram?
  (1) The ocean crust is the same age along line *AB*.
  (2) The oldest ocean crust is located near the continents.
  (3) The age of the ocean crust increases from point *A* to point *B*.
  (4) Most of the ocean crust along line *AB* fromed in the Paleozoic Era.

79. The age of formation of the ocean crust along line *AB* suggests that the United States and Africa are moving
    (1) eastward   (3) closer together
    (2) westward   (4) farther apart

80. Which diagram most closely represents the cross section of the ocean floor along line *AB*?

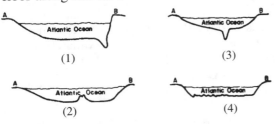

    (1)   (3)

    (2)   (4)

81. According the the diagram, the width of the Cretaceous rock east of the mid-Atlantic ridge along line *AB* is approximately
    (1) 1,000 km   (3) 1,600 km
    (2) 1,200 km   (4) 4,000 km

82. On which map do the marks best represent the distribution of the frequency of earthquakes for this area?

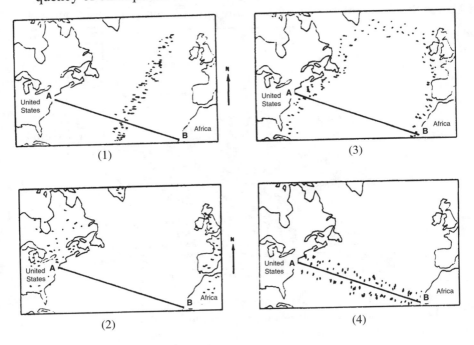

    (1)   (3)

    (2)   (4)

Base your answers to questions 83 through 87 on your knowledge of earth science, the *Earth Science Reference Tables,* and the diagrams below. Diagram I is a map showing the location and bedrock age of some of the Hawaiian Islands. Diagram II is a cross section of an area of the Earth illustrating a stationary magma source and the process that could have formed the islands.

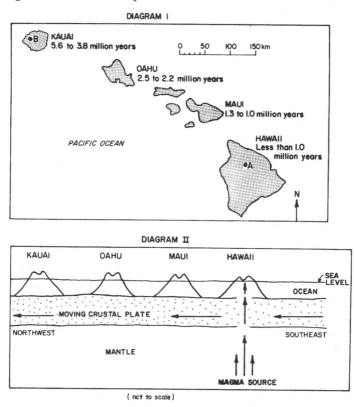

DIAGRAM I

KAUAI
5.6 to 3.8 million years

0   50   100   150km

OAHU
2.5 to 2.2 million years

MAUI
1.3 to 1.0 million years

PACIFIC OCEAN

HAWAII
Less than 1.0
million years

N

DIAGRAM II

KAUAI   OAHU   MAUI   HAWAII

SEA LEVEL

OCEAN

MOVING CRUSTAL PLATE

NORTHWEST   SOUTHEAST

MANTLE

MAGMA SOURCE

(not to scale)

83. If each island formed as the crustal plate moved over the magma source in the mantle as shown in Diagram II, where would the next volcanic island most likely form?
(1) northwest of Kauai
(2) northeast of Hawaii
(3) southeast of Hawaii
(4) between Hawaii and Maui

84. Compared to the continental crust of North America, the oceanic crust in the area of the Hawaiian Islands is probably
(1) thinner and similar in composition
(2) thinner and different in composition
(3) thicker and similar in composition
(4) thicker and different in composition

153

85. Volcanic activity like that which produced the Hawaiian Islands is usually closely correlated with
(1) nearness to the center of a large ocean
(2) sudden reversals in the Earth's magnetic field
(3) frequent major changes in climate
(4) frequent earthquake activity

86. Which of the Hawaiian Islands has the greatest proability of having a volcanic eruption?
(1) Kauai   (3) Maui
(2) Oahu   (4) Hawaii

87. Which graph best represents the ages of the Hawaiian Islands, comparing them from point *A* to point *B*?

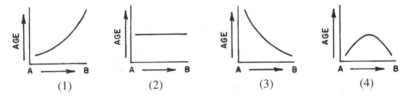

Base your answers to questions 88 through 91 on your knowledge of earth science, the *Earth Science Reference Tables,* and the diagram below. The diagrams represent geologic cross sections of the upper mantle and crust at four different Earth locations. In each diagram, the movement of the crustal sections (plates) is shown by arrows and the locations of frequent earthquakes are indicated by symbols as shown in the key. Diagrams are not drawn to scale.

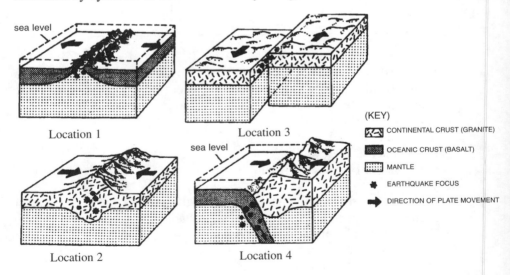

154

88. According to the diagrams, the most probable cause of the shallow earthquakes at location 3 is
    (1) the collision of two oceanic plates
    (2) the spreading of two oceanic plates
    (3) the sinking of an oceanic plate under a continental plate
    (4) the horizontal shifting of two continental plates

89. Compared to the oceanic crust, the continental crust at location 4 is
    (1) thicker and less dense   (3) thinner and less dense
    (2) thicker and more dense   (4) thinner and more dense

90. Which location could be found along the west coast of South America?
    (1) location 1   (3) location 3
    (2) location 2   (4) location 4

91. At a depth of 2,500 kilometers below location 1, what is the actual temperature of the rock?
    (1) 1,000°C   (3) 4,600°C
    (2) 6,800°C   (4) 1,000,000°C

UNIT FOUR _____

Weathering, Erosion and Deposition

KEY IDEAS In preceding units, we explored the mechanisms by which the materials that comprise the Earth's crust were formed. In this unit we will discuss the processes that shape the material exposed at the surface of the Earth's crust—a dynamic, changing system of minerals, rocks, weather, water, ice, wind, and living organisms.

KEY OBJECTIVES

Upon completion of this unit, you should be able to:

- Define weathering as the processes that result in the physical and chemical breakdown of crustal material and describe the products of weathering.
- Define erosion as the process by which sediments are transported and describe the agents of erosion.
- Recognize the factors that affect erosion and relate them to landforms produced by erosion.
- Explain how the characteristics of sediments and of the media carrying them affect their deposition.
- Describe some of the common landforms resulting from deposition by mass wasting, moving water, glaciers, and wind.

A. WHAT ARE THE PROCESSES BY WHICH WEATHERING HAS CHANGED THE SURFACE OF THE EARTH'S CRUST?

The surface of the Earth's crust, or lithosphere, is constantly exposed to the changing weather of the atmosphere's. This environment is very different from the one in which most rocks and minerals were formed. As these materials adjust to their new environment, they change. They break down into smaller pieces, and chemical reactions change them into new substances. Since the processes that produce these changes result from exposure to the weather, they are called *weathering. Weathering is the breakdown of rocks into smaller particles by natural processes.* Whenever rocks are exposed to the air, water, and living things at or near the Earth's surface, weathering occurs. Weathering processes are divided into two general types: physical and chemical.

A-1. Physical Weathering

Some of the changes caused by weathering involve form only. Weathering may break a large, solid mass of rock into loose fragments varying in but identical in composition to the original rock. Only the size and shape of the rock have changed. Processes that break down rocks without changing their chemical compositions are called *physical weathering*.

Frost Action

Frost action is the result of an unusual property of water. Most materials expand when heated and contract when cooled. This is true of water except that when water is cooled from 4°C to 0°C, it *expands*. It expands most when, at 0°C, it solidifies into ice; then its volume increases by 9%. The expansion of water as it cools and solidifies can exert huge forces on anything confining it—forces measuring tens of thousands of pounds per square inch!!

Water, in the form of rain, melting snow, or condensation, seeps into any cracks or pores in rock. When the temperature drops below the freezing point of the water in these cracks and pores, the water changes into ice. The expanding ice exerts tremendous pressure against the confining rock. Acting like a wedge, it widens and extends the opening. Then, when the ice thaws, the water seeps deeper into the opening. When the water refreezes, the process is repeated. In this way, the alternate freezing and thawing of water, or *frost action*, breaks rocks apart (see Figure 4.1).

Frost action is particularly effective where bedrock is directly exposed to the atmosphere, moisture is present, and the temperature fluctuates frequently above and below the freezing point of water. These conditions often exist during the winter in temperate climates such as New York State's, and can occur also on mountain tops and at high elevations in spring or fall.

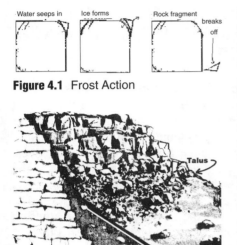

Figure 4.1 Frost Action

Daytime temperatures rise above freezing, causing snow and ice to melt, only to drop below freezing again at night, producing frost action.

Frost action on cliffs of bare rock breaks loose fragments that fall to the base of the cliff. When this takes place rapidly, a pile of fragments called a *talus slope* accumulates at the base of the cliff (see Figure 4.2).

The potholes that threaten motorists in many of our northern states are caused by frost action on exposed road surfaces. Farmers in New England have to clear their fields every spring of rocks and boulders pushed up out of the ground by frost action in soil. As you might expect, frost action is almost nonexistent in states such as Florida and Hawaii.

Figure 4.2 A Talus Slope Forms at the Base of a Cliff. The angle of repose is the steepest slope at which the fragments remain stable.

However, through much of the northern United States, it is probably the primary weathering process, and throughout the world, probably the most significant physical weathering process.

Abrasion

Rocks can also be broken down by *abrasion*, that is, by rubbing against each other. Rock abrasion occurs mainly when fragments are being carried along by agents of erosion. A typical example occurs in streams. As the fragments are carried along by the water, they bounce off and rub against each other. This abrasion breaks smaller pieces off the surface, and the fragments become still smaller and also more rounded (see Figure 4.3). The fragments abrade the bedrock beneath the stream as well. Wind also weathers rock by abrasion. Anyone who has sat on a sandy beach on a windy day can attest to the abrasive power of wind-driven sand. Abrasion is a major physical weathering process.

Time Increases ⟶

Figure 4.3 Rounding of Particles Due to Abrasion.

Exfoliation

Exfoliation is the scaling off, or peeling, of successive shells from the surface of rocks. Exfoliation generally occurs in coarse-grained rocks that contain the mineral feldspar. Whenever the surface of such rocks becomes wet, moisture penetrates pores and crevices between the mineral grains and reacts with the feldspar. A chemical change occurs, producing a new substance: kaolin. This clay has a greater volume than the feldspar it replaces. The expansion pries loose the surrounding mineral grains, and a thin shell of surface material flakes away.

Figure 4.4 Exfoliation

(Note that this is a physical process caused by a chemical change.)

With successive wettings of the rock surface the process is repeated. Since more surface is exposed at the corners of cracks or joints, these split off at the fastest rate. The result is a rounding in the shape of the rock, or the formation of rounded "exfoliation knobs," as seen in Figure 4.4. Feldspar is a common rock-forming mineral, therefore, exfoliation is a significant weathering process.

Plant and Animal Action

Plants and animals interact with rock in a number of ways that cause the rock to break down into smaller pieces. When a rock develops cracks, small particles of rock and soil are washed into the cracks by rain or blown in by wind. If a seed finds its way into the crack, it can germinate and begin to grow. As the plant grows, it sends tiny rootlets deeper into the crack in search of water. The growing rootlets thicken and press against the sides of the crack. Like the expansion of ice in frost action, the growing roots widen and extend the crack (see Figure 4.5a). Eventually the rock is broken apart. The roots of tiny plants like mosses and lichens produce a rock-dissolving acid as they grow and decay, thereby further accelerating the breakdown of the rocks.

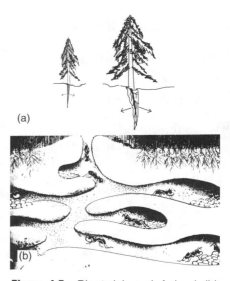

Figure 4.5 Plant (a) and Animal (b) Action on Rocks.

Although animals (with the exception of humans) do not directly attack rock, they contribute indirectly to its weathering in many ways. More than 100 years ago, Charles Darwin calculated that on 1 acre of land earthworms bring as much as 10 metric tons of particles to the surface each year. Exposed to the atmosphere again, these particles are subjected to further breakdown by weathering processes.

Ants (see Figure 4.5b), termites, woodchucks, moles and other burrowing animals also play a role in weathering. Furthermore, the burrows they create allow air and water to penetrate deeper beneath the surface and weather the underlying bedrock.

Humans, however, have probably contributed more to the physical weathering of rock than any other single species. Roadcut building, rock quarrying, and strip mining are just a few examples of the human activities that break up rock. In addition, these activities expose vast quantities of fresh rock to other weathering processes.

The contribution of each individual organism to the weathering of the Earth's crust may seem small. However, consider the number of living things on the Earth and the length of time they have been at work. Taken collectively, the total amount of rock they have affected is tremendous.

Changes in Temperature

Rocks are often exposed to large temperature changes. As rocks heat up during the day, they expand; as they cool off at night, they contract. You might expect this constant change to cause rocks to crack and break up, but, experiments have indicated that this is not generally the case. Only extreme temperature changes, such as those resulting from forest and brush fires, cause rocks to crack or flake off at the surface.

Pressure Unloading

Some rocks form deep beneath the Earth's surface. There they are under great pressure—millions of pounds per square inch. As the rocks form, stresses build up inside them that cannot be released because of the pressure. When forces within the Earth bring these rocks to the surface, the pressure is reduced. The resultant expansion and release of stress cause the rocks to develop large cracks, or joints, at weak points in their structure—a phenomenon known as *pressure unloading*. Pressure unloading can also occur when glaciers melt away and the pressure exerted by the weight of the ice on underlying rock is released.

A-2. Chemical Weathering

Chemical weathering breaks down rocks by changing their chemical compositions. Most rocks form in an environment that is very different from that at the surface of the Earth. Many of the substances in the atmosphere, for example, were not present in the environment where the rocks were formed. When the minerals in a rock are exposed to these substances, they may react with them to form new compounds with properties different from those of the original minerals. Such changes almost always weaken the structure of the rock, so that it either falls apart or is more easily broken down by physical weathering. Oxygen, water, and carbon dioxide are chiefly responsible for the chemical weathering of rocks.

Oxidation

The atmosphere is about 21 percent oxygen. Oxygen combines with many substances in a reaction called *oxidation*, which is an important chemical weathering process. Oxygen reacts most readily with minerals containing iron, such as magnetite, pyrite, hornblende, and biotite. Oxidation of iron produces compounds of iron and oxygen, iron oxides, of which hematite (Fe_2O_3) and magnetite (Fe_3O_4) are common examples. Since the iron in hematite and magnetite can be extracted economically, these compounds are also considered iron ores.

If water is present during oxidation, another reaction can occur; a compound of iron, oxygen, and water called *goethite* can form. Goethite has a yellowish brown color. If goethite is dehydrated, hematite forms. Many reddish or yellow-brown soils and weathered rocks get their color from the hematite or goethite they contain.

Oxidation of iron oxide in the presence of water to form goethite:

$$4FeO \quad + \quad 2H_2O \; + \quad O_2 \quad \rightarrow \quad 4(FeO\text{-}OH)$$

| iron oxide | water | oxygen | goethite |

Dehydration of goethite to form hematite:

$$2(FeO\text{-}OH) \quad \rightarrow \quad Fe_2O_3 \; + \quad H_2O$$

| goethite | hematite | water |

How does oxidation break down or weather rocks? When oxygen combines with iron, the chemical bonds between the iron and other substances in the rock are broken, thereby weakening its structure. Compare the strength of a rusty nail with that of a new one. The effect on rock is similar. In addition to iron, other elements in rock can combine with oxygen, such as aluminum or silicon. The effects are much the same: an oxide is formed, the structure is weakened, and the rock disintegrates.

Hydration, Hydrolysis, and Solution

Most of the Earth's surface is covered with water. Trillions of gallons fall on the Earth as rain every day. With such abundance it is not surprising that water is an important agent of chemical weathering.

When water combines with another substance, the reaction is called *hydration*. For example, the hydration of anhydrite forms gypsum.

Hydration of anhydrite to form gypsum:

$$CaSO_4 \quad + \quad 2H_2O \quad \rightarrow \quad CaSO_4 \cdot 2H_2O$$

| anhydrite | water | gypsum |

Water can also form hydrogen ions (H^+) and hydroxide ions (OH^-). When these ions replace the ions of a mineral, the reaction is called *hydrolysis*. Feldspar, hornblende, and biotite are common minerals that can undergo hydrolysis, which causes them to swell and crumble. The product is a powder of clay minerals that are insoluble in water. For example, the hydrolysis of feldspar forms kaolinite, a clay mineral.

Hydrolysis of feldspar to form kaolinite:

$$4\ KAlSi_3O_8\ +\ 4\ H^+\ +\ 2H_2O\ \rightarrow\ Al_4Si_4O_{10}(OH)_8\ +\ 8SiO_2$$

K-feldspar   hydrogen   water   kaolinite   silica
               ions

(Note that the K does not appear on the right. It is released as a positive ion into the soil water.)

Water also weathers rock simply by dissolving it. This process is called *solution*. So many materials dissolve, at least to some extent, in water that water it is often called the universal solvent. Halite (rock salt) and gypsum are good examples of water-soluble minerals. Water will slowly dissolve these minerals out of a rock, thereby exposing the surrounding minerals to further weathering. In some cases the rock's structure may be so weakened by the empty spaces created that the rock just crumbles. In solution, the dissolved minerals may react with each other to form new substances. If these substances are insoluble in water, they precipitate out.

Carbonation

Carbon dioxide (CO_2) is a colorless, odorless gas that comprises 0.03–0.04 percent of the Earth's atmosphere. The chemical combining of carbon dioxide with another substance is called *carbonation*. As a gas, carbon dioxide has almost no effect on rocks. When carbon dioxide comes in contact with water, though, the following carbonation reaction takes place:

Production of carbonic acid by solution of carbon dioxide:

$$CO_2\ +\ H_2O\ \rightarrow\ H_2CO_3$$

The product, H_2CO_3, is carbonic acid. Although a fairly weak acid, it attacks common rock minerals. Carbonic acid reacts readily with minerals containing the elements sodium, potassium, magnesium, and calcium. The compounds formed when these elements react with carbon dioxide are called *carbonates*.

Carbonic acid is most destructive to the mineral calcite, which is completely dissolved by carbonic acid. Rocks, such as limestones, that are almost entirely composed of calcite are

Figure 4.6 Carbonation Caverns form as rainwater containing carbonic acid seeps into cracks in limestone.

totally dissolved away by carbonic acid in rainwater and groundwater. Spectacular caverns can form (see Figure 4.6) as groundwater containing carbonic acid seeps through bedrock composed of calcite and eats huge holes in it.

Other Chemical Factors

In addition to carbonic acid, other naturally occurring acids attack rocks and minerals. Some of these acids are produced by lightning or the decay of organic material; others, as waste products of certain plants and animals. These acids dissolve in rainwater as it falls through the atmosphere and seeps through the soil. Upon reaching bedrock, they dissolve some rocks and cause others to crumble.

An increasingly important source of acids that weather rock is human activity. Factories, homes, and automobiles release vast quantities of waste gases and other pollutants into the atmosphere. Many, such as the oxides of nitrogen and sulfur, react with water to form very strong, reactive acids.

In some areas surrounding large cities and industrial complexes, the concentration of these acids has reached alarming levels. Rainfall in these areas contains so much acid that it is called *acid rain*. Acid rain accelerates the weathering of rock and the breakdown of manmade structures, and it damages plant and animal life.

A-3. Factors Affecting Weathering

How long does it take for a rock to be broken down by weathering processes? The answer is complex since many factors influence the rate at which a rock will weather. Among the more important factors affecting weathering are climate, particle size, exposure, mineral composition, and time.

Climate

Climate, which is the single most important factor affecting weathering, is the average condition of the atmosphere in a region over a long period of time. It is typically expressed in terms of two factors: temperature and moisture (see Figure 4.7). Both factors influence the type and the rate of weathering. Warm climates favor chemical weathering, cold climates, physical weathering, principally frost action. In both cases, the more moisture present, the more pronounced the weathering.

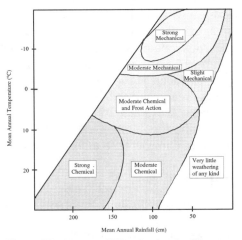

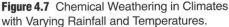

Figure 4.7 Chemical Weathering in Climates with Varying Rainfall and Temperatures.

Chemical reactions tend to occur at a faster rate as temperature increases. Many of these reactions require water, which is a reactant in hydration and carbonation and provides a medium in which acid reactions can occur. Hot, moist climates also support increased biological activity, ranging from burrowing to the production of humic acid as plant matter decomposes. Thus, a hot, moist climate is the ideal environment for rapid chemical weathering.

In cold climates, physical weathering by frost action predominates. It is most effective in cold climates that alternately freeze and thaw. Here again, moisture is a key factor. Without water, ice cannot form and frost action cannot occur. A cold, moist climate will therefore experience strong frost action.

Temperate climates, such as New York's, combine aspects of both extremes. Hot, moist summers followed by moist winters with many freeze-thaw cycles result in an environment in which rocks weather rapidly. A classic example of the destructiveness of our climate is its effect on Cleopatra's Needle, a granite obelisk with hieroglyphics cut into its surface that was moved from Egypt to Central Park in 1880. In Egypt's hot, dry climate the obelisk stood almost unchanged for 3,000 years. In the moist, changing climate of New York City, however, it began to rapidly deteriorate. Today, its surface cracked, worn, and discolored, the hieroglyphics almost unreadable, it stands as mute testimony to the effect of climate on weathering.

Particle Size

The size of rock particles greatly affects the rate at which chemical weathering occurs. Under the same conditions, the smaller the pieces of a particular rock, the faster they will weather. The reason is that a given volume of small particles has more surface area than the same volume of large particles (see Figure 4.8).

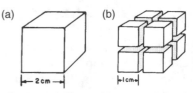

Figure 4.8 As a particle is broken into smaller pieces, its total surface area increases. (a) This cube has a surface area of **24 cm²**. (b) Each of these eight small cubes has a surface area of 6 cm², for a total of **48 cm²**.

A reaction can take place only when the rock comes into contact with the chemical weathering agent. The more rock surface exposed, the more rock is able to react, and the faster the rate of the reaction.

Size is also a factor in what happens to rock particles after they are produced by weathering. Small particles may be transported to a new location that has a different climate, or to a body of water that contains a variety of reagents.

Exposure

Exposure determines the degree to which a rock comes into contact with weathering agents. Soil, ice and vegetation can cover a rock and thereby decrease its contact with weathering agents. Rocks thus protected tend to weather more slowly than those completely exposed at the surface. Also, the weathered surface of a rock can itself shield fresh material underneath, thereby slowing the rate of weathering. Another factor that affects exposure is the slope of the land. On steep slopes, loose materials move downhill because of gravity or are carried downhill by erosion, thus continually exposing fresh rock.

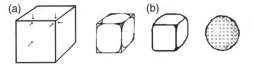

Figure 4.9 Results of Exposure. (a) Corners and edges weather fastest (b) resulting in a rounded particle.

Exposure also has its effects on each individual particle. As you can see in Figure 4.9a, the corners of a cube are exposed on three sides to weathering, the edges on two sides, and the faces on one side. The net effect is that the corners of a particle weather away most quickly and the faces most slowly. Over time the result is a spherical particle, as shown in Figure 4.9b.

Mineral Composition

The mineral composition determines the physical and chemical properties of a rock and thus its susceptibility to weathering. Mineral composition is of greatest importance in chemical weathering. Rocks composed of minerals that react readily with acids, water or oxygen will weather more rapidly than those composed of less reactive minerals. For example, limestone, which is mostly calcite, is dissolved by even mildly acidic rainwater, while granite, which is mostly silicates, is almost unaffected.

Mineral composition also affects physical weathering (see Table 4.1). Softer rocks will abrade more readily than harder rocks. Solid, crystalline rocks have fewer openings into which water can penetrate than rocks composed of cemented-together particles.

TABLE 4.1 RELATIVE RESISTANCES OF THE MAJOR ROCK-FORMING
MINERALS TO WEATHERING

Mineral	Resistance to Weathering
Olivine	Least resistance
Pyroxene	
Hornblende	
Biotite	↓
Plagioclase feldspars	
Muscovite	
Orthoclase	
Quartz	Most resistant

Time

Weathering is a slow process. The 3.5-billion-year-old gneisses in Greenland are scarcely weathered. Even easily weathered rocks may take hundreds of years to be completely broken down. The longer a rock is exposed to weathering processes, however, the more it deteriorates and disintegrates. Eventually, all rocks exposed at the Earth's surface are completely broken down.

WHAT ARE THE PRODUCTS OF WEATHERING?

Since the Earth formed, the rocks exposed at its surface have been attacked by forces that disintegrates them. Whether these forces are physical or chemical, the result is the same—solid rock is broken into fragments. These fragments range from the tiniest particle dissolved in water to the largest boulders. The fragments produced form sediments, mineral deposits, and soil.

B-1. Sediments

The fragments or particles of rock produced by weathering are called *sediments*. Sediments are named by their size (Table 4.2).

TABLE 4.2 SEDIMENT SIZES AND NAMES

Particle Diameter (cm)	Name
0.00001–0.0004	Clay
0.0004–0.006	Silt
0.006–0.2	Sand
0.2–6.4	Pebbles
6.4–25.6	Cobbles
>25.6	Boulders

B-2. Soils

Soil is the accumulation of loose, weathered material that covers much of the land surface of the Earth. Soil varies in depth, composition, age, color, and texture. Although its chief component is weathered rock, a true soil also contains water, air, bacteria, and decayed plant and animal material (humus).

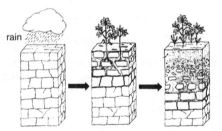

Figure 4.10 The Development of a Residual Soil Weathering opens bedrock to air, water, and dust. Plant and animal life accelerate the breakdown of rock and allow air and water to penetrate ever more deeply. Eventually a deep layer of soil develops atop the bedrock.

The rock from which a soil forms is called the *parent material*. Soil that forms directly from the bedrock beneath it is *residual soil* (see Figure 4.10). If the soil forms from material that was transported to the location by erosion, it is *transported soil*. Transported soil may have a different mineral composition from the underlying bedrock.

As a soil forms, the processes of weathering and plant growth develop recognizable layers, or horizons, in the soil. These horizons differ in structure, composition, color, and texture. A soil that has been forming long enough to have developed distinct horizons is called a *mature soil*. In *immature soils*, horizons are indistinct or altogether lacking.

A *soil profile* is a cross section of a soil from surface to bedrock. Figure 4.11 shows the profile of a typical mature soil with a description of each horizon.

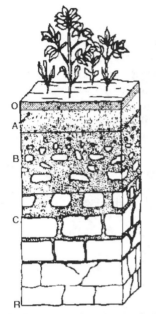

Horizon O (organic horizon)
The material in this horizon is commonly called topsoil and supports plant life. This horizon contains fresh to partly decomposed organic matter. Its color varies from dark brown to black.

Horizon A (leached horizon)
The top of this horizon is highly decomposed organic matter mixed with minerals. Its particles, exposed since the soil began to form, are sand sized or smaller. In a process called *leaching*, soluble minerals and tiny clay particles are carried down to lower layers as water seeps downward through this layer. Leaching is one of the processes that cause horizons to form. This horizon ranges from brown to gray in color.

Horizon B (accumulation horizon)
Materials leached out of horizon A are deposited here. This horizon is richer in clay, has less organic material, and has more and larger particles of bedrock than horizon A. As a result it is less fertile. The clay in this horizon gives it a clumpy or blocky consistency compared to loose, sandy horizon A. The leached materials deposited here (clays and iron oxides) give this horizon a reddish brown or tan color.

Horizon C (partly weathered horizon)
This horizon consists of partly weathered bedrock. Sometimes referred to as subsoil, it is the cracked, broken surface of the bedrock. It marks the final transition from topsoil to unaltered bedrock or parent material.

Parent material (R)
Unaltered bedrock

Figure 4.11 Profile of a Mature Soil

Parent material greatly influences the type of soil that forms, especially when the soil is just starting to form or is immature. Over a long period of time, however, climate has a greater influence on the type of soil that forms. Whatever the parent material, the profiles of mature soils that formed in similar climates are very much alike.

Figure 4.12 Two General Types of Soil in the United States

In the United States, two general types of soils have developed (see Figure 4.12). In the western half of the country, where rainfall is less than 63 centimeters yearly, the soils that formed are mainly a type called *pedocals*. Pedocals are rich in calcium compounds and tend to be slightly alkaline, a perfect growing environment for grasses. In the eastern United States, where rainfall exceeds 63 centimeters yearly, the soils are mainly *pedalfers*. Pedalfers are rich in aluminum and iron compounds produced when water and oxygen react with common rock-forming minerals, and soluble calcium compounds are washed away by successive rainfalls.

167

C. HOW HAS EROSION CHANGED THE SURFACE OF THE EARTH'S CRUST?

What happens to the sediments produced by weathering? In most cases they are moved, often great distances from where they originated. Any process that moves sediments from one place to another on the Earth's surface is called *erosion*. As you read these words, erosion is changing the Earth in countless ways. Waves crashing against shores are scouring away sand, reshaping the coastline. In deserts, hot winds are moving towering dunes of sand grain by grain. On cold, barren mountaintops fragments of rock broken loose by weathering are tumbling to the ground. Glaciers creeping downhill are tearing huge boulders from the ground and carrying them along. Streams, from tiny trickles to raging torrents, are carving their way into the crust.

Together, weathering and erosion wear away the earth's crust. Weathering breaks down solid rock; erosion carries away the pieces. In this way fresh rock is exposed to weathering, and the cycle repeats itself. Sometimes erosion is rapid and unmistakeable, as in a landslide. Usually, though, the process is gradual, and a long time passes before it is evident that erosion is taking place.

C-1. Evidence of Erosion

Any sediment moved from its source is evidence of erosion. At times, this evidence is striking. The owner of a house returning after a mudslide knows that the house was not built with one end hanging over the edge of a cliff. The lack of soil under the house is evidence of erosion. In much the same way, valleys and canyons are evidence of erosion. The Grand Canyon is one of the most spectacular examples of erosion in the United States. Imagine how much sediment had to be carried away to form it!

Quite often, though, the evidence of erosion is subtle. A boulder sitting on bedrock may not immediately seem to indicate erosion. If the boulder is granite and the bedrock is limestone, however, the boulder certainly wasn't derived from the bedrock and had to be carried there from some other source. Similarly, layers of sediment can often be found overlying bedrock that has an entirely different composition. This, too, is evidence that erosion has occurred. Since the sediment did not come from the bedrock, it must have been carried there from some other source.

C-2. Agents of Erosion

Erosion moves sediments. To move anything, a force is needed. The primary driving force of erosion is gravity. Gravity can move sediments by acting on them directly. On cliffs and steep slopes, sediments broken loose by weathering move downhill under the influence of gravity. Gravity can also move sediments by acting on them indirectly, through agents of erosion. For example, water runs downhill under the direct influence of gravity. The running water, in turn, can exert a force on sediments in its path causing them to move. Thus the running water is an agent of erosion. Some other agents of erosion are waves, currents, winds and glaciers.

An agent of erosion, together with the driving force (usually gravity) that sets the agent in motion that picks up and transports sediment, comprises an *erosional system*. Erosional systems are like natural conveyor belts, moving sediments from higher to lower elevations.

C-3. Mass Wasting

Mass wasting is the downhill movement of sediments under the direct influence of gravity. On most slopes, some kind of downhill movement of sediments is going on all the time, because on a slope, gravity acts as if it has two parts, or components. One part, the normal force, pulls downward perpendicular to the surface causing friction, and the other pulls downhill parallel to the surface (see Figure 4.13). As a slope gets steeper, more and more of the gravity acts in a downhill direction. When the downhill force is greater than the friction caused by the normal force, objects move downhill.

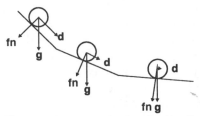

Figure 4.13 Mass Wasting. In the diagram, fn = normal force, g = force of gravity, d = downhill force.

The steepest slope angle at which a particular sediment remains stable is called its *angle of repose* (see Figure 4.14). The size, shape, and density of a rock particle affect its angle of repose. Thus, sand, gravel, and clay have different angles of repose. When their angles of repose are exceeded, sediments move downhill and mass wasting occurs. Depending on the slope and the

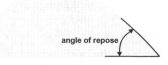

Figure 4.14 Angles of Repose in a Pile of Dry Sand.

angles of repose of the sediments, mass wasting may be rapid or slow.

Rapid Mass Wasting

Rockfalls (see Figure 4.15a) occur when rock fragments broken loose by weathering fall from cliffs or bounce by leaps down steep slopes. Rockfalls are the most rapid of all mass-wasting processes. "Fallen Rock Zone" is a common sign near roadcuts on many highways or in mountainous areas where rockfalls occur frequently. In populated areas or on heavily traveled roads, rockfalls can be very dangerous and have resulted in loss of life. Some

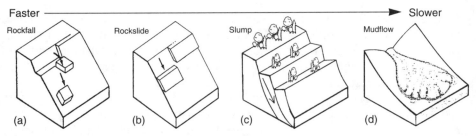

Figure 4.15 Four Rapid Mass-Wasting Processes

localities have spent much money cutting back the rocky slopes of roadcuts or covering them with steel cable nets to prevent rockfalls.

Rockslides (see Figure 4.15b) occur on less steep slopes when rock masses or debris slide downhill. Most rockslides are triggered by heavy rains or earthquakes. Water acts as a lubricant between particles. The shock of an earthquake knocks particles apart, decreasing the friction between them and the underlying surface. The downhill component of gravity is then greater than the friction holding the particles in place, and the particles move downhill. Rockslides can move enormous amounts of material, often millions of cubic meters. In 1959, twenty-seven people were killed when an earthquake caused an entire mountainside to slide into the Madison River gorge in Montana.

Slump (see Figure 4.15c) is a mass-wasting process in which a huge mass of bedrock or soil slides downward from a cliff in one piece. The mass, or slump block, slides along a curved plane of weakness as shown in the diagram. The slump block rotates and comes to rest with its upper surface tilting toward the cliff. Slump is common where ocean waves or streams undercut cliffs.

A *mudflow* (see Figure 4.15d) is the rapid, downhill flow of a fluid mixture of rock, soil, and water. Mud flows generally occur after heavy rains saturate soil that has no protective covering of vegetation, and usually occur in semiarid regions or on slopes denuded by construction or lumbering. The rain mixes with the soil to form a thin mud, which flows downhill, thickening as more soil and debris are picked up and increasing in speed. Since the mudflow is much heavier and thicker than water, the impact when it hits something in its path is devastating. Entire houses, cars, and trees have been swept away by mudflows. A mudflow comes to rest when it reaches the bottom of the slope or when it has picked up enough material so that it thickens to a point where it can no longer move. The four rapid mass-wasting processes described above are diagramed in Figure 4.15.

Slow Mass Wasting

In humid regions, slopes are usually covered with vegetation. The vegetation protects the surface from the impact of raindrops, and its root system holds the soil together, inhibiting downhill movement. However, mass wasting occurs even on slopes covered with vegetation, albeit slowly.

170

On vegetated slopes, rainwater entering the soil loosens the particles and makes them slippery, decreasing their angles of repose. In this condition, an earthflow may occur. In an *earthflow* (see Figure 14.6a), a shallow layer of soil and vegetation, saturated with water, slowly slides downhill. An earthflow may take several hours to ooze its way down a slope. Earthflows are a common cause of road and rail blockages. While they are usually not life threatening since they move at a snail's pace, they can cause considerable property damage.

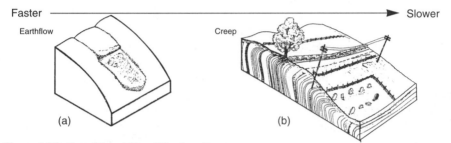

Figure 4.16 Two Slow Mass-Wasting Processes

The slowest of all mass-wasting processes is soil creep. *Soil creep* (see Figure 4.16b) is the invisibly slow, downhill movement of soil, carrying vegetation with it. The tilting of old poles, fenceposts, or tombstones and bulging or broken retaining walls are all evidence of creep.

Any disturbance of the soil on a slope causes creep. Freezing and thawing, wetting and drying, and trampling or burrowing by animals are just a few of the things that can disturb the soil on a slope. When disturbed, the soil particles shift and settle. As they do so, gravity moves them downhill. Creep is common in regions where the soil alternately freezes and thaws frequently.

Large objects on or in a creeping slope are carried along by the moving soil. Boulders that have crept down a hillside may accumulate at the base of the slope, forming a boulder field. Trees growing on creeping slopes are often misshapen. The tree grows straight up, but the soil that its roots are in slowly creeps downhill.

The two slow mass-wasting processes discribed above are diagramed in Figure 4.16.

Mass wasting is usually only the first step in the eroding of the earth. It delivers sediments to the base of slopes. There, agents of erosion, such as streams, can pick up the sediments and carry them farther.

C-4. Erosion by Moving Water

Raindrops and Runoff

Raindrops may fall for thousands of feet before they hit the ground, and the force of falling raindrops can move sediments. When raindrops strike the ground, their impact causes a geyserlike splashing of loose soil. In a process called *splash erosion*, each raindrop impact lifts some soil and drops it in a new position. On one field observed during a rainstorm, splash erosion moved 225 metric tons of soil per hectare.

Runoff is precipitation that does not evaporate or sink into the ground. Under the influence of gravity, runoff flows downhill. On even the gentlest slopes, runoff may flow downhill in a thin sheet called *overland flow.* Overland flow exerts a dragging force on the ground. Depending on its speed, it can exert a force great enough to move sediments ranging from fine clay to coarse sand or gravel. Erosion by overland flow is called *sheet erosion.*

Vegetation greatly decreases both splash and sheet erosion. Plants leaves and stems absorb the impact of falling raindrops, and their network of roots anchors the soil in place. Plant stems also block and slow the downhill movement of water, thereby decreasing the force it exerts on sediments in its path. In arid regions or on slopes denuded of vegetation, however, splash and sheet erosion can remove vast quantities of loose soil.

Splash erosion and sheet erosion are serious problems on farms. They carry away the rich topsoil essential for healthy plant growth. Contour plowing, terracing, strip cropping, and crop rotation help lessen soil erosion.

On steep, unprotected slopes, erosion by runoff can become intense. Numerous tiny grooves, or rills, form as the water runs downhill, carrying away sediment. If erosion continues, the rills may widen and deepen into a gully. Gullies act as funnels for runoff. They concentrate the force of the flowing water, thus increasing the rate of erosion.

Streams

A *stream* is any body of running water that moves under gravity, in a relatively narrow but clearly defined channel, to progressively lower levels. A stream may be small enough to step across, or so wide that the other side is barely visible. Streams are the most important agent of erosion because they affect more of the Earth's surface than any other agent. Most of the sediment carried downhill by runoff ends up in streams.

Streams form as runoff from several slopes that drain into low-lying areas between the slopes. These natural depressions fill with water, forming a puddle, a pond, or even a lake. As runoff continues, they may eventually overflow. Then the overflow continues flowing downhill, along natural passageways or depressions in the surface, to lower and lower levels. As the water flows along these passageways, it cuts into the surface, forming a clearly defined path, or *channel,* along which it flows. Once established, the same channel provides a pathway for all later runoff.

Figure 4.17 shows a typical stream and its parts.

Stream Transport

Water flowing downhill can be very powerful. Have you ever seen whitewater rafters shooting a rapids? The water flowing downhill exerts quite a force! Imagine trying to paddle upstream through the rapids. It would be almost impossible because the force of the flowing water is so great. Now imagine a grain of sand or a pebble in the path of such a stream. Any loose sediments are likely to be carried away.

Material carried by a stream is called its *load*. A stream's load consists mainly of sediments, which can vary in amount and type, depending on the speed and volume of the stream. The steeper the slope, or gradient, of a stream's channel, the faster the water flows. Water can carry more and larger sediments as its speed increases. At any speed, the more water there is in a stream, the more load it can carry.

Streams transport different sediments in different ways. Large sediments, such as pebbles, cobbles, and boulders, are heavy. Even the fastest flowing streams may not be able to pick them up. But sediments can be moved without picking them up! Do you have to pick up a

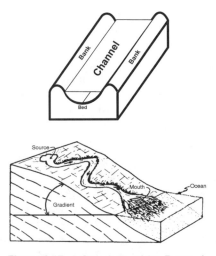

Figure 4.17 A Stream and Its Parts. A stream begins at its **source**, flows along a path called a **channel**, and ends at its **mouth**. The channel consists of sides called **banks** and a bottom called the stream **bed** whose slope is called the stream's **gradient**.

car that has run out of gas in order to move it? Of course not; you can push or drag the car so that it rolls along the road. In much the same way, large particles can be carried downhill by streams. The force of the flowing water in a stream can push or drag large particles downhill by rolling or sliding in a motion called *traction*. If the sediment particles move in a series of bounces, hops, or leaps along the streambed, the motion is called *saltation*.

Smaller particles, such as silt, sand, and clay, are lighter in weight and therefore require less force to move. Although they are heavier than water, the flowing water can pick up these small sediments; and, whenever they begin to sink, the force of the turbulent water pushes them back up again— they are carried downhill *suspended* in the water.

Some sediments dissolve in water. Water that has combined with carbon dioxide or other gases in the atmosphere to become acidic (see Chemical Weathering, page 160) is especially effective at dissolving sediments. Sediments that are dissolved are carried downhill *in solution*.

Stream Erosion

Streams not only transport sediments, they also change the sediments they carry, wear away the surfaces over which they flow, and create new sediments. Pure water flowing over rock wears it away very little, but water carrying sediment acts like a cutting tool. When the sediments being carried hit rock in the stream channel, chips of rock break off. Rolling and sliding peb-

173

bles, cobbles, and boulders collide and knock chips off each other. They crush and grind up smaller particles between them. The process of crushing, grinding, and wearing away rock by the impact of sediments is called *abrasion*. Abrasion wears away a stream's channel and causes the sediments carried by the stream to become rounded.

By abrasion, streams can cut through bedrock over which they flow. The potholes shown in Figure 4.19 were carved out of bedrock by sediments in a swiftly flowing stream. The waterfall formed where a stream flowed over hard bedrock onto soft bedrock. The soft bedrock wore away faster, lowering the stream bed, and a waterfall was formed. Prolonged erosion can carve deep canyons and valleys into the Earth's surface. The Grand Canyon is an amazing example of what sediment-laden water can do.

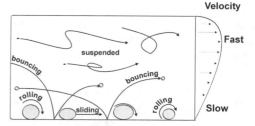

Figure 4.18 Methods of Particle Transport. Larger sediments are transported in streams by rolling, sliding, and bouncing along the stream bed. Smaller sediments are carried in suspension or may even be dissolved in the water of the stream.

A Stream's Life Cycle

As a stream cuts into the surface and carries away sediment, its channel becomes wider, deeper, and longer. Over time the shape and behavior of the stream change as the surface over which it flows is changed.

Youth. Newly formed streams generally flow down sleep slopes. The water flows quickly, often forming rapids. Over time, the stream wears a long, deep groove, or valley, in the surface. Stream-cut valleys typically have steep sides that form a distinctive V-shape. A young stream valley is shown in Figure 4.20a.

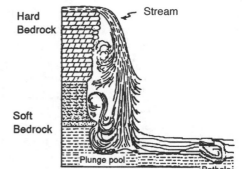

Figure 4.19 Waterfall and Pothole. Waterfalls may form where streams flow from hard bedrock onto softer bedrock. As the easily eroded soft bedrock wears away, the level of the stream bed becomes lower than on the adjacent hard bedrock. The falling water creates turbulence that swirls sediments against the stream bed forming potholes.

Maturity. As a stream erodes the slope over which it flows, the slope becomes less steep. As a result, the water flows more slowly and sediment accumulates at the bottom of the valley, making it flat. As mass wasting and runoff wear away the sides of the valley, the V-shape becomes wider and less steep. The slower moving water flows around obstacles instead of tumbling over them, and the stream's path forms curving loops called *meanders*. When runoff increases suddenly, as after a torrential downpour or sudden spring

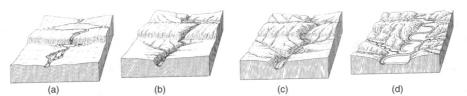

Figure 4.20 The Three Stages in the Life Cycle of a Stream: (a) the relief of the land is steep and the stream has lakes, rapids, and falls. (b) continued downcutting of the stream bed eliminates falls and cutting back of the valley walls allows a narrow flood-plain to form. (c) further downcutting and erosion of the valley walls enlarges the floodplain and allows the stream to meander widely between the valley walls. Shifting meanders result in cutoofs and oxbow lakes.
Source: *The Earth Sciences.* © by Arthur N. Stahler, 1971.

thaw, the stream overflows. Spreading out over the valley floor, the stream slows down and deposits sediments to form a broad, flat *floodplain* adjacent to the stream. Figure 4.20b shows a typical stream at the mature stage.

Old Age. Eventually, the slopes around a stream are worn away almost completely. A flat lowland of gently rolling hills, called a *peneplain*, is formed. Since the slope of the land is very gentle, the water in the stream flows slowly. Its meanders become highly curved and winding and at times may actually loop over one another. During floods, water may gush over the land between meanders, forming a *cutoff*. If the new channel cuts deep enough, part of the meander may become isolated, forming an *oxbow lake*. Figure 4.20c shows a typical stream at the old stage.

Whatever the stage of development, the water in most streams eventually reaches the oceans.

C-5. Ocean Erosion

Ocean water is constantly in motion. Waves, currents, and tides are examples of ocean water movement. Moving ocean water can transport sediments and erode the land.

Waves

Waves can be very powerful. A single wave sends tons of water crashing against a shore. The force of waves can break up rock. Loose fragments are stirred up and carried along in the turbulence of breaking waves.

Wave erosion forms a number of shoreline features, some of which are shown in Figure 4.21. When waves erode steep rocky shores, they remove material from the base of the slopes. A notch is formed, which undercuts the slope, causing overhanging rocks and soil to break away. The result is a steepsided *sea cliff*. When wave erosion penetrates deeply into a cliff, a large hole, or *sea cave* may form. As weaker rock is worn away from the cliff, pillars of resistant rock, called *stacks*, may be left behind. As the erosion continues, fragments broken from the cliff are ground up by abrasion in the turbulent waves. This buildup of material forms a flat platform, or *sea terrace*, at the base of the cliff.

Currents

Currents are movements of water. On gently sloping, sandy shores, waves mainly shift sediment around. When waves strike perpendicularly to the shore, sediments are moved in and out with the waves. When waves strike the shore at other than 90°, however, they are reflected at an angle and interfere with each other, forming a longshore current (see Figure 4.22) parallel to the shore. Longshore currents can move sediments along just as a stream can.

Figure 4.21 Shoreline features formed by Wave Erosion of a Rocky Coast: **P**-platform of abraded sediments, **N**-a wave-cut notch at the base of a cliff, **R**-a crevice eroded back into the rock, **B**-a beach of sediments eroded out of the cliff, **C**-a sea cave, **A**-an arch, and **S**-a sea stack.
The Earth Sciences. © by Arthur N. Stahler, 1971.

Along some coasts, tides flowing into and out of narrow openings form swift currents. These, too, can move large amounts of sediments.

Erosion by moving water affects our lives in many ways. Overland flow may wash away the topsoil your family just bought and spread on the lawn. Splash and sheet erosion of topsoil on farmland decreases fertility and lowers crop yield, resulting in higher prices at the supermarket. Erosion and flooding by streams leave thou-

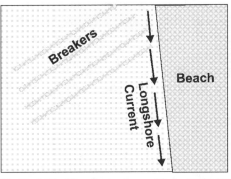

Figure 4.22 The formation of a longshore current as waves approach a shoreline at an oblique angle.

sands of persons homeless annually. Wave erosion threatens the homes of those living near the shore. Only by understanding how water erosion occurs can we attempt to deal with it effectively.

C-6. Wind Erosion

Wind that blows fast enough can pick up and carry loose particles of rock. Wind is also an agent of erosion, though not an important one in most regions. Where rainfall is abundant, plants growing on the surface protect underlying sediments from the full force of the wind. Where the ground is moist, sediment particles tend to stick together and are more difficult for the wind to dislodge and move. (Think of the difference between blowing on a handful of dry, powdery clay and blowing on a handful of wet, sticky clay.) In arid regions, however, there is little plant cover, and the soil, loose sediments, and weathered bedrock are dry and exposed. Under such conditions, wind becomes an important agent of erosion.

Deflation and Abrasion

Wind erodes the land in two ways: deflation and abrasion. In *deflation*, the wind picks up and carries away loose particles much as a stream carries sediments. In *abrasion*, exposed rock is worn away by wind-driven particles. The rock particles worn away by abrasion are then carried away by the wind.

Wind is like a stream of air, and wind carries sediments in much the same way that a stream of water does. However, winds are usually able to carry only small particles, such as sand, silt, clay, and dust, because the wind's medium, air, is not very dense. It exerts less force on a particle than does a denser medium, such as water, moving at the same speed. Imagine the difference between being hit by a balloon filled with air and a balloon filled with water!

Wind carries very small particles, such as clay and dust, in suspension. The finest particles may be carried many meters up into the atmosphere. When volcanoes erupt, dust and ash are often spewed high into the atmosphere, where winds carry them for hundreds or even thousands of miles before they finally settle to the ground. Some particles of dust and ash may reach the jet streams of the stratosphere and remain suspended for years after they are picked up by these fast-moving winds.

Wind carries larger particles closer to the ground. They slide, roll, and bounce, skipping along the surface. As they do so, they collide with exposed bedrock or other particles. In the process of abrasion, these collisions scratch the surface of the exposed rock and cause tiny chips to break off. Like a sandblaster, the wind-borne particles cut and polish the rock surfaces they impact. Over time, rock exposed to wind-driven particles is worn away. The surfaces of both the exposed rock and the particles take on a frosted appearance because of the many tiny scratches inflicted on them.

The amount and the type of erosion caused by wind depend on several factors: the speed of the wind, the size of the particles it carries, and the length of time the wind blows. The faster the wind blows, the larger the particles it can transport. The larger the particles carried by the wind, the greater the abrasion of exposed surfaces. The longer the wind blows, the more deflation and abrasion that can occur.

Features Produced by Wind Erosion

As deflation removes layer after layer of loose surface materials, a shallow depression, known as a *deflation hollow*, is formed. The semi-arid Great Plains of the midwestern United States are dotted with tens of thousands of deflation hollows. In wet years they are covered by grass; and when rainfall is unusually high, they may fill with runoff, forming shallow lakes. In dry years, however, the grass dies off and the wind continues to deflate the bare soil.

When deflation lowers the land surface to the level of the water table, erosion is stopped. The exposed groundwater holds the soil particles together and enables plants to grow. The plants' roots help hold the loose soil in place, and their leaves and stems block the wind, ending any further erosion. A patch of vegetation surrounded by arid wasteland is called an *oasis*.

Deflation usually removes only fine particles. If the soil is composed of a mixture of particle sizes, the larger particles, which are too heavy for the wind to move, are left behind. These larger particles shield the finer particles beneath them from deflation. As exposed fine particles are removed, coarse particles form an increasing percentage of the particles left at the surface. Eventually, a continuous layer of pebbles, gravel and other coarse particles, called a *desert pavement*, may form, as shown in Figure 4.23.

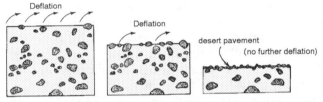

Figure 4.23 Formation of a Desert Pavement. As deflation removes finer sediments, larger sediments that are left behind eventually form a continuous layer called a desert pavement which blocks further deflation.
Physical Geology, 2nd. Edition, by Richard Foster Flint, Brian J. Skinner, John Wiley and Sons, 1977.

Abrasion by wind-blown particles produces spectacular features in arid regions. Caves, arches, and bridges may be cut into exposed bedrock. Unusually shaped rock formations may result when rocks of different hardnesses are exposed to abrasion simultaneously.

Loose rock particles, too heavy to be transported by the wind, can also be abraded by wind-blown particles. Over time, they develop flat surfaces on the side facing the prevailing winds (see Figure 4.24). These unusual particles are called *ventrifacts*, from the Latin word *ventrus*, meaning "wind."

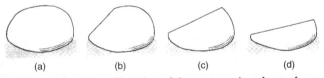

(a)   (b)   (c)   (d)

Figure 4.24 Ventrifact Formation. Abrasion of the exposed surface of a rock fragment by wind-born particles forms a flat-faced ventrifact.
Physical Geology, 2nd. Edition, by Richard Foster Flint, Brian J. Skinner, John Wiley and Sons, 1974, 1977.

Humanity and Wind Erosion

Wind erosion presents humans with many problems. In arid regions, poles carrying electric power and telephone lines have been worn through by abrasion, causing them to fall and interrupt service. In times of drought, the precious topsoil of farmland stripped of vegetation must be protected from deflation. Vehicles used in arid regions require special features to keep abrasive particles out of their moving parts and to prevent particles suspended in the air from clogging fuel, air, and cooling lines.

Awareness of wind erosion and its causes enables us to guard against it. Farmers plant windbreaks of drought-resistant trees. In many arid regions, off-road vehicles such as dirt bikes and dune buggies are banned because they destroy the fragile vegetation that protects the land from wind erosion.

C-7. Erosion by Glaciers

Glaciers form where snowfall exceeds melting over extended periods of time and, as a result, snow accumulates to great depths. As snow at the base is buried deeper and deeper beneath successive snowfalls, the pressure of the overlying material compresses it and the individual ice particles eventually fuse into a solid mass of ice. Ice under pressure behaves as a fluid. The ice in a glacier flows slowly downhill and outward under the influence of gravity.

As a glacier moves along, loose rock may freeze into the ice at the *bottom* of the glacier. The glacier is then like a huge piece of sandpaper. It wears down and smoothes the rock it moves over as sandpaper wears down wood. The smoothed surface of the rock is said to display *glacial polish.* Pebbles and other large particles frozen into the ice scratch deep, parallel grooves, called *glacial striations,* into the surface of the rock over which the glacier moves. If the glacier freezes to a protruding knob of bedrock over which it flows, that rock may actually be torn from the surface, in a process *plucking,* as the glacier moves along. A distinctive feature formed by this process is a *roche moutonée* (French, "fleecelike rock"). As a glacier flows over a small, protruding knob of bedrock, the upstream side is smoothed and elongated while the downstream side is plucked, leaving it steep, rough, and hackly. The roughened surface of the rock is said to resemble a sheep's back both in texture and in general shape—hence its name.

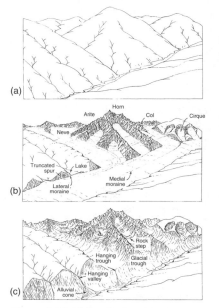

Figure 4.25 Valley Before, During, and After Glaciation. Top–A landscape eroded by streams into gently rolling hills before glaciation. Middle–the same landscape during glaciation. Numerous small glaciers fill the valleys and cut back slopes forming sharp ridges and horns. Smaller glaciers feed into a larger main glacier. Bottom–after the glacier retreats, the sculpted landscape of cirques, u-shaped glacial troughs and hanging valleys is exposed.
The Earth Sciences. © by Arthur N. Strahler, 1971.

As a glacier flows down a valley, it scrapes against the walls of the valley. Rock and other debris break loose and fall onto the *top* of the glacier. In this way trees, soil, and huge boulders may accumulate atop the glacier and be carried along like boxes on a conveyor belt. One glacier in Alaska accumulated so much soil on its surface that trees took root and grew. For years now, it has had a forest growing on it—a forest that is being carried to the sea atop a glacier! Fine, wind-blown sediments may also accumulate on top of a glacier.

A glacier also acts as a plow. Loose rocks and soil are pushed along in front of the glacier and along its sides. When a glacier moves down a V-shaped valley, the shape of the valley is changed to a U by the plowing action of the glacier. Figure 4.25 shows a valley before, during, and after glaciation.

As a glacier moves along, the ice churns and swirls like flowing water, but at an incredibly slow pace. Particles frozen into the sides or bottom of the glacier are drawn into the ice and are carried along "suspended" in it. In the process, rock material in the ice is crushed and ground into a fine powder called *rock flour*. Together, all of the material carried by a glacier is called *drift*. Drift is unsorted and may consist of particles ranging from clay to boulders.

As a glacier moves outward or downhill, it eventually reaches warmer regions and the ice begins to melt. If the ice flows outward faster than it melts back, the glacier's leading edge advances. If the ice melts faster than it flows outward, the leading edge retreats. As climates have changed, glaciers have advanced and retreated over much of the Earth's surface.

D. WHAT ARE THE FACTORS THAT AFFECT THE DEPOSITION OF SEDIMENTS?

At some point, all agents of erosion begin to deposit sediment. *Deposition* is the process by which transported sediment is dropped in a new place. The same agents that erode sediment also deposit it. Several factors influence when deposition will begin.

D-1. Speed of Medium

The *medium* is the substance that carries sediment. In streams the medium is water, in winds it is air, and in glaciers it is ice. Generally, the slower a medium is moving, the less its carrying power. If the carrying power drops below the levels needed to transport a particle, that particle will drop to the ground. Less carrying power results in deposition.

You can see this for yourself. Imagine you are flying a kite and the wind slows down. The slower the wind blows, the less carrying power it has. The kite will begin to drop. If the wind stops blowing, the kite will settle to the ground. Deposition occurs in much the same way. As the medium slows, sediment begins to settle out.

180

Figure 4.26 shows the sediment that water carries at different speeds. Notice that water carries more sediment sizes when it is moving fast. As water slows down, some sediment is deposited. Large, dense particles will settle first, and light, small particles last. The result is a separation of different particles into layers.

The separation of particles during deposition is called *sorting*. Figure 4.27 shows the sorting that occurs when a stream flows a into larger body of water, such as a lake or ocean. As the stream flows into the larger body, it slows down. Coarse gravel and pebbles settle first. Farther from the shore, the speed of the water is slower and sand settles out. Still farther from shore, in very slow moving water, the finest particles of silt and clay settle. Sorted layers of particles result.

D-2. Characteristics of Particle

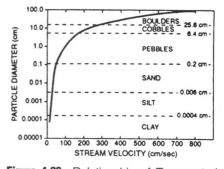

Figure 4.26 Relationship of Transported Particle Size to Water Velocity.
Source: The State Education Department, *Earth Science Reference Tables*, 1994 ed. (Albany, New York; The University of the State of New York).

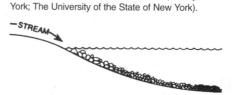

Figure 4.27 Horizontal sorting of Particles Carried by a Stream. As a stream decreases in velocity, the heavier particles settle first and the lighter particles are carried farther downstream before settling.

Particles may vary in size, shape, and density. Each of these characteristics influences the rate at which settling occurs. To learn how particle size influences settling rate, look at Figure 4.28a. Notice that, when all the rock particles are deposited, sorting has occurred. How can that be? The answer is that some particles must settle before others. The particles on the bottom settled fastest, and those on top slowest. Since the larger particles are on the bottom, they settled fastest. The smaller ones, on top, settled slowest.

In general, the larger the particle, the faster it will settle. Very small particles, such as fine clay, take a long time to settle. The muddy appearance of many rivers and lakes is due, in part, to particles that have not yet settled. To see how particle shape affects settling rate, look at Figure 4.28b. Notice that, when all the particles are deposited in water, a bottom layer of round particles results. The round particles must have settled first. Also notice in the figure a top layer of flat particles. These particles must have settled last.

In general, round particles settle faster than flat ones. To understand how particle shape affects settling rate, imagine dropping a crumbled paper and a flat paper to the ground at the same time. The crumbled paper, like a rounded particle, settles faster. Friction between the large surface area of the flat paper and the air slows its fall. A fiat shape in rock particles has the same

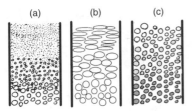

(a) All other factors being the same, large particles settle faster than smaller ones.

(b) All other factors being the same, particles with rounded shapes settle faster than flat-shaped particles.

(c) All other factors being the same, particles composed of denser materials settle faster than those composed of less dense materials.

Figure 4.28 Settling of Different Particles in a Column. When sediments settle in a quiet medium, they form a series of layers. The fastest settling particles reach the bottom first and form the lowest layer. They are followed by slower settling particles which form the next layer and the slowest-settling particles form the top layer.

effect in water. Friction between the flat particle and the water will decrease the rate at which the particle settles.

Finally, to see how the density of a particle affects settling, look at Figure 4.28c, Notice, once again, that sorting has taken place. The densest particles form the bottom layer; the least dense particles, the top layer, Thus it can be concluded that the densest particles settled fastest.

To summarize, several factors affect the deposition of sediment. The speed of the medium determines when particles will be deposited. The size, shape, and density of a particle determine the rate at which it will settle. Differences in the settling rates of sediments result in sorting during deposition.

D-3. Bedding of Sediment

As a result of sorting during deposition, sediment forms layers consisting of different types of particles. Sediment deposited in a quiet body of water will usually form layers similar to those shown in Figure 4.29. Each layer of sediment is called a *bed* and each bed usually represents a period of deposition. Beds of sediment are commonly found in sedimentary rock.

Deposition can create practical problems. For example, materials that settle out of water may clog drain pipes. As a result, most drain pipes have traps, as shown in Figure 4.30. The trap slows down the flow of water much as a lake slows the flow of a stream. As the water slows down, materials settle in the trap, preventing particles from clogging pipes. Clogged pipes can result in water backing up into a house.

Hurricane, September
No rainfall, August
Summer storm, July
Periodic gentle rain, May-June
Heavy rains, May
No rainfall, April
Heavy rainfall and spring thaw - March

Figure 4.29 Bedding of Sediment. Rapid deposition results in **graded bedding**, a vertical sorting. Each sequence represents a depositional event.

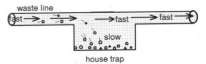

Figure 4.30 Drain Pipe with Trap. Most homes have a "trap," a deep, box-like section, in the waste water line where the water slows down causing solid particles to settle into the trap rather than clogging up the line. The trap must be periodically cleaned.

E. HOW HAS THE DEPOSITION OF SEDIMENTS BY VARIOUS FORCES RESULTED IN LANDFORM DEVELOPMENT?

E-1. Deposition by Gravity

During mass wasting, sediment is pulled downhill by gravity and is deposited, when it stops moving, at the base of a slope. Sediment deposited by gravity does not settle through a medium, as does sediment deposited by wind or water. Thus, little, if any, sorting takes place. Most deposits resulting from mass wasting are unsorted and do not show any distinct layering. In unsorted deposits, all types of particles are arranged without any order, as shown in Figure 4.31.

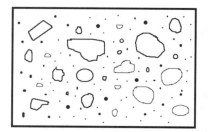

Figure 4.31 Unsorted Deposit. Unsorted sediment contains rock fragments of all sizes randomly mixed together.

Deposits by Rapid Mass Wasting

Rapid mass wasting deposits sediment. As deposition takes place, landforms develop. These landforms are evidence that deposition has occurred.

Rockfalls

During rockfalls and rockslides, fragments slide downhill until they reach the base of a slope, forming a talus. A *talus* is a pile of broken rock that builds up at the base of a cliff or steep slope. As the talus weathers, soil may form on the surface. Then, if the talus is not being constantly covered by new rock fragments, plants may take root and grow.

Mudflow

During a mudflow, mud moves downhill until it reaches the base of a slope. The mud then spreads out in a thin, wide sheet. Huge boulders carried by the mudflow may be left to stand isolated on the gently rolling land.

Deposits by Slow Mass Wasting

Slow mass wasting also deposits sediment. Earthflows, slump, and soil creep all produce sediment deposition from which landforms may develop.

Earthflow

An earthflow deposits a thick pile of sediment. Earthflow deposits are generally thicker than deposits produced by mudflows. The difference between the two can be demonstrated by spilling a jar of thick honey and a jar of thin tomato sauce next to each other. The thick honey, like an earthflow, produces a small, thick deposit. The thinner sauce, like a mudflow, produces a thin, spread.

Slump

Slump forms a deposit that looks like an apron spread on the ground. Notice in Figure 4.15c the steplike appearance of the deposit. The steps are the tops of blocks of material that broke loose from the cliff. Very often, earthflows begin as slump.

Soil Creep

As stated earlier, soil creep is a very slow process. Material deposited by soil creep is often hard to recognize because it is usually covered with plants, and looks like part of the hill. Only a person trained to recognize deposition will perceive the series of ripples at the base of the gentle slope as evidence of deposition by soil creep (see Figure 4.16b).

Deposition by mass wasting along roadsides can cause serious problems. Poorly planned roads may be blocked by deposits. In fact, material deposited by slides, flows, and slump may make a road impassable. Interstate 40, near Rockwood, Tennessee, is one such road; one stretch has been blocked more than 20 times by major slides. Clearing such roads costs taxpayers millions of dollars each year. In addition, these poorly planed roads pose dangers to motorists.

E-2. Deposition by Moving Water

More than 70 percent of the Earth's surface is covered by water. Therefore, it is not surprising that most deposition occurs in water. In this section, you will learn how sediment carried by moving water is deposited and how deposition changes the shape of the land.

Deposition by Streams

Rivers and other streams carry large amounts of sediment. Water from the Colorado River is used to irrigate crops; canals carry the water to the fields. In just one of these canals, up to 5,000 metric tons of sediment has to be removed from the water every day.

Streams deposit sediment when they slow down or decrease in volume. Such slowdown or decrease in volume occurs when the slope decreases, runoff and seepage from groundwater decrease, or the stream enters a standing body of water, such as a lake or ocean. Deposition by streams results in layers that are sorted according to particle size, shape, and density. In addition, these deposited sediments tend to be smoothed and rounded as a result of abrasion while being carried in the stream.

Many landforms develop as a result of stream deposition. These landforms are found in almost all parts of the United States.

Oxbow lakes

In a preceding section you learned how meanders are formed. As meanders become more curved, their ends move closer together. Eventually, the stream erodes its way through the land, separating the meanders (see Figure 4.32). Most of the water then flows through this steeper, straighter shortcut. The slower moving water in the old meander deposits its load. When deposition blocks off the entrance to the old meander and it becomes separated from the stream, an oxbow lake is formed.

Floodplains and Levees

Because of its increased speed and volume, a stream in flood carries much more than the normal amount of sediment. Also, when a stream is

(a) A meander in river

(b) Meamder base eroded

a Narrow neck of land

(c) Meamder bypassed

(d) Meamder abandoned

sediment "dam" | C | b | C sediment "dam"

Oxbow lake

Figure 4.32 Formation of an Oxbow Lake.
From: Earth Science on FIle. © Facts on File, Inc., 1988.

in flood, it overflows its banks. Outside of the channel, the water is slower moving and shallower. As soon as it leaves the channel, the flood water begins to deposit its sediment load over the surrounding land to form a *floodplain*. The larger particles settle out first, building a *levee*, that is, a low, thick, ridge-like deposit along the banks of both sides of the stream (see Figure 4.33).

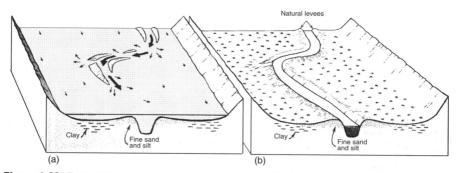

Natural levees

Clay | Fine sand and silt

(a)

Clay | Fine sand and silt

(b)

Figure 4.33 Formation of a Floodplain and Levee. During a major flood, the stream's floodplain becomes a lake. Water in the main channel flows rapidly. However, water escaping the channel immediately slows down and deposits fine sand and silt forming a natural levee. As the water continues to spread out and slow down clay is deposited on the surrounding lower land. When the flood ends, the levees remain as low ridges along the sides of the channel and the floodplain is covered with wet clay and swampy areas.

Physical Geology, 2nd Edition, Richard Foster Flint, Brian J. Skinner, John Wiley and Sons, New York 1974, 1977.

185

Floodplains make good farmland; sediment deposited after each flood renews the soil. For this reason, farmers have been able to raise crops on the floodplains of the Nile River for over 4,000 years!

Deltas and Alluvial Fans

When a stream flows into a quiet body of water such as an inland sea, ocean, or lake, it stops moving. The point at which this occurs is called the *mouth* of the stream. Here most of the stream's sediment is deposited. The resulting landform is a *delta*, as seen in Figure 4.34. A delta is a large, flat, fan-shaped

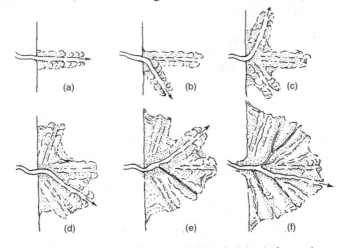

Figure 4.34 Stages in the Formation of a Delta. A delta is formed as a stream flows into a standing body of water such as a lake or ocean.
The Earth Sciences. © by Arthur N. Strahler, 1971.

pile of sediment at the mouth of a stream. It is composed of sand, silt, and clay and is shaped like a triangle. It gets its name from the Greek letter delta, which is written as Δ.

When streams flowing down steep mountain valleys reach flat, open land, they slow down and deposit much of their coarser load. The resulting landform is an *alluvial fan*, that is, a large mound of coarse sediment deposited by a stream onto open, flat land (see Figure 4.35).

Although an alluvial fan is similar in appearance to a delta, there are some differences. An alluvial fan has steeper sides and is made of coarser sediment than a delta.

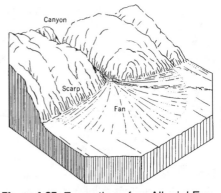

Figure 4.35 Formation of an Alluvial Fan. An alluvial fan forms where a stream flows from a steep gradient onto a gentle gradient.
The Earth Sciences. © by Arthur N. Strahler, 1971.

Deposition by Oceans

Oceans also deposit sediment. Whenever waves and currents slow down, they deposit their loads. Deposition by waves and currents builds new landforms that change the shape of shorelines (see Figure 4.36).

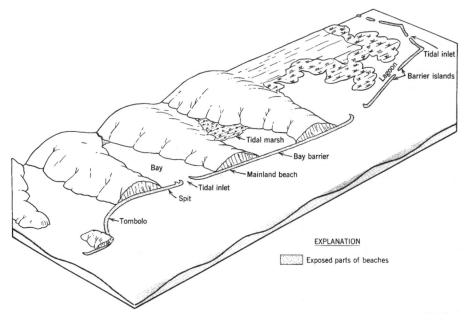

Figure 4.36 Deposition by Oceans. Some coastal landforms created by deposition of sediments by waves and currents.

Physical Geology, 2nd Edition, by Richard Foster Flint, Brian J. Skinner, John Wiley and Sons, New York 1974, 1977.

Beaches

A *beach* is a deposit of sediment along the shoreline of a body of water. Beaches are formed as waves wash up against the land and slow down. Beaches form an almost continuous fringe around the continents and most islands. Although most people think of a beach as a strip of fine, white sand along the seashore, beaches may consist of anything from fine clay to boulders.

Sandbars and Spits

A *sandbar* is a long, narrow pile of sand deposited in open water. Sandbars may form where receding waves wash beach sediments into deeper, quieter waters and also where the shoreline curves away from a longshore current. As the current curves away from the shoreline, it flows into deeper, quieter water and deposits its sediments. Most bars formed in this way are attached at one end to the mainland and are called *spits. Tombolos* are sandbars that connect an island to the mainland.

187

Barrier Bars and Lagoons

Eventually, waves deposit enough sand on a bar so that it is above water. Large waves add more material and in time the bar is built up well above sea level. Such large bars are known as *barrier bars*. Barrier bars may completely block off a bay forming a *tidal marsh*, or may parallel the coast, broken here and there by tidal inlets into elongated *barrier islands*. The calm, protected bodies of water between barrier bars and the mainland are called *lagoons*.

Problems Associated with Deposition by Moving Water

Deposition by moving water can create problems in harbors, which must be deep and have a clear opening to the sea to enable large ships to pass through. Occasionally, sandbars and spits block a harbor's entrance. In addition, sediment from streams and rivers deposited in harbors can create areas of shallow water. Ships would then be in danger of running aground. To prevent this from happening, sediment must be removed from harbors by dredging.

E-3. Deposition By Wind

Wind is moving air. When a wind slows or stops moving, the particles it is carrying settle to the ground and are deposited. Wind deposits generally contain fine sediments and result in certain surface features.

Loess

Loess is a thick, unlayered deposit of very fine, buff-colored sediment deposited by wind. Loess deposits may vary in thickness from a few centimeters to several hundred meters. Loess is found throughout the world. There are large deposits in the central United States, Europe, and parts of China. Loess is a source of excellent soil because of its ability to hold large amounts of water.

Dunes

Dunes are mounds of sand deposited by wind. They are often found in barren, desert regions and can also occur on large, sandy beaches. A dune often forms when wind-blown sand meets an obstacle such as a rock or a bush.

Figure 4.37 shows how a dune develops. On the windward side of the dune, that is, the side facing the wind, sand that strikes an obstacle falls to the ground. In time, the sand forms a mound that slopes gently up toward the tip of the obstacle. Wind then pushes sand grains up this slope to the crest, or top, of the dune.

On the leeward side of the dune, that is, the side sheltered from the wind, winds are very slow. As a result, the sand grains rolling over the crest fall to the ground because the winds can no longer support them. A deposit of sand forms, creating a steep-sided mound. As sand continues to be deposited, this

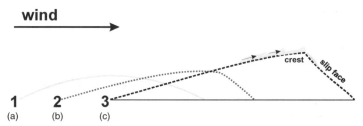

wind

1 2 3

(a) (b) (c)

Figure 4.37 The Formation of a Dune. (a) Wind-blown sand accumulates in low mounds. (b) Steady winds carry sand up the side of the dune and it rolls over the crest onto the sheltered side. (c) As sand is picked up on the gently sloping side facing the wind and deposited on the steep slip face, the entire dune moves.

mound may become unstable. Indeed, sand may slip or slide downhill. For this reason, the steep, leeward side of a dune is called the *slip face*.

As winds blow sand from the windward side of a dune and deposit it on the leeward side, the entire dune moves in the direction in which the wind is blowing. In this way, dunes travel along a desert. Some desert dunes may move from 10 to 25 meters in a year. Moving dunes have been known to engulf forests, farmlands, and buildings.

Sand dunes have a variety of heights, shapes, and patterns. They may reach heights of 30–100 meters depending on wind speed and the amount of sand in the area. Three common shapes and patterns of sand dunes are shown in Figure 4.38. The type of dune that forms in an area depends on the prevailing wind patterns and the amount of sand.

In the United States, the dunes of our coastal areas are an important line of defense for inland properties from storm waters. From 1936 to 1940, the National Park Service built nearly 1,000 kilometers of fencing along the coast of North Carolina. The fenc-

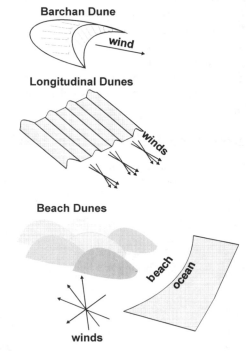

Barchan Dune

wind

Longitudinal Dunes

winds

Beach Dunes

beach *ocean*

winds

Figure 4.38 Common Shapes and Patterns of Sand Dunes.

ing formed an obstacle to the wind-blown sand, trapping it and thereby forming dunes. These dunes were then planted with grass, shrubs, and over two million trees! Today, the resulting barrier of dunes protects inland regions from serious damage by hurricanes.

E-4. Deposition by Glaciers

Types of Glacial Deposits

Part of a glacier's load is carried frozen within the ice or scattered on its surface. The rest is pushed up in piles around the edges of the glacier and carried along as the glacier moves over the Earth's surface. When the glacier reaches warm regions, or the climate changes and becomes warm enough, the ice begins to melt. As the glacier melts, the sediments it was carrying are deposited. These deposits may be sorted or unsorted, and each type is formed differently.

Unsorted Deposits

At the melting edge of the glacier, sediments within the ice, as well as those carried on top of it, are released and drop to the ground. Particles of all shapes and sizes fall to the ground in a confused jumble, forming an unsorted deposit. In addition, as the ice melts back, piles of sediment that were pushed up around the edges of the glacier are left behind. These sediments, too, are unsorted. They were tumbled about and mixed as the ice scraped them from the ground and pushed them along.

Sorted Deposits

Some glacial sediment is deposited into streams that originated from a melting glacier. This material is sorted by the running water and then deposited in layers.

Landforms Resulting from Glacial Deposition

Glacial deposition results in a variety of landforms. (see Figure 4.39.) Some of these were formed during the Ice Age. An area having any of these landforms was once covered by glaciers.

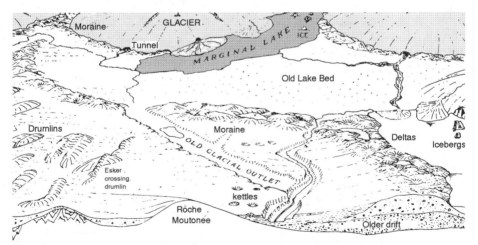

Figure 4.39 Landforms Resulting from Glacial Deposits.
Educational Leaflet #28, The New York State Museum, Albany, New York.

Moraines

Moraines are deposits of till. A moraine that forms a thin, widespread layer of till is called a *ground moraine*. Ground moraines form gently rolling hills and valleys. Piles of till deposited around the side of a glacier are called *lateral moraines*. Deposits of till at the front of the glacier are *terminal moraines*. Terminal and lateral moraines form long, parallel ridges. The ridges mark the boundaries where a glacier once existed.

Drumlins

Drumlins are long, low mounds of till that have a rounded, teardrop shape, as seen in Figure 4.39. By looking at the figure, you can see the direction in which the glacier moved. The rounded part of the drumlin points in the direction in which the glacier advanced. Drumlins are thought to be molded by the ice of a glacier as it slides over previously deposited piles of sediment.

Erratics

Erratics are large, isolated boulders deposited by a glacier. Many erratics are more than 3 meters in diameter and weigh thousands of metric tons. Such boulders are much too large to be carried by wind or water. Glaciers often carry these boulders far from their source. When the ice melts, the boulder comes to rest on a surface having a composition different from its own.

Outwash Plains

At the leading edge of a glacier, some melting is almost always taking place. Streams of glacial meltwater carry sediments out beyond the glacier and deposit them in sorted layers. Over time, a broad outwash plain is built up in front of the glacier by deposition of sediments from glacial meltwater. Such a landform is called an *outwash plain* because the sediments it is made of were "washed out" beyond the glacier by meltwater streams.

Kettles and Kettle Lakes

A *kettle* is a pit found in a glacial deposit. There are several steps in the formation of a kettle. First, a chunk of glacial ice breaks loose. Next, sediment from the glacier covers the chunk of ice. After a while, the ice melts, the overlying sediment sinks, and the kettle is formed. Kettles may later fill with glacial meltwater, rainwater, or groundwater, forming a *kettle lake.*

Glacial Lakes

In addition to kettle lakes, glaciers may form lakes in two other ways. A glacier may gouge out a large depression in the Earth's surface that may fill with water. Alternatively, a *glacial lake* may form when moraines block glacial meltwater. In this instance, the moraine acts like a dam to block the flow of water, thereby creating a lake.

Glacial lakes impact society in several ways. They form some of this country's most spectacular scenery. In addition to being beautiful, glacial lakes have both recreational and commercial uses. For example, ships travel along the Great Lakes, a system of five glacial lakes, carrying freight from port to port along the lakes. In addition, fishing, boating, and swimming, as well as other activities, are enjoyed at glacial lakes throughout the United States.

REVIEW QUESTIONS FOR UNIT 4

1. At high elevations in New York State, which is the most common form of physical weathering?
   (1) abrasion of rocks by the wind
   (2) alternate freezing and melting of water
   (3) dissolving of minerals into solution
   (4) oxidation by oxygen in the atmosphere

2. Which property of water makes frost action a common and effective form of weathering?
   (1) Water dissolves many earth materials.
   (2) Water expands when it freezes.
   (3) Water cools the surroundings when it evaporates.
   (4) Water loses 80 calories of heat per gram when it freezes.

3. Which is the best example of physical weathering?
   (1) the cracking of rock caused by the freezing and thawing of water
   (2) the transportation of sediment in a stream
   (3) the reaction of limestone with acid rainwater
   (4) the formation of a sandbar along the side of a stream

4. Which weathering process tends to form spherical-shaped boulders?
   (1) carbonation   (3) frost action
   (2) exfoliation   (4) oxidation

5. Which substance has the greatest effect on the rate of weathering of rock?
   (1) nitrogen   (3) water
   (2) hydrogen   (4) argon

6. Which type of climate causes the fastest chemical weathering?
   (1) cold and dry   (3) hot and dry
   (2) cold and humid   (4) hot and humid

7. The weathering of earth materials is most affected by
   (1) topography   (3) altitude
   (2) longitude   (4) climate

8. Which factor has the *least* effect on the weathering of a rock?
   (1) climatic conditions
   (2) composition of the rock
   (3) exposure of the rock to the atmosphere
   (4) the number of fossils found in the rock

9. Chemical weathering will occur most rapidly when rocks are exposed to the
   (1) hydrosphere and lithosphere
   (2) mesosphere and thermosphere
   (3) hydrosphere and atmosphere
   (4) lithosphere and atmosphere

10. The four limestone samples illustrated below have the same composition, mass, and volume. Under the same climatic conditions, which sample will weather fastest?

11. Two different kinds of minerals, *A* and *B,* were placed in the same container and shaken for 15 minutes. The accompanying diagrams represent the size and shape of the various pieces of mineral before and after shaking What caused the resulting differences in shapes and sizes of the minerals?

    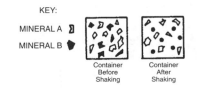

    (1) Mineral *B* was shaken harder.
    (2) Mineral *B* had a glossy luster.
    (3) Mineral *A* was more resistant to abrasion.
    (4) Mineral *A* consisted of smaller pieces before shaking began.

12. For a given mass of rock particles, which graph best represents the relationship between the size of rock particles exposed to weathering and the rate at which weathering occurs?

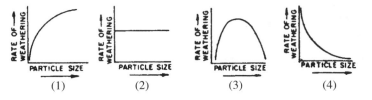

13. The adjacent diagram represents a sedimentary rock outcrop.

    Which rock layer is the most resistant to weathering?
    (1) 1   (3) 3
    (2) 2   (4) 4

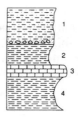

14. Two tombstones, *A* and *B*, each have been standing in a cemetery for 100 years. The same style and size of lettering is clear on *A* but not on *B*. Which is the most probable reason for the difference?
    (1) *B* was more protected from the atmosphere.
    (2) *A*'s minerals are more resistant to weathering than those in *B*.
    (3) *A* is more porous than *B*.
    (4) *B* is smaller than *A*.

15. Why will a rock weather more rapidly if it is broken into smaller particles?
    (1) The mineral structure of the rock has been changed.
    (2) The smaller particles are less dense.
    (3) The total mass of the rock and the particles is reduced.
    (4) There is more surface area exposed.

16. Which factors most directly control the development of soils?
    (1) soil particle sizes and method of deposition
    (2) bedrock composition and climate characteristics
    (3) direction of prevailing winds and storm tracks
    (4) earthquake intensity and volcanic activity

17. A variety of soil types are found in New York State primarily because areas of the state differ in their
    (1) amount of insolation
    (2) distances from the ocean
    (3) underlying bedrock and sediments
    (4) amount of human activities

18. Which change would cause the topsoil in New York State to increase in thickness?
    (1) an increase in slope
    (2) an increase in biologic activity
    (3) a decrease in rainfall
    (4) a decrease in air temperature

19. The chemical composition of a residual soil in a certain area is determined by the
(1) method by which the soil was transported to the area
(2) slope of the land and the particle size of the soil
(3) length of time since the last crustal movement in the area occurred
(4) minerals in the bedrock beneath the soil and the climate of the area

20. Particles of soil often differ greatly from the underlying bedrock in color, mineral composition, and organic content. Which conclusion about these soil particles is best made from this evidence?
(1) They are residual sediments.
(2) They are transported sediments.
(3) They are uniformly large-grained.
(4) They are soluble in water.

21. Which is indicated by a deep residual soil?
(1) resistant bedrock
(2) a large amount of glaciation
(3) a long period of weathering
(4) a youthful stage of erosion

22. On Earth, which agent of erosion is responsible for moving the largest amount of material?
(1) ground water
(2) glaciers
(3) running water
(4) wind

23. Which is the best evidence that erosion has occurred?
(1) a soil rich in lime on top of a limestone bedrock
(2) a layer of basalt found on the floor of the ocean
(3) sediments found in a sandbar of a river
(4) a large number of fossils embedded in limestone

24. For which movement of Earth materials is gravity *not* the main force?
(1) sediments flowing in a river
(2) boulders carried by a glacier
(3) snow tumbling in an avalanche
(4) moisture evaporating from an ocean

25. What is the best explanation for the shape of the cliff in the diagram?

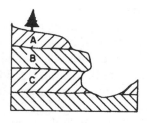

(1) Rocks A and C are made of larger particles than rock B.
(2) The particles in rocks A and C are more firmly cemented than those in rock B.
(3) The minerals in rock A and C erode faster than those in rock B.
(4) Rocks A and C have not been exposed to weathering as long as rock B.

Note that question 26 has only three choices.

26. If the rate of erosion in a particular landscape on the Earth's surface increases and the uplifting forces remain constant, the elevation of that landscape will
(1) decrease
(2) increase
(3) remain the same

27. Which erosional force acts alone to produce avalanches and landslides?
(1) gravity   (3) running water
(2) winds   (4) sea waves

28. Abrasive action is *least* effective in the process of erosion caused by
(1) glaciers   (3) rivers
(2) ground water   (4) waves

29. Which characteristic of a transported rock would be most helpful in determining its agent of erosion?
(1) age   (3) composition
(2) density   (4) physical appearance

Note that question 30 has only three choices.

30. As the amount of plant life in a region is decreased by the activities of humans, the amount of erosion will probably
(1) decrease
(2) increase
(3) remain the same

31. In which area will surface runoff most likely be greatest during a heavy rainfall?
(1) sandy desert   (3) level grassy field
(2) wooded forest   (4) paved city street

32. Which occurs as a stream is gradually uplifted?
(1) Its ability to erode will probably increase.
(2) Its potential energy will probably decrease.
(3) Its amount of streambed deposits will probably increase.
(4) Its stream discharge will probably decrease.

33. Which graph best shows the relationship between the slope of a streambed and the velocity of stream flow?

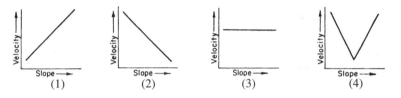

34. Which agent of erosion forms U-shaped valleys?
(1) ground water   (3) rivers
(2) winds   (4) glaciers

35. Which agent of erosion forms water gaps?
(1) glaciers   (3) rivers
(2) ocean waves   (4) wind

36. The accompanying diagram represents a winding stream. At which location is stream erosion most likely to be greater than stream deposition?

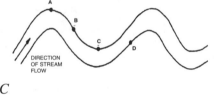

(1) *A*   (3) *C*
(2) *B*   (4) *D*

37. In the two adjacent diagrams, the length of the arrows represents the relative velocities of stream flow at various places in a stream. Diagram 1 shows the different water velocities across the surface, Diagram 2 shows the different water velocities at various depths.

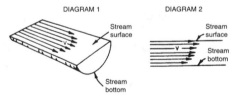

At which location in the stream is the water velocity greatest?
(1) at the center along the bottom
(2) at the center near the surface
(3) at the sides along the bottom
(4) at the sides near the surface

197

Note that questions 38 and 39 have only three choices.

38. As the volume of water in a section of a stream channel increases, the average velocity of the water
(1) decreases
(2) increases
(3) remains the same

39. As the velocity of a stream decreases, the rate of downcutting by the stream will probably
(1) decrease
(2) increase
(3) remain the same

40. Which of the following shows the most likely sequence of events leading to a mature soil?

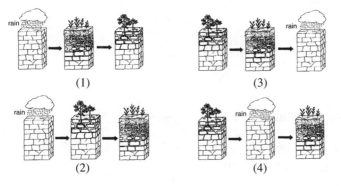

41. According to the *Earth Science Reference Tables*, what is the approximate minimum stream velocity needed to keep a 6.4-centimeter particle moving?
(1) 100 cm/sec   (3) 225 cm/sec
(2) 175 cm/sec   (4) 315 cm/sec

42. According to the *Earth Science Reference Tables,* which stream velocity would transport cobbles, but would not transport boulders?
(1) 50 cm/sec   (3) 200 cm/sec
(2) 100 cm/sec   (4) 400 cm/sec

Note that question 43 has only three choices.

43. As the volume of a stream increases the amount of material that can be carried by the stream generally
(1) decreases
(2) increases
(3) remains the same

44. Based on the diagrams of rock fragments below, which shows the *least* evidence of erosion?

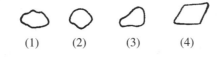

(1)   (2)   (3)   (4)

45. Which material could best be carried in solution by a stream?
   (1) granite   (3) gabbro
   (2) quartz   (4) salt

46. A river transports material by suspension, rolling, and
   (1) solution   (3) evaporation
   (2) sublimation   (4) transpiration

47. Which graph best shows the relationship between the size of a rock particle transported by a stream and the stream velocity?

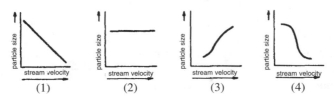

(1)   (2)   (3)   (4)

48. Which rock material was most likely transported by wind?
   (1) large boulders with sets of parallel scratches
   (2) jagged cobbles consisting of intergrown crystals
   (3) irregularly shaped pebbles which contain fossils
   (4) rounded sand grains which have a frosted appearance

49. A contour (topographic) map indicates that a stream is flowing across the landscape. If the stream has a constant volume, where on the map would the stream most likely have the highest velocity?
   (1) as the stream moves parallel to two contour lines
   (2) as the stream moves through a large region that has no contour lines
   (3) as the stream moves across several closely spaced contour lines
   (4) as the stream moves across several widely spaced contour lines

50. The adjacent diagram shows a geologic cross section of the rock layers in the vicinity of Niagara Falls in western New York State.

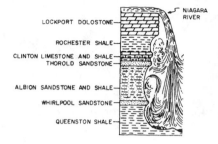

Which statement best explains the irregular shape of the rock face behind the falls?
  (1) The Lockport dolostone is an evaporite.
  (2) The Clinton limestone and shale contain many fossils.
  (3) The Thorold sandstone and the whirlpool sandstone dissolve easily in water
  (4) The Rochester and Queenston shale and the Albion sandstone and shale are less resistant to erosion than the other rock layers.

51. The adjacent diagram represents an aerial view of a stream which is generally flowing from north to south. Which pair of numbers indicates areas where erosion normally occurs more rapidly than deposition?

  (1) 1 and 2   (3) 1 and 4
  (2) 1 and 3   (4) 2 and 3

52. In which location is erosion usually greater than deposition?
  (1) in a stream channel that is being deepened
  (2) along a coast where a sandbar is being enlarged
  (3) at a point where a stream enters a lake
  (4) at the base of a cliff where a pile of rock fragments is accumulating

53. Why do the particles carried by a river settle to the bottom as the river enters the ocean?
  (1) The density of the ocean water is greater than the density of the river water.
  (2) The kinetic energy of the particles increases as the particles enter the ocean.
  (3) The velocity of the river water decreases as it enters the ocean.
  (4) The large particles have a greater surface area than the small particles.

54. A mixture of sand, pebbles, clay, and silt, of uniform shape and density, is dropped from a boat into a calm lake. Which material most likely would reach the bottom of the lake first?
  (1) sand   (3) clay
  (2) pebbles   (4) silt

55. If all the particles below have the same mass and density, which particle will settle fastest in quiet water? [Assume settling takes place as shown by arrows.]

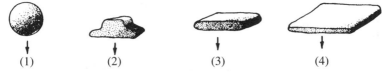

(1)   (2)   (3)   (4)

56. Which rock particles will remain suspended in water for the longest time?

(1) pebbles   (3) silt
(2) sand   (4) clay

Base your answers to questions 57 and 58 on your knowledge of earth science and on the diagrams below, which represent four columns of sedimentary deposits found at different locations.

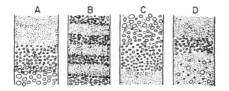

57. Which column best represents deposits from a stream that had a continuously decreasing water velocity?

(1) *A*   (3) *C*
(2) *B*   (4) *D*

58. Which column most clearly indicates that there were cyclic changes in the environment?

(1) *A*   (3) *C*
(2) *B*   (4) *D*

59. The accompanying diagram represents a vertical cross section of sediments deposited in a stream.

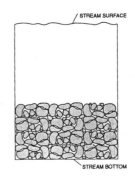

Which statement best explains the mixture of sediments?

(1) The velocity of the stream continually decreased

(2) The stream discharge continually decreased.

(3) The particles have different densities.

(4) Smaller particles settle more slowly than larger particles.

Note that question 60 has only three choices

60. As sediment deposition increases, the pressure on lower layers will
(1) decrease
(2) increase
(3) remain the same

61. The rate at which particles are deposited by a stream is *least* affected by the
(1) size and shape of the particles
(2) velocity of the stream
(3) stream's elevation above sea level
(4) density of the particles

62. The accompanying diagram represents a top view of a river emptying into an ocean bay. *AB* is a reference line along the bottom of the bay. Which characteristic would most likely decrease along the reference line from *A* to *B*?

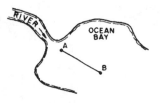

(1) the amount of salt in solution
(2) the size of the sediments
(3) the density of the water
(4) the depth of the water

63. A river carrying pebbles, sand, silt, and clay flows into the ocean. The sediments are sorted by size as they are deposited at different distances from shore. Which sedimentary rock will most likely form from the sediment deposited farthest from shore? [Refer to the *Earth Science Reference Tables*.]
(1) conglomerate   (3) siltstone
(2) sandstone   (4) shale

64. Which deposit is formed along rivers?
(1) coastal plain   (3) lake plain
(2) flood plain   (4) till plain

65. How will the formation of a cutoff change a meandering stream?
(1) It will reduce the stream's velocity.
(2) It will straighten the stream's course.
(3) It will widen the stream's course.
(4) It will increase the stream's rate of deposition.

66. The particles in a sand dune deposit are small and very well sorted and have surface pits that give them a frosted appearance. This deposit most likely was transported by
    (1) ocean currents   (3) gravity
    (2) glacial ice   (4) wind

67. The accompanying map represents a winding stream. At which location is stream deposition most likely to be greater than stream erosion?

    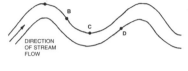

    (1) A   (3) C
    (2) B   (4) D

68. Granite pebbles are found on the surface in a certain area where only sandstone bedrock is exposed. Which is the most likely explanation for the presence of these pebbles?
    (1) The granite pebbles were transported to the area from a different region.
    (2) Some of the sandstone has been changed into granite.
    (3) The granite pebbles were formed by weathering of exposed sandstone bedrock
    (4) Ground water tends to form granite pebbles within layers of sandstone rock.

69. The diagram shows a cross section of soil from New York State containing pebbles, sand, and clay.

    The soil was most likely deposited by
    (1) an ocean current
    (2) the wind
    (3) a river
    (4) a glacier

70. Which agent was responsible for the deposition of fine-grained, angular particles found in a sandstone?
    (1) wave action   (3) running water
    (2) ground water   (4) wind

71. Which landscape features are primarily the result of wind erosion and deposition?
    (1) U-shaped valleys containing unsorted layers of sediment
    (2) V-shaped valleys containing well-sorted layers of sediment
    (3) terraces of gravel containing unsorted layers of sediment
    (4) cross-bedded sand deposits containing finely sorted layers of sediment

Base your answers to questions 72 through 76 on your knowledge of earth science. the *Earth Science Reference Tables,* and on the graph below which was prepared from the results of a study of four different types of cemetery stones. The graph shows the relationship between the ages of four cemetery stones and the percentage of each stone that had weathered away.

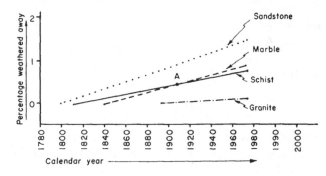

72. Which rock was found to have been exposed to weathering for the greatest number of years?
(1) granite (3) marble
(2) schist (4) sandstone

73. In this study, which rock was the most resistant to weathering?
(1) marble (3) granite
(2) schist (4) sandstone

74. What total percentage of the schist should have weathered away by the year 2000?
(1) 1.0% (3) 0.5%
(2) 2.0% (4) 1.5%

75. Point *A* on the diagram represents the time at which
(1) equal percentages of the marble and schist had weathered away
(2) the marble and schist weathered away at the same rate
(3) climatic conditions changed the weathering rates
(4) industrial pollutants changed the weathering rates

Note that question 76 has only three choices.

76. Studies have shown that pollutants added to the atmosphere in recent years are accumulating to cause an increase in the rate of weathering of marble. This factor should cause the line in the graph for marble in the future to
(1) decrease in slope (curve downward)
(2) increase in slope (curve upward)
(3) remain at the same slope

Base your answers to questions 77 through 79 on your knowledge of earth science and the diagram below, which represents the dominant type of weathering for various climatic conditions.

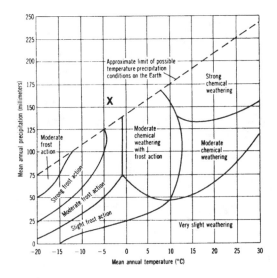

77. Which climatic conditions would produce very slight weathering?
   (1) a mean annual temperature of 25°C and a mean annual precipitation of 100 mm
   (2) a mean annual temperature of 15°C and a mean annual precipitation of 25 mm
   (3) a mean annual temperature of 5°C and a mean annual precipitation of 50 mm
   (4) a mean annual temperature of –5°C and a mean annual precipitation of 50 mm

78. Why is no frost action shown for locations with a mean annual temperature greater than 13°C?
   (1) Very little freezing takes place at these locations.
   (2) Large amounts of evaporation take place at these locations.
   (3) Very little precipitation falls at these locations.
   (4) Large amounts of precipitation fall at these locations.

79. There is no particular type of weathering or frost action given for the temperature and precipitation values at the location represented by the letter *X*. Why is this the case?
   (1) Only chemical weathering would occur under these conditions.
   (2) Only frost action would occur under these conditions.
   (3) These conditions create both strong frost action and strong chemical weathering.
   (4) These conditions probably do not occur on the Earth.

80. Four samples of the same material with identical composition and mass were cut as shown in the diagrams below. When subjected to the same chemical weathering, which sample will weather at the fastest rate?

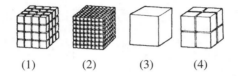

   (1)   (2)   (3)   (4)

81. Assume that the rate of precipitation throughout the year is constant. Which graph would most probably represent the chemical weathering of most New York State bedrock?

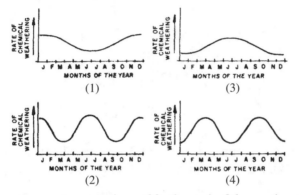

Base your answers to questions 82 through 86 on the *Earth Science Reference Tables*, the information and diagrams below, and your knowledge of earth science.

A mixture of colloids, clay, silt, sand, pebbles, and cobbles is put into stream I at point *A*. The water velocity at point *A* is 400 centimeters per second. A similar mixture of particles is put into stream II at point *A*. The water velocity in stream II at point *A* is 80 centimeters per second.

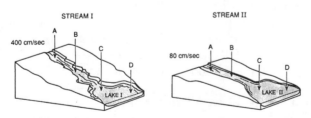

82. Which statement best describes what happens when the particles are placed in the streams?
(1) Stream I will move all particles that are added at point *A*.
(2) Stream II will move all particles that are added at point *A*.
(3) Stream I cannot move sand.
(4) Stream II cannot move sand.

83. Which statement is the most accurate description of conditions in both streams?
(1) The greatest deposition occurs at point *B*.
(2) Particles are carried in suspension and by bouncing along the bottom.
(3) The particles will have a greater velocity than the water in the stream.
(4) The velocity of the stream is the same at point *B* as at point *C*.

84. If a sudden rainstorm occurs at both streams above point *A,* the erosion rate will
(1) increase for Stream I, but not for Stream II
(2) increase for Stream II, but not for Stream I
(3) increase for both streams
(4) not change for either stream

85. What will most likely occur when the transported sediment reaches Lake II?
(1) Clay particles will settle first.
(2) The largest particles will be carried farthest into the lake.
(3) The sediment will become more angular because of abrasion.
(4) The particles will be deposited in sorted layers.

Note that question 86 has only three choices.

86. In Lake I, as the stream water moves from point *C* to point *D*, its velocity
(1) decreases
(2) increases
(3) remains the same

Base your answers to questions 87 through 91 on the information in the table below, which shows the distribution of particles in the load of a certain stream before the load reaches the stream's mouth.

Particles (% of total load)	Miles from Stream's Source		
	100	300	500
Gravel	25%	5%	0%
Coarse sand	30%	18%	1%
Medium sand	30%	50%	5%
Fine sand	14%	20%	25%
Silt	0.8%	5%	50%
Clay	0.2%	2%	19%

87. The decrease in the sizes of the particles carried by the stream as it flows away from its source is most probably caused by the
(1) decreasing velocity of the stream
(2) decreasing amounts of material carried in solution
(3) increasing volume of the stream
(4) increasing gradient of the stream

88. How is most of the load probably carried at a point 500 miles from the stream's source?
(1) in suspension
(2) in solution
(3) by action of wind and convection currents
(4) by rolling action of the particles along the bottom of the stream bed

89. If the sediments were produced from granite, then which mineral would most likely form the coarsest material in the load at 500 miles from the source?
(1) biotite   (3) feldspar
(2) hornblende   (4) quartz

90. What kind of rock will be formed from sediment deposited in the area between 300 and 500 miles from the source?
(1) conglomerate   (3) sandstone
(2) limestone   (4) shale

91. The graph below represents the distribution of which type of particle?

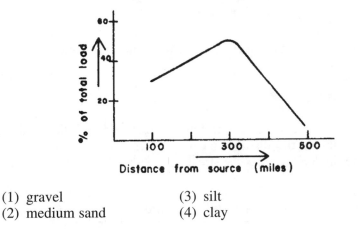

(1) gravel   (3) silt
(2) medium sand   (4) clay

Base your answers to questions 92 through 96 on your knowledge of earth science, the *Earth Science Reference Tables,* and the information below.

A group of students collected rounded, well-sorted mineral particles from a stream that flowed over only coarse-grained igneous bedrock. They sorted the particles by mineral type and then mixed equal volumes of all four minerals together and poured the mixture into a tube of water. The data table below lists the minerals. Figure *A* shows the deposit formed on the bottom of the tube as a result of the deposition of the particles.

DATA TABLE

MINERAL	AVERAGE PARTICLE DIAMETER
Plagioclase feldspar	0.2 cm
Quartz	0.2 cm
Hornblende (Amphibole)	0.2 cm
Olivine	0.2 cm

Figure A

92. As shown in figure *A*, which mineral appears to have the fastest settling rate?
(1) plagioclase   (3) hornblende
(2) quartz   (4) olivine

93. When the mineral particles were collected from deposits on the stream bed, the stream velocity at the time of deposition was approximately
(1) 50 cm/sec   (3) 150 cm/sec
(2) 100 cm/sec   (4) 200 cm/sec

94. The pattern resulting from the deposition of the mineral particles, as shown in figure *A*, is best explained by the fact that the particles have different
(1) volumes   (3) circumferences
(2) densities   (4) surface areas

95. The mineral particles collected by the students were most likely weathered from
(1) rhyolite rocks, only
(2) rhyolite and granite rocks
(3) gabbro rocks, only
(4) gabbro and granite rocks

96. The experiment was repeated using a second plagioclase sample with the original samples of the other minerals. What difference between the first and second samples of plagioclase would best explain the change in the pattern of deposition shown in Figure *B*?

(1) Particles from the second sample of plagioclase had greater density.

(2) Particles from the second sample of plagioclase had greater total volume

(3) Particles from the second sample of plagioclase had flatter shapes.

(4) Particles from the second sample of plagioclase were larger.

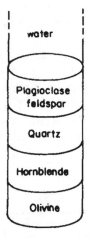

Figure B

OPTIONAL/EXTENDED

TOPIC C _____

Oceanography/ Coastal Processes

> **KEY IDEAS** The ocean is a valuable resource that we are only beginning to understood and utilize. The sediment layers of the ocean bottom are a record of Earth's geologic history, and the fossils they contain are a record of its biological history. The ocean is an important source of chemical and mineral resources. Much of the solar energy that reaches the Earth is stored in the ocean, and the ocean therefore influences weather and climate. The oceans also shape the coastlines of all the continents through erosion and deposition by ocean waves and currents.

KEY OBJECTIVES
Upon completion of this unit, you should be able to:

- Give examples of oceanic resources.
- Describe the composition of ocean water and the mechanisms that led to its present makeup.
- Construct profiles of the ocean bottom and identify its main topographic features.
- Explain the process of seafloor spreading and the evidence that led to its discovery.
- Describe the origins of ocean currents and waves.
- Describe the coastal processes of erosion and deposition that shape the shorelines of the continents.

A. WHY IS IT IMPORTANT TO INVESTIGATE THE OCEANS?

A-1. A Source of Information

The oceans represent a vast source of information and natural resources that has not yet been fully tapped because the ocean environment is hostile and not readily accessible. The oceans contain information in the sediment layers and fossils that line their floors. The core samples that have been retrieved by oceanographic research vessels reveal much information about the Earth's geologic history, and the fossils found in these sediments record a history of life in the oceans. The information obtained from ocean sediments about past climates and the ways they have changed helps us to understand our present climate. Careful study of ocean bottoms also revealed the mechanism for continental drift, which led to the theory of plate tectonics.

A-2. A Source of Natural Resources

The ocean is also an important source of valuable chemical and mineral resources such as bromine, iodine, magnesium, manganese, oil, natural gas, phosphorite, and heavy, metal-rich muds. Moreover, as desalination plants are made more efficient, the oceans are becoming a valuable source of fresh water in arid regions. The oceans are also a critical source of food, particularly animal protein, and of atmospheric oxygen, produced during photosynthesis by phytoplankton.

The resources of the ocean can be divided into renewable and nonrenewable types. Renewable resources, such as food, are replaced by the growth of plants and animals. Nonrenewable resources are present in fixed amounts and once used cannot be replenished. Mineral deposits, such as oil, natural gas, and coal, are examples of nonrenewable resources.

Biological (Renewable) Resources

Fish, crustaceans, mollusks, and other aquatic plants and animals are valuable sources of protein in the diets of people around the world. Fish populations are supported by the growth of phytoplankton and are most abundant in surface waters and places where upwelling brings nutrient-rich water to the surface (see Table C.1). Although aquaculture, or "farming" the oceans, is still in its infancy, it is becoming an increasingly important source of animals such as clams, oysters, and shrimp and plants such as kelp.

TABLE C.1 TOTAL FISH PRODUCTION IN VARIOUS PARTS OF THE OCEAN

Part of Ocean	Percentage of Ocean	Total Fish Production (tons)
Open ocean	90.0	1,600,000
Coastal zone	9.9	120,000,000
Upwelling zone	0.1	120,000,000

Mineral (Nonrenewable) Resources

It is difficult to assess the mineral resources of the oceans because they are underwater, and usual methods of mineral exploration and recovery cannot be used. However, the estimated value of the mineral resources of the oceans is staggering. Some of the resources we are currently aware of are discussed below.

Oil and Natural Gas

The most valuable resources that we now know about are the oil and natural gas found in the sediments of the continental shelf. The value of oil and gas obtained from the oceans yearly is greater than the combined value of all biological resources taken from the ocean.

Hydrothermal Mineral Deposits

The rising of magma in the ocean ridges and the movements of magma near subduction zones create fluids enriched in minerals, which are then deposited on the ocean floor. The deposits found include such minerals as copper, iron, zinc, manganese, nickel, vanadium, lead, chromium, cobalt, silver, and gold.

Phosphorite Deposits

The element phosphorus is important for healthy growth of plant roots, and phosphates are important fertilizers. In many places around the world, rocks and sands rich in phosphorite are exposed on the ocean floor. Phosphorite is usually deposited where cold, nutrient-rich waters are brought to the surface. As these waters warm up and their pH values rise because of carbon dioxide dissolving in from the atmosphere, the phosphates in them precipitate out and are deposited. Rich accumulations have formed where land sediments have not diluted the phosphate precipitates. Phosphorite deposits off the eastern coast of the United States are estimated to contain several billion tons of phosphate.

Muds, Oozes, and Nodules

The deep ocean floor is covered with fine-grained muds and oozes. The brown clay that covers many ocean floors is rich in aluminum (9%) and iron (6%), along with lesser amounts of copper, nickel, cobalt, and titanium. Some of these clays have more metal in them than rocks that are mined on land. Calcareous oozes are 95 percent calcium carbonate, which is used to

Optional/Extended
Topic C **OCEANOGRAPHY/COASTAL PROCESSES**

manufacture cement. Siliceous oozes could be mined for their silica, which is used to make glass and insulation and also serves as a soil additive.

The Red Sea has muds that are particularly rich in heavy metals such as copper, zinc, and silver. The estimated value of the top 10 meters of sediment is several *billion* dollars (see Table C.2).

TABLE C.2 GROSS VALUE OF METALS IN THE UPPER 10 METERS OF SEDIMENTS IN THE ATLANTIS II DEEP OF THE RED SEA, BASED ON 1994 METAL PRICES.

Metal	Average Assay (%)	Quantity (tons)	Value (dollars)
Zinc	3.4	2,900,000	3,393,000,000
Copper	1.3	1,060,000	2,889,560,000
Lead	0.1	80,000	54,400,000
Silver	0.0054	4,500	748,800,000
Gold	0.0000005	45	554,400,000
Total			7,640,160,000

Manganese nodules are common to all oceans. They are rounded spheres, about 1–20 centimeters in diameter, that form by the slow deposition of minerals (about 1 millimeter per million years) from seawater. Manganese nodules contain manganese, iron, nickel, and cobalt in almost pure form. These nodules are typically found scattered across the surface of abyssal plains, where sedimentation is very slow and does not bury them.

Dissolved Elements

Seawater contains almost all known elements (Table C.3). The elements currently removed from seawater economically are salt, bromine, and magnesium. Other elements are present in quantities too small to make their recovery economically feasible at this time. For example, the average concentration of gold in seawater is 0.000004–0.000006 milligram per liter, or about 0.0005 cent worth of gold per ton of seawater.

TABLE C.3 SUMMARY OF MINERAL RESOURCES AND THEIR SOURCES*

Source	Mineral Resource
Seawater	Boron, bromine, calcium, magnesium, potassium, sodium, sulfur, uranium
Sediments (continental shelf and slope)	Sand, gravel, phosphorite, glauconite, lime, silica, heavy minerals (magnetite, rutile, zircon, cassiterite, chromate, monazite, gold)
Heavy-metal muds	Copper, lead, zinc, silver, gold
Subsurface (continental shelf, slope, and rise)	Oil, gas, sulfur
Deep sea	Manganese nodules (copper, nickel, cobalt, manganese)

*After David A. Ross, *Introduction to Oceanography*, Prentice-Hall, 1970.

B. WHAT IS THE COMPOSITION OF OCEAN WATER?

Ocean water is a complex mixture of dissolved inorganic matter, dissolved organic matter, dissolved gases, and particulate matter. *Salinity* is the total amount of dissolved material in ocean water. Salinity is measured in parts per thousand by weight in 1 kilogram of ocean water.

B-1. Inorganic Matter

By weight, ocean water is about 96.5 percent pure water and about 3.5 percent dissolved inorganic substances. Six elements make up more than 90 percent of the materials dissolved in ocean water; they are: chlorine, sodium, magnesium, sulfur, calcium, and potassium. The minor elements include strontium, bromine, and boron. The elements in seawater are almost always present as components of chemical compounds, most of which are salts. While the salinity of ocean water varies from 33 to 37 parts per thousand, depending on where in the ocean the sample is taken, the ratio of the major dissolved substances remains relatively constant (see Figure C.1).

B.2. Organic Matter

The dissolved organic matter in ocean water comes mainly from excreta and dead organisms, is usually present in small amounts (0–6 parts per million), and is highly variable. Dissolved organic matter includes organic carbon, carbohydrates, proteins, amino acids, organic acids, and vitamins. Organic compounds containing nitrogen and phosphorus oxidize to form nitrates and phosphates.

B-3. Dissolved Gases

Nitrogen, oxygen, and carbon dioxide (CO_2) are the major gases dissolved in ocean water. Dissolved gases should not be confused with gaseous elements that are part of compounds. For example, water is a compound of hydrogen and oxygen, both of which are gases. However, the oxygen in a water molecule is not a dissolved gas and is not available to organisms for respiration. Other gases dissolved in ocean water include helium, neon, argon, krypton, and xenon.

With the exception of oxygen and carbon dioxide, gases in ocean water usually enter it from the atmosphere. The amount of a gas that dissolves depends on the temperature of the water and air, the pressure of the gas in the atmosphere, and the salinity of the water. The main controlling factor is temperature. The higher the water temperature, the less gas can dissolve in the ocean water. Low pressure and high salinity also decrease the amount of gas that can dissolve.

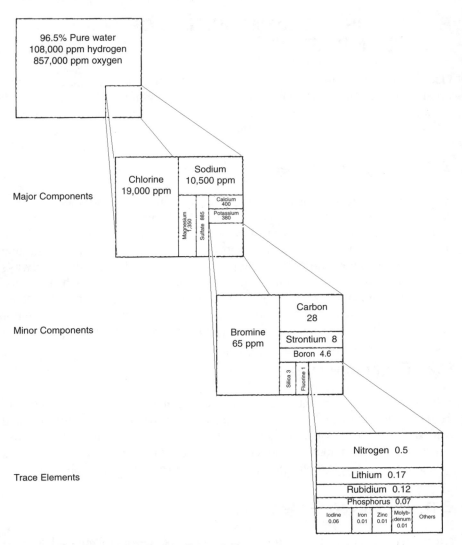

Figure C.1 Materials Dissolved in Seawater.
From *Introduction to Oceanography*, 3rd Ed., by David A. Ross, by permission of Prentice-Hall, 1982.

Ocean water has two sources of oxygen: the atmosphere and the plants that live in the ocean. Oxygen dissolves directly into ocean water from the atmosphere. *Photosynthesis* by plants living in the upper layers of the ocean (mainly phytoplankton) releases oxygen into ocean water.

Photosynthesis (occurs only in surface waters):
$$6CO_2 + 6H_2O + \text{Nutrients} + \text{Radiant energy} \rightarrow \text{Glucose } (C_6H_{12}O_6) + \text{Oxygen } 6(O_2)$$

Since sunlight penetrates only about 200 meters into the ocean, and photosynthesis depends on sunlight, photosynthesis results in an increase in oxygen only in surface waters.

Whereas photosynthesis adds dissolved oxygen to ocean water, *respiration* by organisms consumes dissolved oxygen. In deeper waters, where there is no contact with the atmosphere and light is insufficient for photosynthesis, dissolved oxygen removed from deeper ocean water by respiration is not replaced and a zone depleted of oxygen can form.

Respiration (occurs throughout the ocean):
Glucose $(C_6H_{12}O_6)$ + Oxygen $(O_2) \rightarrow CO_2 + H_2O$

Whether or not an oxygen-depleted zone is found in a particular part of the ocean depends on the degree of mixing of surface and deeper waters.

Some carbon dioxide in ocean water comes from the atmosphere. However, CO_2 is found in higher concentrations in ocean water than in the atmosphere! One reason is that 1 liter of water can absorb more CO_2 than 1 liter of air. Another reason is that ocean water is slightly basic, and excess magnesium and calcium ions in the water combine with CO_2 to form carbonates and bicarbonates. As a result, ocean water is a rich reservoir of CO_2 for photosynthetic reactions and can support huge populations of plants.

B-4 Particulates

Particulates are solid particles of material that are not dissolved in ocean water, but are suspended or settling through it. Aside from living organisms, particulate matter consists mainly of fine-grained minerals and decaying organic material. Particulate matter in ocean water varies greatly, depending on ocean currents, winds, and nearness to rivers and streams.

C. HOW CAN OCEAN SEDIMENTS BE CLASSIFIED?

Ocean sediments can be divided into two main types, terrigenous and pelagic, according to their origins. *Terrigenous sediments* originate on land, and *pelagic sediments* originate in the ocean. The structure and thickness of a pile of ocean sediments can be determined by using *echo sounding* with low-frequency sound waves (see Figure C.2). The composition of sediments can be learned by obtaining cores or by other direct sampling methods.

C-1. Terrigenous Sediments

Terrigenous sediments are found mainly near or on the continental margins, close to land. They are transported into the ocean by streams and, to a lesser extent, by winds and glaciers. Terrigenous sediments may contain quartz, feldspar, and clay minerals common to continental rocks.

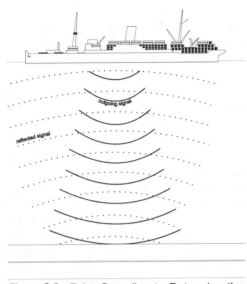

Figure C.2 Echo Sounding to Determine the Topography of the Ocean Floor and Subsurface Layers

Terrigenous sediments make their way into the deep ocean in several ways. In some places there is hardly any continental shelf, and streams deposit their loads directly into the deep ocean, as with the present-day Congo River in Africa and the Magdalena River in Columbia. During glacial periods, sea level is lower and the continental shelves are exposed because so much water is confined to land in the form of ice. Streams flowing over the steep continental shelves during past glacial periods must have carried greater sediment loads, and emptied them directly into the deep ocean.

Another way in which geologic sediments are carried into the deep ocean is by turbidity, or density currents. Sediment-laden water is denser than other water and will sink to the bottom and move downslope. On the 4–6° slope of the continental shelves, sediment-laden water moves downslope at speeds fast enough to erode canyons in the ocean bottom! Turbidity currents thus move sediments across the continental shelves and into the deep oceans.

Many parts of the continental shelf have glacial sediments, which may have been deposited during glacial periods and later inundated when the sea level rose. They may also have been deposited by meltwater streams draining into the ocean. Glacial sediments can also be found in the deep sea. These were probably ice-rafted out over the deep ocean; that is, they were carried frozen in ice that broke off a glacier and were released when the ice melted.

C-2. Pelagic Sediments

Pelagic sediments are usually found in deep ocean bottoms far from land. They consist of the skeletal material of plants and animals, fine-grained clays, or a mixture of the two.

Biological Sediments, or Oozes

Many marine organisms grow "shells" from materials extracted from ocean water (see Figure C.3). Microscopic plankton such as the plantlike coccoliths and animal-like foraminiferans and pteropods grow shells of calcium carbonate. When these organisms die, their shells slowly sink to the ocean bottom. On the way down, the shells dissolve; but if the water is shallow

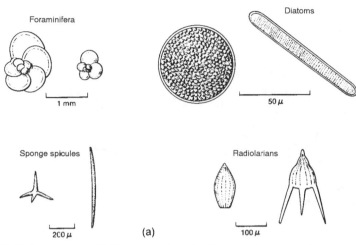

Figure C.3 Calcareous and Silicaceous Shells. Foraminifera have calcareous shells; the others are silicaceous.

From *Introduction to Oceanography*, by David A. Ross, by permission of Prentice-Hall, 1982.

enough or is saturated with carbonate, they may reach the bottom almost intact. The higher pressures and lower temperatures at greater depth increase the solubility of calcareous shells. Thus, they are rarely found at depths greater than 5,000 meters because they completely dissolve before reaching the bottom. Sediments rich in these calcium carbonate shells are called *calcareous oozes*.

Microscopic radiolarians and diatoms grow shells of silica (silicon oxide). Since silica is much less soluble than calcium carbonate, silica skeletons can survive settling through even the deepest waters relatively intact. The abundance of these skeletons in sediments depends on the abundance of the organisms in the surface waters and the silica concentration of the deep water. Sediments rich in silica skeletons are called *siliceous oozes*.

Inorganic Deposits

Fine-grained clays are found on many of the deep ocean bottoms far from land. Little is known about the origin of these clays, but it has been suggested that they were carried in by winds or ocean currents. Other possible sources include ash from volcanoes and dust from meteorites.

Precipitates and Nodules

Some sediments precipitate directly from ocean water. Chemical reactions in ocean water that produce insoluble materials may produce such sediments. Manganese nodules form by precipitation from ocean water directly onto a rock or other small object, such as a shark's tooth or fish bone.

Volcanic Deposits

Sediments of volcanic origin are common on the deep ocean bottoms. Volcanic ash is deposited by both land and submarine volcanoes directly into

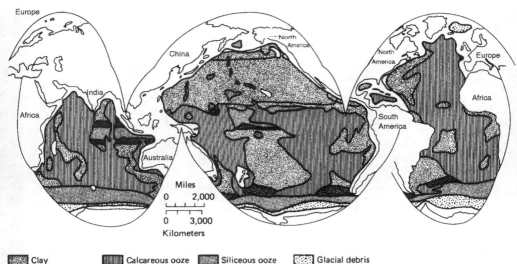

Clay   Calcareous ooze   Siliceous ooze   Glacial debris

Figure C.4 Worldwide Distribution of Deep-sea Sediment Types. Adapted from *Oceanography*, by Drake, Imbrie, Kraus, and Turekian, Holt, Rinehart, and Winston, 1978.

the ocean. Large-scale eruptions have deposited volcanic dust and ash so widely and at such a fast rate that distinct ash layers cover large parts of the ocean bottoms.

Figure C.4 shows the distribution of deep-sea sediments of various types.

D. WHAT IS THE STRUCTURE OF THE OCEAN FLOOR?

The ocean floor can be divided into two main parts, the continental margin and the ocean basin, based on the depth and structure of the crusts. The *continental margin* includes the coastal region and the continental shelf, slope, and rise. The *ocean basin* includes the abyssal plain and the midocean ridge.

The main features of those two parts of the ocean floor are shown in Figure C.5.

D-1. The Continental Margin

The Coastal Region

The *coastal region* is the part of a continent that is adjacent to the ocean. It includes the coast, shorelines, beaches, marshes, estuaries, lagoons, and deltas. Although it is a very narrow zone, it is one of the most important parts of the ocean. Seventy percent of the population of the United States live within 300 kilometers of the coastal zone. Estuaries are good locations for ports and are breeding grounds for many marine species. Beaches are valuable as recreational areas and as sources of sand and gravel.

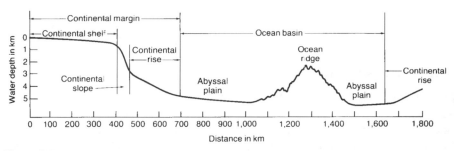

Figure C.5 Diagramatic Profile Showing the Main Features of the Continental Margin and Ocean Basin.
From *Introduction to Oceanography*, by David A. Ross, by permission of Prentice-Hall, 1982.

The Continental Shelf

The *continental shelf* is the shallow part of the ocean floor right next to the coastal zone. It has a relatively smooth bottom that slopes gently toward the ocean basins. The continental shelf ends with an abrupt change in slope at the shelf edge, which marks the beginning of the continental slope. The continental shelves cover a large area, almost one-sixth of the Earth's surface. The topography of the continental shelves is irregular, consisting of many small hills, valleys, and depressions.

The Continental Slope

The *continental slopes* connect the continental shelves with the deep ocean floor. The average pitch of the continental slope is about 4° or about 70 meters in 1 kilometer. Most continental slopes show evidence of slumping, probably due to the steep slope.

The Continental Rise

The continental slope ends in a gently sloping feature called the *continental rise*. The continental rise slopes at an angle of less than 0.5° and is relatively smooth, being interrupted only by submarine canyons and seamounts. Seismic profiles of continental rise areas show that most are wedgelike piles of sediment layers that thin out toward the ocean basins.

D-2. The Ocean Basins

The Midocean Ridges

The most striking feature of the ocean basins are the *midocean ridges*. These rugged, mountainous features often rise to within 1,000 meters or so of the surface. In some cases, the peaks of the ridge actually break the surface and form islands such as Iceland and the Azores. The midocean ridges are a continuous feature of the oceans that can be traced for more than 50,000 kilometers through the Atlantic, Pacific, Indian, and Arctic oceans. Many ridges are split by a central rift zone, a deep valleylike structure that can plunge as much as 2,000 meters below the surrounding peaks. Using submersibles, oceanographers have studied the rift zones and found much evidence of volcanic activity and recent faulting.

The Abyssal Plains

On either side of the midocean ridges are wide, flat areas called *abyssal plains*. The flatness of the abyssal plains is due to the accumulation of sediments. Sediments deposited by turbidity currents or other sediment-laden currents fill in the irregularities of the ocean floor and smooth them out. Seismic profiles have revealed that beneath the sediments the topography is more irregular.

The Deep-Sea Trenches

Deep-sea trenches are deep, V-shaped valleys with narrow, sometimes flat floors. The floor of a trench is seldom more than a few kilometers wide, and is flat where sediments have been deposited in the bottom of the trench. The deepest trench is the Marianas trench, which plunges to 11,022 meters (Table C.4).

TABLE C.4 DIMENSIONS OF SOME OCEANIC TRENCHES*

Trench	Depth (m)	Length (km)
Marianas	11,033	2,550
Tonga	10,882	1,400
Kuril-Kamchatka	10,542	2,200
Philippine	10,497	1,400
Puerto Rico	9,219	1,550
Peru-Chile	8,055	5,900
Sunda	7,455	4,500
Middle America	6,662	2,800

*After David A. Ross, *Introduction to Oceanography,* Prentice-Hall, 1982.

Trenches are zones of seismic activity. The depth of earthquake foci deepens on the continental side of a trench, and the boundary between the crust

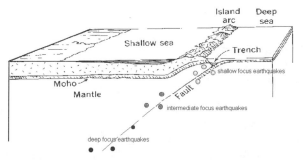

Figure C.6 A Plot of Earthquake Foci. This reveals a plunging boundary called a subduction zone, where the ocean plate is being pushed beneath the continental plate.
From *The Earth Sciences.* © by Arthur N. Strahler, 1971.

and the mantle deepens under the trenches. These important facts helped scientists understand that trenches were places where ocean crust was plunging beneath continental crust in a process called *subduction* (see Figure C.6).

D-3. Ocean-Floor Spreading and Plate Tectonics

The symmetrical shape of the Atlantic Ocean on either side of the mid-Atlantic ridge, and the jigsaw-puzzle fit of the adjacent continents, fascinated scientists for many years. As you have learned, the idea that the continents were once connected and at some point in the past split apart, forming the Atlantic, Indian, and Arctic Oceans and reducing the size of the Pacific, is called *continental drift*. The continental drift concept was first proposed in the early 1800s and was refined over the years, culminating with Wegener's hypothesis in 1912. Wegener argued that the continents float on deeper, denser mantle material and are moved around by convection currents in the mantle. Although the idea had been around for more than a century, it was not widely accepted, mainly because there was little evidence to support it at the time. The main problem was the question "How can a continental region become an ocean floor, and vice-versa?" The mechanism by which continents could move and ocean floors could form in their wake was not yet understood.

With the exploration of the ocean basins, however, the evidence in support of continental drift became obvious and the mechanism of continental drift and ocean floor formation was recognized. The exploration of the ocean basins revealed the following facts:

- The ocean ridges are sites of frequent earthquakes and volcanic activity. New magma is constantly welling up along the center of the ridges, and this upwelling lifts the crust and pushes it outward, causing faulting and earthquakes.
- There are long, thin magnetic patterns in the ocean floor that run parallel to the midocean ridges. These patterns show a sequence of reversals in polarity that is mirrored on both sides of a ridge.
- The rocks of the ocean floor increase in age as one moves away from the ridge, and in the Atlantic none are more than about 200 million years old. This is in sharp contrast to rocks on the continents, many of which are several billion years old.
- The trenches are also sites of seismic activity, and the foci of earthquakes beneath the trench deepen toward the continental side of the trench. Many trenches also show high heat flow and have volcanic islands adjacent to them.

Taken together, this evidence led to the concept of seafloor spreading, which was first proposed in 1960. According to this concept, convection currents in the mantle are pushing magma (containing magnetic minerals) up in the ridge areas. The magma solidifies with its magnetic minerals aligned with the Earth's magnetic field. Rocks that solidify at times when the earth's magnetic poles are reversed will have a reversed pattern of magnetic miner-

223

als. As new magma is pushed up, the rocks break apart and are pushed sideways away from the ridge and the new magma solidifies in the breach. As this sequence of events repeats itself, new ocean floor is created and the older ocean floor is carried sideways away from the ridge in a conveyor belt-like motion (see Figure C.7).

If seafloor is constantly being created along the ridges, the Earth should be getting larger, but it is not. Therefore, seafloor is being destroyed at the same rate at which it forms, but where? The answer lies in the trenches. The plunging ocean floor and deepening pattern of earthquake foci beneath the trenches are consistent with a place where sinking convection currents in the mantle are pulling the ocean floor downward into the mantle, where it melts and is destroyed in a process called *subduction*. With this final piece of evidence the picture is complete, and we have not only an explanation for the features and formation of the ocean floors, but also a mechanism for continental drift. Shortly thereafter, continental drift and seafloor spreading were unified into the theory of plate tectonics.

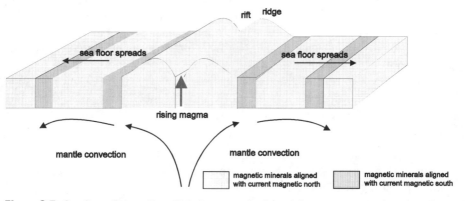

Figure C.7 Seafloor Spreading. This is caused when rising magma pushes into fractures in the ocean crust and forces it apart. When the magma cools and hardens, magnetic minerals solidify in alignment with the existing magnetic field.

E. WHAT IS THE NATURE OF OCEAN WAVES AND CURRENTS?

Movement of the water in the oceans is due almost entirely to differences in density within the ocean and interactions with the atmosphere.

E-1. Density Differences in the Oceans

Differences in the density of ocean waters are largely the result of differences in salinity and temperature.

Salinity

Salinity is the total amount of dissolved material in 1 kilogram of seawater, expressed in parts per thousand by mass. Salinity is usually determined by measuring the electrical conductivity of seawater. In most parts of the ocean, salinity ranges from 33 parts per thousand to 37 parts per thousand, with 75 percent of all ocean water having a salinity between 34.5 parts per thousand and 35.0 parts per thousand. The higher the salinity of seawater, the higher its density. Seawater of high salinity tends to sink in seawater of lower salinity at the same temperature.

The major factors that affect the salinity of seawater are evaporation and precipitation. To a lesser extent, the freezing and melting of ice and runoff from the land also affect salinity. Evaporation and freezing remove the water from seawater but leave behind the salts, causing an increase in salinity. Thus, in areas of high temperatures and low rainfall, the salinity of ocean water is greatest. In the Red Sea it rises to as much as 40 parts per thousand. Runoff from land and melting ice add water to the seawater, thereby causing a decrease in salinity. Near the poles, low evaporation rates and melting ice cause salinity to drop.

Temperature

The surface of the ocean is heated by solar radiation, the conduction of heat through contact with the atmosphere, and the condensation of water vapor. The ocean surface cools by radiating heat back into the atmosphere, conducting heat to the atmosphere, and the evaporation of water. The surface temperature of the oceans is closely tied to the latitude and the season of the year.

The higher the temperature of seawater, the lower its density. Warm water tends to "float" atop cooler water with little mixing unless stirred up by winds and currents. Differences in the temperature and density of seawater have resulted in a stratification of the water in oceans into three main layers. Near the surface is a warm, well-mixed layer about 10–500 meters thick. Beneath it is a layer called the *thermocline*, in which the temperature sharply drops. Beneath the thermocline is a cold, relatively uniform layer that extends to the bottom and in which the temperature slowly decreases.

Effects of Density Differences

The density of seawater is controlled by salinity, temperature, and pressure. In general, density increases with decreasing temperature, increasing salinity, and increasing depth (pressure). Colder, deeper, and more saline water is usually the densest. Gravity tends to cause more dense water to sink toward the ocean bottom and less dense water to rise to the surface.

If a situation develops in which denser water overlies less dense water, the layers will overturn and the denser water will sink until it reaches a zone of similar density. This can occur when high evaporation rates at the surface cause cooling and an increase in salinity. It can also occur when the atmosphere is colder than the ocean, as in winter months, and cools the ocean surface by conduction.

E-2. Interactions with the Atmosphere

The circulation of the oceans is mainly the result of two interactions with the atmosphere: heating of the ocean and wind. The ocean is able to store more heat than the atmosphere or the land because of the high specific heat of water. Differences in the intensity of solar radiation received at the Equator and near the poles result in an uneven distribution of heat in the oceans. The oceans contain more heat near the Equator than near the poles.

Convection currents

The heat in the ocean near the Equator is transferred toward the poles by *convection currents*, or the movement of seawater. Dense, cold water formed near the poles sinks and slowly flows toward the Equator. This cold water is oxygen rich; without it, bottom waters would quickly become oxygen depleted by the oxidation of organic matter that sinks through it. As the cold water moves towards the Equator, it displaces warmer water upward, causing *upwelling*. Upwelling brings nutrients and oxygen to the surface, thereby supporting a rich growth of plants and animals.

Wind-driven currents

Uneven heating of the atmosphere causes similar circulations in it. These circulations, modified by the Coriolis force, result in the pattern of planetary wind belts described in Unit 7. Winds exert a force on the ocean surface and produce wind-driven currents that follow roughly the pattern of surface winds (see Figure C.8).

Throughout the Earth, the trade winds, which blow from the east, form westward-moving currents near the Equator. In the Atlantic and Pacific oceans, these westward-moving currents are obstructed by landmasses and are deflected north and south along the western boundary of the oceans. These deflected currents are among the largest and strongest in the ocean. The Gulf Stream, a western boundary current in the Atlantic Ocean, runs north along the eastern coast of the United States and transports more than 100 times the output of all the rivers in the world.

When the western boundary currents reach the midlatitudes, the prevailing winds, which blow from the west, push them back east across the ocean. The result is a large, circular pattern of motion called a *gyre*. The North and South Atlantic and Pacific, as well as the Indian Ocean, all have large-scale gyres. The northern polar regions of the Atlantic and Pacific have smaller gyres.

One of the unfortunate effects of the large scale gyres is that they carry pollutants from the continental coastlines far out into the oceans. For example, many of the largest cities in the United States are located along the east coast. The sewage, garbage, and industrial waste from these cities have long been dumped into the ocean. The Gulf Stream then carries these wastes far out into the mid-Atlantic. These practices may have had little impact when human populations were small, but today they represent serious threats to the health of ocean life and the future well being of our oceans.

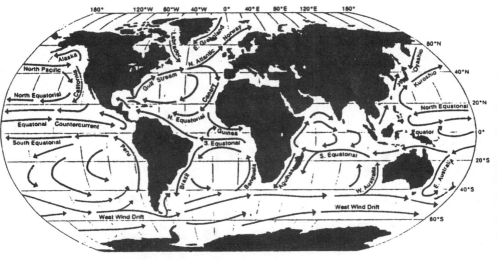

Figure C.8 Major Surface Ocean Currents.

Wind-Generated Waves

Wind forms waves on the surface of the ocean by transferring energy from the moving air to the water. The transfer of energy is not completely understood, but appears to involve pressure differences caused by the deflection of wind over the waveform (much as air flowing over a wing provides lift). It also seems to be related to a resonating movement of the surface caused by wind pushing against the surface in turbulent conditions (like a hand pushing down repeatedly on the surface of the water). The result in either case is a transverse waveform, which moves through the surface water of the ocean. As is discussed in Topic B a transverse waveform has the characteristics shown in Figure C.9.

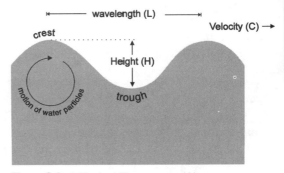

Figure C.9 A Typical Transverse Wave.
From *An Introduction to Oceanography,* by David A. Ross, Prentice-Hall, 1970.

As the waveform moves through the seawater, the water particles move in a circular pattern, as shown in Figure C.10. Near the surface, the movement of water particles is almost perfectly circular. However, the diameter of the circular motion decreases quickly with depth. This decrease is due to a loss of energy during transfer to deeper water. At a depth of one-fourth the wavelength, the circle is only one-fifth of its diameter near the surface. At greater depths the motion is little more than a back-and-forth rocking. Below 100 meters, little if any motion is felt from most wind-generated waves.

227

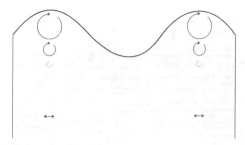

Figure C.10 Movement of a Waveform Through Seawater. Water particles move in circles as a wave passes through them. Circles become smaller and more elliptical with depth, fading out to a gentle back-and-forth rocking motion.

The height and the period of wind-generated waves are affected by three factors: the speed of the wind, the length of time it blows, and the *fetch*, or distance of water, over which it blows. The higher the wind velocity and the longer the duration of the wind, the higher the wave height. Fetch is important in determining wavelength. The longer the fetch, the longer the wavelength.

Wind-generated waves can be divided into three categories: sea waves, swells, and surf.

Sea waves are directly affected by the wind. They are irregular and have no pattern of motion. They differ in height and length, and travel in various directions. When we say the ocean is choppy, we are describing the effect of sea waves.

Swells form as sea waves leave the area of turbulent winds that formed them. Long waves travel faster than short ones (velocity = wavelength/period). The waves sort out into regions of waves of similar characteristics. The waves in these uniform patterns are swells. As waves travel farther from their source, their height decreases but their length remains constant. Waves showing such patterns can travel across entire oceans.

Surf occurs near shore when a wave breaks. The difference between surf, or breaking waves, and sea waves or swells is that the water in breaking waves is not moving in a circular pattern but is actually moving toward the shore. As a result a lot of energy is directed at the beach.

Tsunami

Tsunami are waves caused by undersea movements of the crust due to earthquakes, slumping, or volcanic eruptions. The sudden up-and-down movement of the bottom creates a wave that has a long length and travels at high speeds. In deep water, tsunami may have wavelengths up to 700 kilometers and travel at speeds over 350 kilometers per hour, yet may only be several centimeters high. When tsunami reach a coastline, though, they form huge waves that break against the coast and cause great destruction.

F. HOW CAN WE DESCRIBE COASTAL PROCESSES?

F-1. Breaking Waves

Breaking waves, or surf, form as follows. When waves enter shallow water, the bottom of the waveform comes into contact with the ground at a depth equal to one-half the wavelength. This contact slows the bottom of the waveform and forces the wave upward, while the top continues at its original speed. When the top of the wave outpaces the rest of the wave, it is no longer supported and falls over (see Figure C.11).

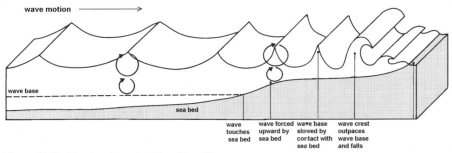

Figure C.11 Mechanics of a Breaking Wave.
From: Earth Science on File. © Facts on File, Inc., 1988.

F-2. Wave Refraction

Waves that enter shallow water at an angle to the beach are refracted; their direction of travel is changed. This refraction occurs because one part of the wave reaches shallow water and slows down while the rest of the wave is still in deep water and moving fast. Like a rolling log whose one end hits a tree, causing the whole log to swing around, the faster moving end of the wave swings around when the end in shallow water slows down.

On irregularly shaped coastlines, waves converge on headlands and diverge at depressions in the shoreline, as shown in Figure C.12. The results are erosion of the headland by the force of the waves directed against it and the deposition of sediments in the calmer zone of divergence within the inlets. Over time, the irregularly shaped shoreline is straightened out.

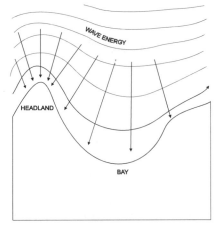

Figure C.12 Wave Refraction by Headlands and Bays. The wave energy is converged on the headland and diverges in the bay.

F-3. Longshore Currents

When waves break, the water is carried into the surf zone and travels toward the beach. The water that washes up against the beach runs back along the incline under the surf zone. When waves approach the beach at an angle, water moves up at an angle and falls straight back. The result is a movement of water parallel to the beach called a *longshore current* (see Figure C.13).

The longshore current and the up-and-back motion of water washing along the beach move sand along the shore. The net result is to transport sed-

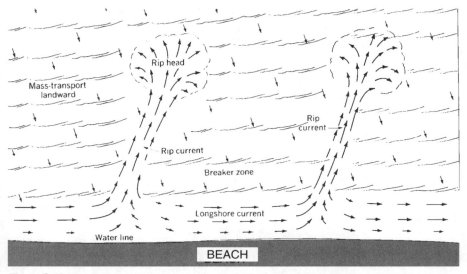

Figure C.13 Formation of Longshore Current. These currents are formed by the zigzag movement of the water as waves approaching the beach at an angle wash up and back out again.
Source: *The Earth Sciences.* © by Arthur N. Stahler, 1971.

iment along the beach and erode it inland. In populated areas, people try to impede this erosion by building structures such as piers and jetties in the surf zone to block the longshore current and trap beach sediments from being washed away (see Figure C.14). These attempts are only moderately successful in the long run. Also, the structures often increase erosion down-current as the longshore current carries away sediment but brings none in from up-current.

F-4. Shoreline Erosion

Waves can be very powerful. A single wave can send tons of water crashing against a shore. The force of waves breaks up rocks on the shore. Loose fragments are stirred up and carried along in the turbulence of the breaking waves.

Wave erosion produces a number of shoreline features. When waves erode steep, rocky shores, they remove mate-

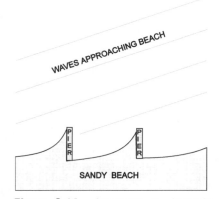

Figure C.14 Attempts to Impede Erosion. Structures such as piers and jetties are sometimes built in an attempt to trap sediment that is being eroded by longshore currents.

rials from the base of the slope. The result is a notch that undercuts the slope. Overhanging rock and soil break loose and slump or fall, leaving a steep-sided *sea cliff*. Pillars of resistant rock, called *sea stacks*, are left behind as

230

weaker rock is worn away. As the erosion continues, fragments broken from the cliff are abraded by wave action, and a flat platform, or *sea terrace*, is formed at the base of the cliff. When wave erosion penetrates deeply into a cliff, a *sea cave* may be formed.

Loose sediments carried into the sea by waves are transported along the shore by longshore currents. Along some coasts, tides flowing in and out of narrow openings form swift currents that also can move large amounts of sediment.

F-5. Shoreline Deposition

Beaches

Beaches are deposits of loose sediments along the shoreline of a body of water. Beaches may be made of earth material of almost any size or composition. The nature of beach sediments depends on the source material and the weathering processes at work in the area. Beaches are shaped by waves and are changed daily and monthly by tides.

Figure C.15 shows the general features of a beach. *Berms* are flat areas formed by wave action. The breaking wave loosens sediments upslope and drags them back toward the water as the wave recedes.

During stormy winter months, waves tend to be high and to have short periods. The sand picked up by one wave does not have a chance to settle to the bottom before the next wave sets it in motion again. As a result, the sand stays in suspension and is carried away from the beach by the backrush of water until it reaches deeper, calmer water and can settle, forming an offshore sand bar.

In summer months, the weather is calmer and the waves that reach the shore tend to be swells of length. The breaking waves pick up sand in shallow water and carry it up onto the beach. Most of the sand is then carried back toward the sea by the backrush of water along the beach. There is time during the backrush, however, for some particles to settle before the next wave again carries sand toward the beach. The overall result is a shifting of sand toward the beach.

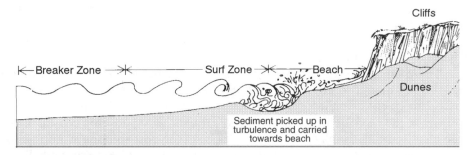

Figure C.15 Formation of a Berm. The usual direction in which breaking waves carry sand is toward the beach. This sand, together with sediments eroded from coastal cliffs, builds a flat area, or a berm. From *Physical Geology*, by Allan Ludman and Nicholas K. Coch, McGraw-Hill, 1982.

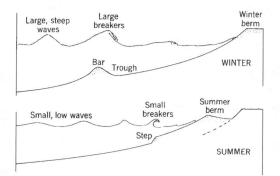

Figure C.16 Winter and Summer Beach Profiles. High waves of short period generally occur in winter months and drag sediments offshore, forming a sandbar. In summer, waves generally have long periods and move sediment onto the beach.
Source: *The Earth Sciences.* © by Arthur N. Stahler, 1971.

Figure C.16 shows the changing shoreline conditions in winter and summer.

Sandbars and Spits

Sandbars are long, narrow piles of sand deposited where waves or currents that are carrying sediments slow down. Typically, this happens where they reach deeper, quieter water. *Spits* are sandbars that are attached to the mainland at one end. They form where longshore currents flow past a sharply curving shoreline into deeper, quieter water.

Barrier Bars and Lagoons

When a *bar* accumulates enough material so that its top is above water level, it forms a barrier to waves and currents. The slowdown of waves and currents results in deposition and the bar is built up further. This large bar, whose top may be well above sea level, is called a *barrier bar*. Between a barrier bar and the mainland, the water is calm and protected. This protected body of water is called a *lagoon*.

F-6. The Erosion "Problem"

Erosion is considered one of the most important problems in the coastal zone, mainly because 85 percent of the shoreline is privately owned and the rising sea level is eroding all shorelines. Erosion is most pronounced on beaches and coastal cliffs, where smaller sediments are easily moved by waves and currents. It tends to be less of a problem on rocky coastlines or beaches that are made of pebbles, which are harder to move than sand.

There are really only three ways people can fight erosion of shoreline areas: (1) dredge sediments to fill in places where erosion has occurred, (2) dam rivers or build coastal structures such as jetties or breakwaters to decrease the force of eroding waves and currents, or (3) develop coastal dune areas as barriers to inland erosion. Of course, the best solution would be to recognize that coastal erosion and deposition are natural processes and allow them to happen without interference. Perhaps the coastal zone is not an environment that is appropriate for private ownership.

1. Beneath the oceans, the Earth's crust is composed primarily of
   (1) basalt
   (2) limestone
   (3) gneiss
   (4) granite

2. The accompanying graph shows the percentage distribution of the Earth's surface elevation above and depth below sea level.

   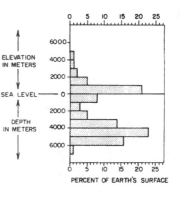

   Approximately what total percentage of the Earth's surface is below sea level?
   (1) 30%
   (2) 50%
   (3) 70%
   (4) 90%

3. What is the average salinity of ocean water?
   (1) 0.035%
   (2) 0.35%
   (3) 3.5%
   (4) 35%

4. What is the probable arrangement of sediments on the continental shelf from the shoreline outward?
   (1) gravel, sand, silt
   (2) gravel, silt, sand
   (3) sand, gravel, silt
   (4) silt, gravel, sand

5. What is the major cause of convection currents in the ocean?
   (1) rotation of the Earth
   (2) dilution of surface water by rainfall
   (3) friction between water and air
   (4) density differences due to changing temperature and salinity

6. What is the chief cause of the surface currents of the ocean?
   (1) tides
   (2) the Moon.
   (3) prevailing winds
   (4) rivers

Base your answers to questions 7 and 8 on the accompanying graph. The graph shows changes in ocean level (tides) at a beach.

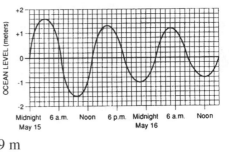

7. On May 15, what is the difference in ocean level between low tide at 10 A.M. and high tide at 4 P.M.?
   (1) 1.0 m
   (2) 2.0 m
   (3) 2.9 m
   (4) 3.9 m

8. What causes this periodic change in ocean level?
   (1) changing wind and storm paths
   (2) motions of the Earth and Moon
   (3) frequent earthquakes and tsunamis
   (4) motions of the crust and mantle

9. Compared to the Southern Hemisphere, the Northern Hemisphere has a greater annual temperature range because it
   (1) contains the greater proportion of land areas
   (2) contains the greater proportion of ocean areas
   (3) is inclined toward the Sun in summer
   (4) is closer to the Sun in winter

10. Many of the atolls in the Pacific are believed to have been caused by
    (1) slow sinking of volcanoes ringed by coral reefs
    (2) erosion of lava on the ocean floor
    (3) faulting of the ocean floor
    (4) breaks in mountain chains on the ocean floor

Base your answers to questions 11 through 17 on the adjacent diagram. The diagram represents a shoreline in New York State along which several general features have been labeled. Letter *B* identifies a location on the shoreline.

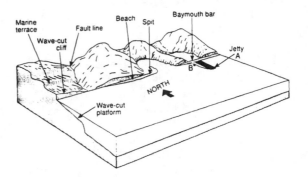

11. After the formation of the baymouth bar, the jetty (a structure made of rocks that extends into the water) labeled *A* was constructed perpendicular to the shoreline. Which statement best describes the result of the construction of this jetty?
    (1) Water current velocity at location *B* decreased.
    (2) Water current velocity at location *B* increased.
    (3) Sand deposition at location *B* decreased.
    (4) Sand deposition was not affected.

12. What is the most likely source of the waves approaching this coastline?
    (1) variations in water temperature
    (2) density differences within the water
    (3) the rotation of the Earth
    (4) surface winds

13. Past movements along the fault line most likely caused the formation of
    (1) the baymouth bar   (3) longshore currents
    (2) tsunamis   (4) tidal currents

14. Which statement best describes the longshore current that is modifying this coastline?
    (1) The current is flowing northward at a right angle to the shoreline.
    (2) The current is flowing southward at a right angle away from the shoreline.
    (3) The current is flowing eastward parallel to the shoreline.
    (4) The current is flowing westward parallel to the shoreline.

15. The marine terraces represent former positions of the wave-cut platform. These terraces indicate that
    (1) crustal uplift has occurred
    (2) crustal sinking has occurred
    (3) the coastline was subjected to many tidal waves
    (4) the coastline was affected by strong winter storms

16. Which feature was formed because more erosion took place than deposition?
    (1) spit   (3) baymouth bar
    (2) wave-cut cliff   (4) beach

17. Often, sediments that are carried down submarine canyons by turbidity currents settle out over the ocean floor. Which cross section represents the pattern of deposition of these sediments?

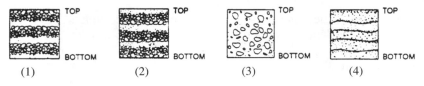

    (1)   (2)   (3)   (4)

Base your answers to questions 18 through 20 on the diagram below, which shows the movement of water particles in ocean waves. The particles are represented by black dots. Letters *A*, *B*, *C*, and *D* are points of reference.

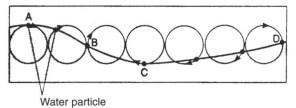

Water particle

18. The passage of a wave will cause a particle of water at the surface to
    (1) move horizontally, only   (3) move in a circular pattern
    (2) move vertically, only   (4) remain stationary

19. The wave pattern shown in the diagram would occur most often in
    (1) a longshore current parallel to a coastline
    (2) shallow water in the breaker zone of a beach
    (3) a turbidity current
    (4) deep water

20. Why do waves form breakers as they move from deep water into shallow water?
    (1) The speed of the waves increases.
    (2) The wave crests are slowed by the air.
    (3) Waves collapse as the bottom of the wave is slowed by friction.
    (4) Waves collapse as the bottom of the wave moves faster.

21. The table shows how density of sea water changes with variations in salinity of the water.

Water Mass	Salinity (0/00)	Density (g/cm³)
A	36.3	1.026
B	34.0	1.024
C	35.3	1.025

If the three water masses were found in the same region of the ocean,
    (1) *C* would be above *A* but below *B*
    (2) *A* would be above both *B* and *C*
    (3) *B* would be below both *A* and *C*
    (4) *C* would be below *A* but above *B*

22. The major source of the dissolved minerals that affect the salinity of ocean water is
    (1) submarine volcanoes   (3) deep ocean sediments
    (2) mostly land erosion   (4) shells of sea animals

Base your answers to questions 23 through 26 on the adjacent profile. Four zones within the profile are labeled. The profile shows ocean waves approaching a beach.

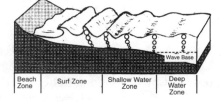

23. In which zone is sand most likely to be moving along the shore?
    (1) deep water zone   (3) surf zone
    (2) shallow water zone   (4) beach zone

24. Which statement best describes the waves in the surf zone?
    (1) The waves are unaffected by the ocean bottom.
    (2) The wave height is decreasing.
    (3) The speed of the wave bottom is decreasing.
    (4) The waves collapse as their wave heights increase.

25. The sand on this beach has its origin from the weathering of granite bedrock. According to the *Earth Science Reference Tables*, a mineral that is likely to be found in the sand is
    (1) quartz   (3) halite
    (2) calcite   (4) olivine

Note that question 26 has only three choices.

26. If a strong coastal storm with high wind speeds moves into the region, the wave heights are likely to
    (1) increase
    (2) decrease
    (3) remain the same

27. Base your answer to this question on your knowledge of earth science, the *Earth Science Reference Tables*, and the accompanying diagram. Which conditions describe the ocean current that would be found at position X in the diagram?

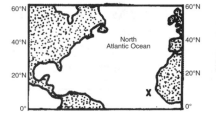

    (1) warm and flowing towards the north
    (2) warm and flowing towards the south
    (3) cold and flowing towards the north
    (4) cold and flowing towards the south

28. What is the primary cause of the major surface currents found in the north Atlantic Ocean?
    (1) planetary winds
    (2) mantle plumes beneath ocean plates
    (3) undersea earthquakes
    (4) high winds from storms

29. The source of much of the sediment found on the deep ocean bottom is
    (1) erosion of continental rocks
    (2) submarine landslides from the mid-ocean ridges
    (3) icebergs that have broken off continental glaciers
    (4) submarine volcanic eruptions

30. What do mid-ocean ridges and hot spots beneath ocean plates have in common?
    (1) origins due to rising magma
    (2) locations along crustal plate boundaries
    (3) earthquakes that originate at great depths
    (4) neither is associated with plate motions

237

TOPIC D _____

Glacial Geology

> **KEY IDEAS** During the Pleistocene epoch, glaciers advanced and retreated across much of the Northern Hemisphere, including nearly all of New York State. Glaciers are responsible for much of New York State's current topography.

KEY OBJECTIVES
Upon completion of this unit, you should be able to:

- Explain how glaciers form and how glacial ice moves under the influence of gravity.
- Identify the two main types of glaciers.
- Describe the processes of glacial erosion and deposition.
- Cite evidence of glaciation in New York State.
- Identify major glacial landscape features of New York State and explain how they formed.
- Discuss the economic resources of New York State that are the results of glaciation.

A. WHAT ARE GLACIERS?

A *glacier* is a large mass of ice, formed on land by the compaction and recrystallization of snow, that is moving downhill or outward under the force of gravity. Modern glaciers are found only at high elevations and in polar regions where snowfall exceeds melting over an extended period of time.

A-1. Formation of Glaciers

For a glacier to form, the average amount of incoming snowfall must be greater than the average amount lost by melting or evaporation year after year. (Another term for glacial ice lost by melting or evaporation is *ablation*.)

When incoming snowfall exceeds ablation, the net amount of snow remaining on the ground increases each year. Freshly fallen snow is a fluffy mass of delicate ice crystals that consists mostly of pockets of trapped air and therefore has a low density. The situation quickly changes, however, as snow piles up on the ground and sits there for weeks or months. Because of evaporation the ice crystals become smaller and rounder and form a denser, *granular snow*. As more and more snow builds up, melting and refreezing, together with the weight of overlying layers, compact the granular snow into an even denser mass called *firn*. Firn is usually a year or more old and has little or no pore spaces. Eventually, the ice crystals in firn grow together into a solid mass of ice (see Figure D.1). When snow is converted to ice in this way, some of the dust and gases of the atmosphere are trapped in the ice. Scientists can sample the trapped air and compare it with the current atmosphere to determine what, if any, changes have occurred.

When ice has accumulated to a thickness such that its own weight deforms the ice and it moves slowly under pressure, a glacier has formed. Usually this stage is reached when the ice is 100 meters or more thick.

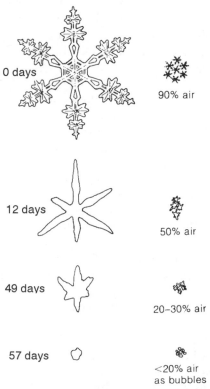

0 days

90% air

12 days

50% air

49 days

20–30% air

57 days

<20% air
as bubbles

Figure D.1 Transformation of Snow Crystals into Glacial Ice. Because of greater surface area, a snowflake's intricate edges melt and evaporate faster than its center. Over time, the snowflake becomes a rounded, compact pellet of ice. From *Earth*, 4th Ed., by Frank Press and Raymond Seiver, W.H. Freeman, © 1974, 1978, 1982, 1986, by W.H. Freeman.

A-2. Movement of Ice

Under pressure, ice behaves like a very thick fluid. The ice *flows* downhill over the Earth's surface, like thick syrup poured on a tilted plate. If there is just a little bit of syrup, it spreads out into a thin layer and moves very slowly. The rate of flow of syrup can be increased in two ways. If you add more syrup, the layer gets thicker and flows more rapidly. You can also get the syrup to flow faster by increasing the tilt of the plate. In much the same way, glaciers flow faster where the glacial ice is thickest and where the slope is steepest.

Variations in the rate of flow of ice cause a glacier to thicken and also to thin out. Where the slope is steep, the ice thins out because it is flowing downslope faster than it is being replenished. The rapid downhill flow exerts a tension on the ice that causes cracks called *crevasses* to form. On extreme slopes the ice may break apart into blocks that cascade downslope in a rock fall. On gentler slopes the ice slows down. The ice from upstream pushes against the slower moving ice, and the compression causes the glacier to bulge and thicken.

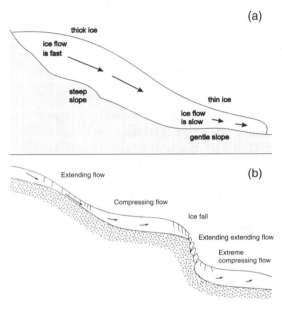

These variations in the rate of flow of ice are illustrated in Figure D.2.

Because the ice in a glacier is flowing, the glacier moves in a unique way. Glaciers do *not* move over the Earth's surface like a brick sliding down a board. A brick slides down an incline along its base as a single unit; nothing moves around *inside* the brick. In ice, both sliding, along the base of the glacier, and flowing, throughout the body of the ice, occur. See Figure D.3 for a digrammatic comparison of the two types of motion.

The amount of a glacier's

Figure D.2 Glacial Flow Rates
(a) Ice flows fastest on steep slopes and in places where the ice is thickest.
(b) On steep slopes, the ice thins out and may even split, forming crevasses, because it is flowing downhill faster than it is being replenished from farther upstream. On gentle slopes, the ice thickens as ice from upstream piles up behind slower moving ice. *Living Ice: Understanding Glaciers and Glaciation*, by Robert P. Sharp, Cambridge University Press, 1988.

movement that is due to the base sliding along the ground, as compared to the amount caused by internal flow, depends on the temperature of the glacier. Temperature affects the behavior of ice in two key ways.

- Ice that is near its melting point is less rigid and is therefore more easily deformed than ice that is well below its melting point. For example, ice at 0°C deforms 10 times faster than ice at -22°C.
- Increasing the pressure on matter causes an increase in temperature. Pressure exerted on ice that is near its melting point may produce enough warming to result in melting.

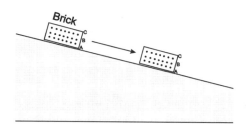

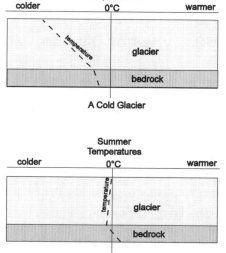

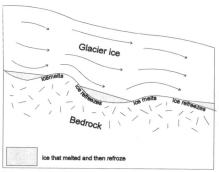

Figure D.3 A Brick and a Glacier in Motion. A glacier both slides along its base and flows like a fluid. The relative position of particles inside the body of the ice change as the glacier flows downhill.

Figure D.4 Temperatures Within a Cold Glacier and a Warm Glacier. From *Living Ice: Understanding Glaciers and Glaciation*, by Robert P. Sharp, Cambridge University Press, 1988.

Temperatures within a warm glacier and a cold glacier are compared in Figure D.4.

Warm glaciers, whose ice is near its melting point, slip along their bases more readily than cold glaciers. When warm glaciers encounter an obstacle, the pressure of the ice against the obstacle causes the ice to deform and flow around the obstacle. The glacier may also bypass the obstacle by melting on one side and refreezing on the other. The pressure on the upstream side of the obstacle may cause the ice to melt and then refreeze on the downstream side, where pressure is lower. Meltwater produced by pressure can also fill in cavities and act as a lubricant, allowing the ice to move over the ground more easily (see Figure D.5).

Figure D.5 Basal Meltwater. High pressure on the upstream side of an obstacle causes ice to melt. The meltwater refreezes on the downstream side of the obstacle, where pressures are lower. Adopted from *Glaciers and Landscape: A Geomorphological Approach*, by David E. Sugden and Brian S. John, A Halsted Press Book, John Wiley, 1976.

In cold glaciers moving over bedrock, almost none of the movement is due to slipping of the base along the ground. The bond between the ice and the rock to which it is frozen is greater than the internal strength of the ice. Therefore, under pressure, the ice will deform before it breaks loose from the bedrock.

A-3. Types of Glaciers

There are two main types of glaciers: valley glaciers and the much larger continental glaciers. *Valley glaciers* are constrained by topography, and their form and flow are strongly influenced by the shape of the land. *Continental glaciers* submerge the land to the extent that the size and shape of the glacier, rather than the shape of the land, control glacial form and flow.

Valley Glaciers

Valley glaciers (see Figure D.6) form between mountain slopes at high elevations. They start as snowfields that accumulate in bowl-shaped hollows called *cirques*. As the snowfield deepens and ice forms from firn, the ice flows out of the cirque and down the valley, confined by the surrounding slopes.

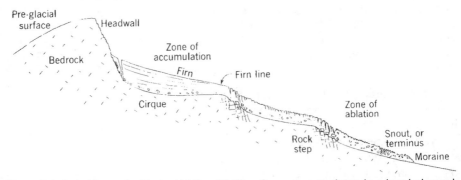

Figure D.6 Side View of a Typical Valley Glacier. Snow accumulates in a bowl-shaped cirque and is compacted into firn. The firn recrystallizes into ice and moves downhill; the firn line marks the beginning of the glacial ice. When the glacier reaches warmer temperatures, ablation consumes the ice. The glacier ends in a thin snout edged by a pile of debris called a moraine. From *The Earth Science*, 2nd Ed., by Arthur N. Strahler, Harper & Row, Copyright 1963, 1971 by Arthur N. Strahler.

As the glacier moves down the valley, it may meet other glaciers from adjacent valleys and the glaciers may merge, forming a larger glacier. Unlike streams, which mix when they merge, glaciers do not mix but rather "weld" together and flow downslope side by side. A large valley glacier may have been fed by dozens of smaller tributary glaciers flowing from high mountain valleys.

Valley glaciers end at a point where there is a balance between ablation and accumulation. This point can be a sensitive indicator of changes in climate. A general warming trend will cause the end of the glacier to melt back up the valley, or retreat. A cooling trend will cause the end of the glacier to move down the valley, or advance.

When a glacier ends on land, streams of meltwater flowing from the glacier often form a river. Large blocks of ice that fall off the face of the glacier may be buried by sediment. A glacier that ends in a body of water forms steep cliffs of ice. Chunks of ice that break off and fall into the water will float since ice is less dense than liquid water. These floating masses of ice, or icebergs, then drift with the currents and gradually melt.

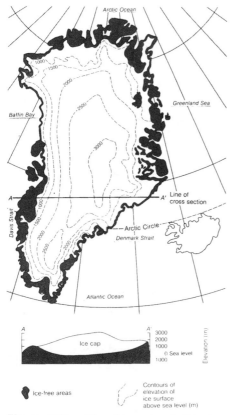

Figure D.7 The Greenland Glacier Contour lines show the elevations of the ice surface in meters, and a cross-sectional view reveals the lenslike shape of the glacier. From *Earth*, 4th Ed., by Frank Press and Raymond Seiver, W.H. Freeman, © 1974, 1978, 1982, 1986 by W.H. Freeman.

Continental Glaciers

Continental glaciers form at high latitudes where temperatures are always cold. These glaciers are immense masses of ice that spread over the entire land surface, rather than being confined to valleys. Two continental glaciers exist today: one covers Greenland, and the other Antarctica. The Greenland glacier (see Figure D.7) covers more than 1.7 million square kilometers and is 3.2 kilometers thick at its center. The Antarctic glacier covers more than 12 million square kilometers (about 1.5 times the size of the mainland U.S.'s 48 states) and is 4 kilometers thick in places. It even extends over the ocean along the margins of the continent of Antarctica, forming several ice shelves.

A continental glacier has a lenslike shape; it is thickest at the center and thinner at the edges. The thick central area is where snow accumulates in a huge firn field and slowly turns to ice. The ice then flows outward in all directions under the pressure of its own weight. The huge weight of a continental glacier actually depresses the surface of the land, and the topography of the land is extensively changed by the movement of the glacier over its surface.

B. HOW HAS GLACIAL EROSION AFFECTED THE EARTH'S SURFACE?

B-1. Processes of Glacial Erosion

Glaciers erode the surfaces over which they move by abrasion and plucking. *Abrasion* is the wearing, grinding, scraping, or rubbing away of rock surfaces by friction. *Plucking* is the process by which rock fragments are loosened, picked up, and carried away by glaciers.

Abrasion

Abrasion is accomplished mainly by fragments of rock frozen in the ice at the bottom of a glacier. These rub and scrape against the underlying bedrock as the base of the glacier slips downslope. Sharp, angular fragments of hard materials such as quartz are very effective abraders. Clay is a good polishing material but doesn't abrade the bedrock very much.

Plucking

Plucking occurs in several ways. The water that forms when pressure on the ice causes melting can seep into cracks and pores in the bedrock. When this water refreezes, it shatters the bedrock and the broken pieces are picked up and carried away by the glacier. Plucking that occurs in this way is probably the most important because it goes on continuously over much of the glacier's bed as it moves along.

Plucking can also occur when glaciers move over bedrock that is broken into large pieces by jointing. Joints are fractures in rock and usually occur in three planes, forming large, rectangular blocks. These blocks freeze to the base of the glacier and are pushed out of the ground and carried along as the ice moves downslope.

A third type of plucking may occur where the glacier is thick and the bedrock is brittle. In such places, the pressure exerted by a glacier may be great enough to crush, shear off, or shatter brittle bedrock, forming fragments that are carried away by the glacier.

B-2. Features Resulting from Glacial Erosion

The processes of abrasion and plucking leave behind unmistakeable evidence of a glacier's passage over bedrock. Numerous features, ranging from scratched and polished rock surfaces to basins excavated by plucking to streamlined hills of solid rock, are formed by glacial erosion.

Glacial Polish

Fine materials like clays and silt that are frozen into the ice at the base of a glacier rub against bedrock as the glacier moves. These particles smooth the rock surface as sandpaper smooths wood. The smooth rock surfaces produced by this type of abrasion are said to be *glacially polished*.

Glacial Striations

Glacial striations (see Figure D.8) are small scratches in the surface of the bedrock. They are created when fragments frozen into the bottom of a glacier scrape against the bedrock as the glacier slides over the bedrock. On smooth surfaces the striations are usually straight and parallel, but on uneven surfaces with protrusions the striations may curve because the ice deformed as it flowed around these obstacles. A striation may also thicken or thin out, depending on what happened to the abrading particle as it scraped against the bedrock. The long axis of a glacial striation indicates the direction of glacial movement.

Figure D.8   Glacial Striations and the Particles That Made Them. Striations vary in size and shape, depending on the nature of the rock particles that formed them and the way the particles were dragged over the bedrock.

Glacial Grooves

Glacial grooves are long, deep, U-shaped grooves in bedrock with smooth bottoms and sides and rounded edges. Most grooves are 10–20 centimeters deep, twice as wide, and tens of meters long. Really large grooves, however, can be more than a meter deep and hundreds of meters long. Grooves are shaped by abrasion and often form along fractures, soft layers, or other weak zones in the bedrock. Once started, a groove acts as a channel for further ice flow and concentrates the abrasive force of the glacier in its channel.

Friction Cracks

If you were to press down and push sideways on the surface of a bowl of Jell-O, it would crack and split apart. In much the same way, the great weight of overlying ice, coupled with the sideward pressure of fast-moving ice, can exert enough force on brittle bedrock to cause it to crack. The most common glacial *friction cracks* are crescent-shaped and several centimeters deep.

Whalebacks

Whalebacks are smooth, elongated bedrock features whose surfaces are often polished and striated. These features get their name from their resemblance to the back of a whale breaking the surface of the ocean. Usually 1–2 meters high, 3–4 meters wide, and 5–10 meters long, whalebacks are formed as ice flowing over bedrock abrades its surface.

Roche Moutonnées

Roche moutonnées are formed by both abrasion and plucking. Abrasion creates a smooth, elongated shape on one side; plucking results in a steep, jagged, clifflike shape along the other side. Roche moutonnées range in size from small knobs to huge masses hundreds of meters high.

Rock Drumlins

Drumlins are hills of loose rock and soil molded into a streamlined shape by glaciers moving over them. Rock drumlins are large outcrops of rock that have been abraded into a streamlined shape by glaciers. Rock drumlins are smoothly abraded like whalebacks, but they are much larger in size, often measuring 100 meters in length.

Glacial Basins

Glacial basins are depressions formed when large quantities of material are carried away by a thick, fast-moving glacier. Well-jointed bedrock may be deeply excavated by plucking, leaving behind a glacial basin that usually ends up holding a lake.

U-Shaped Valleys

Glaciers move from high elevations to low elevations along the easiest paths. Those paths are usually stream valleys (see Figure D.9a) that were eroded before the glacier formed. Stream valleys in mountainous regions tend to be steep-sided and V-shaped and to follow a narrow, twisting path between adjacent slopes. When a glacier moves down a stream valley, it alters its shape. The ice does not meander back and forth as readily as water, so it

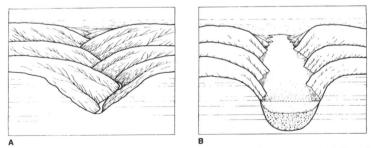

A   B

Figure D.9 A V-shaped Stream Valley (a) and a U-shaped Glacial Valley (b). From *Geology of New York: A Simplified Account*, by Y. Isachsen et al., eds., New York State Museum Geological Survey, The State Education Department, The University of the State of New York, 1991.

often breaks through obstacles to form a straighter path. The ice also cuts back the walls of the valley, making them steeper. The result is a wider, straighter, steeper-sided, *U-shaped valley* (Figure D.9b).

Hanging Valleys

A *hanging valley* (Figure D.10c) is a valley of a tributary glacier whose floor is above, or hanging over, the floor of the main glacier into which it feeds. The difference in the level of the valley floors can be hundreds of meters, and if a stream flows down the tributary valley, it may form a spectacular waterfall where it plunges to the floor of the main valley. Hanging valleys may be formed if the main glacier cuts back the valley of a tributary, or if it simply erodes faster and deeper than the tributary.

Cirques, Arêtes, and Horns

A *cirque* is the open, half-bowl-shaped hollow high on a mountain-side at the head of a valley glacier. Cirques form when snowfields on mountain slopes melt and refreeze during spring and fall. The repeated freezing and thawing breaks apart the rocks. The rock fragments creep downlsope or, when broken into small enough pieces, are carried away by meltwater. In this way, the snow-field creates a hollow in the land on which it rests. As the hollow grows, so does the depth of the snowfield,

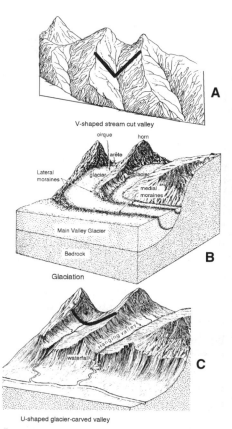

Figure D.10 Landscape Before, During, and After Glaciation. (a) Valley before glaciation. (b) Glacial erosion forms cirques in hollows between slopes. Erosion in adjacent cirques forms sharp ridges called arêtes and converts rounded peaks into sharp horns. Eroded material carried along the ice margins forms moraines. (c) Hanging valleys form where tributary glaciers flowed into the main glacier. From *Earth*, 4th Ed., by Frank Press and Raymond Seiver, W.H. Freeman, © 1974, 1978, 1982, 1986.

eventually becoming large and deep enough to form a glacier. Once formed, the glacier plucks rock from the base of the cirque, further deepening it.

Where cirques from adjacent slopes meet, a narrow, knife-edged ridge called an *arête* is formed between them. Where three or four cirques on the flanks of a mountain peak meet, the central peak takes on a pyramidal shape called a *horn*.

These three features are shown in Figure D.10b.

C. HOW HAVE GLACIAL TRANSPORT AND DEPOSITION ALTERED THE EARTH'S SURFACE?

C-1. Glacial Transport

Glaciers transport materials suspended in the ice, along the sides and bottom of the ice, on the surface of the ice, and in the area directly alongside the ice. Although it is tempting to picture a glacier as a huge block of ice pushing up a pile of debris ahead of it, most glaciers transport very little material by "bulldozing" ahead of the ice. The majority of material carried by a glacier is suspended in the ice.

One of the characteristics that affect the ability of any medium to carry materials in suspension is the viscosity, or resistance to flow, of the medium. The higher the viscosity, the greater is the medium's ability to hold materials in suspension. Ice is millions of times more viscous than liquid water and can carry just about anything in suspension. A boulder that couldn't be budged by the fastest moving stream of water, no less remain suspended in the water, can easily be carried along suspended in the ice of a glacier (see Figure D.11).

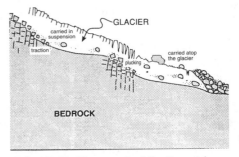
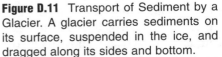

Debris can become suspended in glacial ice in several ways. Many glaciers originate in snowfields high in mountainous regions. Landslides and rockfalls from the steep slopes

Figure D.11 Transport of Sediment by a Glacier. A glacier carries sediments on its surface, suspended in the ice, and dragged along its sides and bottom.

that surround the glacier deposit debris on top of the snow. When more snow accumulates, the debris is buried and eventually encased in the ice that forms. Also, rock protrusions around which a glacier flows shed fragments that become incorporated into the ice. Where two glaciers merge, the debris along the edges of the glaciers becomes incorporated into the body of the larger glacier that forms. Material also becomes incorporated into the bottom and sides of a glacier as the ice melts and refreezes around loose rock fragments. The 10–100 centimeter layer along the base of a glacier is filled with suspended debris.

Material along the bottom of the glacier is carried along by traction. In *traction*, the particles are not encased in the ice but are dragged along, slipping, sliding, and rolling beneath the moving ice. Material along the sides of the glacier is also moved by traction, and rock fragments may slide or roll off the ice onto ground next to the glacier. Meltwater-soaked debris with the consistency of wet concrete also flows off the sides of the glacier.

Most material carried on the top of a glacier is *ablation debris*, material that was suspended in the ice but has been exposed as melting occurred. Thick debris, however, shields the ice beneath it from melting and results in high spots on the ice surface. Blocks of rock that fall onto a glacier from surrounding slopes can be carried for great distances, conveyor-belt style, atop the glacier.

C-2. Glacial Deposition

As mentioned earlier, ice has much more transporting power than liquid water. When a glacier melts, there is a sudden, sharp decrease in transporting power, which results in deposition of the material carried by the glacier.

Processes of Glacial Deposition

Some deposition takes place beneath the ice. The base of a warm glacier is continuously melting because heat radiating from the ground and heat created by friction with the ground. As particles in the base of the glacier are freed by melting, some are rolled or dragged along until they become lodged in the ground. Refreezing of large particles, which jut out more than finer ones, also leaves behind deposits enriched in fine particles. The pressure and stress under which this material is deposited often cause any remaining large particles to become aligned in the direction of the ice flow, and the fine particles take on a compact, streamlined shape. Deposits of this type are called *lodgement till*.

Materials on top of a glacier as deposited in two ways: either meltwater and debris slides carry them off the glacier, or they are set down on the ground as the ice beneath them melts. When meltwater deposits these materials, they are well sorted and somewhat rounded. Materials that slide or fall off the glacier, however, are jumbled masses of large and small angular fragments. By the time a glacier reaches its end, it has already undergone much melting. The remaining material on its surface is dropped off the end of the glacier as the ice beneath it melts, forming a pile called an *end moraine*.

Meltwater carries debris from the glacier beyond the end of the glacier and deposits it over the land. In summer, glacial meltwater streams are in flood and carry huge amounts of sediment, including abundant fine particles called *rock flour*. Rock flour is formed by the crushing pressure and abrasion between particles as they are carried along inside the glacier. If numerous streams spread out over the land and deposit glacial material in a wide sheet, the result is called an *outwash plain*. If the meltwater streams are confined in a valley, they may deposit their load on the floor of the valley, forming a *valley train*, which is an elongated deposit.

Meltwater often fills closed depressions, forming glacial lakes. If the lake forms along the edge of the glacier, debris can fall directly into the lake. If the water undercuts the ice face, icebergs fall into the lake and drift across its surface. As the iceberg melts, it rains *ice-rafted* debris onto the lake bed. During the winter, the lake freezes over and deposition ceases except for the settling out of the very finest particles of clay. In the spring when the ice thaws, meltwater carries in coarser material, which is deposited atop the clay. The result is a pair of layers known as a *varve*. Each varve represents one seasonal cycle—a year.

Features Resulting from Glacial Deposition

Glacial drift is any material deposited as a result of glacial activity, including material deposited by meltwater and debris slides. *Glacial till* is material deposited directly from ice. The main difference between the two is sorting; drift may be sorted because of the action of meltwater, whereas till is unsorted. *Erratics* are large boulders that were transported by the glacier and deposited on the ground when the ice melted. Erratics have usually been transported far from their source and therefore differ in composition from the underlying bedrock.

When glacial drift and till are deposited, they form a number of distinctive landscape features.

Moraines

Moraines (see Figure D.12) are ridges of material deposited by a glacier. Moraines form along the margins of glaciers. Elongated piles of till form along the edge of the melting ice or are deposited along the margins of the glacier by meltwater. *End moraines* form along the end of the glacier, *lateral moraines*, along the sides. *Medial moraines* form where two glaciers flowed together. When the glacier melts, an elongated pile of till is left behind. An end moraine that marks the farthest point to which the glacier advanced is called a *terminal moraine*. As the glacier retreats, there may be brief episodes of advance or pauses in which a new end moraine is built. A series of end moraines forms a *recessional moraine*.

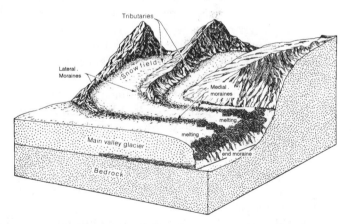

Figure D.12 How Moraines Form. Lateral moraines consist of materials carried along the sides of a glacier. When two glaciers merge, their lateral moraines are welded together between the glaciers, forming a medial moraine. End moraines form where debris melts out of the ice at the end of the glacier. From *Earth*, 4th Ed., by Frank Press and Raymond Seiver, W.H. Freeman, © 1974, 1978, 1982, 1986 by W.H. Freeman.

Ground Moraine

Ground moraine is a blanket of till that settles onto the ground as the ice melts. The surface of ground moraine is uneven and often covered with a pattern of parallel hollows and ridges. The ridges are formed when till is pressed into grooves that were carved into the bottom of the glacier as it moved over rock outcrops.

Ground moraine often contains elongated hills of glacial till that have a streamlined shape and are called *drumlins*. Drumlins are shaped like teardrops, with the blunt end facing upstream. Drumlins occur in groups, or swarms, of thousands and are thought to form near the edge of a glacier carrying a heavy load of debris. As the debris melts out of the glacier, it is shaped by the ice moving over it.

Kettles, Kames and Eskers

A *kettle* is a closed depression, partly filled with water that was formed when a buried block of ice melted (see Figure D.13). The more deeply the ice was buried, the rounder and shallower the kettle that forms. The difference in the depression formed by melting both a shallow and a deeply buried ice block is illustrated in Figure D.13.

Kames are isolated, conical hills of till. A kames forms when a hole in the ice that penetrates to the bed of the glacier is filled with debris by meltwater streams. When the ice melts from around this column of debris, it spreads out until it reaches its angle of repose, forming a conical hill.

If a large meltwater stream runs along the side of a valley glacier, it can spread the debris that would normally form a lateral moraine into a wide, flat streambed. When the glacier shrinks in size as it melts, a flat-topped pile called a *kame terrace* is left along the wall of the valley. A kame terrace looks almost like a giant stair step along the wall of a valley.

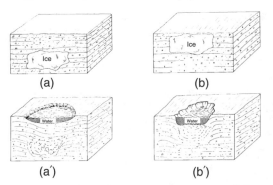

Figure D.13 Formation of a Kettle. Kettles form when buried blocks of ice melt and the overlying sediments collapse, forming a hole. The melting of deeply buried blocks of ice forms rounder and shallower kettles. In (a) and (a′), the depression formed by melting a deeply buried ice block is seen, whereas (b) and (b′) shows the depression formed by melting a shallowly buried ice block. From *Living Ice: Understanding Glaciers and Glaciation*, by Robert P. Sharp, Cambridge University Press, 1988.

Eskers are long, narrow piles of debris deposited from meltwater. Eskers are thought to form as a glacier is melting back at the end of a glacial period. Meltwater streams beneath the glacier, flowing in pipelike tunnels in the ice, become choked with debris. Finer material is carried away by the water, and coarser material is left behind. When the ice melts, the debris that choked the

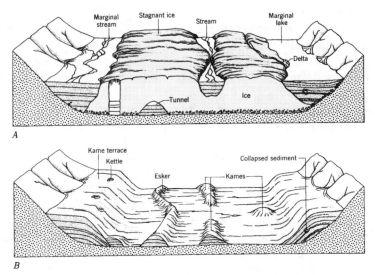

Figure D.14 Features Formed by Deposition Around and Under Melting Ice. In (a), a glacier is in the process of melting. In (b), the glacier has completely melted and the landforms previously buried can be seen. From *Physical Geology*, 2nd Ed., by Richard Foster Flint and Brian J. Skinner, John Wiley, 1974, 1977.

pipe is left behind as a long, snaking ridge of well-sorted gravel. Eskers are prized sources of gravel for roadbuilding.

Figure D.14 shows glacial meltwater features.

Glacial Lakes

As already mentioned, cirques and other basins created in bedrock by plucking often become occupied by lakes. Many such basins are rimmed by moraines, which further increase their capacities. Some glacial lakes are created when a moraine dams the course of a stream. This situation is common in glaciated mountain ranges, and some of the world's most picturesque lakes were formed as a result.

Other glacial lakes are created when the glacial ice itself blocks the course of a stream. Such ice-dammed lakes can cause disastrous floods when the ice dam melts and releases the lake water. There is much evidence that catastrophic floods caused by ice damming occurred repeatedly during the last ice age, 10,000–20,000 years ago.

Glacial Wind Deposits

A glacial outwash plain is much like a desert in that it is barren and has strong winds, as well as a constantly renewing source of fine sediments from the glacial outwash. These conditions lead to transport of fine particles by the wind and, in turn, to the formation of sheets of sand dunes bordering the outwash plain. Silt and clay that become windborne can travel for great distances before settling to the ground. In some places, stable wind patterns have resulted in thick beds of fine wind-blown sediments called *loess*.

D. WHAT OTHER EFFECTS HAS GLACIATION HAD ON EARTH?

In addition to glacial erosion and deposition, there are a number of indirect effects of glaciation, such as changes in sea level, deformations of the crust, fluctuations in ocean temperatures, and the shifting of climate belts, along with their associated plant and animal life.

Most of the water that precipitates as snow and accumulates in glaciers comes from the oceans. In the distant past, as more and more water was tied up as ice and snow, less and less flowed back to the oceans as runoff from rivers. As a result, the balance of the hydrologic cycle was upset and the level of water in the world's oceans lowered. During the height of the Pleistocene, sea level dropped by about 130 meters (425 feet) worldwide, exposing land that had previously been underwater. In some places, the newly exposed coastal plains were more than 100 kilometers wide.

As massive ice sheets accumulated on the land, their weight caused the crust to subside beneath them. Since ice is about one-third as dense as crustal rocks, an ice sheet 3 kilometers thick can cause the crust to subside as much as 1 kilometer. When glaciers melt, their weight is removed from the land and the crust slowly rises. The Hudson Bay region of Canada is still rising as the crust "rebounds" in response to the removal of the ice load from the last ice age.

As the climate cools during glacial periods, the temperature of the ocean decreases. When water evaporates from cold oceans, water containing the lighter oxygen-16 isotope evaporates more easily than water with the heavier oxygen-18 isotope. As a result, oceans lose more oxygen-16 than oxygen-18 and the level of oxygen-18 in the oceans rises. Therefore, plants and animals

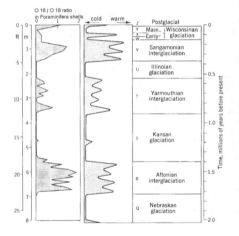

Figure D.15 Correlation Between Quantity of the Isotope O_{18} Present in Deep-Sea Core Material, Temperature, and Glacial Activity.
Source: *The Earth Sciences.* © by Arthur N. Stahler, 1971.

that live in the oceans during glacial periods have more oxygen-18 in their bodies than those that live in the oceans during warmer times. This enrichment can be detected in seafloor sediments containing the remains of glacial-period organisms. Analysis of deep-sea cores has revealed a cyclic pattern of oxygen-16/oxygen-18 ratios, indicating that cyclic changes in ocean temperature coincide with evidence of cyclic glaciation on land.

Figure D.15 shows the fluctuations in the oxygen-18/oxygen-16 ratio with ice ages.

E. WHAT IS THE GLACIAL HISTORY OF NEW YORK STATE?

There is much evidence that New York State has been repeatedly covered by thick ice in the recent geological past.

E-1 Changes in New York State's Landscape During the Pleistocene Epoch

An *ice age* is a time during which conditions exist that cause ice sheets more than 1 million square kilometers in area and thousands of meters thick to develop on nonpolar continents. Since ice ages are times of extensive glacial activity, they are sometimes called *glacial epochs*. The most recent glacial epoch was the Pleistocene. During an ice age there are often times during which glaciers retreat and even disappear. These times are known as *interglacial intervals*. Even though glaciers may disappear during interglacials, they are still considered part of the ice age because they return. An intriguing question is whether or not we are presently living in an interglacial interval.

Figure D.16 The Northern Hemisphere During the Pleistocene Epoch. From *Geology of New York: A Simplified Account*, Y. Isachsen et al., Eds., New York State Museum/ Geological Survey, The State Education Department, The University of the State of New York, 1991.

The Pleistocene epoch, which began 1.6 million years ago, was a time when climates grew colder worldwide. The exact reasons for this cooling are not yet clear, but it resulted in huge ice sheets that advanced and retreated several times across the Northern Hemisphere, including New York State.

During the Pleistocene, snow remained on the ground year round as far south as central New Jersey. Huge continental glaciers formed in arctic and subarctic regions and flowed outward (south) merging with smaller valley glaciers that had already formed in mountainous regions. At the height of the Pleistocene Ice Age, nearly a third of the Earth's land was covered by ice (see Figure D.16).

The glacier that advanced over New York State originated in the region encompassing the Laurentian Mountains of Quebec, Canada, and is therefore called the *Laurentide Ice Sheet*. The Laurentide Ice Sheet made four major advances across the northern United States during the Pleistocene. In order of

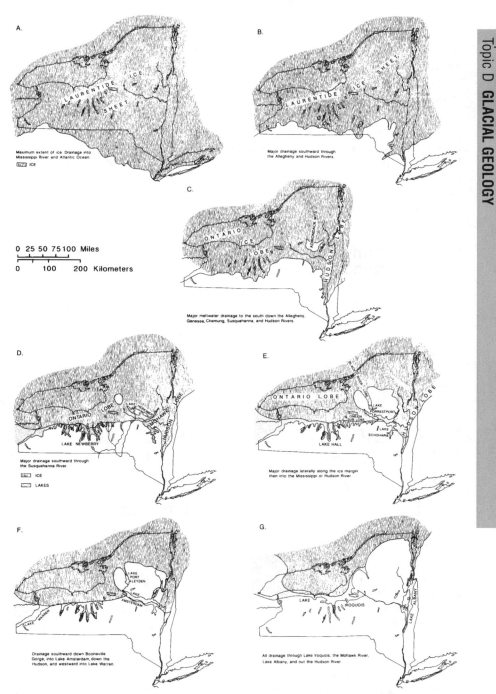

Figure D.17 The Retreat of the Laurentide Ice Sheet. (a) 21,750 years ago; (b) 14,000 years ago; (c) 12,000–13,800 years ago; (d), (e), (f) 11,000–13,000 years ago; (g) 11,000 years ago. From *Geology of New York: A Simplified Account*, Y. Isachsen et al., eds., New York State Museum/Geological Survey, The State Education Department, The University of the State of New York, 1991.

255

increasing age they are the Wisconsin, Illinoian, Kansan, and Nebraskan stages (named after the states in which glacial till and other features of that age were extensively exposed and studied). In New York State the most recent advance, which occurred during the Wisconsin stage, covered almost the entire state and removed nearly all traces of earlier glacial stages.

About 21,500 years ago, the climate began to warm. The Laurentide Ice Sheet started melting back faster than it was flowing southward, and the ice front began to retreat (see Figure D.17). By 10,000 years ago, the Laurentide Ice Sheet had retreated out of New York State, and it had melted completely by 7,000 years ago. Today, all that remains of the great ice sheets of the Pleistocene are the glaciers of Greenland, Antarctica, and the arctic islands of North America and Eurasia.

E-2. Plant and Animal Life During the Pleistocene Epoch

During the Pleistocene, both colder and warmer climates were shifted southward. Near the ice front, the natural environment of New York State might have looked like the barren tundra of northern Canada, Alaska, and Siberia. South of the ice sheet, though, life-forms that were adapted to cold weather flourished. Forests of evergreen trees that could withstand cold, elephantlike woolly mammoths and mastodons, and other animals such as the ground sloth, musk ox, giant beaver, caribou, deer, elk, and bison were common.

F. HOW DID GLACIERS INFLUENCE THE LANDSCAPE AND ECONOMY OF NEW YORK STATE?

F-1. Effects of Glaciation on the Landscape

Nearly all of New York State displays evidence of glaciation. The soils that cover most of the state are composed of weathered till. Superimposed layers of weathered till indicate that there have been several major periods of glaciation in the recent geological past.

Throughout New York State, exposed bedrock shows evidence of glacial erosion, including grooves, striations, rock drumlins, and subglacial potholes. Occurrence of these features on summits in the Catskills and the Adirondacks helps us to estimate the thickness of the glaciers that caused them. For example, there are glacial grooves and striations on peaks in the Adirondacks that are 1.6 kilometers high. Their presence tells us that the ice was more than 1.6 kilometers thick.

Many of the Adirondack Mountains display glacial features such as cirques, arêtes, and horns. Whiteface Mountain, a typical example, was shaped by three cirques into a pyramidal shape with aretes radiating from its peak. Furthermore, the many north-south valleys in New York State bespeak a glacial origin.

Central and western New York State south of Lake Erie display the classic features of a ground moraine, with numerous erratics, drumlin fields, eskers, kames, kettles, and moraine-dammed lakes. The lowlands between Syracuse and Rochester are studded with more than 10,000 drumlins. The Great Lakes are glacial basin lakes, and the Finger Lakes fill moraine-dammed valleys.

The northern coast of Long Island consists of two superimposed terminal moraines; the rest of Long Island is the remainder of an outwash plain. One of the largest lakes on Long Island, Lake Ronkonkoma, is a kettle lake.

The Hudson River valley is a classic U-shaped glacial valley that was flooded when sea level rose after the last ice age, making it a fjord. It extends well into the present Atlantic Ocean, and the ocean tides are felt halfway to Canada along the Hudson River. Many of the river's tributaries occupy hanging valleys and now enter the Hudson River by flowing over a series of waterfalls or cascading down rapids.

F-2. Creation of Natural Resources by Glaciation

The retreating glaciers of the last ice age left behind an abundance of sand, gravel, and clay as a natural resource. To understand the significance of this resource, consider that there are about 2,000 mines in New York State and 85 percent of them produce sand and gravel. The glaciers also left behind a picturesque landscape that attracts much tourism.

Clay

Clay is an extremely fine-grained sediment. Much of New York State's clay accumulated in beds at the bottom of deep glacial lakes. Clay is used in making bricks, pottery, and concrete. Since layers of clay are impermeable to water, this subtance is also used as a cap and lining material for landfills to protect groundwater from contamination.

Sand and Gravel

Sand and gravel are naturally broken rock of many kinds. Sand and gravel made of igneous rock, limestone, dolostone, or sandstone are strong enough to be used in construction. This type of bedrock, which underlies much of New York State, was eroded and deposited by glaciers. The commercially valuable sands and gravels of New York State come from glacial deposits. Sand and gravel are used in the manufacture of concrete for building and in road construction.

Some sands deposited in glacial lakes contained shale, which has weathered into clay. The clay content of these glacial lake sands helps them to hold together and makes it possible to mold them into complex shapes. These molding sands are used to make metal castings in a variety of shapes. Liquid metal is poured into the sand molds, where it hardens, and the metal is then removed and machined to a desired finish.

Water

The glacial till that overlies much of New York State forms a permeable layer in which goundwater can accumulate. The numerous glacial lakes that dot the landscape are natural reservoirs of fresh water. These combine to provide New York State with an abundance of accessible fresh water for residential, commercial, and industrial uses.

F-3. Recreation, Tourism, and Agriculture

Glaciation has left behind spectacular landscapes and abundant lakes, rivers and streams. Consider these few examples. The rugged, glaciated peaks of the Adirondacks, which are reminiscent of the Alps, draw hikers, campers, and skiers year round. The deep, sheltered waters of the Hudson fjord have enabled New York City harbor to become one of the world's largest ports. The Finger Lakes are a year-round playground for fishing and boating enthusiasts. The shores of Long Island, with their white, sandy beaches and rich offshore fishing grounds, draw visitors by the millions each year. The gently rolling hills of fertile, weathered glacial till, dotted with lakes and ponds, constitute a large agricultural region. New York State wines and Long Island potatoes enjoy a national reputation, and New York is a top producer of milk and cheese.

Truly, the glaciers of the ice ages have left New Yorkers with a rich abundance of natural resources.

REVIEW QUESTIONS FOR TOPIC D

1. Which geologic evidence best supports the inference that a continental ice sheet once covered most of New York State?
    (1) polished and smooth pebbles; meandering rivers; V-shaped valleys
    (2) scratched and polished bedrock; unsorted gravel deposits; transported boulders
    (3) sand and silt beaches; giant swamps; marine fossils found on mountaintops
    (4) basaltic bedrock; folded, faulted, and tilted rock structures; lava flows

2. Which is the best evidence that more than one glacial advance occurred in a region?
    (1) ancient forests covered by glacial deposits
    (2) river valleys buried deeply in glacial deposits
    (3) scratches in bedrock that is buried by glacial deposits
    (4) glacial deposits that overlay soils formed from glacial deposits

3. Which landscape characteristic best indicates the action of glaciers?
    (1) few lakes
    (2) deposits of well-sorted sediments
    (3) residual soil covering large areas
    (4) polished and scratched surface bedrock

4. Which feature is more likely to be formed by a valley glacier than by a continental glacier?
   (1) drumlin
   (3) medial moraine
   (2) esker
   (4) till plain

5. Which landscape features are primarily the result of erosion by continental glaciers?
   (1) U-shaped valleys with unsorted layers of soil
   (2) V-shaped valleys containing swiftly flowing streams
   (3) mountains with rugged slopes and shallow layers of soil
   (4) well-established stream drainage patterns with few lakes

6. The formation of the Finger Lakes of central New York State and the formation of Long Island are both examples of
   (1) climatic changes resulting in a modification of the landscape
   (2) uplifting and leveling forces being in dynamic equilibrium
   (3) soil associations differing in composition depending upon the bedrock composition
   (4) activities of man altering the landscape

7. A continental ice sheet which covered New York State most likely originated in which area?
   (1) Alaska
   (3) Greenland
   (2) Labrador
   (4) the Antarctic

8. The accompanying diagram represents a landscape area.

   Which process is primarily responsible for the shape of the surface shown in the diagram?
   (1) crustal subsidence
   (3) glacial action
   (2) wave action
   (4) stream erosion

9. Which diagram best represents a cross section of a valley which was glaciated and then eroded by a stream ?

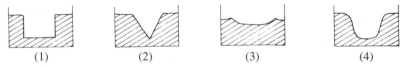

   (1)   (2)   (3)   (4)

10. Many elongated hills, each having a long axis with a mostly north-south direction, are found scattered across New York State. These hills contain unsorted soils, pebbles, and boulders. Which process most likely formed these hills?
    (1) stream deposition
    (3) wave deposition
    (2) wind deposition
    (4) glacial deposition

11. An area contains numerous winding ridges, cone-shaped hills, and small circular lakes. This area was most likely formed by
(1) an arid climate
(2) folding of rock layers
(3) continental glaciation
(4) marine flooding

12. Which rock material was most likely transported to its present location by a glacier?
(1) rounded sand grains found in a river delta
(2) rounded grains found in a sand dune
(3) residual soil found on a flat plain
(4) unsorted loose gravel found in hills

13. Which statement best describes the major heat flow associated with an iceberg as it drifts south from the Arctic Ocean into warmer water?
(1) Heat flows from the water into the ice.
(2) Heat flows from the ice into the water.
(3) A state of equilibrium exists, with neither ice nor water gaining or losing energy.
(4) Heat flows equally from the ice and the water.

Base your answers to questions 14 through 18 on your knowledge of earth science, the *Earth Science Reference Tables*, and the adjacent diagram. The diagram represents two branches of a valley glacier. Points A, B, G, and H are located on the surface of the glacier. Point X is located at the interface between the ice and the bedrock. The arrows indicate the general direction of ice movement.

Scale (km)

14. Which type of weathering most likely is dominant in the area represented by the diagram?
(1) biologic activity
(2) frost action
(3) acid reactions
(4) chemical reactions

15. Which force is primarily responsible for the movement of the glacier?
(1) ground water
(2) running water
(3) gravity
(4) wind

16. The sediment deposited by the valley glacier at position X is best described as
(1) sorted according to particle size
(2) sorted according to particle density
(3) sorted according to particle texture
(4) unsorted

17. Metal stakes were placed on the surface of the glacier in a straight line from position *A* to position *B*. Which diagram best shows the position of the metal stakes several years later?

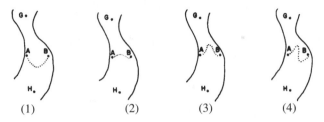

(1)   (2)   (3)   (4)

18. Which cross section best represents the valley shapes of this landscape area after the glacier melts?

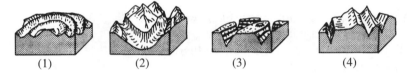

(1)   (2)   (3)   (4)

Base your answers to questions 19 through 23 on the *Earth Science Reference Tables*, the diagram below, and your knowledge of earth science. The diagram represents a glacier moving out of a mountain valley. The water from the melting glacier is flowing into a lake. Letters *A* through *F* identify points within the erosional/depositional system.

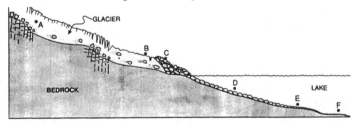

19. Deposits of unsorted sediments would probably be found at location
(1) *E*   (3) *C*
(2) *F*   (4) *D*

20. An interface between erosion and deposition by the ice is most likely located between points
(1) *A* and *B*   (3) *C* and *D*
(2) *B* and *C*   (4) *D* and *E*

21. Colloidal-sized sediment particles carried by water are most probably being deposited at point
(1) *F*   (3) *C*
(2) *B*   (4) *D*

261

22. Which characteristic would form as the glacier advances from point *A* to point *B*?
(1) V-shaped valleys   (3) layers of salt and other evaporites
(2) a thick, well-sorted soil   (4) scratched and polished bedrock

23. Which graph best represents the speed of a sediment particle as it moves from point *D* to point *F*?

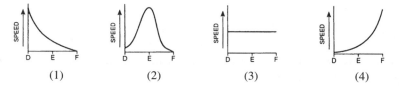

    (1)   (2)   (3)   (4)

24. A glacial deposit would most likely consist of
(1) particles in a wide range of sizes
(2) particles the size of pebbles and larger
(3) sediments in flat, horizontal layers
(4) sediments found only in the bottoms of stream valleys

25. When were large parts of North America covered by ice sheets?
(1) only once, early in the geologic history of the Earth
(2) only once, in the recent geologic past
(3) once early in the geologic history of the Earth, and once in the recent geologic past
(4) many times during the geologic history of the Earth

26. Which event would most likely cause a new ice age in North America?
(1) a decrease in the energy produced by the Sun
(2) a decrease in the light reflected by the surface of the Earth
(3) an increase of carbon dioxide in the Earth's atmosphere
(4) an increase in the westward drift of the North American continent

27. Which statement presents the best evidence that a boulder-sized rock is an erratic?
(1) The boulder has a rounded shape.
(2) The boulder is larger than surrounding rocks
(3) The boulder differs in composition from the underlying bedrock.
(4) The boulder is located near potholes.

28. The direction of movement of a glacier is best indicated by the
(1) elevation of erratics
(2) alignment of grooves in bedrock
(3) size of kettle lakes
(4) amount of deposited sediments

29. The general direction of continental glacial advance in New York State was from
    (1) south to north   (3) west to east
    (2) north to south   (4) east to west

30. Another ice age would probably result in a change in
    (1) sea level
    (2) Moon phases
    (3) the speed of the Earth in its orbit
    (4) the time between high tides

Base your answers to questions 31 through 33 on the diagrams below. The diagrams represent glacial events in the geologic history of the Yosemite Valley region in California.

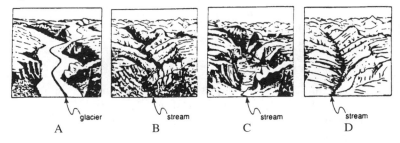

31. Which diagram most likely represents the most recent stage of landscape development ?
    (1) A   (3) C
    (2) B   (4) D

32. Which features currently found in this region are the result of glaciation?
    (1) intrusions of volcanic rock
    (2) deep U-shaped canyons with steep sides
    (3) fossils of marine organisms
    (4) thick soils covering all rock surfaces

33. Which natural resource of economic value would most likely be found in this region?
    (1) rock salt   (3) natural gas
    (2) petroleum   (4) sand and gravel

Base your answers to questions 34 through 38 on the diagrams below. Diagram I represents a section of the northeastern United States and Canada. Five different source regions, *A* through *E*, are shown along with the positions of glacial deposits containing boulders which originated from each source region. Diagram II represents the appearance of the surface of a typical boulder from any of the deposit locations.

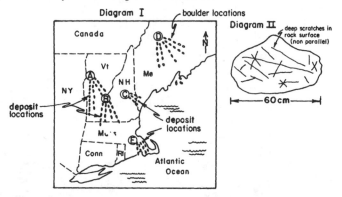

34. The force that caused the deposits to be distributed in the pattern shown in Diagram I most likely came from which general direction?
(1) northwest   (3) southwest
(2) northeast   (4) southeast

35. Which characteristic do all of the deposits most likely have in common?
(1) They have the same chemical composition.
(2) They were eroded from source region *A*.
(3) They are composed of unsorted sediments.
(4) They are found at the ends of large rivers.

36. According to the *Earth Science Reference Tables*, during which geological epoch were the deposits most likely transported to the locations shown in Diagram I?
(1) Mesozoic   (3) Jurassic
(2) Pleistocene   (4) Paleocene

37. The scratches in the boulder shown in Diagram II were most likely caused by the
(1) internal arrangement of the minerals in the boulder
(2) splitting of a large boulder into two smaller boulders
(3) erosion of the boulder by running water
(4) movement of the boulder over bedrock

38. About how many times larger is the actual boulder than the model shown in Diagram II?
(1) 5 times   (3) 30 times
(2) 14 times   (4) 60 times

39. The graph below shows the average temperature of the Earth during the past 250,000 years and the beginning and end of the most recent glacial and interglacial stages.

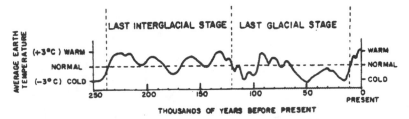

According to the graph, the duration of the last glacial stage was from
(1) 250,000 years ago to 240,000 years
(2) 120,000 years ago to 10,000 years ago
(3) 240,000 years ago to 120,000 years ago
(4) 10,000 years ago to the present

40. Base your answer to the following question on the accompanying photograph, which shows the surface of limestone bedrock exposed in New York State.

The scratches on the exposed bedrock were most likely caused by
(1) the internal arrangement of minerals in the bedrock
(2) sand blasting by wind-blown sediments
(3) erosion of the bedrock by running water
(4) movement of gravel frozen into ice over the bedrock

41. Which force is primarily responsible for the movement of a glacier?
(1) ground water   (3) gravity
(2) running water   (4) wind

42. Lateral moraines of a valley (alpine) glacier are principally composed of rock debris
(1) deposited by glacial meltwater
(2) freed from the enclosing ice by frost heaving
(3) derived from the weathering of the valley walls
(4) deposited by streams flowing along the ice margins

43. If the front of an active glacier is observed to be stationary, it is correct to infer that the ice in the glacier is
(1) not advancing at all
(2) melting as fast as it advances
(3) advancing faster than it melts
(4) melting faster than it advances

265

Base your answers to questions 44 through 48 on the map of New York State and surrounding areas shown below. The short lines indicate the location and direction of streamline features such as drumlins and glacial striations.

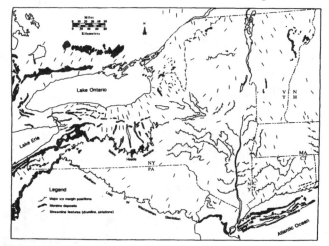

44. The glacial ice that caused the deposits in the region south of Lake Ontario and north of the Finger Lakes most likely came from which general direction?
(1) north   (3) southwest
(2) southeast   (4) west

45. According to the *Earth Science Reference Tables*, during which geologic time period were the glacial deposits in New York State most likely formed?
(1) Mesozoic   (3) Pleistocene
(2) Jurassic   (4) Paleocene

46. Which characteristic do all of the deposits most likely have in common?
(1) They have the same chemical composition.
(2) They are all composed of sorted sediments.
(3) They are all composed of unsorted sediments.
(4) They are found at the ends of large rivers.

47. In which two landscape regions of New York State do most of the moraine deposits seem to be located?
(1) Adirondack Mountains and the Catskill Mountains
(2) Appalachian Plateau and the Atlantic Coastal Plain
(3) Tug Hill Plateau and the Erie-Ontario Lowlands
(4) The Atlantic Coastal Plain and the St. Lawrence Lowlands

48. Of the several states shown on the map, which would show evidence of glacial deposits over the smallest percent of its surface area?
(1) Connecticut (CT)   (3) Pennsylvania (PA)
(2) Vermont (VT)   (4) Massachusetts (MA)

266

UNIT FIVE

Earth's History

KEY IDEAS   Until the nineteenth century, nearly everyone believed that the Earth was only a few thousand years old. Scientific evidence indicates, however, that some of the rocks on the Earth's surface are several billion years old. Observations of patterns in rock layers and the locations of various kinds of fossils allow inferences concerning the relative ages of rocks and the events that formed them. The absolute age of a rock can be determined from the relative amounts of a radioisotope and its decay products in the rock.

The rock record and fossil evidence reveal that life-forms and environments have changed over time.

KEY OBJECTIVES

Upon completion of this unit, you should be able to:

- Determine the relative ages of the rock layers in a series, and of any igneous intrusions, faults, folds, or fossils they may contain, based on the principle of superposition.
- Explain the significance of index fossils and volcanic ash deposits in correlating widely separated rock layers.
- Interpret the geologic time scale.
- Determine the absolute age of a rock, given the relative amounts of a radioisotope and its decay product in the rock, and the half-life of the radioisotope.
- Explain how a study of the fossil record shows that life-forms have evolved through geologic time.
- Describe how fossils provide evidence of past environments.

A. HOW CAN THE ORDER IN WHICH GEOLOGICAL EVENTS OCCURRED BE DETERMINED?

Much of what is known about the Earth's history has been learned by studying rocks. Rock layers contain traces of events that occurred in the past—clues to the origin of the rocks and the environment in which they formed. With careful logic, inferences can be drawn about the Earth's past from data obtained from rocks. For example, the fact that a rock layer contains sorted sand grains that are rounded implies that they were carried by running water. The size of the grains can indicate how fast the water that deposited them was moving. If the layer contains shells, the type of shell may show that the rock formed in a lake, not an ocean. The shells can also provide clues to the depth and temperature of the water. Microscopic pollen grains in the rock can serve to identify plants living around the lake when the rock formed. Still other rock layers may contain evidence of glaciers or deserts or volcanic eruptions.

Such inferences, however, are based on an assumption proposed by James Hutton in 1795—uniformitarianism. *Uniformitarianism* is the idea that the Earth's features, such as mountains, valleys, and rock layers, have formed gradually by processes still underway, not by instantaneous creation. It assumes that the Earth has essentially always behaved in the same the way it does now; that water, for example, ran downhill in streams carrying sand and silt 10 million years ago just as it does today. This idea can be summarized as "the present is the key to the past."

According to uniformitarianism, every rock layer contains a record of a short part of the Earth's long history. By working out the age of each layer in a series, paleontologists can determine the order in which the layers formed and can arrange the events they represent in the order in which they occurred. In this way a history of the Earth can be pieced together. There are two ways of expressing the age of Earth materials or geologic events—relative age and actual age.

Relative age is the age of a rock, fossil, geologic feature, or event relative to the age of another rock, fossil, geologic feature, or event, rather than in terms of years. Relative age does not express an exact age in years, but indicates only that one thing is older or younger than another. To say that you are older than a first grader but younger than your parents is to give your relative age.

Actual age is the age of a rock, fossil, geologic feature, or event given in units of time, usually years. A person who states, "I am 15 years old," is giving his or her actual age. The actual ages of earth materials and events are usually determined by analyzing the radioisotopes in objects.

B. HOW CAN THE RELATIVE AGE OF A ROCK BE DETERMINED?

Two key ideas in determining relative age are the law of original horizontality and the principle of superposition. *The law of original horizontality* states that sediments deposited in water form flat, level layers parallel to the Earth's surface. *The principle of superposition* is the idea that the bottom layer of a series of sedimentary layers is the oldest, unless it has been overturned or older rock has been thrust over it. Each layer forms atop the one that is already there. (Layers of sediment do not form in midair, nor do they hover halfway between the

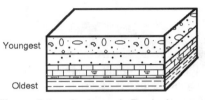

Figure 5.1 Series of Rock Layers, Showing Oldest and Youngest

ocean surface and the ocean bottom!) Therefore, each layer is younger than the one under it and older than the one on top of it (see Figure 5.1). The relative ages of any two layers can be determined based on which layer is on top and which is on the bottom.

B-1. Applying the Principles of Superposition

The principle of superposition must be used with care because certain events can disturb the positions of rock layers. Forces within the earth may tilt, fold, or fault rock layers. Older layers may be pushed on top of younger ones. In such cases, it is necessary to work out the original positions of the rock layers before applying the principle of superposition.

In general, a rock layer is older than any joint, fault, or fold that appears in it; the rock had to already exist in order to be folded or faulted. By unfolding or unfaulting the rock layers, one can determine their positions before they were disturbed. The principle of superposition can then be applied to determine relative ages.

For example, consider Figure 5.2. If we were to look only at present posi-

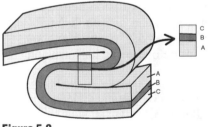

Figure 5.2

tion, we would conclude that layer *C* is the youngest and *A* is the oldest because *C* is on top and *A* is on the bottom. If we look at the entire rock structure, however, we see that three layers of rock have been folded into a syncline. Since the fold is a syncline, we can assume that the sides have been pushed upward. By straightening the limbs of the fold, we place the layers in their original positions and see that layer *A* is the *youngest* while *C* is the *oldest*.

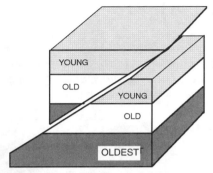

Figure 5.3 Fault, Showing Older Rocks Pushed Atop Younger Ones

The same is true of faults, that is, fractures along which the rocks on either side have moved. If rock layers fracture and then move along that fracture, they are displaced. Any rock displaced by a fault is older than the fault. During faulting, underlying rock layers may be pushed up so that they are found on top of younger rock (see Figure 5.3). Again, to obtain true relative ages, one must work backwards to determine the positions the rock layers were in before the fault offset them.

Igneous intrusions or extrusions are often found in association with other types of rock. Igneous intrusions form when molten rock forces its way into preexisting rock, cools, and hardens. Thus, an intrusion is younger than any rock it cuts through. Extrusions form during volcanic eruptions when molten rock flows out onto the Earth's surface as lava and hardens, or is blown into the atmosphere and settles on the ground, forming a blanket of volcanic rock particles. Thus, extrusions are younger than the rocks beneath them, but older than any layer that may form over them. Therefore, if an igneous body is found in rock, one must first determine whether it is an intrusion or an extrusion before relative ages can be established.

Figure 5.4 shows layers of rock before and after intrusion.

B-2. Recognizing Unconformities

Layers of rock are generally deposited in an unbroken sequence. However, if forces within the Earth uplift rocks, deposition ceases. Erosion may wear away many layers of rock before the land is low enough for another layer to be deposited.

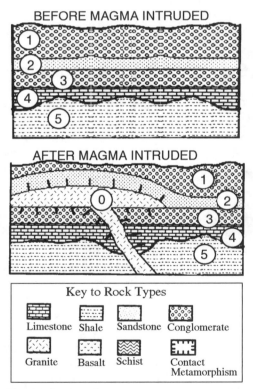

Figure 5.4 Five Rock Layers Before and After Intrusion of Magma. Rock (1), which formed from the cooling magma, is younger than the five rock layers into which it was intruded. (0 = youngest, 5 = oldest)

The result is an *unconformity*, a break, or gap, in the sequence of a series of rock layers. Thus, the rocks above an unconformity are quite a bit younger than those below it. There are three types of unconformity.

Angular unconformities form when rock layers are tilted or folded before being eroded. When new layers are deposited, they form horizontally and the layers below the unconformity are at an angle to those above it.

Disconformities are irregular erosional surfaces between parallel layers of rock. Disconformities occur when deposition stops and layers are eroded, but no tilting or folding occurs. These surfaces are not easy to discern that are often found when fossils of very different ages are discovered in adjacent layers.

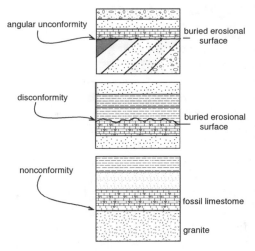

Figure 5.5 Angular Unconformity, Disconformity, and Noncomformity

Nonconformities are places where sedimentary layers lie on top of igneous or metamorphic rocks.

Figure 5.5 illustrates the three types of unconformities.

C. HOW CAN ROCKS AND GEOLOGIC EVENTS IN ONE PLACE BE MATCHED WITH THOSE IN OTHER LOCATIONS?

With care and logic the relative ages of rock layers in any particular location can be worked out. However, that information provides only a small part of the overall picture of the Earth's history, and unconformities may mean that even more of the picture is missing.

C-1. The Correlation Process

To broaden the scope of the picture, rocks in one place need to be correlated with rocks in other locations. The process of *correlation* involves determining that rock layers in different areas are the same age. In this way, rocks in one location may fill in gaps in the record for another location. Correlation also allows the relative ages of rocks in widely separated outcrops to be determined.

C-2. Methods of Correlation

Walking the Outcrop

Rock layers can sometimes be followed from one location to another by *walking the outcrop*, that is, by physically following the layers from one place to another (see Figure 5.6). In this way, rock layers in two different places, such as two adjacent mountains or ridges, can sometimes be correlated.

Unfortunately, rock layers are rarely continuously visible for any distance. Most rocks are hidden beneath *regolith*, the loose fragments of rock and soil that blanket the Earth's surface. Rock outcrops poke out here and there like islands. As a result, geologists are often faced with the problem of correlating rocks in widely separated outcrops.

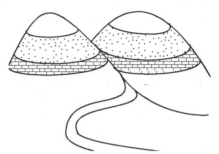

Figure 5.6 Walking the Outcrop

Matching Physical Characteristics

Rocks can sometimes be correlated on the basis of distinct similarities in physical characteristics such as composition, color, thickness, and fossil remains. Distinctive rocks or sequences of rock types can also be used to correlate rocks.

For example, Alfred Wegener used this type of correlation as part of his evidence for continental drift:

"…igneous rocks in Brazil and Africa [have] no less than five parallels: 1) the older granite, 2) the younger granite, 3) alkali-rich rocks, 4) volcanic Jurassic rock and intrusive dolerite, and 5) kimberlite, alnoite, etc. … The last of the rock groups (kimberlite, alnoite) is best known since both in Brazil and South Africa the beds yield the famous diamond finds. In both these regions the peculiar type of stratification known as 'pipes' occurs. There are white diamonds in Brazil in Minas Geraes State and in South Africa north of the Orange River only."

This type of correlation must be done with care, though, because different formations can look almost identical.

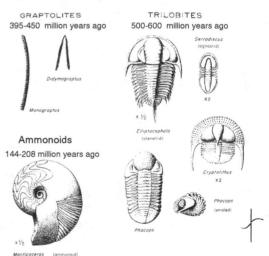

Figure 5.7 Graptolites, Trilobites, and Ammonites, Labeled with Ages

Studying Index Fossils

Index fossils are remains of organisms that had distinctive body features, were common and abundant, and had a broad, even worldwide range, yet existed only for a short period of time. The best index fossils include swimming or floating organisms that evolved rapidly and were distributed widely, such as graptolites, trilobites, and ammonites (see Figure 5.7). Their distinctive bodies and broad distribution make index fossils easy to find in widely separated rock layers (see Figure 5.8), and their short existence pinpoints the time period during which the rock layer was formed.

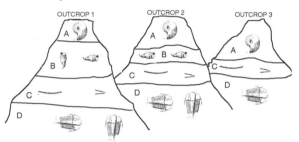

Figure 5.8 Correlation by Fossils of Strata in Several Locations

Examining Key Beds

Key beds are well-defined, easily identifiable layers or formations that have distinctive characteristics or fossil contents that allow them to be used in correlation. Key beds can be readily identified and are not easily confused with other layers. Materials like volcanic ash layers that are rapidly deposited over a wide area make excellent key beds.

One well-known key bed is the iridium-rich layer of rock discovered in Italy, Denmark, and many other places around the world. Iridium is an extremely rare element on Earth, but not in meteorites. The iridium-rich layer is thought to have formed when a meteorite impact 66 million years ago created a dust cloud that encircled the Earth and then slowly settled to the ground.

D. HOW CAN THE EARTH'S GEOLOGIC HISTORY BE SEQUENCED FROM THE FOSSIL AND ROCK RECORD?

D-1. The Geologic Column

Even though the rock record in any one place is incomplete, correlation has made it possible to piece together a fairly complete record by examining rocks in different locations. By the nineteenth century, geologists had correlated rocks worldwide into a single sequence called the *geologic column*. Rocks are still being added to the geologic column as more outcrops are mapped and described.

D-2. The Geologic Time Scale

Geologists have divided the Earth's history into a sequence of time units called the *geologic time scale*, based on the fossil record found in rocks of the geologic column. At first, these time units were based only on the relative ages of fossils in the rocks. Now, however, radioisotope dating has enabled geologists to assign actual ages to the time units of the geologic time scale, as shown in Figure 5.9.

Geologic time is the entire time that the Earth has existed. As stated above, the geologic time scale divides geologic time into units and subunits based on the fossil record. Geologic time is divided into eons, eons are subdivided into eras, eras are subdivided into periods, and periods are subdivided into epochs.

Eons

Eons are the largest time units on the geologic time scale. The oldest is the *Archean* (Gr., ancient); it covers the time

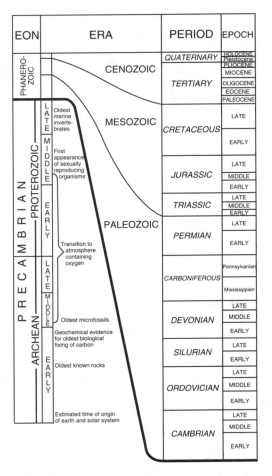

Figure 5.9 The Geologic Time Scale

from the Earth's formation to the appearance of multicelled organisms. Archean rocks are the oldest eon known on earth and contain microscopic fossils of single-celled bacterialike organisms. The *Proterozoic* (Gr., earlier life) is the next oldest eon, and its rocks contain traces of multicelled organisms that had no preservable hard parts. The most recent eon is the *Phanerozoic* (Gr., visible life), for which an abundant fossil record of preserved hard parts exists. Before the traces of soft-bodied organisms and microfossils were found in Archean and Proterozoic rocks, these two eons were lumped together under the general term *Precambrian*. The oldest known rocks that contain fossils of the hard parts of organisms are called Cambrian because they were found in Wales, which was once known as Cambria. Therefore, rocks that are older than Cambrian rocks are pre-Cambrian, or Precambrian.

Eras

Eons are subdivided into *eras,* based on the fossil record. Since there are so few fossils from the Archean and Proterozoic, these eons have not yet been formally divided into eras. During these eons, simple forms of life such as bacteria, microorganisms, and later algae and soft animals (e.g., worms and jellyfish) predominated. The Phanerozoic eon, however, is subdivided into the *Paleozoic* (old life), *Mesozoic* (middle life), and *Cenozoic* (recent life) eras, based on the types of life-forms that predominated during those time intervals. In general, life-forms developed in complexity over time. During the Paleozoic era, marine invertebrates dominated and the earliest land plants and animals developed. The Mesozoic was dominated by reptiles, such as the dinosaurs, and saw the first mammals develop. The Cenozoic has been dominated by mammals, including humans, and flowering plants have become dominant. Notice, though, that human existence has been very brief in comparison with the total expanse of geologic time.

Periods

Eras are subdivided into *periods*, which have a much less organized terminology. The names for the time periods are based on the names of rock formations in England, Germany, Russia, the United States, and many other countries. Some are named for locations, such as the Permian, for the province of Perm in Russia, and the Pennsylvanian and Mississippian, after those places in the United States. Others are named for characteristics of the rock layers where rocks of this age were first studied, such as the Cretaceous, from the Latin word for chalk.

Epochs

The smallest unit of time on the geologic time scale is the epoch. *Epochs* are subdivisions of periods, usually into early, middle, and late. The epochs of the Tertiary period, however, are designated by a jumble of Greek words describing degrees of recentness, for example, Holocene (wholly recent); Miocene (less recent); Eocene, (dawn of the recent).

In Figure 5.10, the Geologic History of New York State at a Glance, the geologic time scale is shown. Next to, and correlated with, the time scale are a series of columns that contain additional information.

The *Life on Earth* column describes the types of life that existed during the different time periods, based on the fossil record.

The *Record in New York* column shows the time periods for which there are rock and fossil records in New York State. A thick, black line indicates that rock or fossil evidence from that time period can be found. If there is no line, there is no rock or fossil record for that time. These gaps reflect unconformities in the rocks of New York State.

The *Important Fossils of New York* column gives the names and pictures of important fossils found in New York State. They range from stromatolites of the Proterozoic to mastodons of the Tertiary.

The *Tectonic Events Affecting Northeast North America* column uses thick, black lines to indicate the time periods during which major plate inter-

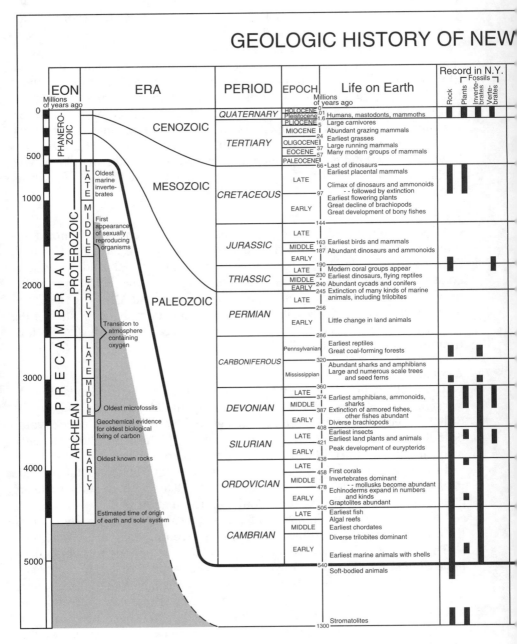

Figure 5.10 Geologic History of New York State at a Glance

YORK STATE AT A GLANCE

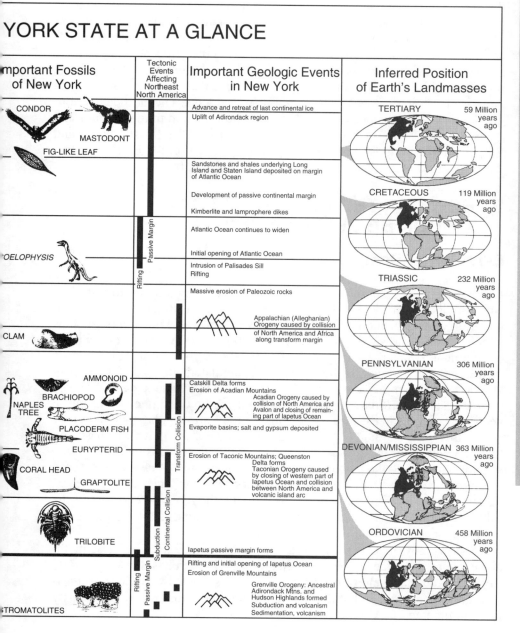

Important Fossils of New York	Tectonic Events Affecting Northeast North America	Important Geologic Events in New York	Inferred Position of Earth's Landmasses
CONDOR, MASTODONT, FIG-LIKE LEAF		Advance and retreat of last continental ice / Uplift of Adirondack region	TERTIARY — 59 Million years ago
	Passive Margin / Rifting	Sandstones and shales underlying Long Island and Staten Island deposited on margin of Atlantic Ocean / Development of passive continental margin / Kimberlite and lamprophere dikes	CRETACEOUS — 119 Million years ago
COELOPHYSIS		Atlantic Ocean continues to widen / Initial opening of Atlantic Ocean / Intrusion of Palisades Sill / Rifting	TRIASSIC — 232 Million years ago
CLAM		Massive erosion of Paleozoic rocks / Appalachian (Alleghanian) Orogeny caused by collision of North America and Africa along transform margin	PENNSYLVANIAN — 306 Million years ago
AMMONOID, BRACHIOPOD, NAPLES TREE, PLACODERM FISH, EURYPTERID, CORAL HEAD, GRAPTOLITE	Transform Collision	Catskill Delta forms / Erosion of Acadian Mountains / Acadian Orogeny caused by collision of North America and Avalon and closing of remaining part of Iapetus Ocean / Evaporite basins; salt and gypsum deposited / Erosion of Taconic Mountains; Queenston Delta forms / Taconian Orogeny caused by closing of western part of Iapetus Ocean and collision between North America and volcanic island arc	DEVONIAN/MISSISSIPPIAN — 363 Million years ago
TRILOBITE	Subduction / Continental Collision	Iapetus passive margin forms	ORDOVICIAN — 458 Million years ago
STROMATOLITES	Rifting / Passive Margin	Rifting and initial opening of Iapetus Ocean / Erosion of Grenville Mountains / Grenville Orogeny: Ancestral Adirondack Mtns. and Hudson Highlands formed / Subduction and volcanism / Sedimentation, volcanism	

actions occurred. The lines are labeled with the type of plate tectonic event that produced the mountain building.

The *Important Geologic Events in New York* column explains in more detail how the plate tectonic events shown in the preceding column affected New York State. It describes the origins of well-known features of New York State, such as the Palisades sill, which was intruded during the late Triassic, and the Adirondacks, which were uplifted during the Pliocene.

The *Inferred Position of Earth's Landmasses* column shows the changing positions of landmasses due to plate movements. On the maps, North America is shown in black to highlight its movements. The latitude and longitude lines show the positions of landmasses in relation to the Equator and poles, as well as to each other. Notice that, during the Mississippian and Pennsylvanian periods, North America was centered on the Equator, a fact that explains how coalbeds formed from tropical rain forests.

E. HOW CAN THE ACTUAL AGE OF A ROCK BE DETERMINED?

E-1. Early Attempts

Early attempts at determining the age of the Earth based on uniformitarianism were inaccurate at best. For example, careful measurements of the accumulation of sediments in a shallow sea over objects of known age, such as wrecked ships) show that it takes about 15,000–30,000 years to form a layer of sediment 1 meter thick. Layers of sedimentary rock exist that are more than 2,000 meters thick. If the sediment accumulated at a rate of 1 meter in 20,000 years, then the layers would have taken 2,000 meters × 20,000 years per meter, or *40 million years*, to form! If this method is applied to the entire geologic column, an estimate for the age of the Earth can be obtained.

Another early method used to estimate the age of the Earth was based on the salt contents of the oceans. Salt is carried into the oceans by rivers. First, scientists measured the amounts of salt in all the oceans. Then they measured the amounts of salt that all the world's rivers were carrying into the oceans. Finally they calculated how long it would take the rivers to carry in all the salt now in the oceans and, on the basis of these results, estimated the age of the Earth at several hundred million years.

We now know that such figures are much too low. Dating by these methods is not reliable because they depend too heavily on rates that vary widely from place to place and through time. What was needed was a way to measure time by a process that does not vary, a process that runs continuously through time and leaves a record with no gaps in it.

E-2. Radioactive Decay

The discovery of radioactivity by A. H. Becquerel in 1896 provided the needed process.

All elements are made up of tiny bits of matter called *atoms*. Also, most elements have a number of isotopes. *Isotopes* are varieties of the same element whose atoms differ slightly in mass. For example, most carbon atoms have a mass of 12 units, and this isotope is called carbon-12. Some carbon atoms, however, have a mass of 14 units, and this isotope is carbon-14.

The most common isotopes of elements are stable, meaning that their atoms do not change. However, some isotopes are unstable. In a process called *radioactive decay*, the atoms of unstable isotopes break apart. During this process the atoms go through a series of changes. They give off energy and some of the small particles that make up their nuclei. Finally, they form a stable isotope of a new element that is not radioactive. This new substance is called the *decay product*. For example, uranium-238 is a radioactive isotope. It decays slowly into the decay product lead-206.

Radioactive decay takes place at a steady, constant rate. It is not affected by outside factors such as changes in temperature, pressure, or chemical state.

The minerals in certain types of rocks contain radioactive isotopes. They start to decay when the rock forms. Therefore, the decay process may be used as a clock to determine the actual age of these rocks. The decay of a radioactive isotope occurs at a statistically predictable rate known as its half-life. The *half-life* is the time required for one-half of the unstable radioisotope to change into a stable decay product. Table 5.1 gives the half-lives of commonly used radioisotopes.

TABLE 5.1 RADIOACTIVE DECAY DATA

Radioactive Isotope	Disintegration	Half-life (years)
Carbon-14	$C^{14} \rightarrow N^{14}$	5.7×10^3
Potassium-40	$K^{40} \rightarrow Ar^{40}$ $\rightarrow Ca^{40}$	1.3×10^9
Uranium-238	$U^{238} \rightarrow Pb^{206}$	4.5×10^9
Rubidium-87	$Rb^{87} \rightarrow Sr^{87}$	4.9×10^{10}

Example

A rock contains 50 grams of potassium-40 and 50 grams of its decay product, argon-40. How old is the rock?

Since all of the decay product was originally radioactive isotope, the rock originally contained 100 grams of potassium-40 (50 grams of the radioisotope that still exist plus the 50 grams that decayed into argon-40). Thus, exactly one-half of the original radioisotope has decayed; the rock is one half-life old. From Table 5.1 we find that the half-life of potassium-40 is 1.3×10^9, or 1.3 billion years. Therefore, the rock is 1.3 billion years old.

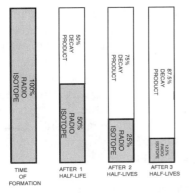

TIME OF FORMATION   AFTER 1 HALF-LIFE   AFTER 2 HALF-LIVES   AFTER 3 HALF-LIVES

Figure 5.11 Radioisotope/Decay Product Ratios at Time of Formation and After One, Two, and Three Half-Lives

If the half-life of a radioisotope is known, the age of a rock can be determined from the relative amounts of that radioisotope and its decay product in the rock. Every half-life, the percentage of decay product will increase and the percentage of radioisotope will decrease. Thus, the ratio of radioisotope to decay product tells how many half-lives have gone by. This, in turn, indicates how many years the decay process has been going on. In this manner, geologists are able to find the absolute ages of rocks. Figure 5.11 shows how the relative amounts of radioisotope and its decay product change as each half-life elapses.

There are two reasons why a certain isotope may be used to date a rock: because it is present in the minerals of the rock you wish to date or because its half-life is especially long or short. Radioisotopes with long half-lives are used to date very old rocks; those with short half-lives, to date something young. For example, potassium-40 can be used to date rocks between 100,000 and 4.6 billion years old. On the other hand, carbon-14 is best for dating objects between 100 and 50,000 years old.

Carbon-14 is especially useful because it can be used to date the remains of living things. All living things contain carbon, and some of that carbon is carbon-14. Carbon-14 is present also in air, water, and food. As long as an organism is alive, the amount of carbon-14 in its body remains constant; whatever decays is quickly replaced from the surroundings. When the organism dies, however, the carbon-14 that decays is not replaced. The longer the organism has been dead, the less carbon-14 remains.

Carbon-14 can be also used to date fossils such as wood, bones, and shells. However, carbon-14 has a short half-life. After about 50,000 years the amount of carbon-14 left in a fossil is too small to be measured. Therefore, other isotopes, such as potassium-40, must be used to date older rocks.

Table 5.2 lists six radioisotopes and gives helpful information about them.

Example

Suppose a wood branch is found buried beneath layers of sediment in a lake. The wood is found to contain one-fourth the amount of carbon-14 present in contemporary wood. How old is the layer of sediment in which the wood is found?

After one half-life, a sample would contain one-half the normal amount of carbon-14. After a second half-life, half of that would decay, leaving one-quarter the normal amount of carbon-14. Therefore, two half-lives have elapsed since the tree died. From the Table 5.1 we know that the half-life of carbon-14 is 5.7×10^3, or 5,700 years. Since two half-lives have elapsed, the wood is $5,700 \times 2$, or 11,400 years old.

TABLE 5.2 SIX RADIOISOTOPES USED IN DATING*

Isotopes				
Parent	Decay Product	Half-life (years)	Dating Range (years)	Minerals (or other materials that can be dated)
Uranium-238	Lead-206	4.5 billion	10 million–4.6 billion	Zircon,
Uranium-235	Lead-207	710 million	10 million–4.6 billion	uraninite
Uranium-232	Lead-208	14 billion	10 million–4.6 billion	
Potassium-40	Argon-40 Calcium-40	1.3 billion	50,000–4.6 billion	Muscovite, biotite, hornblend
Rubidium-87	Strontium-87	47 billion	10 million–4.6 billion	Muscovite, biotite, potassium feldspar
Carbon-14	Nitrogen-14	5,730 ± 30	100–70,000	Wood, peat, grain, charcoal, bone, tissue, cloth, shell, stalacites, glacier ice, ocean water

*From *The Dynamic Earth*, by Brain J. Skinner and Stephen C. Porter, John Wiley, 1992.

F. HOW CAN ROCK RECORDS AND FOSSILS REVEAL CHANGES IN PAST LIFE AND PROVIDE EVIDENCE OF ANCIENT ENVIRONMENTS?

Many thousands of layers of sedimentary rock contain evidence of the long history of the Earth and the changing life-forms, whose fossils are found in the rocks. Fossils show that a great variety of plants and animals lived on Earth in the past. Fossils can be compared to one another and to living organisms in terms of their similarities and differences. Some fossil organisms are similar to existing organisms, but many are quite different. A study of the fossil record reveals that more recently deposited rock layers are more likely to contain fossils resembling existing life-forms and that most life-forms of the distant geologic past have become extinct.

F-1. Changes in the Past Life

Evolution

The history of life on Earth depicted by the fossil record is one of change. The record shows that life-forms have changed gradually over time, or evolved, so that current life-forms differ significantly from the earliest. Biological evolution is the idea that existing life-forms have developed from earlier, different life-forms.

Natural Selection

The theory of natural selection provides a scientific explanation for the evolution of life-forms seen in the fossil record. Natural selection proposes the following mechanism for evolution:

1. Individual organisms of the same species have different characteristics, and sometimes these differences give one organism an advantage in surviving and reproducing.
2. Offspring that inherit the advantage are more likely to survive and reproduce.
3. Over time the proportion of individuals with the advantageous characteristic will increase.

In this way, natural selection leads to organisms that are well suited for their environments. Changes in the environment can affect the survival value of some inherited characteristics. The characteristics advantageous for survival may change, so changes in the environment lead to changes in organisms. Small differences between parents and offspring can accumulate over many generations so that descendants may become very different from their ancestors.

Evolution by natural selection does not imply long-term progress toward a goal or have a set direction. Nor does evolution always occur gradually. Much recent information indicates that evolution occurs in spurts brought on by sudden, large-scale changes in the environment. A good analogy for evolution is the growth of a hedge. Some branches exist from the beginning with little or no change. Other branches grow and then die out. Still others grow a little, then branch apart repeatedly. The end result is a complex network of large and small branches.

From the fossil record, life on Earth is thought to have begun as simple, one-celled organisms about 4 billion years ago. During the first 2 billion years, only these one-celled organisms existed. Then, about 1 billion years ago, cells with nuclei began to develop. Since then, increasingly complex multicellular organisms have evolved.

One important reason to preserve life on Earth is that evolution builds on what exists. The more variety there is, the more variety can exist in the future. Human behavior that results in the extinction of organisms decreases the variety of life-forms on Earth and may have far-reaching effects in the future.

F-2. Evidence of Ancient Environments

Rocks in which fossils are found provide evidence of ancient environments. Many fossils are clear indicators of particular types of environment. For example, fossil coral that is almost identical to existing corals can be found in some rocks. Corals that exist today can live only in shallow, tropical seas. Therefore, we can infer that the rocks containing fossil coral existed in a similar environment. Oak and birch trees inhabit moist, temperate climates. The

presence of abundant oak and birch pollen in a sedimentary layer implies that, when the rock layer formed, the climate was moist and temperate. The fossils in coal represent tropical rain forest plants and animals, and layers of coal imply a tropical rain forest environment. Inferences of this type, together with the inferences that can be drawn from the composition of a sedimentary layer, enable geologists to piece together a model of the environment that existed when any particular rock layer formed. Figure 5.12 shows rock layers in a series together with pictures of the environment at the time of their formation inferred from fossil evidence contained in the rock.

Figure 5.12 Fossil-bearing Rock Layers in a Series. With illustrations of environments inferred from each layer's fossil evidence.

REVIEW QUESTIONS FOR UNIT 5

1. Unless a series of sedimentary rock layers has been overturned, the bottom rock layer usually
   (1) contains fossils
   (2) is the oldest
   (3) contains the greatest variety of minerals
   (4) has the finest texture

2. If the vertical cross section, shown on the accompanying diagram, represents sedimentary rock layers which have *not* been overturned, then the oldest rock layer is most probably indicated by which letter?
   (1) *A*   (3) *C*
   (2) *B*   (4) *D*

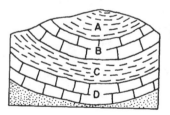

3. Tilted layers of sandstone found in the Rocky Mountains indicate that
   (1) sandstone was deposited at the tops of the mountains
   (2) the oceans were once higher than they are now
   (3) sandstone layers were formed and then displaced
   (4) the oceans rose, deposited sandstone, and receded

4. Older layers of rock may be found on top of younger layers of rock as a result of
   (1) weathering processes   (3) joints in the rock layers
   (2) igneous extrusions   (4) overturning of rock layers

5. Which feature in a rock layer is older than the rock layer?
   (1) igneous intrusions   (3) rock fragments
   (2) mineral veins   (4) faults

6. The diagram below represents layers of rock.

Rock layer *A* is inferred to be older than intrusion *B* because
   (1) layer *A* is composed of sedimentary rocks
   (2) parts of layer *A* were altered by intrusion *B*
   (3) layer *B* is located between layer *A* and layer *C*
   (4) parts of layer *C* were altered by intrusion *B*

Base your answer to question 7 on the diagram below, which shows a geologic cross section and landscape profile of a section of the Earth's crust.

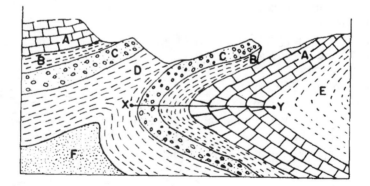

7. Which graph best represents the age of the rocks along line *XY*?

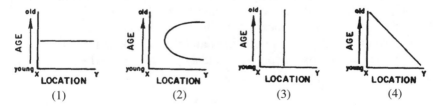

8. The adjacent diagram shows a sample of conglomerate rock.

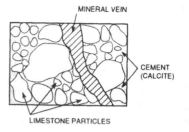

The oldest part of this sample is the
(1) conglomerate rock sample
(2) calcite cement
(3) limestone particles
(4) mineral vein

9. Erosion has created gaps in the geologic record preserved in rocks. According to the *Earth Science Reference Tables*, in New York State there are no rocks of the which ages?
(1) Permian and Tertiary
(2) Ordovician and Cretaceous
(3) Ordovician and Cambrian
(4) Triassic and Jurassic

10. What process most directly caused the formation of the feature shown by line AB in the geologic cross section shown here?

(1) erosion   (3) igneous intrusion
(2) faulting   (4) folding

11. The best method for the correlation of sedimentary rock layers several hundred kilometers apart is comparing the
(1) index fossils in the layers
(2) layers by walking the outcrop
(3) thickness of the rock layers
(4) color of the rock layers

12. Why can layers of volcanic ash found between other rock layers often serve as good geologic time markers?
(1) Volcanic ash usually occurs in narrow bands around volcanoes.
(2) Volcanic ash usually contains index fossils.
(3) Volcanic ash usually contains the radioactive isotope carbon-14.
(4) Volcanic ash usually is rapidly deposited over a large area.

13. What characteristics of fossils are most useful in correlating sedimentary rock layers?
(1) limited geographic distribution, but found in many rock formations
(2) limited geographic distribution, and limited to a particular rock formation
(3) wide geographic distribution, but limited to a particular rock formation
(4) wide geographic distribution, and found in many rock formations

285

14. The geologic columns *A*, *B*, and *C* in the accompanying diagrams represent widely spaced outcrops of sedimentary rocks. Symbols are used to indicate fossils found within each rock layer. Each rock layer represents the fossil record of a different geologic time period.

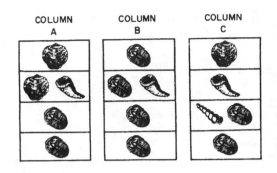

According to the diagrams for all three columns, which would be the best index fossil?

15. The adjacent diagram represents two geologic rock columns. The color and environment of deposition of each sedimentary rock are indicated beside the rock layers. Which rock layer in the West geologic column is most likely the same as rock layer *X* in the East column?
(1) *A*
(2) *B*
(3) *C*
(4) *D*

16. Which is the best method of determining the relative ages of a layer of sandstone in western New York State and a layer of sandstone in eastern New York State?
(1) Compare the thickness of the two layers.
(2) Compare the colors of the two layers.
(3) Compare the sizes of sand particles of the two layers.
(4) Compare the index fossils in the two layers.

17. Unconformities (buried erosional surfaces) are good evidence that
(1) many life-forms have become extinct
(2) the earliest life-forms lived in the sea
(3) part of the geologic rock record is missing
(4) metamorphic rocks have formed from sedimentary rocks

Base your answers to questions 18 and 19 on the cross-sectional diagram of a folded region shown below.

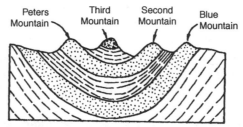

Peters Mountain   Third Mountain   Second Mountain   Blue Mountain

18. Where would a person normally expect to find the same type of fossil remains as are present at the top of Peters Mountain?
    (1) in the sandstone on Blue Mountain
    (2) in the sandstone on Second Mountain
    (3) in the conglomerate on Third Mountain
    (4) in the shale between Second Mountain and Blue Mountain

19. The present surface features known as Peters, Blue, Third, and Second mountains were mainly the result of
    (1) intrusion of igneous material
    (2) faulting
    (3) different erosion rates
    (4) metamorphism

20. Geologic time is divided into units based upon
    (1) erosion rates   (3) surface topography
    (2) rock types   (4) fossil evidence

21. Using the information in the *Earth Science Reference Tables*, students plan to construct a geologic time line of the Earth's history from its origin to the present time. They will use a scale of 1 meter equals 1 billion years. What should be the total length of the students' time line?
    (1) 10.0 m   (3) 3.8 m
    (2) 2.5 m   (4) 4.5 m

22. A timeline is made on a strip of paper to illustrate the Earth's history. A length of 1 centimeter is used to represent 10 million years. According to the *Earth Science Reference Tables*, what distance should be used to best represent the length of the Mesozoic Era?
    (1) 0.18 cm   (3) 18 cm
    (2) 1.8 cm   (4) 180 cm

23. According to the *Earth Science Reference Tables*, which era represents the shortest amount of geologic time?
    (1) Precambrian   (3) Mesozoic
    (2) Paleozoic   (4) Cenozoic

287

24. For which segment of the Earth's geologic history are fossils rarely found?
    (1) Cenozoic   (3) Paleozoic
    (2) Mesozoic   (4) Precambrian

25. For which geologic period are no fossils found in New York State?
    (1) Ordovician   (3) Devonian
    (2) Silurian   (4) Permian

26. According to the *Earth Science Reference Tables,* which area of New York State has the youngest bedrock?
    (1) the area south of the Finger Lakes
    (2) the area around Mt. Marcy
    (3) the area between Syracuse and Rochester
    (4) the area east of Albany

27. According to the *Earth Science Reference Tables,* most of the surface bedrock found in New York State was formed during which era?
    (1) Precambrian   (3) Mesozoic
    (2) Paleozoic   (4) Cenozoic

28. Which statement best explains why dinosaur fossils have *not* been found in the bedrock in the area around Syracuse, New York? [Refer to the *Earth Science Reference Tables.*]
    (1) Fossils are found only in sedimentary rock.
    (2) In the Syracuse area, dinosaur bones are located deep below the Earth's surface.
    (3) The dinosaurs were mobile and left no remains in the Syracuse area.
    (4) No rock record exists in the Syracuse area for the time period when the dinosaurs lived.

29. According to the *Earth Science Reference Tables,* which rock is most likely the oldest?
    (1) conglomerate containing the tusk of a mastodon
    (2) shale containing trilobite fossils
    (3) sandstone containing fossils of flowering plants
    (4) siltstone containing dinosaur footprints

30. According to the *Earth Science Reference Tables,* a straight line connecting which two cities would cross surface bedrock of only two geologic periods?
    (1) Watertown and Plattsburgh
    (2) Jamestown and Old Forge
    (3) Kingston and New York City
    (4) Binghamton and Syracuse

31. The geologic cross section shown here represents the fossil remains present in several rock layers in the Earth's crust. [If a fossil symbol is not shown in a rock layer, the plant or animal did not exist when the rock layer was formed.]

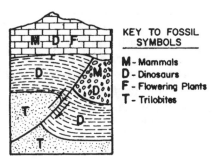

KEY TO FOSSIL
SYMBOLS

M - Mammals
D - Dinosaurs
F - Flowering Plants
T - Trilobites

Based on this diagram and the information in the *Earth Science Reference Tables*, during which geological time did the faulting shown in the diagram take place?
(1) Early Permian Period
(2) Late Cambrian Period
(3) Late Jurassic Period
(4) Early Tertiary Period (Paleocene Epoch)

32. Why are radioactive substances useful for measuring geologic time?
(1) The disintegration of radioactive substances occurs at a predictable rate.
(2) The ratio of decay products to undecayed products remains constant in sedimentary rocks.
(3) The half-lives of most radioactive substances are shorter than five minutes.
(4) Measurable samples of radioactive substances are easily collected from most rock specimens.

Note that question 33 has only three choices.

33. When the quantity of a radioactive material decreases, the half-life of that substance will
(1) decrease
(2) increase
(3) remain the same

34. According to the *Earth Science Reference Tables*, which radioactive element formed at the time of the Earth's origin has just reached about one half-life?
(1) carbon-14   (3) uranium-238
(2) potassium-40   (4) rubidium-87

35. Why is carbon-14 *not* usually used to accurately date objects more than 50,000 years old?
(1) Carbon-14 has a relatively short half-life and too little carbon-14 is left after 50,000 years.
(2) Carbon-14 has a relatively long half-life and not enough carbon-14 has decayed after 50,000 years.
(3) Carbon-14 has been introduced as an impurity in most materials older than 50,000 years.
(4) Carbon-14 has only existed on Earth during the last 50,000 years.

36. An ancient bone was analyzed and found to contain carbon-14 that had decayed for nearly two half-lives. According to the *Earth Science Reference Tables*, approximately how old is the bone?
(1) 1,400 years   (3) 5,600 years
(2) 2,800 years   (4) 11,000 years

37. A rock sample contained 8 grams of potassium-40 (K^{40}) when it was formed, but now contains only 4 grams due to radioactive decay. Based on the *Earth Science Reference Tables*, what is the approximate age of this rock?
(1) 0.7×10^9 years   (3) 2.8×10^9 years
(2) 1.3×10^9 years   (4) 5.6×10^9 years

38. A rock sample containing the radioactive isotope potassium-40 is calculated to be 4.2×10^9 years old. Based on the information in the *Earth Science Reference Tables*, how much of the original potassium-40 would be left in this rock sample?
(1) 0   (3) 1/4
(2) 1/8   (4) 1/2

39. Which radioactive substance shown on the graph below has the longest half-life?

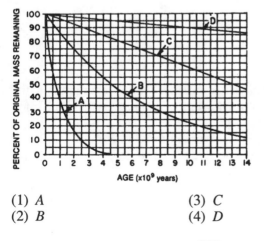

(1) *A*   (3) *C*
(2) *B*   (4) *D*

40. Which fact provides the best evidence for the scientific theory of the evolutionary development of life on the Earth?
  (1) Fossils are found almost exclusively in sedimentary rocks.
  (2) Characteristics of simpler forms of life can be found in more complex forms of life.
  (3) Only a small percentage of living things have been preserved as fossils.
  (4) Most species of life on the Earth have become extinct.

41. Which conclusion can be made based on existing fossil evidence?
  (1) Present life-forms have always existed.
  (2) The Earth's environment has always been the same.
  (3) Many life-forms have become extinct.
  (4) All life-forms will remain the same in the future

42. Fossils of two trilobite species from different geologic periods are illustrated here.

MIDDLE CAMBRIAN TRILOBITE

A comparison of these fossils provides evidence that these species may have
  (1) undergone evolutionary development
  (2) undergone metamorphism
  (3) experienced identical lifespans
  (4) experienced weathering and erosion

SILURIAN TRILOBITE

43. According to the *Earth Science Reference Tables*, studies of the rock record suggest that
  (1) the period during which humans have existed is very brief compared to geologic time
  (2) evidence of the existence of humans is present over much of the geologic past
  (3) humans first appeared at the time of the intrusion of the Palisades sill
  (4) the earliest humans lived at the same time as the dinosaurs

44. The climate that existed in an area during the early Paleozoic Era can best be determined by studying
  (1) the present climate of the area
  (2) recorded climate data of the area since 1700
  (3) present landscape surface features found in the area
  (4) the sedimentary rocks deposited in the area during the Cambrian and Ordovician periods

45. Trilobite fossils found in shale bedrock in the Albany, New York area indicate that this area once
(1) was covered by an ocean
(3) had iron ore deposits
(2) was covered by a large forest
(4) had many land animals

46. According to the *Earth Science Reference Tables*, at which location is the surface bedrock likely to contain fossils of early fishes, coelenterates, and mollusks?
(1) Elmira
(3) Watertown
(2) Long Island
(4) Old Forge

47. In the Earth's geologic past there were long warm periods which were much warmer than the present climate. What is the primary evidence that these long warm periods existed?
(1) United States National Weather Service records
(2) polar magnetic directions preserved in the rock record
(3) radioactive decay rates
(4) plant and animal fossils

48. Earth scientists studied fossils of a certain type of plant. They noted slight differences in the plant throughout geologic time. What inference is best made from this evidence?
(1) When the environment changed, this type of plant also changed, allowing it to survive.
(2) When uplifting occurred, the fossils of this type of plant were deformed.
(3) The processes which form fossils today differ from those of the past.
(4) The fossils have changed as a result of weathering and erosion.

49. Which best explains why fossil remains of sharks have been found in the sedimentary rocks of Wyoming?
(1) Sharks used to live in fresh water basins.
(2) Sharks were once land animals.
(3) The teeth were carried to Wyoming by streams from another area.
(4) The area was once covered by the sea.

50. The diagrams below show geologic cross sections of the same part of the Earth's crust at different times in the geologic past.

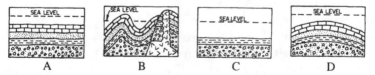

Which sequence shows the order in which this part of the crust probably formed?
(1) *A - B - C - D*
(3) *C - A - D - B*
(2) *C - D - A - B*
(4) *A - C - B - D*

Base your answers to questions 51 through 55 on your knowledge of earth science, the *Earth Science Reference Tables*, and the diagram below, which shows matching geologic columns from three different locations, *A*, *B*, and *C*. The locations are about 5 kilometers apart and the layers have not been overturned.

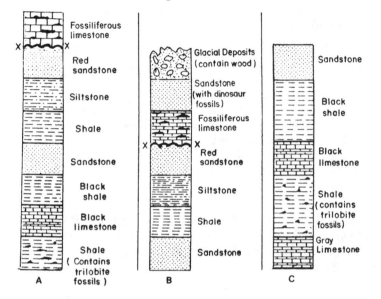

51. Which is the oldest layer shown?
(1) gray limestone
(2) sandstone
(3) glacial till containing wood
(4) shale containing trilobite fossils

52. The shale which contains the trilobite fossils was most likely deposited during which geologic period?
(1) Cretaceous
(2) Tertiary
(3) Triassic
(4) Ordovician

53. The formation of the fossiliferous limestone in column *B* was probably due to
(1) heat and pressure which has metamorphosed the limestone and fossils
(2) deposition of a variety of glacial sediments
(3) compaction and cementation of skeletons and shells of sea organisms
(4) cooling and solidification of molten material containing fossils

54. The feature at *X* is a buried erosional surface. Based on this, what inference can best be supported?
  (1) Faulting has occurred along the boundary between the red sandstone and the fossiliferous limestone.
  (2) No rock layers were ever formed between the red sandstone and the fossiliferous limestone.
  (3) An igneous intrusion has destroyed part of the fossiliferous limestone layer.
  (4) The red sandstone and the fossiliferous limestone do not provide a continuous geologic record.

55. Radioactive carbon-14 would be most useful in determining the age of the
  (1) trilobite fossils in the shale
  (2) wood in the glacial till
  (3) calcite in the black limestone
  (4) iron oxide in the red sandstone

Base your answers to questions 56 through 60 on your knowledge of earth science, the *Earth Science Reference Tables*, and the diagram below. The diagram represents a cross section of the Earth's crust showing several rock layers containing marine fossils. Overturning has not occurred. [Diagram is not to scale.]

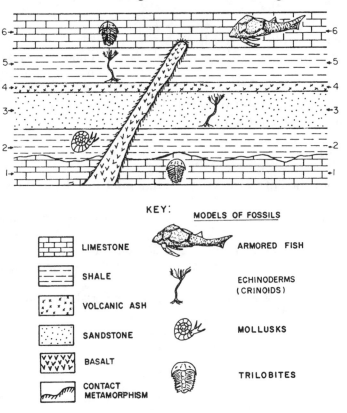

56. Why is layer 4 likely to be a good time marker?
    (1) Volcanic ash is usually a unique gray color.
    (2) Volcanic ash is usually rapidly deposited over a large area.
    (3) Volcanic ash can usually be dated with carbon-14.
    (4) Volcanic ash usually contains index fossils.

57. What could be an approximate age of rock layer 1?
    (1) 110 million years old   (3) 510 million years old
    (2) 210 million years old   (4) 810 million years old

58. Which best describes the order of events for the formation of this section of the Earth's crust?
    (1) intrusion of basalt; deposition of rock layers 1,2,3,4,5, and 6
    (2) deposition of rock layers 1,2,3; intrusion of basalt; deposition of rock layers 4,5, and 6
    (3) deposition of rock layers 1,2,3,4, and 5; intrusion of basalt; deposition of rock layer 6
    (4) deposition of rock layers 1,2,3,4,5, and 6; intrusion of basalt

59. Which is the best explanation for the irregular surface between layers 1 and 2?
    (1) Layer 1 was folded after 2 was deposited.
    (2) Volcanic actions pushed layer 1 up before 2 was deposited.
    (3) Pressure from the layers above pushed layer 2 into layer 1.
    (4) Layer 1 was partially eroded before 2 was deposited.

60. Which rock was formed by the compaction and cementation of particles 0.07 centimeter in diameter?
    (1) limestone   (3) shale
    (2) sandstone   (4) basalt

Base your answers to questions 61 through 65 on your knowledge of earth science, the *Earth Science Reference Tables*, and the graphs below. Graph I shows the population distribution of the fossils from four species of brachiopods (*A, B, C,* and *D*) found in the rocks of New York State. Graph II shows the range of the average shell thickness of adult individuals of each species.

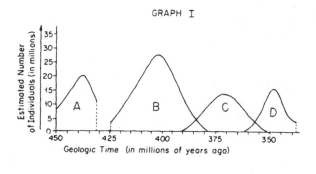

GRAPH I

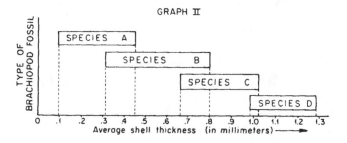

GRAPH II

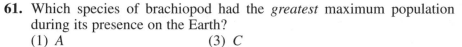

61. Which species of brachiopod had the *greatest* maximum population during its presence on the Earth?
(1) *A*   (3) *C*
(2) *B*   (4) *D*

62. Which species of brachiopod had the *least* variation in shell thickness?
(1) *A*   (3) *C*
(2) *B*   (4) *D*

63. Which would best explain the incompleteness of the fossil records of species *A, B,* and *D,* as seen in Graph I?
(1) the existence of a gap in the preserved rock record for these time periods in New York State
(2) the flooding of large portions of New York State during these time periods
(3) volcanic activity resulting in the intrusion of the Palisades sill
(4) the sudden extinction of all three brachiopod species

64. According to the Generalized Geologic Map of New York State, which map symbols represent rocks in which brachiopod species *B* and *C* are found?

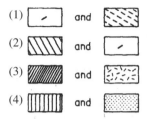

65. An undisturbed sequence of sedimentary rocks contains each of these four types of brachiopod fossils. Which species would be found farthest below the surface?
(1) *A*   (3) *C*
(2) *B*   (4) *D*

Base your answers to questions 66 through 70 on your knowledge of earth science and on the adjacent graph, which shows the relationship between mass and time for a radioactive element during radioactive decay.

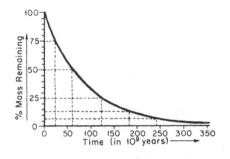

66. What is the half-life of the element?
(1) 20×10^9 years   (3) 125×10^9 years
(2) 60×10^9 years   (4) 180×10^9 years

67. When 35 percent of the original radioactive material is left, the sample will be approximately how many years old?
(1) 5×10^9 years   (3) 100×10^9 years
(2) 60×10^9 years   (4) 120×10^9 years

68. According to the graph, which statement best describes the decay of the radioactive element?
(1) The actual mass of radioactive material that decays in each half-life decreases with time.
(2) The rate of decay is greatest between 150×10^9 years and 250×10^9 years.
(3) There is the greatest loss in mass between 250×10^9 years and 300×10^9 years.
(4) The same number of atoms of radioactive material decay during each half-life.

297

69. Based on the trend indicated in the graph, the radioactive element will be completely decayed
(1) in 60×10^9 years
(2) in 250×10^9 years
(3) in 350×10^9 years
(4) at a time greater than 350×10^9 years

Note that question 70 has only three choices.

70. If the rock containing this radioactive material is buried deeper in the Earth's crust and subjected to an increase in temperature and pressure, the rate of radioactive decay will probably
(1) decrease
(2) increase
(3) remain the same

Base your answers to questions 71 through 75 on your knowledge of earth science and on the block diagram below showing rock layers A, B, C, D, and E with holes drilled at I, II, III, and IV through these rock layers.

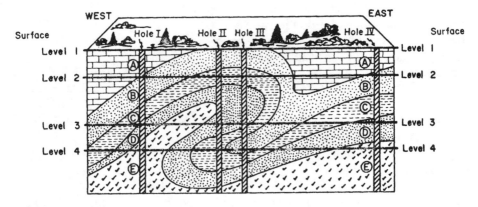

71. Which sequence of rock outcrops would be found on the surface by an observer who walked a straight line from west to east ?
(1) A, B, A (3) B, A, B
(2) A, D, A (4) D, B, A

72. At which level line would all four drill holes show the same rock layer?
(1) 1   (3) 3
(2) 2   (4) 4

73. The best evidence for overturned rock layers could be found by examining and correlating rock samples taken from
(1) Hole I, only   (3) Holes II and III
(2) Holes I and IV   (4) Hole IV, only

74. If rock layer D was formed during the Devonian Period, during which period could rock layer B have been formed? [Use the *Earth Science Reference Tables.*]
(1) Cambrian   (3) Silurian
(2) Ordovician   (4) Devonian

75. If rock layer B is *more* resistant than rock layer A, as the surface weathers and erodes away, the probable surface elevation of Hole III, as compared to Hole 1 and Hole IV, will be
(1) lower than Hole I, but higher than Hole IV
(2) higher than Hole I, but lower than Hole IV
(3) lower than both Hole I and Hole IV
(4) higher than both Hole I and Hole IV

Base your answers to questions 76 through 80 on your knowledge of earth science, the *Earth Science Reference Tables*, and the diagrams below. The diagrams show cross sections of the Earth's crust at four widely scattered locations, *A* through *D*. Numbers 1 through 10 represent fossils located in the rock layers. (The numbers do not represent the relative ages of the fossils.) The rock layers have not been overturned.

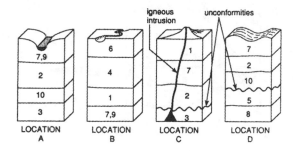

76. What is the most likely cause of the unconformities at locations C and D?
(1) volcanic activity   (3) faulting
(2) human activity   (4) uplift and erosion

77. Which location most likely contains the youngest fossil?
(1) A (3) C
(2) B (4) D

78. What is the relative age of the igneous intrusion at location C?
(1) younger than the layer containing fossil 10
(2) older than the layer containing fossil 7
(3) the same age as the layer containing fossil 1
(4) the same age as the layer containing fossil 9

79. Index fossils such as 7 are useful for correlating rocks because the fossils
    (1) are found only in sedimentary rocks
    (2) contain radioactive carbon-14, which is used for relative dating
    (3) represent organisms that lived for a relatively short period of geologic time in widespread areas
    (4) represent organisms that lived close to the Earth's surface for a relatively long period of time

80. Fossil 8 represents the earliest fish. How many millions of years ago was the rock layer containing this fossil probably formed?
    (1) 560   (3) 300
    (2) 450   (4) 275

Base your answers to questions 81 through 85 on the *Earth Science Reference Tables*, the graph below, and your knowledge of earth science. The graph shows the development, growth in population, and extinction of the six major groups of trilobites, labeled *A* through *F*.

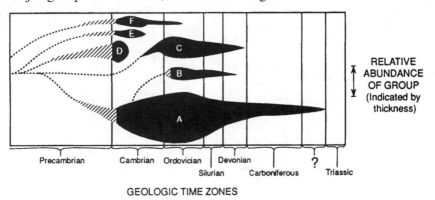

81. The fossil evidence that forms the basis for this graph was most likely found in
    (1) lava flows of ancient volcanoes
    (2) sedimentary rock that formed from ocean sediment
    (3) granite rock that formed from former sedimentary rocks
    (4) metamorphic rock that formed from volcanic rocks

82. Which group of trilobites became the most abundant?
    (1) *A*   (3) *C*
    (2) *B*   (4) *D*

83. During which period did the last of these trilobite groups become extinct?
    (1) Cretaceous   (3) Permian
    (2) Triassic   (4) Carboniferous

84. Which inference is best supported by the graph?
  (1) All trilobites evolved from group *A* trilobites.
  (2) The trilobite groups became most abundant during the Devonian Period.
  (3) Precambrian trilobite fossils are very rare.
  (4) Trilobites could exist in present-day marine climates.

85. The diagrams below represent rock outcrops in which the rock layers have not been overturned. Which rock outcrop shows a possible sequence of the trilobite fossils?

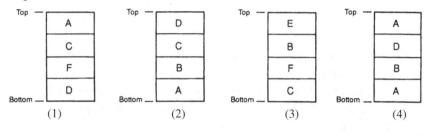

UNIT SIX _____

Meteorology

> **KEY IDEAS** Weather, the present condition of the atmosphere at any given location, is described by the physical characteristics of the atmosphere. The transfer of heat energy from solar radiation and of water vapor into and out of the atmosphere causes regions of different densities to form. The rise or fall of these regions of different densities, due to the action of gravitational force, produces circulation of the air, including winds.
>
> Field maps of atmospheric characteristics reveal the positions and movements of these regions, which are the basis of weather forecasting.
>
> Hazardous weather is the result of the vigorous transfer of energy at atmospheric boundaries.

KEY OBJECTIVES
Upon completion of this unit, you should be able to:

- Define weather and identify the atmospheric characteristics used to describe it.
- Describe the relationships between important atmospheric characteristics.
- Explain how clouds and precipitation form.
- Construct weather maps, given weather information at numerous locations, and identify the positions of air masses and fronts.
- Predict weather and the movement of weather systems based on current weather maps.
- Describe the weather conditions associated with the various types of frontal boundaries.
- Explain how weather information can be used to make forecasts.
- Identify the conditions that typically lead to hazardous weather such as hurricanes and tornadoes.

A. HOW CAN WEATHER BE DESCRIBED?

Weather is the present condition of the atmosphere at any location. Weather changes are mainly the result of unequal heating by solar radiation, of Earth's landmasses, oceans, and atmosphere. As heat energy is transferred at the boundaries between the atmosphere, oceans, and land masses, layers of different temperatures and densities form in the atmosphere. When gravity acts on these layers of different densities, they rise and fall, producing circulation of air in the atmosphere. This circulation causes the air at any one location to constantly be moved away and replaced by air from a different location—with a resulting change in weather.

The constantly changing weather is described in terms of the characteristics of the atmosphere that change, or *atmospheric variables*. The atmospheric variables used to describe weather are measured with weather instruments and include the temperature, pressure, humidity, movement, and transparency of the air. Atmospheric variables are interrelated, so that a change in one is likely to cause a change in another. This interrelatedness makes weather prediction a complex task. However, measurement of atmospheric variables and the creation of field maps based on these data reveal large-scale patterns that can be used to make predictions.

A-1. Air Temperature

The air temperature at any given location is related to the heat energy present in the atmosphere at that location. Solar radiation is the main source of heat energy in the atmosphere; therefore anything that influences solar radiation will ultimately influence air temperature. Although the radiation emitted by the sun is fairly constant, the amount that reaches the Earth's surface and is converted to heat energy in the atmosphere is influenced by many factors. These include the angle at which solar radiation strikes the Earth's surface; the number of hours of solar radiation per day; the reflection, refraction, or absorption of solar radiation by the atmosphere (e.g., cloud cover); and the nature of the surface materials absorbing solar radiation. In general, the more direct the solar radiation and the longer it occurs, the higher the temperature.

The angle at which solar radiation strikes the Earth's surface varies during the course of a day. Because of the Earth's rotation, radiation is received during the day but not at night. Therefore, air temperature often varies in a daily cycle. Similar changes occur as the Earth moves through its orbit around the Sun, resulting in seasonal temperature variations.

Temperature is measured with thermometers. A thermometer works on a simple principle—matter expands when heated and contracts when cooled, and the amount of expansion or contraction depends on the amount of heat gained or lost. A typical thermometer measures the height of a column of alcohol or mercury in a glass tube. Some modern devices, however, use the change in electrical conductivity of certain metals when heated or cooled to measure temperature. Several scales, named for their creators, have been developed over the years to measure temperature: Fahrenheit, Celsius, and Kelvin. Meteorologists in the United States still use the Fahrenheit scale but are changing over to the Celsius.

Meteorologists use several types of specialized thermometers when measuring air temperature. Simple thermometers are mounted in shaded, vented boxes so that they will measure the temperature of the air and not be heated additionally by absorption of sunlight. Maximum-minimum thermometers have pinched tubes and markers that are pushed or dragged back as the fluid in them rises and falls. This feature enables them to record the highest and lowest temperatures during a time period by the positions of the markers. Continuous temperature readings are made with a thermograph, a thermometer in which a coil of heat-sensitive metal moves a pen against a rotating drum.

A-2. Air Pressure

Air is a mixture of gases. A gas consists of many tiny individual molecules that are far apart and moving rapidly. As they whiz to and fro, they are kept from escaping into space by the walls of a container or, in the case of the atmosphere, by Earth's gravity.

Gases exert pressure on surfaces with which they come in contact. Pressure is a force that acts on a surface area. The pressure that a gas exerts on a surface is the result of gas particles colliding with the surface. Since gas particles move randomly in all directions, they exert pressure equally in all directions. Figure 6-1 shows gas molecules speeding around in a container and striking surfaces held at different angles.

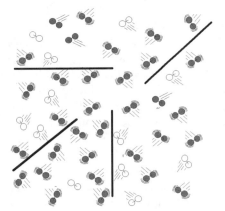

Air pressure is not an atmospheric variable that you can sense, in the way that you feel temperature or see cloud cover. By itself, knowledge of the air pressure in one location is of little use in weather forecasting. However, *field maps of air pressure in many locations* reveal distinct regions within the atmosphere, and *changes* in air pressure signal movements of these large masses of air and their associated weather conditions. Rising pressure usually signals the approach or continuation of fair weather; falling pressure, the approach of a storm.

Figure 6.1 Gas Molecules in a Container. No matter what angle a surface is held at, air molecules collide with it, exerting the same pressure in all directions.

Air pressure is measured with a barometer. Two commonly used barometers are the mercury barometer and the aneroid barometer.

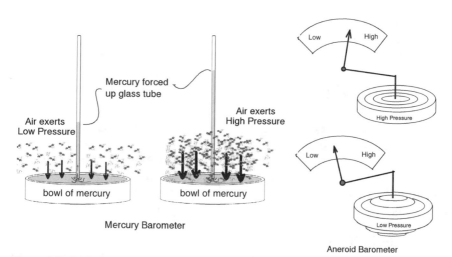

Figure 6.2 A Mercury Barometer and an Aneroid Barometer

In a mercury barometer (see Figure 6.2a), air pressure forces liquid mercury up a tube. The mercury rises in the tube because the tube contains a vacuum and no pressure is exerted on the mercury inside the tube. The higher the air pressure outside the tube, the higher the mercury is pushed up in the tube. Since pressure is measured by how far up the tube the mercury is pushed, pressure is often described as "inches or millimeters of mercury." The word *aneroid* means no liquid, and an aneroid barometer (see Figure 5.2b) is a can with no liquid inside it. The can contains a spring scale. As air pressure on the sides of the can increases or decreases, the spring compresses and expands and the scale records the magnitude of the force.

Pressure is measured in units of force per unit area. A variety of units are used to describe atmospheric pressure. The *atmosphere* (atm) is the average pressure exerted by the Earth's atmosphere at sea level and is equal to 1.013×10^5 newtons per square meter, or 14.7 pounds per square inch. The *bar* is equal to 10,000 newtons per square meter, and the *millibar* is 1/1,000 of a bar or 100 newtons per square meter. The *torr*, the pressure needed to push a column of mercury up 1 millimeter, is equal to 133 newtons per square meter. In meteorology, atmospheric pressure is measured in millibars. Normal atmospheric pressure is 1,013 millibars.

Air pressure is often misunderstood because much of our personal experience with pressure is of solids pressing against us, of our bodies pressing against something because of gravity acting on us. Since the molecules of solids are bonded together in a single unit, they act as a single unit. Under the influence of gravity the entire object is pulled downward and exerts a downward pressure. In gases, on the other hand, each tiny molecule acts independently. Gravity pulls the molecule down; but since it is hurtling through space, the effect is like that of gravity acting on a baseball speeding toward home. The ball sinks in a downward trajectory but still exerts a sizable horizontal force, as any professional catcher will attest. Think of gases as millions of tiny baseballs flying in all directions at the same time. Anything that affects the strength and the number of collisions that gas molecules will make with a surface also affects the pressure they exerts.

A-3. Air Humidity

Humidity is the amount of moisture in the air. Humidity is *not* liquid water that is suspended in the air. Humidity is *water vapor*, a colorless, odorless gas that enters the atmosphere when liquid water evaporates. Humidity is an important weather factor because water vapor in the atmosphere is the source of the liquid water that condenses to form clouds, fog, and precipitation. Humidity is often expressed as absolute humidity or relative humidity.

Absolute humidity is the number of grams of water vapor in 1 cubic meter of air. It is difficult to isolate the water vapor in air and to measure its mass; therefore, absolute humidity is seldom directly measured. Also, knowing the absolute humidity of a given sample of air is of little use in predicting how the air will feel, or whether that humidity will condense to form clouds or rain, because air can hold different amounts of water vapor at different temperatures. Warm air can hold more moisture than cold air, so the same amount of moisture that would saturate cold air will not saturate warm air.

Consider this example. Suppose you were told that a vehicle contained 50 people. You would have no idea whether you would feel crowded in that vehicle unless you knew its capacity. A bus that holds 40 passengers would be overcrowded with 50, but a 747 plane with only 50 passengers aboard would have enough room for each passenger to have a whole row of seats. If, however, you were told that a vehicle was 10 percent full or 100 percent full, you would have a much better idea of what you might experience while traveling in it. Similarly, it is more useful to know whether air is 10 percent or 100 percent full of moisture than to know simply the amount of moisture the air contains.

Relative humidity is the ratio of the water vapor actually in the air to the maximum amount of water vapor the air can hold at that temperature:

$$\text{Relative humidity} = \frac{\text{Amount of moisture in air}}{\text{Air's capacity to hold moisture}} \times 100\%$$

It is expressed as a percent and can vary from almost 0 percent over a desert to as much as 100 percent in rain or thick fog. The way in which air containing moisture will interact with its surroundings depends on its capacity to hold moisture. As you can see in Figure 6.3, the same amount of moisture in air at different temperatures can result in air that behaves very differently.

Relative humidity is a more useful expression of humidity than absolute humidity. Relative humidity can be used to predict how the air will feel to a person and how likely condensation is to occur. The higher the relative humidity, the more uncomfortable people feel because the air has less capacity available to hold moisture evaporating from their bodies. Therefore, less water evaporates from the skin, leaving a wet and "sticky" feeling.

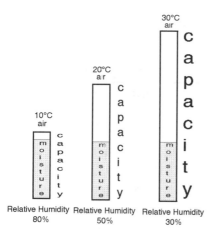

Relative Humidity 80% Relative Humidity 50% Relative Humidity 30%

Figure 6.3 Relative Humidity. These three parcels of air contain the same amount of moisture. As the temperature increases, the capacity of the air to hold moisture also increases, however the relative humidity decreases because the moisture represents a smaller and smaller percentage of the capacity.

High relative humidity also means that the air is holding close to its capacity of moisture, and a decrease in temperature, which in turn decreases capacity, could trigger condensation, which forms clouds, fog, or precipitation. A commonly used value in this connection is the *dew point*, that is, the temperature at which the water vapor in the air will begin to condense into liquid water. It is the temperature at which the moisture in the air fills the air to its capacity.

Humidity is measured with a hygrometer or a psychrometer. A hygrometer consists of strands of human hair attached to a pointer. Human hair lengthens slightly as humidity increases. As the hair lengthens and shortens because of changing humidity, the pointer changes position.

A psychrometer consists of two thermometers, one whose bulb is kept dry and another whose bulb is kept wet. Water evaporating from the wet bulb causes cooling, and the temperature of the wet-bulb thermometer decreases. The dry-bulb thermometer, however, remains unchanged and registers the temperature of the air. The lower the relative humidity, the more water can evaporate from the wet-bulb thermometer and the more it is cooled. Therefore, the difference in temperature between the wet-bulb thermometer and the dry-bulb thermometer is directly related to the relative humidity of the air.

The relationship among dry-bulb temperature, wet-bulb temperature, and relative humidity is shown in Table 6.1. To find the relative humidity, look for the intersection of the dry-bulb temperature and the difference between the wet-bulb and dry-bulb temperatures. The number at the intersection is the percent relative humidity.

Example

Find the relative humidity if the dry-bulb temperature is 10°C and the wet bulb temperature is 5°C.

To find the relative humidity you need the *dry-bulb temperature* and the *difference between the dry-bulb and wet bulb temperatures.*

The dry-bulb temperature is given as 10°C.

The difference between the dry-bulb and wet-bulb temperatures must be calculated. It is 10°C - 5°C = 5°C.

The number where these two temperatures intersect is 43. The relative humidity is 43%.

TABLE 6.1 RELATIVE HUMIDITY (%)

Dry-Bulb Temperature (•C)	Difference Between Wet-Bulb and Dry-Bulb Temperatures (C•)														
	1	2	3	4	5	6	7	8	9	10	11	12	13	14	15
-20	28														
-18	40														
-16	49	0													
-14	55	11													
-12	61	23													
-10	66	33	0												
-8	71	41	13												
-6	74	38	20	0											
-4	77	54	32	11											
-2	79	58	37	20											
0	81	63	45	28	11										
2	83	67	51	36	20	6									
4	85	70	56	42	27	14									
6	86	72	59	46	35	22	10	0							
8	87	74	62	51	39	28	17	6							
10	88	76	65	54	43	33	24	13	4						
12	88	78	67	57	48	38	28	19	10	2					
14	89	79	69	69	59	41	33	25	16	8	1				
16	90	80	71	62	54	45	37	29	21	14	7	1			
18	91	81	72	64	56	48	40	33	26	19	12	6	0		
20	91	82	74	66	58	51	44	36	30	23	17	11	5	0	
22	92	83	75	68	60	53	46	40	33	28	21	15	10	4	0
24	92	84	76	69	62	55	49	42	36	30	25	20	14	9	4
26	92	85	77	70	64	57	51	45	39	34	28	23	18	13	9
28	93	86	78	71	65	59	53	47	42	36	31	26	21	17	12
30	93	86	79	72	66	61	55	49	44	39	34	29	25	20	16

The difference between the wet-bulb temperature and the dry-bulb temperature is also directly related to the dew point of the air. The higher the relative humidity, the closer the air is to being filled to its capacity. Cooling air decreases its capacity to hold moisture. The closer the air is to being filled to capacity, the less it has to be cooled to reach the point of maximum moisture. Table 6.2, which shows the relationship among wet-bulb temperature, dry-bulb temperature, and dew point, can be used to find the dew point. The number at the intersection of the dry-bulb temperature and the difference between the wet-bulf band dry-bulb temperatures is the dew point in degrees Celsius.

Example

Find the dew point if the dry-bulb temperature is 10°C and the wet-bulb temperature is 5°C.

The dry-bulb temperature is given as 10°C.

The difference between the dry-bulb and wet-bulb temperatures must be calculated. It is 10°C – 5°C = 5°C.

The number where these two temperatures intersect is –2. The dew point is –2°C.

TABLE 6.2 DEW POINT TEMPERATURES

Dry-Bulb Tempera-ture (•C)	Difference Between Wet-Bulb and Dry-Bulb Temperatures (C•)														
	1	2	3	4	5	6	7	8	9	10	11	12	13	14	15
-20	-33														
-18	-28														
-16	-24														
-14	-21	-36													
-12	-18	-28													
-10	-14	-22													
-8	-12	-18	-29												
-6	-10	-14	-22												
-4	-7	-12	-17	-29											
-2	-5	-8	-13	-20											
0	-3	-6	-9	-15	-24										
2	-1	-3	-6	-11	-17										
4	1	-1	-4	-7	-11	-19									
6	4	1	-1	-4	-7	-13	-21								
8	6	3	1	-2	-5	-9	-14								
10	9	6	4	1	-2	-5	-9	-14	-28						
12	10	8	6	4	1	-2	-5	-9	-16						
14	12	11	9	6	4	1	-1	-6	-10	-17					
16	14	13	11	9	7	4	1	-1	-6	-10	-17				
18	16	15	13	11	9	7	4	2	-2	-5	-10	-19			
20	19	17	15	14	12	10	7	4	2	-2	-5	-10	-19		
22	21	19	17	16	14	12	10	8	5	3	-1	-5	-10	-19	
24	23	21	20	18	16	14	12	10	8	6	2	01	05	-10	-18
26	25	23	22	20	18	17	15	13	11	9	6	3	0	-4	-9
28	27	25	24	22	21	19	17	16	14	11	9	7	4	1	-3
30	29	27	26	24	23	21	19	18	16	14	12	10	8	5	1

A-4. Air Movements

In the atmosphere, air circulates because of density differences. The vertical movements of rising and sinking air are called *air currents*. When sinking air reaches the Earth's surface, it spreads out horizontally. When rising air expands, it also spreads out horizontally. Horizontal movements of air parallel to the Earth's surface are called *winds*. Wind is described by both its speed and its direction.

Wind speed is measured with an anemometer. An anemometer consists of three or four cups mounted on a vertical axis and driven by the wind causing the axis to spin. The speed of rotation of the axis varies with wind speed. Wind speed is measured in knots, a nautical measure of speed; 1 knot = 1.85 kilometers per hour.

Wind direction is determined with a wind vane. A wind vane is a pointer mounted on an axis that is attached to a compass rose. The tail of the pointer is larger in surface area than the tip. Thus, the wind exerts more pressure on the tail, causing it to swing around so that the tip points in the direction from which the wind is blowing. A wind is named according to the direction from which it blows. Just as a person who comes from the South is called a Southerner, a wind that blows from the south is called a south wind.

A-5. Atmospheric Transparency

All of the gases of which air is composed are transparent. However, there are many substances that become suspended in the atmosphere and decrease its transparency. These substances include dust, volcanic ash, smoke, salt particles, aerosols (droplets of liquids), ice particles, and water droplets. By far, water droplets have the greatest effect on atmospheric transparency. Clouds and fog consist of tiny, but visible, droplets of water. Atmospheric transparency is expressed in three ways: visibility, cloud ceiling, and cloud cover.

Visibility is the horizontal distance through which the eye can distinguish objects. It is usually expressed in miles.

Cloud ceiling is the base height of cloud layers. It is measured with a ceilometer, which uses pulses of light and a photoelectric telescope.

Cloud cover is the fraction of the sky that is obscured by clouds.

B. WHAT ARE THE RELATIONSHIPS AMONG THE SEVERAL ATMOSPHERIC VARIABLES?

Atmospheric variables are interrelated, so that a change in one is likely to cause a change in another. Since the atmosphere is gaseous, this interrelatedness is best understood in light of the kinetic theory of gases. According to the kinetic theory, gases consist of many tiny, individual molecules that do not interact with each other except when they collide. The molecules are far apart and are in constant random motion, and the temperature of the gas is proportional to the speed at which they are moving.

B-1. Air Pressure and Air Temperature

As explained earlier, air pressure is the result of the forces exerted by gas molecules colliding with a surface or each other. Air temperature depends on the speed of the gas molecules. Air pressure and temperature are related because both involve the motion of molecules. The higher the temperature, the faster the gas molecules are moving. The faster the gas molecules are moving, the more force they exert in a collision. Thus, you would expect that an increase in temperature would result in an increase in pressure, but in the atmosphere this is not the case. The reason is that the atmosphere is not contained by rigid walls but is an open system free to expand.

When the temperature of the atmosphere increases, its molecules move faster and collide more vigorously. As a result, the molecules bounce farther apart and the atmosphere expands. Even though the molecules are moving faster, and each individual impact exerts greater force, the total number of collisions decreases because the molecules are spread farther apart. The decrease in collisions far exceeds any increase in force due to faster moving molecules, so the net effect is a *decrease* in air pressure when air temperature increases (see Figure 6.4).

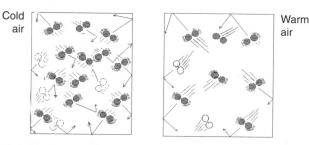

Figure 6.4 Relationship Between Air Temperature and Air Pressure. Even though the molecules are moving slower and collide with less force, the cold air on the left exerts more pressure because its closely spaced molecules collide much more frequently.

Air temperature ↑, Air pressure ↓   Air temperature ↓, Air pressure↑

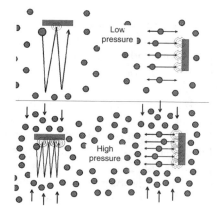

Figure 6.5 Relationship Between Air Pressure and Air Temperature. Each black rectangle represents a unit of surface area. When vertically moving molecules are confined by surrounding molecules as air pressure increases, the number of collision-producing rebounds increases, so more collisions occur per unit of surface area. When horizontally moving molecules are confined, their spacing decreases, so again there are more collisions per unit of surface area.

Now let's consider what happens to temperature if pressure is changed (see Figure 6.5). To increase air pressure, an outside force must be exerted to confine the air. That force, pushing air inward, causes the molecules that collide with the inward-moving air to rebound at a faster speed. Think of the difference between a ball bouncing off a bat held still in a bunt and a ball bouncing from a bat swung hard at the ball. Thus, an increase in pressure causes an increase in temperature. Conversely, decreasing the pressure requires a decrease in the force confining the molecules and the air expands outward. The outward motion causes a decrease in the speed at which the molecules rebound from the outward-moving air. Think of moving the bat *away* from the ball as it strikes the bat, causing it to rebound very little. Thus, a decrease in pressure causes a decrease in temperature.

Air pressure ↑, Air temperature ↑   Air pressure ↓, Air temperature ↓

311

B-2. Air Pressure and Humidity

The greater the mass of the molecules, the more force they exert during a collision. The mass of a molecule in air depends on the element(s) of which it is composed. Nitrogen and oxygen molecules have more mass than water molecules (molecular mass $N_2 = 28$, of $O_2 = 32$, of $H_2O = 18$). As water vapor builds up in the atmosphere, lighter water molecules displace heavier nitrogen or oxygen molecules. Since the lighter water molecules exert less force during a collision than the heavier nitrogen and oxygen molecules, air pressure decreases as humidity increases. (Don't be confused by thinking of moist air as containing *liquid* water, which is denser than any gas in air; moist air contains water *vapor,* which is less dense than most gases in air.) Figure 6.6 shows the displacement of heavier gas molecules by lighter water vapor molecules as humidity increases.

Humidity ↑, Air pressure ↓   Humidity ↓, Air pressure↑

B-3. Air Temperature and Air Humidity

The ability of air to hold water vapor is directly related to air temperature. Warm air can hold more water vapor than cool air. Why is this so? Imagine air over a body of liquid water. In order for molecules of water in the liquid to break free and become a gas, they require energy. Air molecules striking the water molecules on the surface of the liquid water impart their energy to the water molecules, enabling some to break free and become a gas. The higher the temperature, the faster the air molecules are moving, the more energy they impart, and the more water molecules can break free (see Figure 6.7). For every 10°C increase in temperature, air can hold approximately twice as much moisture.

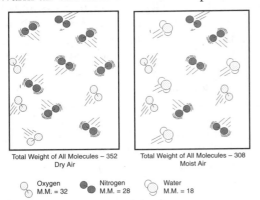

Total Weight of All Molecules – 352
Dry Air

Total Weight of All Molecules – 308
Moist Air

Oxygen M.M. = 32   Nitrogen M.M. = 28   Water M.M. = 18

Figure 6.6 Relationship Between Molecular Mass and Force. A volume of dry air has more mass than an equal volume of moist air because water molecules have less mass than the nitrogen and oxygen molecules they displace. Therefore, moist air is less dense than dry air.

Since the capacity of air to hold moisture changes with temperature, the relative humidity also changes with temperature. If the temperature increases and the moisture in the air remains the same, the relative humidity will decrease. Conversely, if the temperature decreases and the moisture in the air remains the same, the relative humidity will increase. See Figure 6.3.

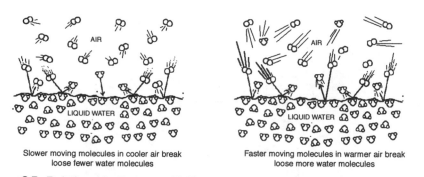

Slower moving molecules in cooler air break loose fewer water molecules

Faster moving molecules in warmer air break loose more water molecules

Figure 6.7 Relationship Between Air Temperature and the Amount of Water Vapor Entering in the Air from a Water Surface. At higher temperatures, faster moving air molecules impart more energy to surface water molecules, enabling them to enter the atmosphere as a gas.

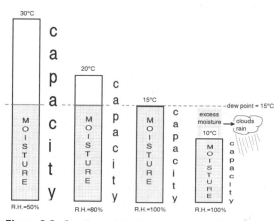

Figure 6.8 Cooling of Air Decreases its Capacity to Hold Moisture Resulting in Condensation. As air is cooled, its capacity to hold moisture approaches the amount of moisture it contains. At the dew point the moisture fills the air to its capacity. If the air is cooled below the dew point, its capacity is less than the moisture in the air and the excess moisture condenses to form clouds and precipitation.

As air temperature decreases, it eventually reaches the dew point, the temperature at which the capacity of the air to hold moisture is equal to the moisture the air contains. At that point the relative humidity is 100 percent. If the temperature decreases further, the capacity of the air to hold moisture falls below the amount of moisture in the air, and moisture is released. The released moisture forms droplets of water or ice crystals that may form clouds or precipitation (see Figure 6.8). Therefore, as air temperature approaches the dew point, precipitation becomes more likely.

B-4. Air Pressure and Winds

Winds blow from regions of high air pressure to regions of low air pressure. This is easy to understand if you think of two people pushing against each other. The person pushing harder will advance against the person pushing with less force. In much the same way, air exerting high pressure will advance against air exerting low pressure. The net movement of the air is from high pressure toward low pressure (see Figure 6.9).

313

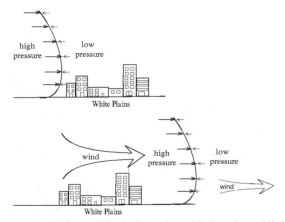

Figure 6.9 Air Pressure and Winds. Winds Blow from High to Low. High-pressure air exerts more force than low-pressure air and pushes the low-pressure air ahead of it, resulting in a horizontal movement of air—a wind.

Land and Sea Breezes

Land and sea breezes illustrate this very clearly. Differences in air pressure between adjacent regions occur wherever land and water meet. During the day, when exposed to sunlight, land surfaces increase in temperature more than water surfaces, (because water has a higher specific heat than soil). As a result, air over land surfaces will have a higher daytime temperature than adjacent air over water surfaces. Since warm air exerts less pressure than cool air, the pressure over the land will be lower and the pressure over the water will be higher. The result is a movement of air from the water toward the land. Since a wind is named for the direction or place from which it is blowing, this wind is called a *sea breeze*.

At night, though, land cools off faster than water. The air over the land becomes cooler than the air over the water. Thus, the air pressure is greater over the land than over the water and air moves from the land toward the water. This wind is called a *land breeze* because it comes from the land.

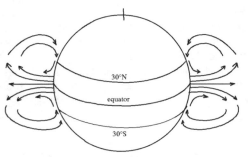

Global Wind Belts

Differences in pressure between adjacent regions of air also occur on a global scale. The Equator is a region of hot, moist air that exerts low pressure. The regions

Figure 6.10 Convection Cells Near the Equator. Air near the Equator is heated by more intense insolation, becomes less dense, and floats upward in the surrounding denser air. As the air rises, it expands and cools, causing it to sink back toward the surface. The result is a circular movement of air, or convection cell.

314

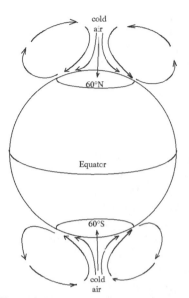

Figure 6.11 Convection Cells Near the Poles. Air near the poles cools because of less intense insolation, becomes more dense, and sinks downward. The sinking air spreads outward at the surface, is warmed as it moves away from the poles, and rises. The result is a circular movement of air, or convection cell.

adjacent to the Equator are cooler and exert higher pressure. Thus, there is a global movement of air from regions surrounding the Equator toward the Equator. As a result the hot air at the Equator is displaced upward. As the air is forced upward, its pressure decreases and the air expands outward and cools, causing it to become more dense and to sinks back toward the surface. The end result is a circular pattern of motion in the atmosphere, as shown in Figure 6.10.

In much the same way, air over the poles is cold and exerts high pressure. In the warmer regions adjacent to the poles, the air exerts less pressure. Thus, there is a global movement of air over the poles outward. This displaces the warmer adjacent air, causing it to rise; and another circular pattern is set up, as shown in Figure 6.11.

The polar circulation pattern and the equatorial circulation pattern cause the air in between to move in a circular pattern as well. The result is three circular patterns of motion between the poles and the Equator, as shown in Figure 6.12.

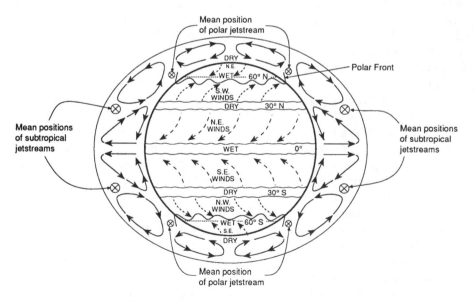

Figure 6.12 Planetary Wind and Moisture Belts

315

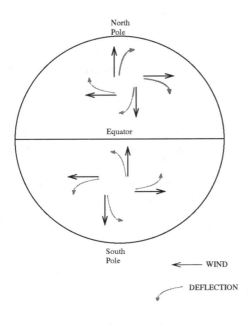

North Pole

Equator

South Pole

←——— WIND

——— DEFLECTION

Figure 6.13 The Coriolis Effect. Winds are deflected to the right from their point of origin in the Northern Hemisphere and to the left from their point of origin in the Southern Hemisphere.

Wind directions are influenced by the rotation of the Earth. Since the Earth rotates from west to east, winds blowing from areas of high to areas of low pressure follow a path over the Earth's surface that is curved. This curving of the path of winds because of the Earth's rotation is called the *Coriolis effect*. Deflection is to the right of the wind's point of origin in the Northern Hemisphere and to the left of its point of origin in the Southern Hemisphere. (remember that someone who is *left*-handed is called a *south*paw), as shown in Figure 6.13.

C. HOW DO CLOUDS AND PRECIPITATION FORM?

Clouds and precipitation form when air is cooled below its dew point and water vapor condenses into tiny water droplets or ice crystals. The conditions under which this cooling takes place determine what type of cloud or precipitation will form.

Condensation is the change of phase from gas to liquid. To condense into a liquid, a gas must have a surface to condense on. In the atmosphere, this surface is provided by particles suspended in the air, such as salt or ice crystals, dust, or smoke, called *condensation nuclei*. Water vapor in the air condenses when the air is cooled to the dew point and there are condensation nuclei on which the condensation can form. *Sublimation* is the change of phase from a gas directly to a solid. Water vapor may sublime to form ice crystals if moisture is released from the air when its temperature is below the freezing point of water.

Air that is warmed by contact with the sun-heated surface of the Earth expands, becomes less dense, and floats upward in the denser surrounding air. As the air rises, it continues to expand, causing a decrease in pressure and temperature. Air will also expand and cool if it is forced upward by moving over a natural obstacle, such as a mountain, or by cooler, denser air pushing its way underneath it. These three conditions that cause air to rise are diagramed in Figure 6.14.

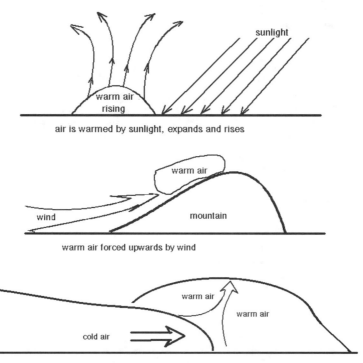

sunlight

warm air
rising

air is warmed by sunlight, expands and rises

warm air

wind

mountain

warm air forced upwards by wind

warm air

warm air

cold air

dense, cold air mass pushes under warm air forcing it to rise

Figure 6.14 Three Conditions That Cause Air to Rise. Air rises when it is heated, forced upward as winds blow against mountains, or forced upward as cold, dense air pushes under it.

The cooling of air as it rises results in condensation when the air temperature drops to the dew point and condensation nuclei are present. This condensation generates a *cloud* made of a great many tiny but visible water droplets or ice crystals. The *cloud base* indicates the elevation at which the rising air reached its dew point. Since this depends on the amount of moisture in the air and the starting temperature, clouds may form at many different elevations.

Fog is a cloud whose base is at ground level. It forms when moist air at ground level is cooled below its dew point. When moist air comes in contact with a surface such as grass, or with a car or other object that has cooled below the dew point because of heat radiating out of it overnight, droplets of water called *dew* condense on the surface. If the temperature is below freezing, tiny ice crystals called *frost* will form instead of dew.

The water droplets or ice crystals in clouds are very tiny (1/50 of a millimeter) and often remain suspended in the air. Thus, all clouds do not form precipitation. Precipitation occurs only when cloud droplets or ice crystals join together and become heavy enough to fall. Table 6.3 summarizes the different types of precipitation and their origins.

317

TABLE 6.3 TYPES OF PRECIPITATION

Name	Description	Origin
Rain	Droplets of water up to 4 mm in diameter	Cloud droplets coalesce when they collide.
Drizzle	Very fine droplets falling slowly and close together	Cloud droplets coalesce when they collide.
Sleet	Clear pellets of ice	Raindrops freeze as they fall through layers of air at below-freezing temperatures.
Glaze	Rain that forms a layer of ice on surfaces it touches	Supercooled raindrops freeze as soon as they come in contact with below-freezing surfaces.
Snow	Hexagonal crystals of ice, or needle-like crystals at very low temperatures	Water vapor sublimes, forming ice crystals on condensation nuclei at temperatures below freezing.
Hail	Balls of ice ranging in size from small pellets to as large as a softball, with an internal structure of concentric layers of ice and snow	Again and again hailstones are hurled up by updrafts in thunderstorms and then fall through layers of air that alternate above and below freezing. Each cycle adds a layer to the hailstone. The more violent the updrafts, the larger and heavier the hailstone can become before falling.

As precipitation falls through the atmosphere, condensation nuclei and other suspended material that adhere to it are brought down to the Earth's surface. In this way dust and many pollutants are removed from the atmosphere by precipitation. However, some pollutants react with the water in precipitation to form harmful solutions such as acid rain, which has had many negative effects on the environment.

D. WHAT INFORMATION IS SHOWN BY WEATHER MAPS?

D-1. Synoptic Weather Maps

One of the most widely used methods of forecasting weather is known as *synoptic forecasting*. Synoptic forecasting is based on looking at a synopsis, or summary of the total weather picture at a particular time.

A synoptic weather map is made by measuring atmospheric variables at thousands of weather stations around the world four times a day. These data are then used to create field maps, which reveal large-scale weather patterns. By looking at a sequence of synoptic weather maps, meteorologists can perceive the development and movement of weather systems and make predictions.

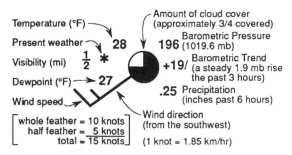

Figure 6.15 A Station Model. The symbol summarizes atmospheric variables in a shorthand notation.

Synoptic weather maps use a symbol called a *station model* to show a summary of the weather conditions at a weather station. Figure 6.15 shows a typical station model and explains what each of its elements represents.

D-2. Field Maps

Station models plotted on a map summarize a wide range of weather data and can be used to create many different field maps. In Unit 1 a *field* was defined as a region of space that has a measurable quantity at every point. Field maps can be used to represent any quantity that varies in a region of space. One way to represent field quantities on a two-dimensional field map is to use *isolines*, which connect points of equal field value. For example, a temperature field map would contain lines connecting points of equal temperature, or *isotherms*. A pressure field map would contain lines connecting points of equal pressure, or *isobars*. Field maps clearly show the weather patterns in the atmosphere.

Figure 6.16 shows a synoptic weather map, followed by the same map with isotherms and isobars drawn in.

As you can see, there are distinct regions in the atmosphere that have similar conditions. For example, a region of high pressure, cooler temperatures, and clear skies is centered near Salt Lake City, Utah. You can also see a region of low pressure, warmer temperatures, and cloudy skies centered near Cincinnati, Ohio.

Within a field, a field value changes as you move from place to place. The rate at which the field value changes is called its *gradient*. Rapid changes in field values, or steep gradients, show up as closely spaced isolines on a field map. In Figure 6.16, you can see that the isobars are widely spaced within each of the two air masses, but are closely spaced between them. This means that there is little change in pressure within each of the regions, but there is a boundary between them where the pressure changes rapidly. You can see the same pattern repeated for temperature: there are regions of the atmosphere with relatively uniform characteristics, which have boundaries separating them from adjacent regions.

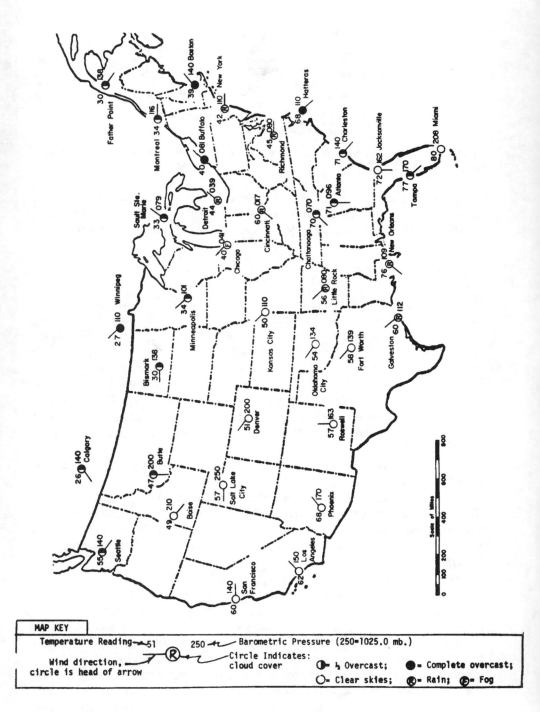

MAP KEY

Temperature Reading ⟶ 51 ____ 250 ⤼ ___ Barometric Pressure (250=1025.0 mb.)

Wind direction, ___⤵ ___ ⟨Ⓡ⟩ ___ Circle Indicates: cloud cover
circle is head of arrow

◖ = ½ Overcast;   ● = Complete overcast;

○ = Clear skies;   Ⓡ = Rain;   Ⓕ = Fog

Figure 6.16a A Synoptic Weather Map of the United States

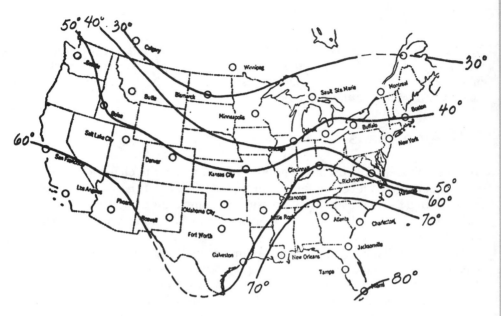

Figure 6.16b Isotherms

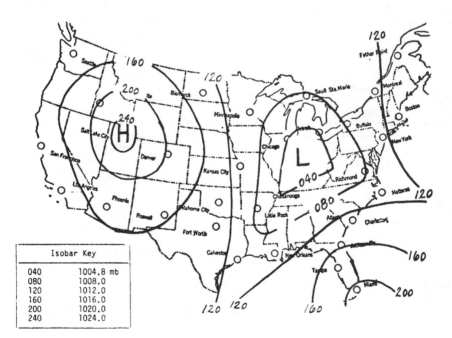

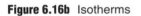

Isobar Key	
040	1004.8 mb
080	1008.0
120	1012.0
160	1016.0
200	1020.0
240	1024.0

Figure 6.16c Isobars

E. HOW CAN WEATHER INFORMATION BE USED TO MAKE FORECASTS?

Most weather forecasts are based on the movements of large regions of air with fairly uniform characteristics, called *air masses*. By examining a series of synoptic weather maps, meteorologists can track the movements of air masses and predict future movements.

E-1. Air masses

As stated above, a region of the atmosphere in which the characteristics of the air at any given level are fairly uniform is called an *air mass*. Air masses are identified by their average air pressure, air temperature, moisture content, and winds. The boundaries between air masses, where rapid changes occur, are called *frontal boundaries*, or *fronts*.

The characteristics of an air mass are the result of the geographical region over which it formed, or its *source region*. Air resting on, or moving very slowly over, a region tends to take on the characteristics of that region. For example, air that sits over the Gulf of Mexico in the summer is resting on very warm water. The air is warmed by contact with the water, and evaporation causes much humidity to enter the air. As a result the air becomes warm and moist, just like the Gulf of Mexico. In general, air masses that form near the poles are cold, and those that form near the Equator are warm. Air masses that form over water are moist; those that form over land, dry. The longer the air remains stationary over a region, the larger it becomes and the closer it matches the characteristics of the region. On weather maps, air masses are named for their source region and are designated by a letter or letters as shown in Figure 6.17.

	arctic A	polar P	tropical T
	Formed over extremely cold, ice-covered regions	Formed over regions at high latitudes where temperatures are relatively low	Formed over regions at low latitudes where temperatures are relatively high
maritime m Formed over water, moist		mP—cold, moist Formed over North Atlantic, North Pacific	mT—warm, moist Formed over Gulf of Mexico, middle Atlantic, Caribbean, Pacific south of California
continental c Formed over land, dry	cA—dry, frigid Formed north of Canada	cP—cold, dry Formed over northern and central Canada	cT—warm, dry Formed over southwestern United States in summer

Figure 6.17 Air Mass Names and Sources

322

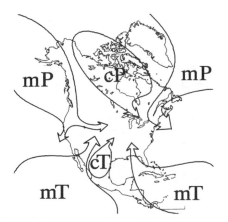

Figure 6.18 Air Masses That influence the United States and Their Source Regions.

Air masses are moved by planetary winds and jet streams (study Figure 6.12 carefully). In general, arctic and polar air masses are moved south and west over the United States by the northeast wind belt near the North Pole and tropical air masses are moved to the north and east by the southwest wind belt between 30° and 60° N. As the air masses are carried over the Earth's surface, they slowly change and take on the characteristics of the regions over which they are moving. For example, a cP air mass moving south out of Canada will gradually become warmer as it moves farther and farther south.

Air masses that affect U.S. weather and the regions in which they form are shown in Figure 6.18.

E-2. Weather Fronts

At any given point in time, several air masses will be moving across the United States. They generally move from west to east, driven by the prevailing southwesterlies. When different air masses meet, very little mixing of the air takes place and a sharp transition zone, called a *weather front*, forms between them. The zone as termed a weather front because it marks the leading edge, or front, of an air mass that is pushing against another air mass. The whole surface along which the air masses meet is called the *frontal surface*, and the line on the ground marking the transition from one air mass to the other is the *front*. Fronts are areas of rapid changes in weather conditions and are often sites of unsettled and rainy weather. Different types of frontal surfaces, such as cold, warm, stationary, and occluded fronts, will form, depending on which type of air mass is advancing against another.

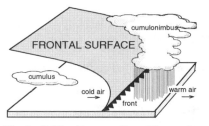

Figure 6.19 Perspective View of a Frontal Surface and a Front Line on the Ground. A *frontal surface* forms along the boundary between two different air masses. The line along which the boundary touches the surface is the *front*.

Cold Fronts

A cold front forms along the leading edge of a cold air mass that is advancing against a warmer air mass. The cold air mass is denser than the warmer air

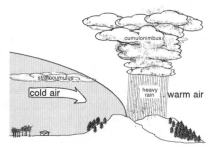

Figure 6.20 A Typical Cold Front

ahead of it, so it pushes against and under it like a wedge. This *forces* the warmer air upward rapidly and results in turbulence, rapid condensation forming heavy, vertically developed clouds, and heavy precipitation or thunderstorms. Cold fronts are often preceded by a zone of thunderstorms in the summer and of snow flurries in the winter. After the front passes, the temperature drops sharply and the pressure rises rapidly.

Warm Fronts

A warm front forms where a warm air mass overrides the trailing edge of a cold air mass ahead of it. The less dense warm air mass rides up and over the denser cold air. As the warm air mass rises above the denser cold air mass, it expands and cools, causing condensation to occur over the wide, gently sloping boundary. The results are thickening, lowering clouds and widespread precipitation. Figure 6.21 shows a typical warm front.

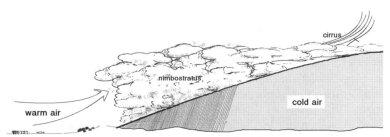

Figure 6.21 A Typical Warm Front

Stationary Fronts

A stationary front forms along the boundary between a warm air mass and a cold air mass when neither moves appreciably in any direction. A stationary front slowly takes on the shape and characteristics of a warm front as the denser cold air slowly slides beneath the less dense, warmer air. Although stationary fronts develop the same widespread rain and cloudiness as do warm fronts, the rain and clouds may persist for a longer period, perhaps many days, until another air mass comes along with enough impetus to get the stalled air masses moving.

Occluded Fronts

Cold air is denser and exerts more pressure than warm air. Therefore, cold air masses and the cold fronts associated with them tend to move faster than warm air masses and warm fronts. If a slow-moving warm front is followed by a fast-moving cold front, the cold front will sometimes overtake the warm front. Then the cold air lifts the warm air entirely aloft, forming an occluded front. The lifting of the warm air mass causes large-scale condensation and precipitation, resulting in widespread rain and thunderstorms. Then, depending on whether the cold air mass coming from behind is colder or warmer than the cold air mass now ahead, a cold-front occlusion or a warm-front occlusion may form, as shown in Figure 6.22.

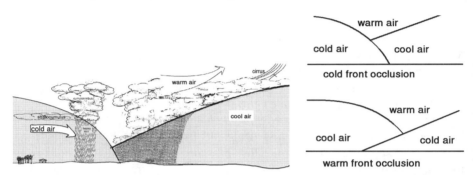

Figure 6.22 An Occluded Front. Two types of occlusions can form, depending on whether the advancing cold air mass is warmer or cooler than the cold air mass ahead of the warm air mass.

E-3. Cyclones, Anticyclones, and Wave Cyclones (Frontal Cyclones)

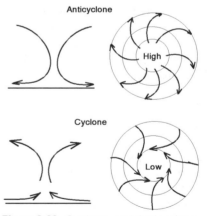

Regions of low pressure surrounded by higher pressure form winds that blow toward the center of the warm air mass but, because of the Coriolis effect, are deflected to the right. The result a is *cyclone*, a flow of air, counterclockwise that moves in a curved path toward the center of the low pressure.

Similarly, regions of high pressure surrounded by lower pressure form winds that blow outwards from the center. The winds are deflected by the Coriolis effect into a flow of air, an *anticyclone,* that is clockwise and outward from the center.

Figure 6.23 Cyclone and Anticyclone

A cyclone and an anticyclone are diagramed in Figure 6.23.

Most of the weather systems that move across the United States are *wave cyclones*, or *frontal cyclones*. Wave cyclones are warm and cold fronts that move in a counterclockwise direction around a low-pressure center. They typically form in the westerly wind belts along the boundary between polar and tropical air, called the *polar front*. (see Figure 6.12) In the United States, this boundary falls just about along our northern border with Canada.

A wave cyclone forms when south-westerly winds push warm, low-pressure air against cold air along the polar front (see Figure 6.24). The result is to lower the pressure along the polar front, and the boundary bends to form a wave of cold air advancing from west to east against the warm air—a cold front. The cold air pushes the warm air ahead of it into the cold air on the other side of it, forming a warm front. The end result

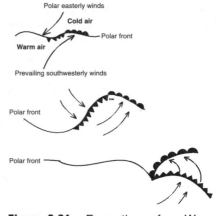

Figure 6.24 Formation of a Wave Cyclone

is a wavelike structure consisting of a cold front overtaking a warm front. The counterclockwise flow in the low-pressure center causes the cold front to swing around the equatorial side of the low-pressure center as the wave moves from west to east. The wave cyclone may even break loose from the polar front and be carried, swirling along, by the jet stream. In the United States, wave cyclones generally move from west to northeast and get carried out over the north Atlantic. The stormy weather associated with a wave cyclone may affect millions of square miles and last for several days.

E-4. Methods of Weather Forecasting

Synoptic forecasting is one of the methods of weather forecasting most commonly used today. The sequence of synoptic charts, or weather maps, used to make predictions can be seen daily on television news programs and in newspapers. Until the 1960s, weather data from stations across the United States were plotted on weather maps by hand and analyzed at local weather offices. Now, computers plot the data, create a wide array of weather maps, and project the future positions of air masses, fronts, and storms. Computer-drawn maps predict temperature, humidity, and wind patterns at different levels in the atmosphere.

Another method of forecasting is *statistical forecasting*, which uses records of past weather to predict future weather. Mathematical equations based on weather records are applied to current conditions to find the probability of occurrence of different weather events. Statistical forecasting is often used to map the probable temperature changes and precipitation associated with weather predicted by synoptic forecasting. Statistical forecasting is most useful for predicting weather 5–90 days in the future.

A third method, *numerical forecasting*, is based on the physics involved with atmospheric behavior. Physical laws such as the gas laws, thermodynamics, and fluid mechanics are used to create mathematical models. These models are then applied to the conditions in the atmosphere in order to predict its behavior. Six basic equations are used: one for each of the dimensions of motion, and one each for the conservation of mass, moisture and heat. Solving these complex equations was not practical before high speed computers become available. Now, however, computers solve the equations to provide near-instantaneous predictions of changes at a grid of points covering the entire United States. Numerical forecasting is very useful for predicting weather up to about 5 days in the future.

Weather data from satellites and computerized analysis of synoptic data have greatly improved the accuracy of forecasts. Satellites provide pictures of clouds that make it possible to detect weather systems from the time they begin. They also make infrared radiation measurements that distinguish warm and cold ocean currents and temperatures at different levels in the atmosphere, and can tell the difference between snow, ice, and rain. In addition, satellites measure the absorption and reflection of solar radiation by the Earth, as well as the amount of ozone in the atmosphere, and they provide wind data based on the movement of clouds.

On the surface of the Earth, radar is a powerful tool for tracking precipitation, thunderstorms, hurricanes, and tornadoes. The National Weather Service has established a network of radar stations that transmit regular reports.

Computer analysis of this wealth of data from satellites and radar has become crucial in making forecasts.

No matter what method of forecasting is used, however, forecasting reliability decreases as the time frame increases. Short-term forecasts (1–3 days) of weather are usually far more accurate than long-term forecasts (a week or more). The decrease in forecast reliability over time is due to a number of factors: the wide spacing of initial data points, unreliability in measurements taken over many areas, and a lack of understanding of the behavior of the atmosphere. A small error in an initial measurement can cause errors in computer-calculated forecasts.

For example, let's make a simple mathematical model that assumes that temperature increases or decreases in direct proportion to the ratio of day to night:

Temperature today × Length of night/Length of day = Temperature tomorrow

Let's compare the temperatures predicted by this model, using a correct temperature reading and one in which an error was made:

Starting Temperature	Prediction When Daylight = 13 hours, Night = 11 hours	Prediction When Daylight = 14 hours, Night = 10 hours
21°C (correct)	21 × 13/11 = 24.8°C	24.8 × 14/10 = 34.7°C
22°C (incorrect)	22 × 13/11 = 26°C	26 × 14/10 = 36.4°C

As you can see, an initial error of 1°C has almost *doubled* as it was compounded in the calculations. Errors like these multiply as forecasts are extended farther into the future, and at some point the forecast becomes useless.

Chaos theory shows that small, unpredictable changes can result in large-scale changes in a system over time. For example, a forest fire set by lightning may cause a localized increase in temperature that may set in motion large-scale changes in air flow patterns. For this reason, chaos theory seems to indicate that long-term forecasts may never be reliable.

F. WHERE AND HOW ARE THE MOST HAZARDOUS WEATHER SITUATIONS LIKELY TO OCCUR?

F-1. Thunderstorms

One of the most common weather hazards is a thunderstorm (see Figure 6.25). A thunderstorm begins as a convection cell in the atmosphere. Warm air rises, expands, cools and then sinks back to the Earth's surface. If the convection in the cell is strong, because of intense heating, the air may rise very rapidly, reaching heights of many kilometers in just a few minutes. A cloud forms in the updraft of the convection cell and grows as more and more warm, moist air is carried upward. The strong updrafts in the rising air support water droplets and ice crystals in the cloud so that they grow in size.

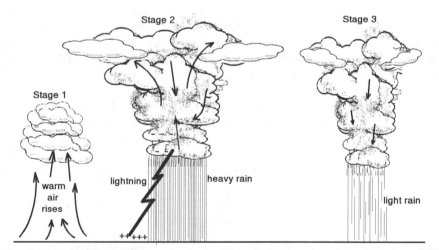

Figure 6.25 Formation of a Thunderstorm. Stage 1: Intense heating causes warm air to rise rapidly, forming tall, vertically developed clouds. Stage 2: Strong updrafts keep precipitation aloft until it is heavy. Falling rain creates downdrafts, and internal friction with updrafts causes buildup of static charge, leading to lightning and thunder. Stage 3: Downdrafts cause cooling of air and updrafts subside, so rain becomes lighter.

When the updrafts can no longer support the moisture, it falls as rain or even hail. The falling rain sets up downdrafts that cause internal friction with updrafts, and the internal friction builds up static electric charges that may discharge as lightning. Thunderstorms are likely to occur wherever and whenever there is strong heating of the Earth's surface.

F-2. Tornadoes

Late in the day, when the Earth's surface is warmest and convection is strongest, a tornado may form. Tornadoes are small (most are less than 100 meters in diameter), brief (most last only a few minutes) disturbances that usually develop over land from intense thunderstorms. When heating is very intense, warm air rises in strong convection currents. The upward movement of the air causes a sharp decrease in pressure. Air rushes into the low pressure region from the sides and, deflected by the Coriolis effect, is given a spin. As air moves toward the center of the updraft, it reaches high speeds because of the big difference in pressure. The rapidly moving air decreases the pressure even more, further feeding the updraft. The whole process spirals upward in intensity, and a funnel forms that eventually touches the

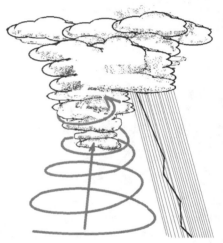

Figure 6.26 Formation of a Tornado

ground. Wind speeds near the center of a tornado may reach 500 kilometers per hour or more.

F-3. Hurricanes

When air over warm oceans is heated by solar radiation, the warming of the air causes a decrease in pressure and the evaporation of water from the ocean adds humidity to the air, decreasing pressure even further. The result is numerous convection cells and thunderstorms. Widespread thunderstorm activity can merge into a large updraft that is part of a huge convection cell. The rising air further decreases the pressure, as does latent heat released when the moisture in the air condenses. As a result a region of very low pressure is created. Winds begin to blow toward the low pressure center as air moves in from surrounding areas of higher pressure. The Coriolis effect deflects the winds, and a cyclone forms. Heat, released when fresh moisture brought in by winds condenses, feeds the convection cell with new energy, and it grows larger and stronger. When the winds in the cyclone exceed 119 kilometers per

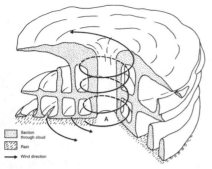

Figure 6.27 Cross Section of a Hurricane.
From: Earth Science on File. © Facts on File, Inc., 1988.

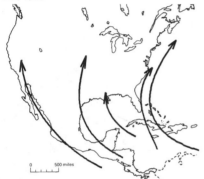

Figure 6.28 Typical Tracks of Tropical Cyclones—Hurricanes—near the United States.

hour, we have a hurricane (see Figure 6.27). Fully formed hurricanes are huge cyclones, often exceeding 500 kilometers in diameter.

Hurricanes tend to form over oceans near the Equator during summer months, when the ocean surface is the warmest. Once the huricane has formed, heat and moisture from the ocean provide the energy that maintains it. Anything that cuts off the supply of heat or moisture will diminish the strength of the hurricane, so hurricanes lose strength as they move over land or cool water. Therefore a hurricane is usually most destructive when it first moves over land. However, even after moving over land, hurricanes can last for many days and cause much damage because of high winds and flooding.

Planetary winds tend to move hurricanes to the west until they reach 30° N latitude; then the prevailing southwesterlies move them from southwest to northeast (see Figure 6.28). However, air masses that the hurricane encounters may deflect its path.

REVIEW QUESTIONS FOR UNIT 6

1. The adjacent diagram represents a map of Western and Central New York State on a day in August. The location of an 18°C isotherm is shown. Why is the 18°C isotherm line farther north over the land than it is over Lake Ontario?

   (1) There is a high plateau at the eastern end of Lake Ontario.
   (2) The air is warmer over the land than over Lake Ontario.
   (3) Isotherms are the same shape as the latitude lines.
   (4) The prevailing winds alter the path of the isotherms.

Note that question 2 has only three choices.

2. In New York State, when do the highest air temperatures for the year usually occur?
   (1) a few weeks before maximum insolation is received
   (2) a few weeks after maximum insolation is received
   (3) at the time that maximum insolation is received

3. The adjacent table shows the noontime data for air pressure and air temperature at a location over a period of one week.

   Based on the data provided, which air pressure would most likely occur at noon on November 15?
   (1) 987 mb
   (2) 1,015 mb
   (3) 1,017 mb
   (4) 1,022 mb

WEATHER DATA RECORDED AT NOON

Date	Nov. 9	Nov. 10	Nov. 11	Nov. 12	Nov. 13	Nov. 14	Nov. 15
Air Temperature (°C)	1	6	0	–2	–4	5	10
Air Pressure (millibars)	1024	998	1015	1021	1030	1013	?

Note that question 4 has only three choices.

4. Two weather stations are located near each other. The air pressure at each station is changing so that the difference between the pressures is increasing. The wind speed between these two locations will probably
   (1) decrease
   (2) increase
   (3) remain the same

Base your answers to questions 5 through 7 on your knowledge of earth science and on the accompanying diagram, which shows air temperatures and dewpoint temperatures recorded during a 24-hour period in New York State.

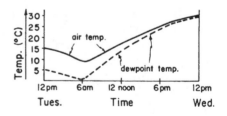

5. The probable changes in pressure for this period are best represented by which graph?

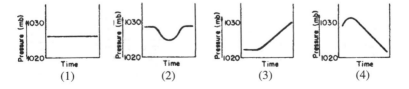

6. Which is the best explanation for the continual increase in temperature after 6:00 P.M. in the evening?
(1) The amount of insolation increased at that time.
(2) A warm airmass moved into the region.
(3) The amount of cloud cover decreased causing more radiative cooling of the Earth.
(4) The burning of fossil fuels to heat homes warmed the night air.

7. According to the *Earth Science Reference Tables*, what is the approximate relative humidity at 12 p.m. Tuesday?
(1) 5%   (3) 50%
(2) 15%   (4) 75%

8. What would an airmass which forms over the land in central Canada most likely be labeled? [Refer to the *Earth Science Reference Tables*.]
(1) cP   (3) mT
(2) cT   (4) mP

9. The accompanying diagram represents a cross-sectional view of airmasses associated with a low-pressure system. The

cold frontal interface is moving faster than the warm frontal interface. What usually happens to the warm air that is between the two frontal surfaces?

(1) The warm air is forced over both frontal interfaces.
(2) The warm air is forced under both frontal interfaces.
(3) The warm air is forced over the cold frontal interface but under the warm frontal interface.
(4) The warm air is forced under the cold frontal interface but over the warm frontal interface.

10. According to the *Earth Science Reference Tables*, which graph best represents the relationship between the altitude and the amount of water vapor in the atmosphere?

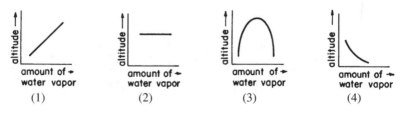

11. The characteristics of an air mass depend mainly upon the
    (1) rotation of the Earth
    (2) cloud cover within the air mass
    (3) wind velocity within the air mass
    (4) surface over which the air mass was formed

12. How does air circulate within a cyclone (low-pressure area) in the Northern Hemisphere?
    (1) counterclockwise and toward the center of the cyclone
    (2) counterclockwise and away from the center of the cyclone
    (3) clockwise and toward the center of the cyclone
    (4) clockwise and away from the center of the cyclone

13. A high-pressure center is generally characterized by
    (1) cool, wet weather
    (2) cool, dry weather
    (3) warm, wet weather
    (4) warm, dry weather

14. On a weather map, an airmass that is very warm and dry would be labeled
    (1) mP   (3) cP
    (2) mT   (4) cT

15. An air mass from the Gulf of Mexico, moving north into New York State, has a high relative humidity. What other characteristics will it probably have?
    (1) warm temperatures and low pressure
    (2) cool temperatures and low pressure
    (3) warm temperatures and high pressure
    (4) cool temperatures and high pressure

16. Which symbol would be used to identify an air mass originating in the Northern Pacific?
    (1) mT   (3) cT
    (2) mP   (4) cP

17. Why do clouds usually form at the leading edge of a cold airmass?
    (1) Cold air contains more water vapor than warm air does.
    (2) Cold air contains more dust particles than warm air does.
    (3) Cold air flows over warm air, causing the warm air to descend and cool.
    (4) Cold air flows under warm air, causing the warm air to rise and cool.

18. Which diagram best represents the air circulation as seen from above in a high-pressure center (anticyclone) in the Northern Hemisphere?

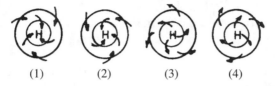

    (1)   (2)   (3)   (4)

19. Precipitation often occurs along a frontal surface because the air along a frontal surface
(1) has a high density
(2) contains condensation nuclei
(3) is rising
(4) is low in humidity.

20. Cities *A*, *B*, *C*, and *D* on the weather map are being affected by a mid-latitude low-pressure system (cyclone).

Which city is located in the warm air mass?
(1) *A*   (3) *C*
(2) *B*   (4) *D*

21. According to the *Earth Science Reference Tables*, what is the air temperature shown on the station model below?
(1) 8°F   (3) 50°F
(2) 21°F   (4) 70°F

22. Which weather station model indicates the greatest probability of precipitation?

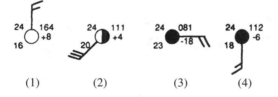

    (1)   (2)   (3)   (4)

23. A weather station records a barometric pressure of 1,013.2 millibars. Which diagram below would best represent this weather station on a weather map? [Refer to the *Earth Science Reference Tables*.]

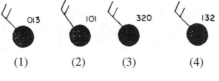

    (1)   (2)   (3)   (4)

Base your answers to questions 24 and 25 on the weather map shown below.

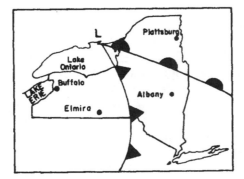

24. Which city is located in the warmest airmass?
    (1) Buffalo   (3) Albany
    (2) Elmira   (4) Plattsburg

25. Which city has most recently experienced a change in wind direction, brief heavy precipitation, and a rapid drop in air temperature?
    (1) Buffalo   (3) Albany
    (2) Plattsburg   (4) Elmira

Base your answers to questions 26 and 27 on the *Earth Science Reference Tables* and the accompanying diagram of a weather station model.

26. The weather forecast for the next 6 hours at this station most likely would be
    (1) overcast, hot, unlimited visibility
    (2) overcast, hot, poor visibility
    (3) overcast, cold, probable snow
    (4) sunny, cold, probable rain

27. The barometric pressure is
    (1) 1,013.0 mb   (3) 130.0 mb
    (2) 913.0 mb   (4) 10.28 mb

28. Which process most directly results in cloud formation?
    (1) condensation   (3) precipitation
    (2) transpiration   (4) radiation

29. Why is it possible for no rain to be falling from a cloud?
    (1) The water droplets are too small to fall.
    (2) The cloud is water vapor.
    (3) The dewpoint has not yet been reached in the cloud.
    (4) There are no condensation nuclei in the cloud.

30. Which atmospheric variable usually increases before precipitation occurs?
(1) insolation   (3) visibility
(2) air pressure   (4) relative humidity

31. Which is a form of precipitation?
(1) frost   (3) dew
(2) snow   (4) fog

32. At what temperature would ice crystals form from air that has a dew-point temperature of -6°C?
(1) 6°C   (3) -2°C
(2) 0°C   (4) -6°C

33. Cities *A*, *B*, *C*, and *D* on the weather map below are being affected by a low-pressure system (cyclone).

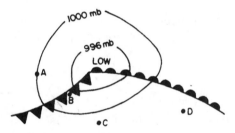

Which city would have the most unstable atmospheric conditions and the greatest chance of precipitation?
(1) *A*   (3) *C*
(2) *B*   (4) *D*

34. Wind moves from regions of
(1) high temperature toward regions of low temperature
(2) high pressure toward regions of low pressure
(3) high precipitation toward regions of low precipitation
(4) high humidity toward regions of low humidity

35. A strong surface wind is blowing from city *A* toward city *B*. City *A* has a barometric pressure of 1,020 millibars. The barometric pressure of city *B* could be
(1) 996 mb   (3) 1,025 mb
(2) 1,020 mb   (4) 1,038 mb

36. Winds appear to curve toward the right in the Northern Hemisphere. This curving to the right is caused by the Earth's
(1) revolution   (3) size
(2) rotation   (4) shape

37. The primary cause of winds is the
    (1) unequal heating of the Earth's atmosphere
    (2) uniform density of the atmosphere
    (3) friction between the atmosphere and the lithosphere
    (4) rotation of the Earth

38. The process by which air flows from one location in the atmosphere to another is called
    (1) absorption   (3) convection
    (2) conduction   (4) radiation

39. To accurately describe the wind, the measurement should include
    (1) a direction, but not a magnitude
    (2) a magnitude, but not a direction
    (3) both a magnitude and a direction
    (4) neither a magnitude nor a direction

40. The accompanying diagram shows the Earth's high and low air pressure belts and direction of prevailing winds for a particular time of the year. The winds do not appear to blow in a straight line from the high-pressure belts to the low-pressure belts. Which statement best explains this observation?

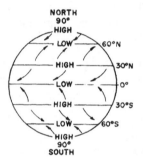

    (1) Wind direction is modified by the Earth's rotation.
    (2) Wind direction is modified by landforms.
    (3) Wind direction is modified by water areas.
    (4) Wind direction is modified by the Sun's motion.

41. Compared to polar areas, why are equatorial areas of equal size heated much more intensely by the Sun?
    (1) The Sun's rays are more nearly perpendicular at the Equator than at the poles.
    (2) The equatorial areas contain more water than the polar areas do.
    (3) More hours of daylight occur at the Equator than at the poles.
    (4) The equatorial areas are nearer to the Sun than the polar areas are.

42. Present-day weather predictions are based primarily upon
    (1) land and sea breezes   (3) ocean currents
    (2) cloud height   (4) air mass movements

43. A prediction of next winter's weather is an example of
    (1) a measurement   (3) an observation
    (2) a classification   (4) an inference

337

44. What form of precipitation results from rain which falls through a freezing air mass?
(1) fog   (3) sleet
(2) snow   (4) frost

45. Which natural process best removes pollutants from the atmosphere?
(1) evaporation   (3) transpiration
(2) precipitation   (4) convection

46. People sometimes release substances into the atmosphere to increase the probability of rain by
(1) raising the air temperature within the clouds
(2) providing condensation nuclei
(3) lowering the relative humidity within the clouds
(4) increasing the energy absorbed during condensation and sublimation

47. Which graph best represents the relationship between water droplet size and the chance of precipitation?

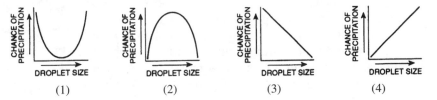

(1)   (2)   (3)   (4)

48. Condensation of water vapor in the atmosphere is most likely to occur when a condensation surface is available and
(1) a strong wind is blowing
(2) the temperature of the air is below 0°C
(3) the air is saturated with water vapor
(4) the air pressure is rising

49. At which location will a low-pressure storm center most likely form?
(1) along a frontal surface between different airmasses
(2) near the middle of a cold air mass
(3) on the leeward side of mountains
(4) over a very dry, large, flat land area

50. During winter, New York City frequently receives rain when locations just north and west of the city receive snow. Which statement best explains this difference?
(1) The snow in the clouds has been depleted by the time the storm reaches New York City.
(2) The ocean modifies New York City's temperatures.
(3) New York City usually receives its weather from the south.
(4) New York City has a higher elevation.

51. Which diagram shows the *usual* paths followed by low-pressure storm centers as they pass across the United States?

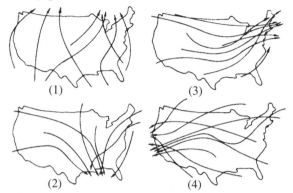

(1)   (3)

(2)   (4)

Base your answers to questions 52 through 56 on your knowledge of earth science, the *Earth Science Reference Tables*, and the diagram below.

Diagram I shows weather systems over a part of North America that includes five weather stations *A, B, C, D,* and *E*. Solid lines represent isobars and dashed lines, *X* and *Y*, represent the horizontal location of frontal surfaces.

Diagram II shows the same weather system in cross-sectional view through stations *A, B, C, D,* and *E*. The relative temperatures of the different air-masses are shown along with their directions of movement. The dashed lines represent the vertical location of frontal surfaces.

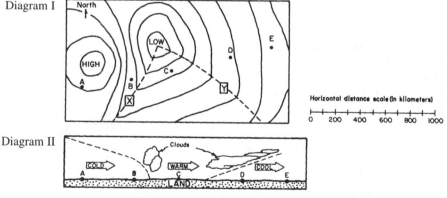

Diagram I

Diagram II

52. According to the information on page 6 of the *Earth Science Reference Tables*, which symbol should be used on a weather map to represent frontal surface (front) *X*?

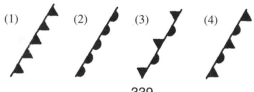

(1)   (2)   (3)   (4)

339

53. Which location is probably experiencing the warmest air temperature?
(1) *A*   (3) *C*
(2) *B*   (4) *D*

54. Wind velocity is probably greatest at which location?
(1) *A*   (3) *C*
(2) *B*   (4) *D*

55. Which diagram best represents the direction of the winds that would normally be associated with these weather systems?

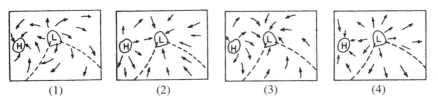

      (1)   (2)   (3)   (4)

56. Which shaded area best represents the portion of the Earth's surface which is most probably experiencing precipitation?

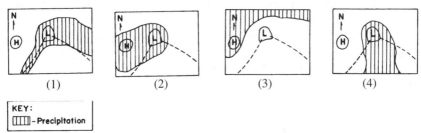

      (1)   (2)   (3)   (4)

KEY:
⊞ - Precipitation

Base your answers to questions 57 through 61 on your knowledge of earth science, the *Earth Science Reference Tables*, and the surface weather map shown here. The map shows weather systems over the United States and weather station data for cities *A*, *B*, *C*, and *D*. Note that part of the weather data for city *C* 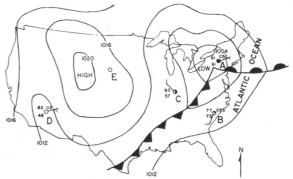 and all of the weather data for city *E* are missing. The pressure field (isobars) on the map has been labeled in millibars.

57. Which city is experiencing the *highest* air temperature?
(1) *A*   (3) *C*
(2) *B*   (4) *D*

58. What type of front extends eastward away from the low-pressure center?
(1) cold   (3) stationary
(2) warm   (4) occluded

59. Which city is probably experiencing a slow, steady rain?
(1) *A*   (3) *C*
(2) *B*   (4) *D*

60. Which weather station model best represents the weather conditions probably existing at city *E*?

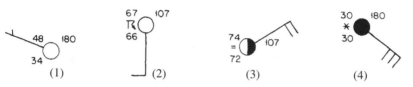

61. If this low-pressure center has followed a normal storm track, which map shows its most likely path over the last two days?

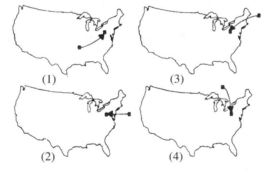

Base your answers to questions 62 through 66 on your knowledge of earth science, the *Earth Science Reference Tables*, and the accompanying chart which contains a summary of weather observations taken at noon during a six-day period.

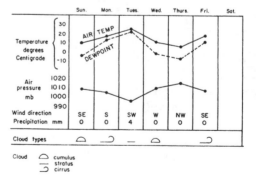

62. Between which two days did the direction of the wind shift 180°?
(1) Sunday and Monday   (3) Wednesday and Thursday
(2) Monday and Tuesday   (4) Thursday and Friday

63. On which day did the greatest chance of precipitation exist?
(1) Monday   (3) Wednesday
(2) Tuesday   (4) Thursday

64. Which weather symbol would most likely have been used to represent this region on Thursday?

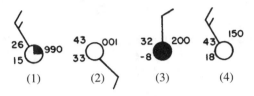

(1) (2) (3) (4)

65. At noon on Friday, the relative humidity was approximately
(1) 0% (3) 50%
(2) 17% (4) 77%

66. If the trends that started on Friday continue, the weather on Saturday most likely will be
(1) cold with clear skies (3) warm with showers
(2) cold with snow (4) hot with clear skies

Base your answers to questions 67 through 71 on your knowledge of earth science, the *Earth Science Reference Tables*, and the diagram below. The diagam shows a common weather condition approaching a section of New York State.

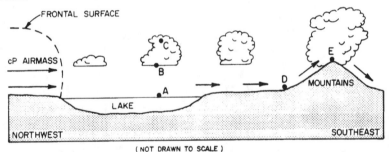

(NOT DRAWN TO SCALE)

67. Which geographic area is the source of the cP airmass shown in the diagram?
(1) Caribbean Sea (3) North Atlantic Ocean
(2) central Canada (4) southwestern United States

68. Which surface weather map below best represents the frontal system shown in the diagram?

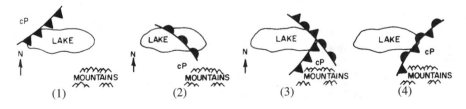

(1) (2) (3) (4)

69. As the air moves from point *D* to point *E*, it will be
   (1) warmed by compression
   (2) warmed by expansion
   (3) cooled by compression
   (4) cooled by expansion

70. In October, the surface water temperature is higher than the surface land temperature because the water
   (1) is heated by the air above it
   (2) is located farther south than the land
   (3) has a higher specific heat than the land
   (4) gains energy during a phase change

71. A strong wind blowing from the northwest toward the southeast would be caused primarily by differences in
   (1) elevation   (3) air pressure
   (2) cloud cover   (4) dewpoint temperature

Base your answers to questions 72 through 76 on your knowledge of earth science and the adjacent diagram. The diagram represents the general circulation of the Earth's atmosphere and the Earth's planetary wind and pressure belts. Points *A* through *F* represent locations on the Earth's surface.

72. The curving paths of the surface winds shown in the diagram are caused by the Earth's
   (1) gravitational field   (3) rotation
   (2) magnetic field   (4) revolution

73. Which location might be in New York State?
   (1) *A*   (3) *C*
   (2) *B*   (4) *D*

74. Which location is experiencing a southwest planetary wind?
   (1) *A*   (3) *C*
   (2) *B*   (4) *F*

75. Which location is near the center of a low-pressure belt where daily rains are common?
   (1) *E*   (3) *F*
   (2) *B*   (4) *D*

343

76. The arrows in the diagram represent energy transfer by which process?
   (1) conduction   (3) convection
   (2) radiation   (4) absorption

77. The United States weather map below shows weather data plotted for a December morning.

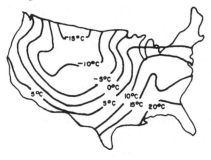

The isolines shown on the map most likely are
   (1) contour lines   (3) isobars
   (2) latitude lines   (4) isotherms

78. The map below shows average annual temperatures in degrees Fahrenheit across the United States.

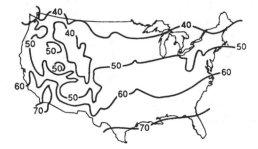

Which climatic factor is most important in determining the pattern shown in the eastern half of the United States?
   (1) ocean currents
   (2) mountain barriers
   (3) elevation above sea level
   (4) latitude

79. What is the air pressure indicated by the diagram of the mercury barometer below?
   (1) 1,028.1 mb
   (2) 1,028.5 mb
   (3) 1,029.5 mb
   (4) 1,031.0 mb

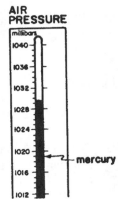

80. A balloon carrying weather instruments is released at the Earth's surface and rises through the troposphere. As the balloon rises, what will the instruments generally indicate? [Refer to the *Earth Science Reference Tables.*]
(1) a decrease in both air temperature and air pressure
(2) an increase in both air temperature and air pressure
(3) an increase in air temperature and a decrease in air pressure
(4) a decrease in air temperature and an increase in air pressure

81. According to the *Earth Science Reference Tables*, an air temperature of 15°C is equal to
(1) -10°F　　　　　　　(3) 59°F
(2) 41°F　　　　　　　(4) 78°F

82. According to the *Earth Science Reference Tables*, an atmospheric pressure of 1,019 millibars is equal to
(1) 31.05 inches of mercury
(2) 30.15 inches of mercury
(3) 30.09 inches of mercury
(4) 30.00 inches of mercury

83. Which graph represents the greatest rate of temperature change?

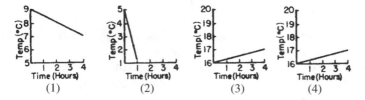

(1)　　　　　　(2)　　　　　　(3)　　　　　　(4)

Base your answers to questions 84 and 85 on your knowledge of earth science, the *Earth Science Reference Tables*, and the diagram below which shows the readings on a psychrometer (wet- and dry-bulb thermometers) after it has adjusted to atmospheric temperatures and humidity.

Wet bulb

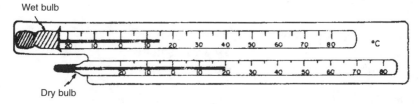

Dry bulb

84. What is the approximate dewpoint temperature?
(1) 5°C　　　　　　　(3) 15°C
(2) 12°C　　　　　　　(4) 20°C

85. Which statement best explains the difference in the readings on the two thermometers?
(1) Evaporation removes heat energy from the wet bulb.
(2) Evaporation absorbs heat energy from the surrounding air.
(3) Condensation removes heat energy from the wet bulb.
(4) Condensation absorbs heat energy from the surrounding air.

86. A container of water is placed in an open area outdoors so that the evaporation rate may be observed. The water will probably evaporate fastest when the air is
(1) cool and humid   (3) warm and humid
(2) cool and dry   (4) warm and dry

87. If the air temperature were 20°C, which dewpoint temperature would indicate the highest water vapor content?
(1) 18°C   (3) 10°C
(2) 15°C   (4) 0°C

88. According to the *Earth Science Reference Tables*, what is the approximate dewpoint temperature if the dry bulb temperature is 24°C. and the wet bulb temperature is 19°C?
(1) 50°C   (3) 18°C
(2) 16°C   (4) 22°C

89. There will most likely be an increase in the rate at which water will evaporate from a pond, if there is a decrease in the
(1) wind velocity   (3) moisture content of the air
(2) temperature of the air (4) altitude of the Sun

90. Rapidly falling barometric pressure readings usually indicate
(1) clearing conditions
(2) approaching storm conditions
(3) decreasing humidity
(4) decreasing temperatures

91. Air pressure is usually highest when the air is
(1) cool and dry   (3) warm and dry
(2) cool and moist   (4) warm and moist

92. Clouds usually form when moist air rises because
(1) the air pressure increases
(2) the dewpoint decreases
(3) the air is cooled to its dewpoint
(4) additional water vapor is added to the air

93. Which graph best represents the effect that heating has on air density in the troposphere?

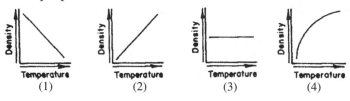

| (1) | (2) | (3) | (4) |

Note that questions 94 and 95 have only three choices.

94. Moisture is evaporating from a lake into stationary air at a constant temperature. As more moisture is added to this air, the rate at which water will evaporate will probably
(1) decrease
(2) increase
(3) remain the same

95. As a stationary airmass is heated, its density will generally
(1) decrease
(2) increase
(3) remain the same

96. Which statement best explains why the air pressure is usually greater over the ocean than over the land during the daytime hours in the summer?
(1) Air temperature is lower over the ocean.
(2) Air is more humid over the ocean.
(3) Prevailing winds are usually blowing from the land.
(4) Water absorbs heat from the land.

97. The graph below shows the changes in air temperature and dewpoint temperature over a 24-hour period at a particular location.

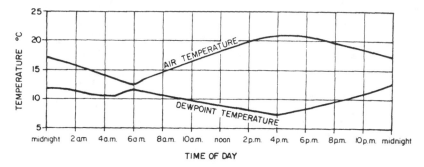

At what time was the relative humidity lowest?
(1) midnight
(2) 6 A.M.
(3) 10 A.M.
(4) 4 P.M.

The phrases in questions 98 through 102 describe various forms of moisture. Write the *number* of the form, *chosen from the list below*, that is described by each phrase. [An answer may be used once, more than once, or not at all.]

Forms of Moisture
(1) Snow
(2) Glaze
(3) Sleet
(4) Hail
(5) Rain
(6) Fog
(7) Dew

98. Most common form of precipitation

99. Formed from a liquid into a solid by strong convectional currents in a cumulonimbus cloud

100. Formed when supercooled cloud droplets strike the wing of an airplane

101. Form of precipitation which has a hexagonal crystal

102. Formed when droplets of suspended moisture become larger and then fall to the Earth

OPTIONAL/EXTENDED

TOPIC E _____

Latent Heat/ Atmospheric Energy

KEY IDEAS The various electromagnetic waves that emanate from the Sun are the main source of energy for weather changes. Energy is absorbed as it passes through the atmosphere, and some is stored in the form of latent heat.

Evaporation and condensation, which result from the interplay of energy and moisture in the atmosphere, are probably the most important meteorological processes. Convection in the atmosphere distributes solar energy over the whole Earth.

Computer analysis of weather data collected from ground stations and satellites has allowed increasingly reliable short-term forecasts, but the complex development of atmospheric variables makes accurate predictions of weather beyond 1–2 weeks in the future unlikely.

KEY OBJECTIVES
Upon completion of this unit, you should be able to:

• Explain why the Sun emits different types of electromagnetic radiation.
• Compare and contrast the different forms of electromagnetic radiation shown in the *Earth Science Reference Tables*.
• Describe the mechanisms by which solar energy is absorbed by the atmosphere.
• Define the terms *latent heat of vaporization* and *latent heat of fusion*, and solve problems involving changes of phase.

- Explain the energy transformations that occur during adiabatic heating and cooling, and describe cloud formation by adiabatic cooling.
- Interpret a diagram of the planetary wind and pressure belts.
- Describe several recent technological advances in meteorology.
- Compare and contrast the accuracy of short- and long-term weather forecasts.

A. HOW DOES THE ATMOSPHERE STORE AND RELEASE ENERGY?

The Sun is the major source of energy for weather changes. Energy from the Sun reaches the Earth in the form of electromagnetic waves.

A-1. Energy from Electromagnetic Waves

A magnet can make a compass needle move from a distance because the magnet is surrounded by an invisible magnetic field. If you move the magnet, the magnetic field will move and the moving magnetic field, in turn, will cause the compass needle to move. In much the same way, a statically charged balloon held near your head will attract your hair from a distance because the balloon is surrounded by an invisible electric field. If you move the balloon, the invisible electric field moves with it and you can feel the effect of the moving field on your hair. Such fields extend outward infinitely in all directions from their sources, but they weaken with distance from these sources (see Figure E.1).

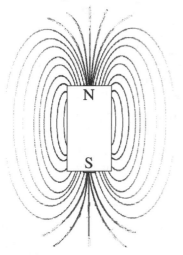

Figure E.1 An Invisible Magnetic Field Surrounding a Magnet. This field decreases in strength with distance.

All matter is composed of atoms. Every atom consists of electrically charged particles such as protons and electrons, which are surrounded by an electric field. Whenever a charged particle moves, a magnetic force is produced and the particle is then also surrounded by a magnetic field. These two fields—an electric field and a magnetic field—exist simultaneously around all electrically charged particles that are moving. Together, they are called an *electromagnetic field*, and they extend outward infinitely in all directions around the particle.

When a particle moves back and forth, its electromagnetic field moves with it (see Figure E.2a). Like a ripple in a pond, this movement spreads out through the field in the form of a transform wave. The way that the wave moves through the electromagnetic field depends on the way in which the particle moves. The faster the particle oscillates, the shorter the wavelength, or distance between successive "ripples" in the electromagnetic field. Every frequency of oscillation produces a wave of a different wavelength (Figure E.2b).

A moving wave contains energy. It can exert forces on matter with which it interacts. Since an electromagnetic field does not require a medium to exist, it can extend through space. Disturbances in electromagnetic fields around particles can travel through space as they move outward, or radiate, through the field. In this way, energy is transferred from the Sun to the Earth without the existence of a physical medium between the two.

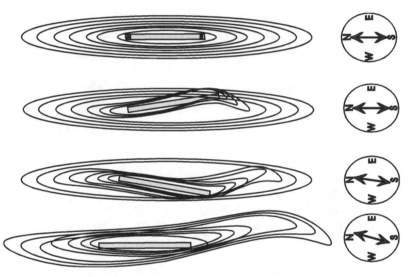

(a) As a magnet moves, its magnetic field moves too. Like a ripple in a pond, the movement of the field spreads outward in all directions at a rate of 3×10^8 meters per second—the speed of light. When the moving field reaches *another* field (such as the one around a compass needle), it exerts a force of attraction or repulsion that can cause motion of the field and even of the object that produced the field. A similar disturbance would travel through the electric field around an electrically charged particle in motion.

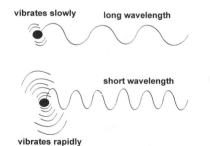

(b) Electromagnetic waves produced by particles vibrating at different rates. The higher the temperature, the faster the particle vibrates, and the shorter the wavelength produced.

Figure E.2 Electromagnetic Waves Are Produced by Vibrating Particles

The Electromagnetic Spectrum

Electromagnetic waves are classified by their lengths, and range from short waves, such as X rays, to long waves, such as radio waves. The *electromagnetic spectrum* (see Figure E.3) is a continuum in which electromagnetic waves are arranged in order, from longest to shortest wavelength. Infrared (heat) waves and visible-light waves fall roughly in the middle range of wavelengths. The wavelength, or distance between "ripples" of visible light waves, falls between 10^{-6} and 10^{-7} meter.

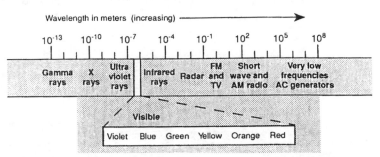

Figure E.3 Electromagnetic Spectrum. This shows the categories into which electromagnetic waves are classified according to their lengths.

Many different waves emanate from the Sun (Figure E.4) because it contains particles moving at many different speeds. However, most waves coming from the Sun are in the visible-light and infrared range. To move at the speed needed to emit waves of this length, particles at the Sun's surface must

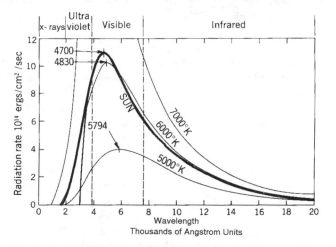

Figure E.4 Makeup of Solar Radiation. The Sun emits electromagnetic radiation ranging in wavelength from infrared to X rays. The majority of the waves are in the visible-light range with a peak at 4,700 angstrom units (1 angstrom unit = 1×10^{-9} meter—one billionth of a meter). This is very close to the theoretical output of a body at a temperature at 6,000K, which is why the Sun's surface temperature is thought to be between 6,000 and 7,000K.

have a temperature between 5,000K and 7,000K. The Earth's magnetic field and the Van Allen belts of charged particles in the upper atmosphere deflect many of the waves with short lengths. This circumstance is fortunate, since most short-wavelength radiation is harmful to living things.

The Global Radiation Budget

The Earth intercepts electromagnetic waves from the Sun as they travel through space. The Sun's electromagnetic waves exert a force on the fields surrounding particles of matter in the atmosphere, hydrosphere, and lithosphere, causing the particles to move. This increased motion shows up as an increase in the level of the Earth's heat energy. At the same time, the Earth's matter radiates energy in the form of electromagnetic waves, causing a decrease in the Earth's heat energy. Over long periods of time, the average level of the Earth's heat energy remains constant. We say that the Earth is in *radiative balance*.

Both incoming solar radiation and outgoing terrestrial radiation pass through the atmosphere. Table E.1 shows the global radiation budget.

TABLE E.1 GLOBAL RADIATION BUDGET*

	Percent	
Incoming Solar Radiation	Gain	Loss
Reflection from clouds to space		21
Diffuse reflection (scattering) to space		5
Direct reflection from Earth's surface		6
Net energy loss back to space		32
Absorbed by clouds	3	
Absorbed by molecules, dust, water vapor and CO_2	15	
Absorbed by Earth's surface	50	
Net incoming energy gained by Earth-atmosphere system	**68%**	
Outgoing Terrestrial Radiation		
Total infrared radiation emitted by the Earth's surface	98	
Absorbed by atmosphere	90	
Lost to space		8
Total infrared radiation emitted by the atmosphere	137	
Absorbed by Earth's surface	77	
Lost to space		60
Net outgoing energy lost from entire Earth-atmosphere		**68%**
Net energy leaving *Earth's surface* (98 emitted − 77 reabsorbed)		21
Net energy leaving the *atmosphere* (137 emitted − 90 reabsorbed)		47

*Source: *The Earth Sciences.* © by Arthur N. Stahler, 1971.

A-2. Energy from Phase Changes of Moisture and Adiabatic Changes in the Atmosphere

Moisture in the atmosphere may exist in three possible states: gas, liquid, and solid. Gaseous water, called *water vapor*, diffuses freely amidst the other gases of the atmosphere. The amount of water vapor in the atmosphere varies from as little as 0.1 to 1.8 percent of air by weight. As a result, there is almost 20 times more water vapor in the air in some places than in others. Liquid water in the atmosphere is found mainly in the form of the tiny water droplets that make up clouds and fog, but liquid water exists also as precipitation and dew. Solid water takes the form of very tiny crystals of ice in the cirrus clouds of the upper troposphere, snow, or frost.

Latent Heat of Fusion and Vaporization

Typically, when matter gains heat energy, its temperature increases; and when matter loses heat energy, its temperature decreases. However, when matter changes its state, heat is given off or absorbed in the process of change. The heat energy is involved in causing the change of state and does not result in a change in temperature. Since this heat does not show up as a temperature change, it is called hidden or *latent heat*. The heat absorbed when solid water melts into liquid water, called the *latent heat of fusion*, amounts to about 80 calories per gram

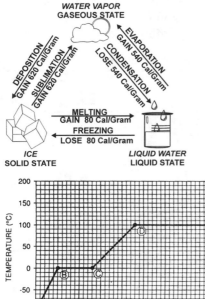

(a) Water gains or loses energy when it undergoes a change in state of matter. Since this gain or loss of heat energy does not result in a change in temperature, it is called hidden or *latent heat*.

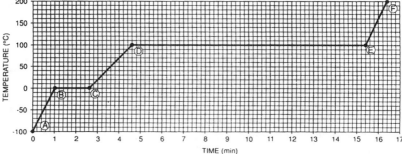

(b) The heating curve for a sample of water heated from a starting temperature of −100°C to a final temperature of +200°C. The same amount of heat is added to the sample every minute. From A to B the ice increases in temperature until it reaches its melting point. From B to C there is a change in state from solid ice to liquid water with no increase in temperature (the curve is flat, like a plateau). From C to D liquid water increases in temperature until it reaches its boiling point. From D to E there is a change in state from liquid water to water vapor, again with no increase in temperature. From E to F water vapor is increasing in temperature. Note that the flat area from B to C is shorter than the flat area from D to E since less energy is required to change water from a solid to a liquid than to change water from a liquid to a gas.

Figure E.5 Changes in State of Water and Heating Curve for Water

of water. The heat absorbed when liquid water evaporates into water vapor, called the *latent heat of vaporization,* amounts to about 540 calories per gram of water. The changes of state of water and the heat given off or absorbed in each process are summarized in Figure E.5.

Solar energy absorbed by water during the process of evaporation is stored in the form of latent heat in water vapor of the atmosphere. Every gram of water vapor in the atmosphere contains the 540 calories of latent heat that was absorbed during the change from a liquid to a gas. When condensation occurs, as during cloud formation, this heat is released back into the atmosphere. As a result, condensation is a warming process that causes the temperature of air to rise as droplets of water in clouds grow in size. The increase in temperature causes the air to expand, become less dense, and rise. Thus, latent heat is an important energy source for violent storms. The release of latent heat during large-scale condensation produces localized heating, which causes updrafts. At the same time, precipitation associated with the heavy condensation causes downdrafts. The result is the violent circulations of air that occur in storms such as hurricanes and tornadoes.

Adiabatic Changes in the Atmosphere

As air rises in the atmosphere, it expands because of a reduction in the air pressure confining it and becomes cooler. Conversely, air that sinks in the atmosphere is compressed into a smaller volume because of an increase in confining pressure and becomes warmer. These temperature changes are called *adiabatic*, because no heat energy is absorbed from outside sources or lost to the surroundings to produce the changes. Rather, the temperature of the air changes because the average energy of its molecules changes.

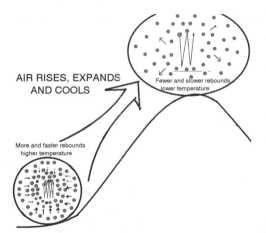

AIR RISES, EXPANDS AND COOLS

Fewer and slower rebounds lower temperature

More and faster rebounds higher temperature

Figure E.6 Adiabatic Cooling. As air rises and expands, there are fewer and less energetic collisions, resulting in a decrease in the average speed of the molecules and a decrease in temperature.

The Mechanism of Adiabatic Heating and Cooling

How does changing the pressure that confines air change the average energy of its molecules? Think of a parcel of air being contained by the pressure of the surrounding air as being in a vessel. If the confining pressure increases, the surrounding air moves inward on the parcel of air, as would the walls of the vessel if pushed toward each other. As the inward-moving air collides with the molecules in the parcel of air, it imparts energy to the molecules as does a bat hitting a ball. Molecules rebounding from the inward-moving air move faster than they did before. The increase in the energy of molecules that collide with the inward moving air raises the average energy of the molecules in the parcel of air.

Conversely, if confining air pressure decreases, the parcel of air expands into a larger space. The surrounding air moves outward as if the walls of the vessel were moving apart. Molecules rebounding from the outward-moving air move slower than before. The decrease in the energy of molecules that collide with the outward-moving air lowers the average energy of the parcel of air.

Molecular changes during adiabatic cooling are illustrated in Figure E.6.

Adiabatic Lapse Rates

In rising or sinking air, the temperature change caused by adiabatic processes is called the *adiabatic lapse rate*. When air that is not saturated with water vapor rises, the air temperature falls 10°C for every 1,000-meter increase in elevation; this is called the *dry adiabatic lapse rate*.

The dew point temperature also falls with elevation because the expanding air contains less water vapor per unit of volume. The less water vapor in a given volume of air, the more the air has to be cooled before its capacity to hold moisture equals the moisture in the air. The dew point falls at 2°C for every 1,000-meter increase in temperature; this is called the *dew point lapse rate*.

As you can see, if the dew point falls only 2°C in 1,000 meters while the air temperature drops by 10°C, the air temperature will approach the dew point at a rate of 8°C per 1,000 meters. Eventually the air will reach its dew point, and condensation will begin to occur. As soon as condensation occurs, latent heat is released into the air and warms it, offsetting the adiabatic lapse rate. As a result, the temperature of air saturated with water vapor, in which condensation is occurring, falls only 6°C per 1,000 meters; this is called the *moist adiabatic lapse rate*.

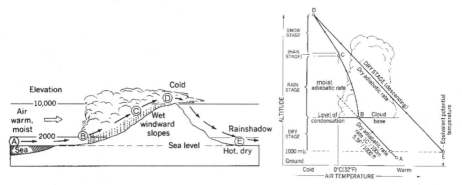

Figure E.7 The Orographic Effect. Warm, moist air forced to rise over the windward side of a mountain range cools adiabatically and loses much of its moisture because of condensation and precipitation. When the dry air sinks on the leeward side, it warms adiabatically. The hot, dry air sweeps over the land, absorbing any moisture and creating an arid region in the lee of the mountain range.

Air that sinks is compressed and increases in temperature. As a result, the air temperature moves away from the dew point and condensation ceases. Therefore, sinking air rises in temperature at the dry adiabatic lapse rate of 10°C per 1,000 meters. Figure E.7 summarizes the temperature changes that take place in an air mass because of adiabatic changes as it is forced to rise over the windward side of a mountain range and then sinks on the leeward side.

The lapse rate chart in the *Earth Science Reference Tables* shows the dry adiabatic lapse rate as bold slanting lines and the dew point lapse rate as dotted slanting lines. It can be used to predict the altitude at which condensation will occur when a parcel of air rises.

Example

A psychrometer is used at ground level in a parcel of rising air. The dry-bulb temperature reads 30°C, and the wet-bulb temperature is 20°C. At what altitude will condensation occur in this parcel of air?

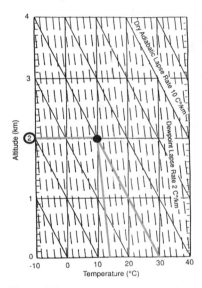

Figure E.8 Lapse Rate Chart. Solid slanting lines represent the lapse rate of dry air; dotted slanting lines, the lapse rate of the dew point. Air that is 30°C with a dew point of 14°C at the ground will undergo condensation and cloud formation when it rises to an altitude of 2 kilometers.

In order to use the lapse rate chart you need to know the *air temperature* and the *dew point.*

The air temperature = 30°C (the dry-bulb temperature reading).

The dew point can be determined by using the dew point temperature chart on page 12 of the *Earth Science Reference Tables.*

Difference between wet-bulb and dry-bulb temperatures = 30°C – 20°C = 10°C.

Finding the intersection of 30°C and 10°C on the dew point chart, determine the dew point.

Dew point = 14°C.

Now, using the lapse rate chart, find the bold slanting line for 30°C. This line marks the dry adiabatic lapse rate.

Now find the dotted line slanting upward to the left from 14°C. This line marks the dew point lapse rate.

Follow these two lines upward until they intersect (see Figure E.8). At their point of intersection, read the altitude from the vertical scale. The altitude is 2 kilometers.

Condensation will begin when the air rises to an altitude of 2 kilometers.

357

Cloud Formation and Cloud Types

Clouds are composed of tiny droplets of water or ice crystals ranging from 0.02 to 0.06 millimeter in diameter. Cloud particles form by condensation and sublimation. Each water droplet or ice crystal forms around a tiny particle called a *condensation nucleus*. Most condensation nuclei are minute particles of salt, dust, or other chemical compound. Condensation nuclei provide surfaces on which condensation or sublimation can occur and are usually *hygroscopic*, that is, they readily absorb moisture. Common table salt is hygroscopic; unless chemically treated, it absorbs moisture and becomes soggy if left out on humid days. Salt particles enter the atmosphere from the oceans when droplets of seawater are blown into the air and evaporate, leaving behind tiny salt crystals that remain suspended in the air. Dust enters the atmosphere from volcanic eruptions or may be picked up by winds and blown aloft.

The tiny water droplets in clouds or fog remain in the liquid state at temperatures far below the freezing point of water. Until the temperature drops below −10°C, almost all cloud droplets are in the liquid state. At temperatures below 0°C, droplets of water are said to be supercooled. Between −10°C and −30°C, clouds contain a mixture of water droplets and ice crystals. Only at temperatures below −40°C are clouds composed entirely of ice crystals. As you can see from Figure E.9, temperature drops steadily with altitude in the troposphere. In the upper portion of the troposphere, temperatures below −40°C are the norm year-round. For this reason, even in the summer, cirrus clouds are composed of tiny ice crystals, and thunderstorms whose cloud tops extend into the upper troposphere may produce hail.

Clouds are important components of the atmosphere for several reasons. They are the producers of precipitation, such as rain, snow, sleet, and hail.

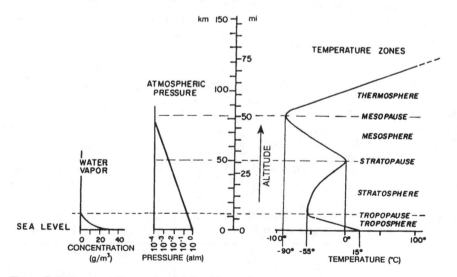

Figure E.9 Physical Properties of the Atmosphere

They are also major reflectors of solar radiation, and key indicators of over-all weather conditions.

As shown in Figure E.10, clouds are categorized by the altitude at which they exist and their degree of vertical development.

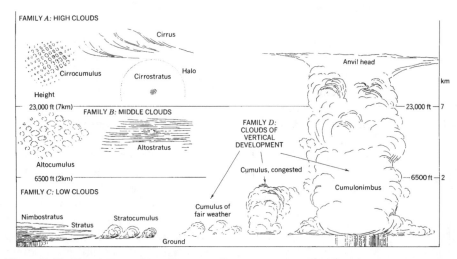

FAMILY A: HIGH CLOUDS

Cirrus

Cirrocumulus

Cirrostratus

Halo

Anvil head

km

Height

23,000 ft (7km)

FAMILY B: MIDDLE CLOUDS

23,000 ft — 7

FAMILY D:
CLOUDS OF
VERTICAL
DEVELOPMENT

Altostratus

Altocumulus

Cumulus, congested

6500 ft (2km)

6500 ft — 2

FAMILY C: LOW CLOUDS

Cumulonimbus

Nimbostratus

Stratus

Stratocumulus

Cumulus of
fair weather

Ground

Figure E.10 Cloud Form Classification. Clouds are classified into families according to altitude and degree of vertical development. From Arthur N. Strahler, *The Earth Sciences*.

A-3. Global Convection Patterns in the Atmosphere

Unequal heating of the Earth's surface at different latitudes results in unequal heating of the atmosphere. The equatorial regions of the Earth are predominantly water and receive more solar energy per year than the polar regions. Air over the Equator becomes warm and moist through contact with the Earth's surface. As was discussed in Unit 6, air that is warm and moist exerts less pressure than cooler, drier air. Thus, the air at the Equator is surrounded by air of higher pressure that pushes inward toward the Equator and displaces the warm, moist equatorial air upward. As this air is forced upward it expands outward and cools adiabatically, resulting in condensation and precipitation. The end result is cooler, dryer air that is denser than the air beneath it and begins to sink back toward the surface. Together, the rising warm, moist air and sinking cool, dry air form a circular pattern of motion called a *convection cell*.

A similar convection cell forms over the poles, although it begins with sinking air, which spreads southward when it reaches the Earth's surface, warms, and rises. In between the polar and equatorial convection cells lies a third cell,

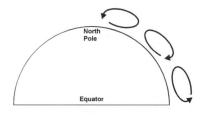

North
Pole

Equator

Figure E.11 Three-Cell Theory of Convection in the Atmosphere.

which is set in motion by the westerly winds of the middle latitudes. This *three-cell theory* of global atmospheric circulation (see Figure E.11) explains how the atmosphere distributes solar energy over the whole Earth.

B. WHAT IS THE CORIOLIS EFFECT?

One of the consequences of the Earth's rotation is a tendency of all matter that is in motion on the Earth's surface to be deflected to the right from its point of origin in the Northern Hemisphere and to the left in the Southern Hemisphere. This is called the *Coriolis effect* after the French mathematician G. G. Coriolis, who first analyzed it in the nineteenth century. The Coriolis effect is not an actual force, but is the apparent effect of a number of different forces acting on any particles in motion.

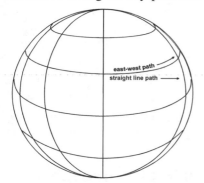

east-west path
straight line path

The main elements of the Coriolis effect are the curvature of the Earth's surface, the rotation of the Earth, and the tendency of objects in motion to remain in motion in a straight line (Newton's first law of motion). The east-west path between any two points on the Earth's surface is not a straight line, but a curve—a segment of a parallel (line of latitude). As a result, any object following a straight path will appear, to an observer on Earth, to curve from its path. We say "appear" because actually it is the observer who is traveling a curved path while the object travels in a straight line. In Figure E.12 the solid line represents the actual path of an observer on the surface of a spherical Earth as it rotates. The dotted line represents a straight path for a moving object. The Coriolis effect is most pronounced where there is the greatest difference between the curvature of a parallel and

Figure E.12 Deflection Due to Curvature of the Earth's Surface. An object moving in a straight line travels along the dashed line. An observer on the Earth's surface moves due east-west as the Earth rotates, following a path that curves. To an observer on the Earth's surface, the straight-line path seems to veer off, or be deflected, to the right in the Northern Hemisphere. In the Southern Hemisphere, the deflection is to the left.

a straight-line path—in other words, near the poles.

A similar deflection in the north-south direction is due to the difference in the speed at which points on the Earth's surface are moving because of rotation. As the Earth rotates, every object on its surface rotates with it. In one 24-hour rotation, objects near the poles travel less distance than objects near the Equator and thus are moving slower (see Figure E.13).

Now, let's suppose a rocket at the Arctic Circle (60° N) is aimed due south at New York City (40° N). Before the rocket is even launched, it is moving west to east at the speed of the Earth's surface—233 meters per second. The

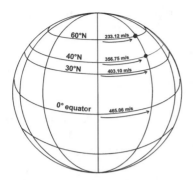

Figure E.13 Deflection Due to Different Velocities at Different Latitudes. As the Earth rotates, objects on its surface move at different speeds, depending on their latitude. Objects at the Equator move faster because they cover more distance in one 24-hour rotation than objects near the poles.

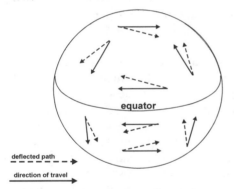

Figure E.14 The Coriolis Effect. Objects are deflected to the right from their points of origin in the Northern Hemisphere and to the left from their points of origin in the Southern Hemisphere.

target, New York City, is also moving west to east at the speed of the Earth's surface—356 meters per second. Note that the target is moving west to east faster than the rocket! When the rocket is fired, it moves due south and west to east; but since the target is moving west to east faster than the rocket, the rocket lags behind the target and actually hits the ground behind the target. To an observer, the rocket appears to follow a path that curves to the *right* of the target. Similarly, a rocket fired from New York City toward the Arctic Circle would land ahead of the target because New York City is moving west to east faster than the target. Nevertheless, the deflection would still be to the *right*. As seen in Figure E.14, no matter what direction the motion, deflection is always to the right in the Northern Hemisphere.

For objects following paths that fall somewhere between due north-south or due east-west, total deflection is a combination of the deflection due to curvature and the deflection due to differences in rotational speed. Of course, frictional forces within the atmosphere and with the Earth's surface modify the Coriolis effect, but it still persists.

C. HOW CAN WE FORECAST THE WEATHER?

Until relatively recently, weather forecasts were based on local observations made directly by human senses. Through first-hand experience and passed-down weather lore, people learned to recognize certain patterns of weather changes associated with cloud types or shifts in wind direction. It was not until the invention of the barometer and the thermometer in the 1600s that accurate measurements of pressure and temperature changes could be made. Even then, forecasting was not practicable until communication among many points by telegraph made possible the rapid collection and sharing of weather data. The first government weather forecasts were issued in the United States in 1870.

C-1. Synoptic Weather Forecasting

The basis of most weather forecasting is the systematic collection of weather data from a network of weather-reporting sites that blanket the globe. At the simplest level, these data are summarized on a synoptic weather map using shorthand symbols called *station models*. At their most sophisticated, computer programs compile, store, analyze, and plot weather data, plot isolines for various weather factors, and can display weather data in a variety of graphic formats.

Synoptic weather maps are made four times a day—at midnight, noon, 6 A.M., and 6 P.M. Greenwich Mean Time. By looking at a sequence of these maps, a meteorologist can detect the development and movement of weather systems. The movements of these systems are projected into the future, and predictions of weather for various locations are made. Computer-drawn maps show wind, temperature, air pressure, and humidity patterns at many different atmospheric levels. Statistical analysis is then used to map projections of weather factors.

Synoptic forecasting is fairly accurate for large scale, short-term forecasts. However, there are always locally unique conditions that modify the general behavior of weather systems. For example, cities tend to absorb and reradiate more heat energy than rural areas, creating urban "heat islands." Local meteorologists must take such specific conditions into account when making forecasts, for they have a significant effect on such factors as the exact path a storm will take locally, the times when rain will begin and end, and the extremes that high and low temperatures will reach. For this reason, most local meteorologists have a network of weather hobbyists who feed them weather data in addition to those collected by official Weather Service stations.

C-2. Statistical Weather Forecasting

Statistical weather forecasting is based on the past behavior of the atmosphere. Mathematical equations are used to predict the probability that the atmosphere will behave in a certain way. Basically the system works something like this: suppose you have math first period on Mondays, Tuesdays, and Wednesdays, and science first period on Thursdays. A person observing you would soon see a pattern in which three days of first-period math are followed by a first period of science. However, there would be exceptions. There might be a holiday on a Monday, so the sequence would be science after *two* maths. Or Thursday might be a holiday, so the sequence would be *five* maths before the next science. However, such deviations in the schedule would probably be the exception rather than the rule. Let's suppose that in

10 weeks there was one deviation. Statistical forecasting would then say that for the next week there was a 9 in 10, or 90 percent, chance that three maths would be followed by a science.

Now let's apply this to weather. Suppose we observed that, nine out of the past ten times the wind shifted to blow from the northeast, we had rain within 6 hours. Then, if the wind shifted to the northeast, statistical forecasting would say that there was a 90 percent chance of rain in the next 6 hours.

The process of statistical forecasting is threefold: maintaining detailed historical records, identifying all previous occurrences of a pattern, and then calculating the probabilities. However, there is no way to anticipate unusual behavior, nor is past behavior a guarantee of future behavior.

C-3. Numerical Weather Forecasting

Numerical forecasting is based on fluid dynamics. Theoretically, if the initial states of the atmosphere, land surfaces, and oceans are known, we should be able to predict the behavior of the atmosphere based on the laws of physics governing heat and moisture transfer and fluid behavior. However, such complete information is not currently available.

The sheer quantity of data, coupled with the number and complexity of the calculations needed to carry out numerical forecasting, made it impracticable until computers capable of high-speed calculations were developed in the 1940s. Now, computers solve six equations simultaneously—one for each of the three dimensions of motion, and one each for the conservation of moisture, heat, and mass. These equations are solved for thousands of points on a map grid, and changes at each point are projected. These changes are then fed back into the loop for another round of calculations, producing a projection further into the future. As new actual data are entered into the equations throughout the day, the revised calculations serve to correct the predictions. Calculations for several levels in the atmosphere, projecting changes over the next 2 days, are made every 12 hours. A 5-day forecast is calculated once a day.

C-4. Long-Range Weather Forecasting

No matter which method of forecasting is used, the reliability of the forecast decreases as its time range increases. Currently, weather cannot be accurately predicted beyond 5 days. This unreliability is due to the wide spacing between points from which data are collected, the unreliability of some measurements, and our lack of understanding of some aspects of atmospheric behavior.

Unlike short-term forecasts, which look at existing weather systems, long-range forecasting deals with predictions of the behavior of weather systems that do not yet exist. Forecasting for time periods ranging from a week to a month in the future is based on the positions of large-scale flow patterns in the atmosphere, such as the convection cells and the waves in the polar front. These flow patterns remain relatively stable over long periods of time and steer lows and highs along preferred pathways. They are somewhat useful in predicting the periods of time during which heat waves, cold spells, droughts, and excessive rain will occur. Beyond a month, however, forecasting is essentially done statistically on the basis of historical records. Large-scale seasonal changes are cyclic and can be predicted in advance, but not with any great degree of accuracy.

Chaos theory indicates that very small anomalies can grow in importance to influence worldwide weather changes. Such small-scale anomalies as a plowed field, which absorbs more heat than a forested area, or a newly built housing development that deflects surface winds, are relatively unpredictable, yet can theoretically lead to large-scale changes in weather systems. Therefore, many meteorologists believe that chaotic relationships make long-term predictions impossible.

REVIEW QUESTIONS FOR TOPIC E

1. At which temperature would an object radiate the least amount of electromagnetic energy?
   (1) the boiling point of water (100°C)
   (2) the temperature at the stratopause (0°C)
   (3) the temperature of the North Pole on December 21 (–60°F)
   (4) room temperature (293°K)

2. The graph below represents the relationship between the intensity and wavelength of the Sun's electromagnetic radiation. Which statement is best supported by the graph?

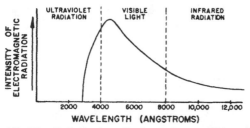

   (1) The infrared radiation given off by the Sun occurs at a wavelength of 2,000 angstroms.
   (2) The maximum intensity of radiation given off by the Sun occurs in the visible region.
   (3) The infrared radiation given off by the Sun has a shorter wavelength than ultraviolet radiation.
   (4) The electromagnetic energy given off by the Sun consists of a single wavelength.

3. An object that is a good radiator of electromagnetic waves is also a good
  (1) insulator from heat
  (2) reflector of heat
  (3) absorber of electromagnetic energy
  (4) refractor of electromagnetic energy

4. In the Earth's atmosphere, the best absorbers of infrared radiation are water vapor and
  (1) nitrogen   (3) hydrogen
  (2) oxygen   (4) carbon dioxide

5. Most of the energy in the Earth's atmosphere comes from
  (1) the rotation and revolution of the Earth
  (2) the rotation of the Earth and wind from the Earth
  (3) radioactive decay of elements and radiation from the Earth
  (4) radiation from the Earth and insolation from the Sun

6. Which model best represents what happens to energy from the Sun as the energy approaches the Earth? [Diagrams are not to scale.]

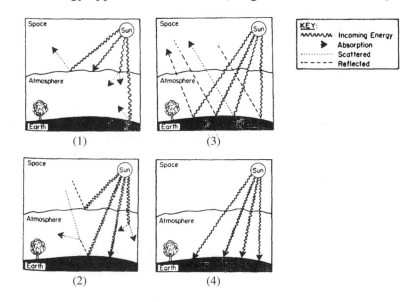

7. Which color is the best radiator of electromagnetic energy?
  (1) red   (3) black
  (2) white   (4) yellow

8. Electromagnetic energy that is being given off by the surface of the Earth is called
  (1) convection   (3) specific heat
  (2) insolation   (4) terrestrial radiation

Note that questions 9 and 10 have only three choices.

9. As the amount of solid aerosols (pollutants) in the air increases, the amount of insolation that reaches the Earth's surface will
   (1) decrease
   (2) increase
   (3) remain the same

10. The environment is in dynamic equilibrium when it is gaining
    (1) less energy than it is losing
    (2) more energy than it is losing
    (3) the same amount of energy as it is losing

11. The adjacent diagram shows part of the electromagnetic spectrum.

    Which form of electromagnetic energy shown on the diagram has the lowest frequency and longest wavelength?
    (1) AM radio   (3) red light
    (2) infrared rays   (4) gamma rays

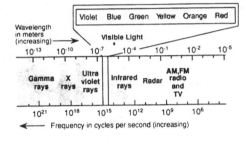

12. A sample of water undergoes the phase changes from ice to vapor and back to ice as shown in the model below. During which phase change does the sample gain the greatest amount of energy?

    (1) A   (3) C
    (2) B   (4) D

Note that question 13 has only three choices.

13. If equal masses of water in various phases (states) are compared, which phase will contain the greatest amount of stored energy (latent heat)?
    (1) solid ice
    (2) liquid water
    (3) water vapor

14. The change from the vapor phase to the liquid phase is called
    (1) evaporation   (3) precipitation
    (2) condensation   (4) transpiration

15. Why is the condensation of water vapor considered to be a process which heats the air?
    (1) Liquid water has a lower specific heat than water vapor.
    (2) Energy is released by water vapor as it condenses.
    (3) Water vapor must absorb energy in order to condense.
    (4) Air can hold more water in the liquid phase than the vapor phase.

16. Based on the *Earth Science Reference Tables,* what is the total amount of energy required to melt 100 grams of ice at 0°C to liquid water at 0°C?
    (1) 5,400 cal   (3) 54,000 cal
    (2) 8,000 cal   (4) 80,000 cal

17. According to the *Earth Science Reference Tables*, how many calories of heat energy must be added to 20 grams of liquid water to change its temperature from 20°C to 30°C?
    (1) 10 cal   (3) 200 cal
    (2) 20 cal   (4) 400 cal

18. An ice cube is placed in a glass of water at room temperature. Which heat exchange occurs between the ice and the water within the first minute?
    (1) The ice cube gains heat and the water loses heat.
    (2) The ice cube loses heat and the water gains heat.
    (3) Both the ice cube and the water gain heat.
    (4) Both the ice cube and the water lose heat.

19. Which process results in a release of latent heat energy?
    (1) melting of ice
    (2) heating of liquid water
    (3) condensation of water vapor
    (4) evaporation of water

20. Which statement best explains why a desert often forms on the leeward side of a mountain range, as shown in the diagram below?

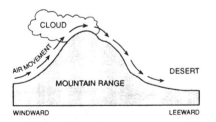

    (1) Sinking air compresses and warms.
    (2) Sinking air expands and warms.
    (3) Rising air compresses and warms.
    (4) Rising air expands and warms.

21. The diagram below shows the flow of air over a mountain from point *A* to point *C*. Which graph best shows the approximate temperature change of the rising and descending air due to the adiabatic process?

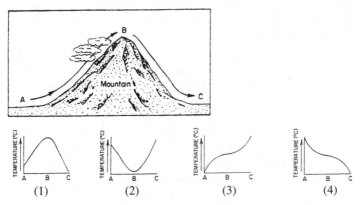

22. The diagram below shows a container of water that is being heated.

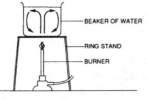

The movement of water shown by the arrows is most likely caused by
(1) density differences   (3) the Coriolis effect
(2) insolation   (4) the Earth's rotation

23. Which drawing best illustrates the general result that the Earth's rotation would have on the direction of the wind as it moves away from the center of a high-pressure system in the Northern Hemisphere?

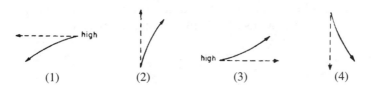

Note that question 24 has only three choices.

24. As a sample of very moist air rises from sea level to a higher altitude, the probability of condensation occurring in that air sample will
(1) decrease
(2) increase
(3) remain the same

25. A weather balloon that rises at the rate of 600 feet per minute disappears into an overcast in 4 minutes. What information about the weather is provided by this observation?
    (1) The cloud type is altostratus.
    (2) The visibility is about ½ mile.
    (3) The wind velocity is about 4 miles per hour.
    (4) The ceiling is 2,400 feet.

26. On a clear, dry day an air mass has a temperature of 20°C and a dew point temperature of 10°C.

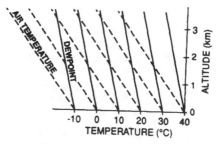

According to the graph, about how high must this air mass rise before a cloud can form?
(1) 1.6 km	(3) 3.0 km
(2) 2.4 km	(4) 2.8 km

27. The diagram below shows a sealed container holding liquid water and clean air saturated with water vapor. (Relative humidity is 100%.) The container has been placed on a block of ice to cool.

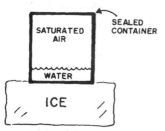

Which statement best explains why a cloud has not formed in the sealed container?
    (1) The air in the container is above the freezing point.
    (2) The ice is cooling the water in the container.
    (3) The air in the container lacks condensation nuclei.
    (4) The water in the container is still evaporating.

369

Base your answers to questions 28 through 32 on your knowledge of earth science and the graph below of measurements of the insolation above the Earth's atmosphere and at the Earth's surface on a clear day. [Note that the graph does not show the entire solar spectrum at the longer wavelengths.]

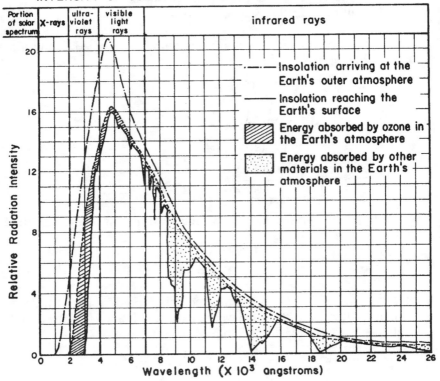

INTENSITY OF SOLAR RADIATION AT DIFFERENT WAVELENGTHS

28. Which of the following types of solar radiation has the longest wavelength?
(1) X-rays
(2) ultraviolet rays
(3) visible light rays
(4) infrared rays

29. The greatest intensity of energy reaching the outer atmosphere of the Earth from the Sun has a wavelength of approximately
(1) 4.5×10^0 angstroms
(2) 4.5×10^3 angstroms
(3) 3.0×10^3 angstroms
(4) 9.1×10^2 angstroms

30. In which portion of the solar spectrum does ozone absorb the greatest amount of energy?
(1) X-rays
(2) ultraviolet rays
(3) visible light rays
(4) infrared rays

31. What quantity is most likely represented by the area between the "dot-dash" curve (.—.) and the "dash" curve (– – –)?
    (1) the amount of radiation that is given off by the Sun
    (2) the amount of radiation absorbed in outer space
    (3) the amount of insolation reflected by the atmosphere back into space
    (4) the amount of insolation absorbed by the Earth's surface

32. According to the graph, in which portion of the solar spectrum was the greatest total amount of energy absorbed by ozone and other materials in the Earth's atmosphere?
    (1) ultraviolet rays and infrared
    (2) X-rays and visible light
    (3) visible light and infrared
    (4) X-rays and infrared

Base your answers to questions 33 through 37 on the *Earth Science Reference Tables*, the diagrams and graph below, and your knowledge of earth science. The diagrams show the general effect of the Earth's atmosphere on insolation from the Sun at middle latitudes during both clear-sky and cloudy-sky conditions. The graph shows the percentage of insolation reflected by the Earth's surface at different latitudes in the Northern Hemisphere in winter.

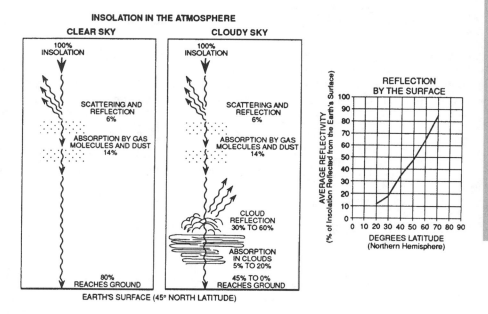

33. Approximately what percentage of the insolation actually reaches the ground at 45° North latitude on a clear day?
    (1) 100%   (3) 60%
    (2) 80%   (4) 45%

34. Which factor keeps the greatest percentage of insolation from reaching the Earth's surface on cloudy days?
(1) absorption by cloud droplets
(2) reflection by cloud droplets
(3) absorption by clear-air gas molecules
(4) reflection by clear-air gas molecules

35. According to the graph, on a winter day at 70° North latitude, what approximate percentage of the insolation is reflected by the Earth's surface?
(1) 50%   (3) 85%
(2) 65%   (4) 100%

36. Which statement best explains why, at high latitudes, reflectivity of insolation is greater in winter than in summer?
(1) The North Pole is tilted toward the Sun in winter.
(2) Snow and ice reflect almost all insolation.
(3) The colder air holds much more moisture.
(4) Dust settles quickly in cold air.

37. The radiation that passes through the atmosphere and reaches the Earth's surface has the greatest intensity in the form of
(1) visible-light radiation (3) ultraviolet radiation
(2) infrared radiation   (4) radio-wave radiation

Base your answers to questions 38 through 42 on your knowledge of earth science, the *Earth Science Reference Tables*, and the graph below. The graph shows the temperatures recorded when a sample of water was heated from –100°C to +200°C. The water received the same amount of heat every minute.

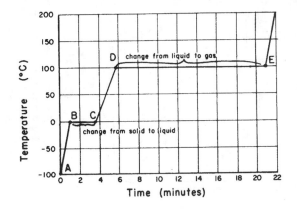

38. The water temperature reached 75°C after the sample had been heated for approximately
(1) 5 min   (3) 6 min
(2) 2 min   (4) 4 min

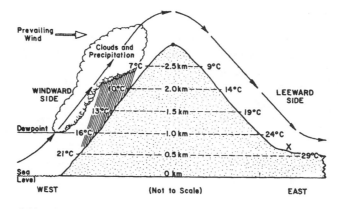

39. The line from *D* to *E* on the graph represents a phase (state) change. Between points *D* and *E*, the water was
(1) condensing
(2) evaporating
(3) freezing
(4) melting

40. The greatest amount of energy was added to the water between points
(1) *A* and *B*
(2) *B* and *C*
(3) *C* and *D*
(4) *D* and *E*

41. How many calories of heat were required to change the temperature of 10 grams of the water from point *C* to point *D*?
(1) 100 cal
(2) 800 cal
(3) 1,000 cal
(4) 5,400 cal

42. Between which two points did the temperature change most rapidly?
(1) *A* and *B*
(2) *B* and *C*
(3) *C* and *D*
(4) *D* and *E*

Base your answers to questions 43 through 47 on your knowledge of earth science and the diagram below, which shows a mountain. The prevailing wind directions and air temperatures at different elevations on both sides of the mountain are indicated.

43. What would be the approximate air temperature at the top of the mountain?
(1) 12°C
(2) 10°C
(3) 0°C
(4) 4°C

44. On which side of the mountain and at what elevation is the relative humidity probably 100%?
(1) on the windward side at 0.5 km
(2) on the windward side at 1.5 km
(3) on the leeward side at 1.0 km
(4) on the leeward side at 2.5 km

373

45. How does the temperature of the air change as the air rises on the windward side of the mountain between sea level and 0.5 kilometer?
(1) The air is warming due to compression of the air.
(2) The air is warming due to expansion of the air.
(3) The air is cooling due to compression of the air.
(4) The air is cooling due to expansion of the air.

46. Which feature is probably located at the base of the mountain on the leeward side (location *X*)?
(1) an arid region   (3) a glacier
(2) a jungle   (4) a large lake

47. The air temperature on the leeward side of the mountain at the 1.5-kilometer level is higher than the temperature at the same elevation on the windward side. What is the probable cause for this?
(1) Heat stored in the ocean keeps the windward side of the mountain warmer.
(2) The insolation received at sea level is greater on the leeward side of the mountain.
(3) The air on the windward side of the mountain has a lower adiabatic lapse rate than the air on the leeward side of the mountain.
(4) Potential energy is lost as rain runs off the windward side of the mountain.

Base your answers to questions 48 through 52 on your knowledge of earth science and on the accompanying diagram. The diagram shows the apparatus used as a model to study large-scale motions within the Earth's atmosphere. The water was heated for several minutes.

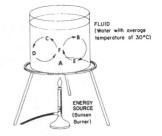

48. What type of energy transfer is indicated by the arrows in the diagram?
(1) conduction   (3) radiation
(2) convection   (4) insolation

49. The density of the water in the beaker is *least* at point
(1) *A*   (3) *C*
(2) *B*   (4) *D*

50. What primary source of energy causes the Earth's atmosphere to move in a manner similar to the motion of the water in the beaker?
(1) ocean currents   (3) the Sun
(2) the Moon   (4) tides

51. The region of the Earth's atmosphere where towering clouds of vertical development could start to form is best represented in the model at
(1) *E*   (3) *F*
(2) *B*   (4) *D*

Note that question 52 has only three choices.

52. As water moves from *A* to *F* to *B*, the water's kinetic energy will generally
(1) decrease
(2) increase
(3) remain the same

Base your answers to questions 53 through 57 on your knowledge of earth science and the graph below, which shows the hourly surface air temperature, dew point, and relative humidity for a 24-hour period.

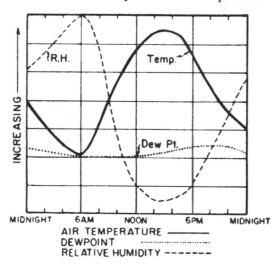

53. The greatest change in air temperature occurred during the period from
(1) midnight to 6 A.M.   (3) noon to 6 P.M.
(2) 6 A.M. to noon   (4) 6 P.M. to midnight

54. Which conclusion concerning the relationship between the dew point and the air temperature is best justified by the data shown?
(1) Changes in air temperature are caused by dew point changes, but not in proportion to these changes.
(2) The dew point increases in proportion to an increase in air temperature.
(3) The dew point increases in proportion to a decrease in air temperature.
(4) Changes in dew point do not necessarily occur when air temperature changes.

375

55. The graph indicates that, as the air temperature increases, the relative humidity
(1) decreases, only
(2) increases, only
(3) sometimes increases and sometimes decreases
(4) remains the same

56. Condensation most likely occurred at approximately
(1) 6 A.M.   (3) 7 P.M.
(2) 9 A.M.   (4) 10 P.M.

57. The greatest difference between the wet- and dry-bulb temperatures on a sling psychrometer would occur at approximately
(1) 6 A.M.   (3) 3 P.M.
(2) 12 noon   (4) 9 P.M.

UNIT SEVEN _____

The Water Cycle and Climates

> **KEY IDEAS** Water is one of the most abundant substances on Earth, covering nearly three-quarters of its surface. Water on Earth exists as a solid in the form of ice and snow; as a liquid in oceans, lakes, streams, and the water droplets in clouds; and as an invisible gas in the atmosphere. The process that keeps the earth supplied with fresh water is the water cycle.
>
> *Climate* is a term that describes the general weather conditions in a given area over a long period of time. Climate is important in determining the habitability of a location and the most reasonable land uses within the region. One of the most important effects of climate is the availability of water at the surface and within the ground. Climate is controlled by a variety of geographic factors.

KEY OBJECTIVES
Upon completion of this unit, you should be able to:

- Identify the forms in which water exists on Earth.
- Analyze the water cycle and describe the transfer of water and energy in and out of the atmosphere at each step in the cycle.
- Explain why fresh water is a limited but renewable resource.
- Describe the factors that determine the amount of solar energy an area receives.
- Differentiate between weather and climate.
- Identify the various factors that control climate.
- Describe the role that the cycling of water and energy in and out of the atmosphere plays in determining climatic patterns.

377

A. WHERE DOES WATER COME FROM?

A-1. The Origin of Earth's Water

The origin of Earth's water is directly linked to the formation of our planet. Meteorites called Type I carbonaceous chondrites provide a clue to the probable origin of Earth's water. These meteorites, which contain 5 percent carbon compounds and 20 percent water tied up in mineral compounds, are believed to be debris left over from the formation of the solar system. As such, they are a type of material thought to have formed while the planets were condensing from a cloud of dust and hydrogen gas.

As Earth cooled and condensed from a cloud of dust and gas, hydrogen reacted with oxygen to form water, which in turn reacted with crystals of solids that were forming. These reactions formed silicate minerals, which contain water, and oxidized iron, so the original form in which water existed on Earth was as part of solid compounds—rocks. The common clay minerals found in soil are an example of this type of compound (refer to Unit 2). As these substances formed, they were drawn inward by gravity and heated. We know that, as rocks are heated above 500°C, the water in their molecules is released in a process called *degassing*. When vented to the Earth's surface, the water is released into the atmosphere and upon further cooling condenses to form liquid water. Considerable evidence supports the idea that much of Earth's water was released to the surface soon after it formed. However, the process continues on a small scale even today through degassing associated with volcanic eruptions.

A-2. The Water Cycle

The water cycle, or *hydrologic cycle*, is the process by which Earth's water cycles in and out of the atmosphere (see Figure 7.1). Let's describe this process in the oceans, which cover nearly three-quarters of the Earth's surface.

1. Each day, solar radiation reaching the surface of the oceans provides energy to the water molecules at the surface. The molecules heat up, and trillions of tons of water change from a liquid to a gas, or *evaporate*. The resulting water vapor enters the atmosphere. In the atmosphere, the water vapor is circulated by winds and carried upward by rising warm air.
2. When air is cooled to its dew point by expansion during rising or by contact with cooler surfaces, the water vapor changes from a gas back to a liquid, or *condenses*, forming tiny water droplets. These water droplets suspended in air form clouds.
3. When the water droplets become too large to remain suspended, they fall back to the surface as *precipitation*. Precipitation may fall on the oceans or the land. Water that falls on the oceans has completed its

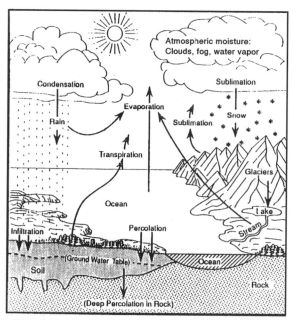

Figure 7.1 The Hydrologic Cycle.
Permission from Thomas McGuire

cycle and may evaporate back into the atmosphere, beginning the entire process anew.

4. A number of things may happen to precipitation that falls on land. It may become *runoff*, water that flows downhill into rivers and streams, which eventually carry it back into the ocean or other body of water. It may seep into the earth to become *groundwater*, water that has infiltrated open spaces in the outer part of the Earth's crust. Groundwater may move slowly back to the oceans through the outer crust; however, most of it seeps into rivers and streams. It is then carried back to the oceans.

5. Some of the water may evaporate again to be carried farther over land, or be blown back over the oceans.

6. Once the water is back in the oceans, the cycle begins all over again.

There are many variations in the water cycle. Sometimes the water that evaporates over the oceans condenses over them and falls directly back into them as rain. Water falling on land may evaporate almost immediately, and in some cases precipitation evaporates before it ever reaches the ground.

Plants play a role in the water cycle as well. The roots of plants take in water that has seeped into the soil. Then the water is transported to the leaves of the plants and released back into the atmosphere by a process called *transpiration*. In an area of abundant vegetation, such as a forest, transpiration returns more than one-third of the precipitation back to the atmosphere as water vapor.

Other variations are possible, but all are part of the endless water cycle, which receives its energy from the Sun. In this way, the water cycle continually renews our supply of fresh water. Each day an estimated 15 trillion liters of water in the form of rain or snow falls on the United States alone. The Earth is a closed system, and the water in use today will eventually be recycled for use by a future generation. In theory, Earth's water will never be exhausted. If, however, persistent, long-lasting pollutants enter the basins and ground from which our water supply is drawn, water that is cleansed by evaporation is polluted as soon as it condenses and falls back into these polluted areas. Consider what happens if the ground is contaminated with nuclear waste, which remains radioactive for tens or even hundreds of thousands of years. For the entire time the material remains radioactive, any water that infiltrates that ground will immediately become unusable.

A-3. Groundwater

Groundwater is an important source of fresh water. There is 37 times as much fresh water beneath the ground as there is on top of it in lakes, streams, glaciers, and other sources.

Porosity

The Earth's surface is not completely solid; it is filled with empty spaces, or *pores*. When precipitation falls, some of it *infiltrates*, or seeps through openings in the surface and into pore spaces. Pore spaces are usually filled with air unless water, oil, or natural gas has forced the air out. The amount of open space in a material is its porosity.

Porosity is expressed as a percentage of the total volume of a given substance. If half of the volume of a sample of rock or sediment is open space, the sample is said to have a porosity of 50 percent. Earth materials differ greatly in porosity; loose sediments such as sand, gravel, and clay usually have the highest porosity (see Figure 7.2). Porosity can be calculated by measuring the amount of water required to fill the pores in a substance and then setting up a ratio of the volume of water to the total volume of the sample being tested.

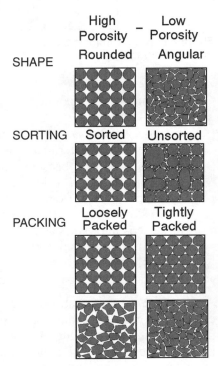

Figure 7.2 Factors That Affect the Porosity of a Material.

380

Example

Water is poured into a 100-cm^3 sample of sand. When 35 mL of water has been poured in, the water is just even with the surface of the sample. What is the porosity of the sand? (1 mL = 1 cm^3)

$$\text{Porosity} = \frac{\text{Volume of pore space}}{\text{Total volume of sample}} \times 100\%$$

$$\text{Porosity} = \frac{35 \text{ cm}^3}{100 \text{ cm}^3} \times 100\% = 35\%$$

Permeability

Permeability is the rate at which water can infiltrate a material. A substance through which water passes rapidly is said to be *permeable*; a substance through which water cannot flow is *impermeable*. The permeability of a substance depends on two factors: the size of its pores and the degree to which they are interconnected (see Figure 7.3). Small pores constrict the flow of water; and if water cannot get from one pore to another, it cannot flow through a substance.

Isolated pore spaces Interconnected pore spaces

Figure 7.3 Isolated and Interconnected Pore Spaces. Permeable materials have interconnected pore spaces.

Knowing the permeability of a material is very useful. From this information the rate at which rainwater will sink into the ground, the speed and direction in which sewage in a septic tank will flow, and the rate at which a well will refill with water after some has been pumped out can be determined.

A-4. The Water Table

When rain falls on land, or snow that has fallen melts, gravity pulls the water down into the ground through interconnected pore spaces. Some of the water clings to particles near the surface, forming a moist layer called the *soil moisture zone*. This is the layer from which plants obtain the water they need to survive. Most of the water that infiltrates continues to trickle downward through the ground until it reaches an impermeable substance. Then the water begins to fill up all the pore spaces from the impermeable material upward. The region of permeable ground in which all of the pore spaces are filled with water is the *zone of saturation*. The top of this zone, known as the *water table*, is the boundary between pores filled with water and pores filled with air. Just above the water table is a narrow region, called the *capillary fringe*, in which narrow pore spaces are filled with water drawn upward by capillary action. The area above the capillary fringe, where the pore spaces are mostly filled with air, is called the *zone of aeration*. The various soil water and groundwater zones are diagramed in Figure 7.4.

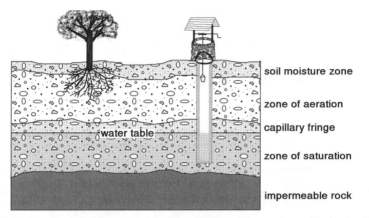

soil moisture zone

zone of aeration

capillary fringe

water table

zone of saturation

impermeable rock

Figure 7.4 Zones of Soil Water and Groundwater. The water table is the boundary between the zone in which pores are filled with air and the zone in which pores are filled with water.

Groundwater can leave the ground in several ways. It can be drawn up through tiny interconnected pore spaces by capillary action and reach the soil moisture zone, where plant roots remove it. Instead, it may evaporate and slowly diffuse out of the ground through the zone of aeration. It may also seep through the ground sideways and slowly move out of the area. If the surface of the ground dips beneath the level of the water table, water can seep out and run off, forming streams, or collect in depressions, forming springs, ponds, or lakes. When the water table in a depression is just above the surface, a swamp may form (see Figure 7.5).

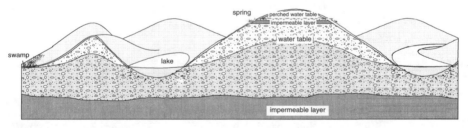

spring   perched water table
          impermeable layer
          water table

swamp

lake

impermeable layer

Figure 7.5 Variations in the Subsurface Position of the Water Table. The water table roughly follows the contours of the surface. Where the surface dips beneath the water table, water seeps out of the ground and can collect in depressions, forming a body of water, such as a lake, or runoff as a stream. Impermeable layers in hills can create a perched water table, which may seep out of the hillside as a spring.

A-5. Aquifers and Wells

Layers of permeable rock or loose sediments whose pore spaces are filled with water are called *aquifers*. Groundwater in aquifers can be tapped by digging a *well*, which is a hole dug into the ground so that it penetrates the water table. Beneath the water table the well acts as a large pore space in which water accumulates and can then be pumped up to the surface.

A-6. Artesian Wells

In artesian rock formations (see Figure 7.6) groundwater can be tapped by wells and the water brought to the surface without pumping. In these rock formations an aquifer is sandwiched between two impermeable layers called *aquicludes*. The aquifer slopes downward away from the area where water seeps into it. The aquicludes act like the walls of a pipe, confining the water as it seeps downslope, and water pressure builds up in the aquifer. If the upper aquiclude is pierced by a well, this pressure may push the water up to the surface or even higher. The amount of pressure in an artesian well depends on the difference in elevation between the well and the water table in the aquifer. Artesian wells get their name from the town of Artois, France, where the first well of this type was dug.

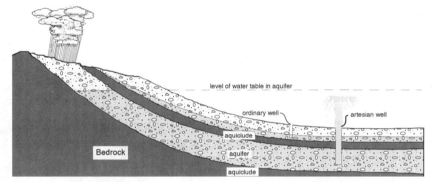

Figure 7.6 An Artesian Formation. The water rises in an artesian well because of hydrostatic pressure created by the weight of the water in the aquifer.

B. WHAT FACTORS DETERMINE THE AMOUNT OF SOLAR ENERGY AN AREA RECEIVES?

The Sun provides nearly all of the energy received at the Earth's surface. Solar radiation reaches the upper atmosphere at a fairly constant rate of about 200 kilocalories per minute per square meter. About one-third of this radiation is reflected back into space, mostly by clouds. As the remaining radiation passes through the atmosphere, some is absorbed by gas molecules. Some is refracted as it crosses boundaries between layers of differing densities. Some is scattered by particles of dust or aerosols in the air. The radiation that reaches the Earth's surface, called *insolation*, short for *in*coming *sol*ar radi*ation*, is either reflected or absorbed.

A number of factors control the amount of solar energy that an area absorbs or reflects, including the angle at which insolation strikes the surface, the length of time each day insolation is received, and the nature of the surface.

B-1. Angle of Insolation

Latitude

Since the Earth is a sphere, insolation does not strike all points on the Earth's surface at the same angle. Near the Equator insolation strikes the surface almost vertically, but near the poles it strikes at a more glancing angle. Therefore, the insolation reaching the tropics is more concentrated than that reaching polar regions (see Figure 7.7).

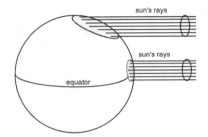

Daily and Annual Cycles

The angle at which insolation strikes the Earth's surface at any location also varies in both daily and annual cycles. The daily cycle starts at dawn with insolation striking the surface at a very low angle; it then becomes increasingly direct throughout the morning, reaching its greatest directness at noon and then becoming increasingly indirect again throughout the afternoon until sunset.

Figure 7.7 Insolation Near the Equator and the North Pole. Note that the two beams of sunlight approaching the Earth are identical, but the angle at which they strike the Earth's surface causes insolation received near the Equator to be more concentrated at the surface than that received near the poles. The result is greater heating at the Equator.

The annual cycles vary with latitude and the Earth's position in its orbit around the Sun. In the United States, insolation is most direct on June 21, the summer solstice, and least direct on December 21, the winter solstice.

Figure 7.8a shows the daily cycle of intensity of solar radiation; Figure 7.8b, the yearly cycle of intensity of solar radiation at noon.

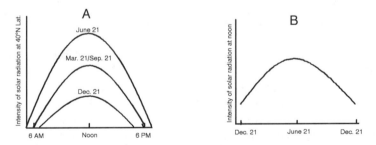

Figure 7.8 Daily and Yearly Cycles of Solar Radiation Intensity. (a) Daily cycle of intensity of solar radiation from 6 A.M. to 6 P.M. at 40° N on the Equinox and solstice days. Note that for each day there is a cyclic change from low intensity in early morning to higher intensity at noon and then back to low intensity in late afternoon. (b) Yearly cycle of intensity of solar radiation at noon for an entire year. Note the cyclic change in intensity from low intensity at noon on the winter solstice to high intensity at noon on the summer solstice and back to low intensity at the next winter solstice.

B-2. Duration of Insolation

The length of time that the surface receives insolation each day, or the *duration of insolation*, depends on the season and the latitude of the location.

Season

The length of daylight varies in an annual cycle with the season (see Figure 7.9). In the United States, the greatest number of hours of insolation is received on June 21, the summer solstice. Each day thereafter, the number of daylight hours decreases, reaching a minimum on December 21, the winter solstice. Then the number of daylight hours increases daily until it peaks again the following June 21. At the fall and spring equinoxes, the number of daylight hours equals the number of hours of darkness. New York State varies from roughly 15 hours of daylight at the summer solstice to only 9 hours at the winter solstice.

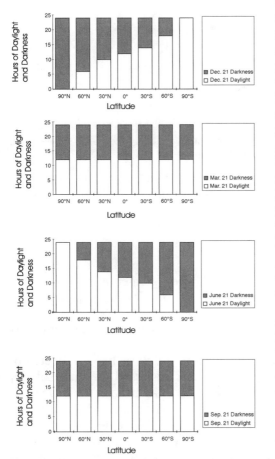

Figure 7.9 Duration of Day and Night at Different Locations on Key Dates Throughout the Year.

Latitude

The length of daylight on any given day also varies with location north or south of the Equator (see Figure 7.10). Since the Earth's axis of rotation is tilted, the circles that form the parallels of latitude are also tilted. At different latitudes, different fractions of the parallels are in daylight and darkness. Thus, observers at different latitudes spend different fractions of each twenty-four hour rotation in daylight and in darkness. When the Northern Hemisphere is tilted toward the Sun, points between the North Pole and the Arctic Circle rotate through an entire 24-hour day without ever entering into darkness. As you move southward, the number of daylight hours decreases.

Although the maximum duration of insolation occurs on June 21 in the Northern Hemisphere, maximum temperatures are reached sometime after this date. Temperatures continue to increase after June 21 because the number of daylight hours still exceeds the number of nighttime hours for many weeks after this date. As long as more energy is received each day than is lost overnight, the temperature will continue to increase.

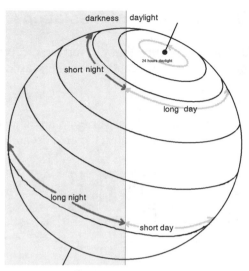

Figure 7.10 Variations in the Lengths of Day and Night Because of the Earth's Tilted Axis of Rotation

B-3. The Nature of the Earth's Surface

The Earth's surface consists of a wide variety of substances, including ice, water, soil of many types, and vegetation. All interact differently with the insolation that strikes them.

Color

Just as a ball is more likely to bounce off a concrete wall than off a pillow, insolation striking different substances will have a greater or lesser tendency to be absorbed or reflected. Which process will predominate depends on the molecular structure of the substance. The color of a substance is a fairly good indicator of whether the substance absorbs more insolation than it reflects or vice-versa.

The lighter the color of a substance, the more light is being reflected by it; the darker the color, the more light is being absorbed. When insolation strikes substances such as sand and snow, their light color indicates that much of the insolation is being reflected. On the other hand, when insolation strikes rich black soil *or* deep green leaves, their dark color indicates that much of the insolation is being absorbed. In general, dark-colored substances absorb more insolation than light-colored ones, and the more insolation a substance absorbs the more it is heated.

Specific Heat Capacity

When different substances absorb equal amounts of heat energy, they do not all change temperature by the same number of degrees. This difference can be thought of as the result of a kind of molecular inertia. Heavy molecules and tightly held molecules require more energy to get them moving than light and loosely held molecules. The temperature of a substance is strictly a result of how fast the molecules are moving. If you put the same amount of energy into two substances, one with tightly held molecules and one with loosely held molecules, the loosely held molecules would move faster and that substance would register a higher temperature.

Table 7.1 shows the specific heat capacities of different Earth materials and also the number of degrees the substance will increase in temperature if 1 calorie of heat energy is added to it. Notice that water increases in temperature much less than rock material such as granite or basalt with the addition of the same amount of heat. Therefore, if the same amount of insolation was absorbed by an ocean and by the sand-sized particles of broken rock that make up the beach next to it, the sand would get much hotter than the water. (This is why we seek out bodies of water to swim in when temperatures on land rise too high.) In general, land increases in temperature more than water when exposed to the same insolation.

TABLE 7.1 SPECIFIC HEAT CAPACITIES OF COMMON EARTH MATERIALS

Material		Specific Heat Capacity (cal/g • °C)	Temperature Increase if 1 Calorie of Heat Is Added to 1 Cubic Centimeter of Material (°C)
Water	Solid	0.5	2
	Liquid	1.0	1
	Gas	0.5	2
Dry Air		0.24	4.2
Granite		0.19	5.3
Basalt		0.20	5
Marble		0.21	4.8
Iron		0.11	9.1
Copper		0.09	11.1
Lead		0.03	33.3

Latent Heat of Fusion and Vaporization

Since water is so abundant and exists in all three phases on Earth, latent heat also has an effect on the temperature changes that occur when insolation strikes water. When water is melting or evaporating, it absorbs energy with no resulting change in temperature. The energy is used to break internal bonds rather than to increase the speed at which water molecules are moving. Since this heat causes no change in temperature, it is called latent heat (latent means "hidden"). As a result, ice that is melting or water that is evaporating from the surface of oceans absorbs insolation without increasing in temperature. Ice-covered land and oceans remain cooler than adjacent land areas. Latent heat of fusion or vaporization is not an issue on land because the substances that land is composed of do not melt or evaporate in the normal range of temperatures at the Earth's surface.

B-4. Radiative Balance

Every object not at a temperature of absolute zero radiates energy. The type of energy radiated depends on the temperature of the object. The Sun, which is at an extremely high temperature, radiates energy of relatively short wavelengths such as ultraviolet and visible light. The Earth, which is at a much lower temperature, radiates heat energy of longer wavelength.

The Earth's temperature depends on the balance between the energy gained by absorbing insolation and the energy lost by radiating heat. Since the average temperature of the Earth's surface does not change over the long term, the Earth as a whole must be in *radiative balance*, that is, reradiating as much energy as it absorbs.

This fact does not mean, however, that all points on the Earth's surface are in radiative balance at all times. Polar regions clearly reradiate more energy than they receive, and regions near the Equator receive more energy than they reradiate. Land areas both absorb and reradiate energy faster than water. This unequal heating and cooling would result in ever-warmer temperatures near the Equator and ever-cooler temperatures near the poles if global winds and ocean currents were not constantly shifting energy from the tropics toward the poles.

There is much evidence that over long periods of time the Earth experiences global warming and cooling trends. These may be triggered by changes

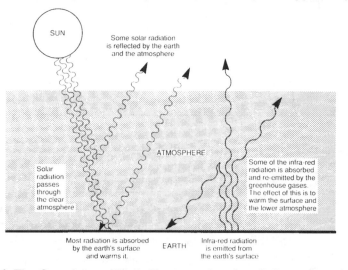

Figure 7.11 The Greenhouse Effect. Short-wavelength radiation passes through the atmosphere and is absorbed by the Earth. The long-wave heat reradiated by the Earth is absorbed or reflected back to the surface by water vapor, carbon dioxide, and other gases in the atmosphere. This sequence of events is called the *greenhouse effect* because the glass in a greenhouse allows light to enter but blocks heat from escaping.
From *Climate Change*, The Intergovernmental Panel on Climate Change, World Meteorological Organization/UN Environmental Programme, Cambridge University Press.

in the Earth's tilt due to precession as it spins on its axis and to increases or decreases in certain gases in the atmosphere.

Water vapor and carbon dioxide both absorb long-wavelength heat energy radiated by the Earth. An increase in carbon dioxide in the atmosphere would prevent the escape of some of the heat energy radiated by the Earth and would cause the Earth to become warmer. This higher temperature could, in turn, increase evaporation and the amount of water vapor in the atmosphere, which would absorb even more heat energy. This chain of events, called the *greenhouse effect* (see Figure 7.11), could result in global warming.

On the other hand, warmer temperatures that increase evaporation would produce more clouds, which in turn, would decrease the amount of insolation reaching the Earth's surface. Recent observations have cast doubts on our ability to predict global warming or cooling.

C. WHAT FACTORS DETERMINE THE TYPE OF CLIMATE THAT AN AREA WILL HAVE?

Climate is the typical weather pattern in a region, averaged over an extended period of time. The main elements of a region's climate are its temperature and precipitation patterns.

C-1. Factors That Determine Climate

Climate is controlled by various factors, including latitude, the distribution of land and water, high- and low-pressure belts and prevailing winds, ocean currents, altitude, relief (the differences in the height of an area's land forms), clouds, and cyclonic storm activity.

Latitude

Latitude is the primary element of temperature control in climate because it determines both the angle and duration of insolation. Much more insolation is received in low latitudes near the Equator than in high latitudes near the poles.

Only the areas that lie between 23½° N (the Tropic of Cancer) and 23½° S (the Tropic of Capricorn) ever receive the vertical rays of the Sun. The average angle of insolation decreases toward the poles, causing the insolation to be spread over a larger area therefore to be less intense. As the angle of insolation decreases, the solar energy passes through more of the atmosphere, so more is reflected or absorbed and insolation is further decreased.

While the average length of daylight, or duration of insolation, is 12 hours everywhere on Earth at the time of the equinoxes, at other times it varies. It is always 12 hours at the Equator; but as one moves north or south of the Equator, the duration of insolation varies in a cyclic manner. It increases toward the pole tilted toward the Sun, ranging from 12 hours at the Equator to 24 hours at the pole. It decreases toward the pole tilted away from the Sun, ranging from 12 hours at the Equator to 0 hour at the pole. The result is that the higher the latitude, the greater the cyclic change in duration of insolation over the course of a year. The poles range from 24 hours of insolation in the summer to no insolation in the winter. At the latitudes of New York State, insolation ranges from about 15 hours in the summer to about 9 hours in the winter. At the Equator the duration is always 12 hours.

Differences in angle and duration of insolation with latitude result in three main temperature zones: the always hot *torrid zone* near the Equator, the always cold *frigid zone* near the poles, and the seasonally changing, intermediate *temperate zone* in between.

Distribution of Land and Water

The irregular distribution of land and water surfaces on the Earth is another major factor controlling climate. As discussed earlier, land surfaces increase in temperature more than water surfaces when insolation strikes them, and they also cool off more rapidly than water surfaces. As a result, air temperatures are warmer in the summer and cooler in the winter over landmasses than they are over oceans at the same latitude. Large bodies of water tend to moderate the temperatures of nearby landmasses by warming them in winter and cooling them in summer. For this reason cities in the interior United States have a greater annual temperature range than coastal cities (see Figure 7.12).

	J	F	M	A	M	J	J	A	S	O	N	D
San Francisco, CA	49	51	53	56	58	61	63	63	64	61	55	50
St. Louis, MO	32	35	43	55	64	74	78	77	70	58	44	35

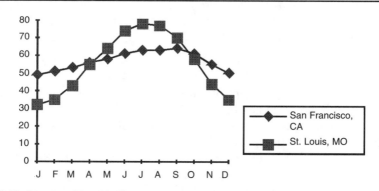

Figure 7.12 Average Monthly Temperatures at St. Louis, Missouri, an Inland City, and San Francisco, California, a Coastal City, Both at Approximately the Same Latitude.

Large areas permanently covered by ice also affect climate. The light color, high specific heat capacity, and latent heat of fusion of ice combine to keep it from melting completely in certain areas. The air over ice surfaces is chilled by contact with the ice, and this cold air, in turn, chills the land and water surrounding these surfaces.

Pressure and Wind Belts

The huge convection cells that form in the atmosphere because of unequal heating give rise to global wind and pressure belts. A belt of low pressure caused by rising warm air lies over the Equator. Low pressure favors evaporation, so the air here tends to be both warm and moist. The rising air spreads outward from the Equator, cools, loses moisture because of condensation, and then sinks back to Earth. This descending air creates a belt of high pressure on either side of the equatorial low between 25° and 30° north or south latitude. As the air sinks toward the ground, it is compressed and warms slightly. This warm, dry air is able to absorb moisture, and land surfaces in these regions tend to be dry. Most of the Earth's deserts are located in these subtropical, high-pressure belts.

The adjacent high- and low-pressure belts give rise to a belt of winds that carry warm, moist air outward from the Equator. These are known as the tropical easterlies or trade winds.

The poles are regions of high pressure because of their frigid, dry air. As the frigid air moves outward from the poles, it warms up and begins to rise, creating the subpolar low-pressure belts at 60° north and south latitude. The 30° high-pressure belts and the 60° low-pressure belts give rise to the moist but cooler prevailing westerlies.

In general, the high-pressure belts create regions of dry air, and the low-pressure belts create regions of moist air. The planetary wind belts carry moist or dry air over landmasses, greatly influencing their climate. These winds also transfer heat energy from the Equator toward the poles, modifying temperature patterns.

Figure 7.13 shows the planetary wind and pressure belts and their climatic effects.

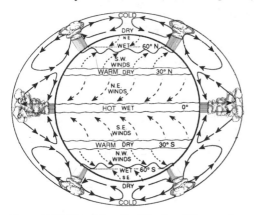

Figure 7.13 The Climatic Effects of Convection Cells in the Atmosphere.

Ocean Currents

Ocean currents control climates by transferring heat from the Equator toward the poles, thereby cooling the equatorial regions, warming the polar regions, and influencing the temperature of adjacent landmasses these currents pass. The major ocean currents (see Figure 7.14) follow the prevailing winds that blow out of the subtropical high-pressure belts, giving rise to a clockwise flow in the Northern Hemisphere and a counterclockwise flow in the Southern Hemisphere.

Wherever the ocean currents flow toward the poles, they carry warm water from the Equator. Wherever they flow toward the Equator, they carry

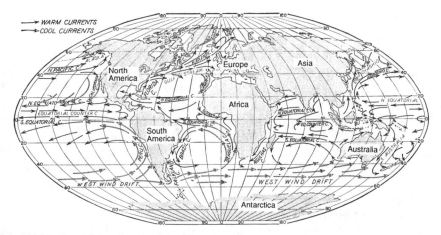

Figure 7.14 The Major Ocean Currents of the World

Source: The State Education Department, *Earth Science Reference Tables*, 1994 ed. (Albany, New York: The University of the State of New York).

cold water from the polar regions. Ocean currents such as the Gulf Stream and the North Atlantic Current carry warm water northward and eastward. The warm air over these ocean currents moderates the climate of the eastern coast of the United States, the Azores, and western Europe. The cold air over the icy waters of the Peru Current modifies the climate all along the western coast of South America, causing it to be much cooler and dryer than places with the same latitudes on the eastern coast of South America, where the Brazil Current and South Equatorial Current bring warm, moist air with their waters.

Altitude

There is a gradual decrease in average temperature with elevation. This is due to the decrease in air pressure with elevation, which causes air to expand and cool. The fact that temperature decreases approximately 1°C for every 100-meter rise in elevation explains why high mountains may have tropical vegetation at their bases but permanent ice and snow at their peaks.

Example

Mount Kilimanjaro in Africa is 5,896 meters high. If the temperature at its base was 35°C (95°F), what would be the temperature at its peak to the nearest degree Celsius?

The temperature drops 1°C for every 100 m of elevation.

Therefore, the temperature drop will equal 5,896 m × 1°C/100 m = 58.96°C, rounded off to 59°C.

And 35°C − 59°C = −24°C (−12°F).

Relief

Relief, or the differences in height of landforms in an area, is another factor that controls climate. For example, mountain ranges serve as barriers to outbreaks of cold air. In this way, the Alps protect the Mediterranean coast and the Himalayas protect India's lowlands.

In a process called the *orographic effect*, the windward side of a mountain facing winds carrying moisture-laden air usually receives much more precipitation than the leeward side. Mountain ranges force the air that is blown over them by winds to rise, and cool by expansion. This decreases the air's capacity to hold moisture and causes condensation, which forms clouds and precipitation. By the time the air reaches the top of the mountain, it has lost much of its moisture. When the air descends on the leeward side of the mountain, it is warmed by compression, its capacity to hold moisture is decreased, and precipitation is less likely. This orographic effect explains why cities along the Oregon coast west of the Cascades (e.g., Tillamook) have much precipitation while cities in the state's interior (e.g., Bend) have a dry climate (see Figure 7.15).

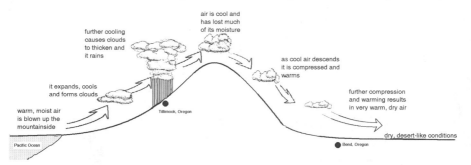

Figure 7.15 The Orographic Effect. On the windward side of the mountains, moist air is forced to rise and then cools by expansion, resulting in clouds and precipitation. On the leeward side, the cool air, which has lost much of its moisture, descends and warms by compression, resulting in warm, dry air that creates desertlike conditions.

Clouds and Cyclonic Storms

Clouds control climate by reducing the amount of insolation gained or lost by a region. During daylight, clouds block sunlight and keep the temperature from rising as high as it would if the skies were clear. At night, the water droplets and water vapor in clouds absorb long-wavelength heat being radiated into space and keeps the region from getting as cold as it would without them.

The warm and cold fronts of wave cyclones that form along the polar front produce the changing weather of middle latitudes. These fronts produce precipitation and carry it over a wide area, causing changes in temperature as both warm and cold fronts pass through.

C-2. Types of Climates

Three Patterns Used to Classify Climates

Climates are classified by moisture, temperature, and vegetation patterns. There are several systems in use today. Table 7.2 summarizes terms often used to describe different patterns of moisture, temperature, and vegetation.

TABLE 7.2 DESCRIPTIVE TERMS FOR CLIMATIC PATTERNS

Moisture	Temperature	Vegetation
Arid	Polar	Desert
Semiarid	Subpolar	Grassland, steppe, taiga
Subhumid	Subtropical	Deciduous forest
		Coniferous forest
Humid	Tropical	Rain forest

Figure 7.16 is a map of the major climates of the world.

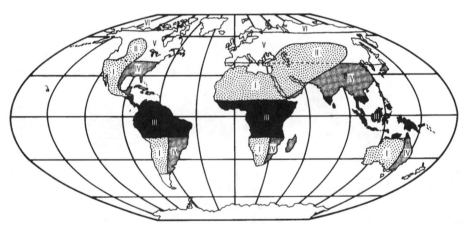

Figure 7.16 Major Climates of the World. I. Dry—hot summers, warm or cool winters. II. Dry—hot summers, cold winters. III. Humid—hot all year. IV. Humid—hot summers, cool winters. V. Humid—warm summers, cold winters. VI. Humid—cold summers, cold winters.

From Part I: *Teacher's Guide—Investigating the Earth*, Earth Science Curriculum Project, Sponsored by the American Geological Institute and the National Science Foundation, Copyright © 1967, American Geological Institute.

C-3. Effects of Climate

Climate has its greatest effect on vegetation. Individual plant species can generally survive only in a certain range of sunlight, temperature, precipitation, humidity, soil type, and wind. There is such a strong relationship between climate types and vegetation that climate regions are often named for their dominant vegetations.

Climate also affects the type of soil that will develop in an area. Warm, wet climates break down rock into soil faster than cool, dry climates. Soils in wet climates tend to be less fertile, though, because heavy precipitation and infiltration dissolve out nutrients needed by plants. Climate also affects soil by determining the types of plants that will become part of the soil when they eventually decay.

Another effect of climate is the shape into which landforms are eroded. In humid climates, where abundant rainfall makes running water the primary agent of erosion, landforms tend to have rounded contours. In arid climates, where wind is dominant and water is likely to erode in violent bursts, the landforms are likely to be more jagged and angular.

The Water Budget

One useful model for determining climate type is the water budget. A water budget tracks the amount of moisture available to plants throughout the year. It combines precipitation and temperature patterns into a single numerical model.

Water Budget Variables

A water budget tracks moisture in the same way a household budget tracks money. A household budget keeps track of dollars and includes variables such as income, bills due, bills paid, bank deposits, and bank withdrawals. A water budget keeps track of moisture, expressed in millimeters of water (the depth to which a particular amount of water will cover an area), and consists of the eight variables described below.

1. **Precipitation (P).** Precipitation is the moisture source for the water budget. Whatever its form (rain, snow, etc.), precipitation is expressed in millimeters of liquid water, the depth to which the precipitation would cover the surface if it did not evaporate, run off, or infiltrate.
2. **Potential evapotranspiration (E_p).** *Evapotranspiration* is a term that combines evaporation and transpiration, the two mechanisms by which liquid water is returned to the atmosphere as water vapor. *Potential evapotranspiration* is the maximum amount of water that could be returned to the atmosphere as water vapor if the water was available. E_p depends on temperature and vegetation. The higher the temperature, the greater the air's capacity to hold moisture, and the greater the evaporation. The more vegetation, the more leaf surface through which water vapor can transpire. Therefore, the highest E_p usually occurs in summer, when temperatures are highest and vegetation is most lush.
3. **The difference between precipitation and potential evapotranspiration ($P - E_p$).** The difference between precipitation and potential evapotranspiration is a measure of how much capacity the air still has to draw moisture from sources other than precipitation, such as the soil and bodies of water. $P - E_p$ may be positive, negative, or zero.

395

4. **Actual evapotranspiration (E_A).** The actual evapotranspiration is the "actual" amount of water that was returned to the atmosphere by evaporation and transpiration. It can never exceed E_P, since by definition E_P is the maximum amount of water that can be lost. However, E_A can exceed P, since precipitation may not be the only water available in the environment to be lost. For example, water in the soil moisture zone may evaporate.

5. **Soil-moisture storage (St).** Soil moisture is water that adheres to the surface of soil particles after seeping into the ground. It is not groundwater, but rather the thin film of water that remains on the surfaces of soil particles after the water has seeped past. This is the water that is absorbed by plant roots and transpired, and it is close enough to the surface to be lost by evaporation. The maximum amount of soil moisture varies, depending on the surface area of the particles, the porosity and permeability of the soil, and the amount of organic matter the soil contains. Soil moisture is considered water in storage because, if it does not evaporate or get drawn into a plant, it can remain in the soil over time. For the purposes of a water budget, 100 millimeters is considered the maximum amount of water that can be held in storage as soil moisture. Soil moisture doesn't exceed the typical maximum of 100 millimeters because once a particle's surface is "wetted" with a film of water, any additional water does not stick to it but continues to seep downward.

6. **Change in storage (ΔSt)—usage or recharge.** If precipitation is less than potential evapotranspiration ($P - E_P < 0$), the air will have the capacity to absorb more moisture. It will absorb that moisture from the soil, causing the amount of soil moisture in storage to drop. This "drying out" of the soil is called *usage*.

   If precipitation exceeds potential evapotranspiration ($P - E_P > 0$), the additional moisture will seep into the soil and replenish the moisture that was lost. This is called *recharge*, because the soil is recharged with water.

7. **Surplus (S).** If precipitation exceeds potential evapotranspiration ($P - E_P > 0$) and the soil is holding all of the moisture it can, a water surplus exists. This additional water will run off out of the area or seep down past the soil and enter the groundwater. Surplus water eventually ends up in streams as runoff or base flow. Base flow is groundwater that seeps out of the banks of the stream, and thereby increases its discharge. If there is no surplus, base flow may lead to depletion of groundwater.

8. **Deficit (D).** When precipitation and soil moisture are both consumed by evapotranspiration ($E_A < E_P$, $St = 0$), and the air still has the capacity to absorb more moisture, a moisture deficit is said to exist. The deficit is the amount of water needed to cover the shortfall between E_A and E_P.

Sample Water Budget

A water budget is calculated based on the precipitation and potential and actual evapotranspiration values in a location. It is then plotted on a graph, and the surplus, deficit, usage, and recharge are identified by shaded areas. When completed, the graph indicates when irrigation might be needed and when flooding might occur. Water budgets are still widely used because they are easy to calculate and they correlate well with actual stream discharge and crop conditions.

Look at the table headed "The Water Budget for Port Gerry" on page 398.

The P and E_p values for the year are given in the first two rows of the table. To complete the rest of the water budget, follow these steps:

1. The E_p values are subtracted from the P values to obtain the $P - E_p$ values entered in row 3.
2. Since you cannot calculate ΔSt or E_A unless you know how much moisture is in soil storage, and the soil storage isn't usually listed for any of the months at this point, you must now make an assumption. Assume that St is 0. Then look at the $P - E_p$ values for a series of months with positive values that add up to more than 100. At the end of that series of months you can be certain that, even if storage started as 0, by the end of this string of months it would be a full 100 mm. This happens in November and December ($43 + 96 = 139$). Put down 100 for St at the end of December, and start calculating the water budget in January. (In a dry location, you may not be able to find a series of months adding up to 100 mm. Then assume that storage is a full 100 mm and look for a string of $P - E_p$ values that adds up to -100 mm. In the month at the end of that series you could be certain that even if storage started out full, it would be depleted and St would be 0. Put down 0 for St in that month, and begin filling out the table in the next month.)
3. The difference between P and E_p is now considered.
   - If $P - E_p$ is negative, moisture will be lost from storage until $St = 0$.
     This first happens in April, $P - E_p$ is -7. The 7-mm shortfall is lost from storage, leaving 93 mm in storage at the end of the month. This will continue as long as $P - E_p$ is negative until no more water remains in storage. If $St = 0$, no more water can be lost from it, and this shortfall of water is listed as a deficit. This first happens in July, when $P - E_p$ is -72 but only 4 mm of water remained in storage at the end of July. The 4 mm is lost from storage, reducing St to 0. The remaining 68-mm shortfall ($72 - 4$ from St) is listed on the D row.
   - If $P - E_p$ is positive, water will be added to storage until $St = 100$ mm.
     This first happens in October, November, and December. In December, $P - E_p = 96$. Since $St = 58$ at the end of November, water is added to storage until $St = 100$ mm; then no more water can be added. That uses up only 42 mm of the 96 mm. The excess 54 mm of water is then listed as in the S row for December.

397

4. The moisture gained or lost by storage is indicated in the ΔSt box.
5. Storage is the amount of water that remains in storage after any gains or losses.
6. E_A is equal to precipitation *plus any moisture taken from storage*. (For example, in April E_A is 55 mm even though P was only 48 mm. The additional 7 mm came from storage.)
7. When E_A is less than E_P, the shortfall is listed as a deficit (D).
8. When P is greater than E_A and soil storage is full, the excess water is listed as a surplus (S).

TABLE 7.3 THE WATER BUDGET FOR PORT GERRY

	J	F	M	A	M	J	J	A	S	O	N	D	TOTALS
P	125	90	77	48	37	32	17	20	35	78	94	127	780
E_P	15	25	37	55	73	85	89	91	77	63	51	31	692
$P - E_P$	110	65	40	-7	-36	-53	-72	-71	-42	15	43	96	
ΔSt	0	0	0	-7	-36	-53	-4	0	0	+15	+43	+42	
St	100	100	100	93	57	4	0	0	0	15	58	100	
E_A	15	25	37	55	73	85	21	20	35	63	51	31	511
D	0	0	0	0	0	0	68	71	42	0	0	0	181
S	110	65	40	0	0	0	0	0	0	0	0	54	269

When the table has been completed, the P, E_P, and E_A values for each month are graphed as shown below:

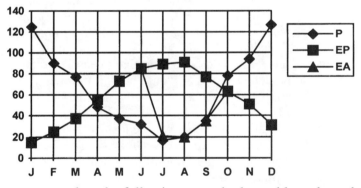

Now try to complete the following water budget table and graph on your own. Since Big Eddy Springs is a dry area with no series of months in which $P - E_P = +100$, look for a series of negative $P - E_P$ values that add up to -100. The total from January to June is -104; therefore St must be 0 at the end of June. Begin filling in the water budget chart for the next month, July. Then add the E_A values (Δ) to the graph.

TABLE 7.4 THE WATER BUDGET FOR BIG EDDY SPRINGS

	J	F	M	A	M	J	J	A	S	O	N	D	TOTALS
P	7	15	27	51	49	28	30	28	25	23	17	18	**317**
E_P	10	18	35	57	78	83	92	81	67	42	21	12	**596**
$P - E_P$	−3	−3	−8	−6	−29	−55	−62	−53	−42	−19	−4	6	
ΔSt													
St						0							
E_A													
D													
S													

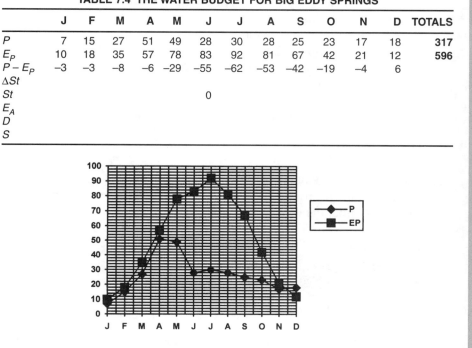

REVIEW QUESTIONS FOR UNIT 7

A diagram of the water cycle is shown below. Letters *A* through *D* represent the processes taking place.

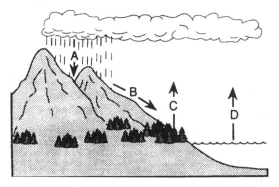

1. Which arrow represents the process of transpiration?
   (1) *A*   (3) *C*
   (2) *B*   (4) *D*

2. The primary source of most of the moisture for the Earth's atmosphere is
   (1) soil-moisture storage   (3) melting glaciers
   (2) rivers and lakes   (4) oceans

3. There will most likely be an increase in the rate at which water will evaporate from a lake if there is
(1) a decrease in the temperature of the air
(2) a decrease in the altitude of the Sun
(3) an increase in the moisture content of the air
(4) an increase in the wind velocity

4. By which process does moisture leave green plants?
(1) convection   (3) transpiration
(2) condensation   (4) radiation

5. All of the glass containers shown below contain the same amount of water and are receiving the same amount of heat energy. In a given amount of time, the most water will evaporate from which container?

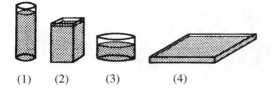

   (1)   (2)   (3)   (4)

6. Which atmospheric condition will cause the greatest amount of evaporation from the surface of a lake?
(1) calm, dry, cold   (3) calm, moist, hot
(2) moist, cold, windy   (4) dry, hot, windy

7. The graph below represents how the rate of evaporation of water is affected by a variable, X. Which variable is most likely represented by X?

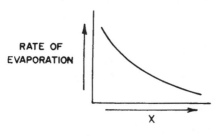

(1) temperature   (3) exposed surface area
(2) wind velocity   (4) moisture content of the air

8. In the diagram below, points *A* and *B* identify two points in the atmosphere above the surface of a body of water.

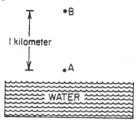

Which graph best represents the vapor pressure (amount of moisture) from point *A* to point *B*?

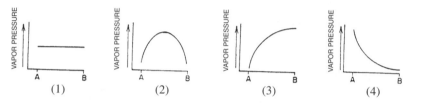

9. When rainfall occurs, the water will most likely become surface runoff if the surface of the soil is
(1) highly permeable   (3) covered with trees
(2) steeply sloped   (4) loose and sandy

10. Water will infiltrate surface material if the material is
(1) impermeable and unsaturated
(2) impermeable and saturated
(3) permeable and unsaturated
(4) permeable and saturated

11. During a rainstorm, when is surface runoff *least* likely to occur?
(1) when the permeability rate of the soil equals the rainfall rate
(2) when the pore spaces of the ground are saturated with water
(3) when rainfall rate exceeds the permeability rate of the soil
(4) when the slope of the surface is too great for infiltration to occur

12. Which graph best represents the relationship between the particle size of a loose material and the loose material's permeability?

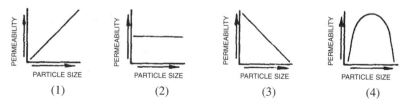

13. Characteristics such as composition, porosity, permeability, and particle size are used to describe different types of
(1) hillslopes   (3) soils
(2) stream drainage patterns (4) landscapes

14. Which is most important in determining the amount of groundwater that can be stored within a rock?
(1) the rock's geologic age   (3) the rock's porosity
(2) the rock's hardness   (4) the rock's color

15. Why does water move very slowly downward through clay soil?
(1) Clay soil is composed of low-density minerals.
(2) Clay soil is composed of very hard particles.
(3) Clay soil has large pore spaces.
(4) Clay soil has very small particles.

Note that question 16 has only three choices.

16. As the temperature of the soil decreases from 10°C to –5°C, the infiltration rate of groundwater through this soil will most likely
(1) decrease
(2) increase
(3) remain the same

17. The diagram below represents two identical containers filled with samples of loosely packed sediments. The sediments are composed of the same material, but differ in particle size. Which property is most nearly the same for the two samples?

(1) infiltration rate   (3) capillarity
(2) porosity   (4) water retention

Note that question 18 has only three choices.

18. As the amount of precipitation on land increases, the depth from the surface of the Earth to the water table will probably
(1) decrease
(2) increase
(3) remain the same

19. Which diagram best illustrates the condition of the soil below the water table?

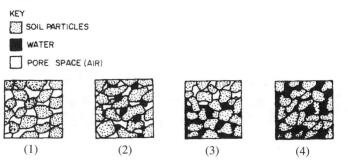

KEY

▨ SOIL PARTICLES

■ WATER

□ PORE SPACE (AIR)

(1) (2) (3) (4)

20. The Earth's surface temperatures are due chiefly to energy received from
(1) insolation from the Sun
(2) radioactivity within the Earth's crust
(3) fusion in the Earth's core
(4) friction between the crust and the mantle

21. By which process does most of the Sun's energy travel through space?
(1) absorption   (3) convection
(2) conduction   (4) radiation

22. Four trays, each containing sand at the same temperature but with different characteristics, were placed on a sunny windowsill. The type of sand in each tray is listed below:

Tray 1—light-colored sand that is dry
Tray 2—light-colored sand that is wet
Tray 3—dark-colored sand that is dry
Tray 4—dark-colored sand that is wet

After 30 minutes, which tray would probably contain the sand that had undergone the *greatest* temperature change?
(1) 1   (3) 3
(2) 2   (4) 4

23. Which latitude on the Earth would receive the highest average yearly insolation per square meter of surface if the atmosphere were completely transparent at all locations?
(1) 90° N   (3) 0°
(2) 23½° N   (4) 23½° S

24. Electromagnetic energy that reaches the Earth from the Sun is called
(1) insolation   (3) specific heat
(2) conduction   (4) terrestrial radiation

25. The length of time that daylight is received at a location during one day is called the location's
   (1) angle of insolation
   (2) intensity of insolation
   (3) duration of insolation
   (4) eccentricity of insolation

26. Which graph best illustrates the relationship between the angle of insolation and the time of day at a location in New York State?

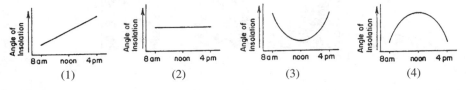

(1)   (2)   (3)   (4)

27. Which graph best represents the relationship between latitude north of the Equator and the length of the daylight period on March 21?

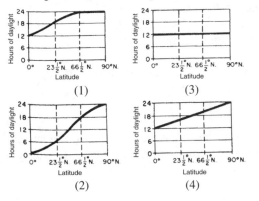

(1)   (3)

(2)   (4)

Note that question 28 has only three choices.

28. The graph below shows the average daily temperatures and the duration of insolation for a location in the mid-latitudes of the Northern Hemisphere during a year.

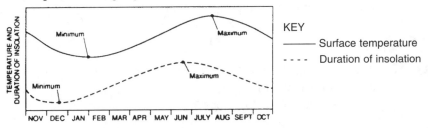

KEY
——— Surface temperature
- - - - Duration of insolation

Compared to the date of maximum duration of insolation, the date of maximum surface temperature for this location is
   (1) earlier in the year
   (2) later in the year
   (3) the same day of the year

29. During which month does the *minimum* duration of insolation occur in New York State?
(1) February   (3) September
(2) July   (4) December

30. What is the usual cause of the drop in temperature that occurs between sunset and sunrise at most New York State locations?
(1) strong winds   (3) cloud formation
(2) ground radiation   (4) heavy precipitation

31. The adjacent diagram represents a portion of the Earth's surface that is receiving insolation. Positions *A*, *B*, *C*, and *D* are located on the surface of the Earth.

At which position would the intensity of insolation be greatest?
(1) *A*   (3) *C*
(2) *B*   (4) *D*

32. Each of the sunbeams in the diagrams below contains the same amount of electromagnetic energy and each sunbeam is striking the same type of surface. Which surface is receiving the greatest amount of energy per unit area where the sunbeam strikes the surface?

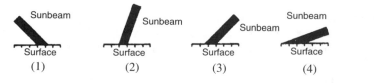

33. The adjacent diagram shows the Sun's maximum altitude (*C*) relative to a vertical stick in New York State on June 21. In which direction from the base of the stick does shadow *C'* point?

(1) north   (3) east
(2) south   (4) west

34. Which two factors determine the number of hours of daylight at a particular location?
(1) longitude and season
(2) longitude and the Earth's average diameter
(3) latitude and season
(4) latitude and the Earth's average diameter

35. Which factor would most affect the total amount of solar energy received by a location on the Earth's surface during a period of 1 year, assuming no cloud cover?
(1) climate
(2) longitude
(3) latitude
(4) distance from a body of water

36. The Earth's axis of rotation is tilted 23½° from a line perpendicular to the plane of its orbit. What would be the result if the tilt was only 13½°?
(1) shorter days and longer nights at the Equator
(2) colder winters and warmer summers in New York State
(3) less difference between winter and summer temperatures in New York State
(4) an increase in the amount of solar radiation received by the Earth

37. For which location and date will the shortest duration of insolation occur?
(1) 60° N latitude on June 21
(2) 23½° N latitude on June 21
(3) 60° N latitude on December 21
(4) 23½° N latitude on December 21

38. A surface will most effectively reflect insolation if it is
(1) rough and light-colored
(2) smooth and light-colored
(3) rough and dark-colored
(4) smooth and dark-colored

39. When visible light strikes a snow-covered, flat field at a low angle, most of the energy will be
(1) absorbed by the snow
(2) refracted by the snow
(3) reflected by the snow
(4) radiated by the snow

40. Bodies of water cool slower than land areas because
(1) some insolation is converted into potential energy as water evaporates
(2) bodies of water have a lower density than land areas
(3) water has a higher specific heat than land
(4) water is a better reflector of sunlight than land

41. Between the years 1850 and 1900, records indicate that the Earth's mean surface temperature showed little variation. This would support the inference that
(1) the Earth was in radiative balance
(2) another ice age was approaching
(3) more energy was coming in than was going out from the Earth
(4) the Sun was emitting more energy

42. An observer noted that even though the ice on a pond was melting, the temperature of the liquid water in the pond remained constant until all the ice melted. Which statement best explains this observation?
(1) The angle of insolation is not great enough.
(2) The daylight hours are longer than the nighttime hours.
(3) The nighttime hours are longer than the daylight hours.
(4) The heat energy is being used to change the solid ice to liquid water.

43. How much energy is lost by water when 1 gram is cooled from 20°C to 19°C? (Use the *Earth Science Reference Tables.*)
(1) 1.0 cal   (3) 19 cal
(2) 0.5 cal   (4) 80 cal

44. Infrared radiation is absorbed in the atmosphere mainly by
(1) nitrogen and oxygen   (3) ice crystals and dust
(2) argon and radon   (4) carbon dioxide and water vapor

45. Which model best represents how a greenhouse remains warm as a result of insolation from the Sun?

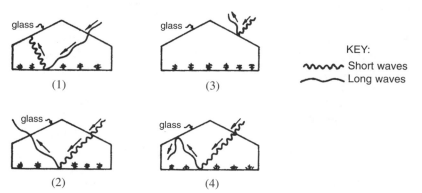

KEY:
〰〰〰 Short waves
——— Long waves

(1)   (3)

(2)   (4)

46. In which region of the electromagnetic spectrum is most of the outgoing radiation from the Earth?
(1) infrared   (3) ultraviolet
(2) visible   (4) X-ray

47. When is an object in radiative balance?
(1) when the radiation emitted by the object is equal to that absorbed by the object
(2) when the radiation emitted by the surroundings is equal to that absorbed by the object
(3) when the radiation emitted by the object is equal to that absorbed by the surroundings
(4) when the wavelength of the radiation emitted by the object is equal to that absorbed by the object

48. The type of climate for a location can be determined by comparing the yearly amounts of
(1) precipitation and potential evapotranspiration
(2) soil storage and potential evapotranspiration
(3) precipitation and infiltration
(4) change in soil storage and stream discharge

49. Two locations, one in northern Canada and one in the southwestern United States, receive the same amount of precipitation each year. The location in Canada is classified as a humid climate. Why would the location in the United States be classified as an arid climate?
(1) The yearly distribution of precipitation is different.
(2) The soil-moisture storage in the southwestern United States is more than that in northern Canada.
(3) The potential evapotranspiration is greater in the southwestern United States than in northern Canada.
(4) The vegetation of the southwestern United States is different from that of northern Canada.

50. Omaha, Nebraska, and Eureka, California, are both at 41° North latitude. Omaha is a midcontinent city and Eureka is a coastal city. Why is Eureka warmer in winter and cooler in summer than Omaha?
(1) The angle of insolation is greater in Eureka than it is in Omaha.
(2) The duration of insolation in Eureka is longer in winter and shorter in summer than it is in Omaha.
(3) The prevailing winds in Eureka are from the west and those in Omaha are from the east.
(4) The ocean water near Eureka is a poorer absorber and reradiator of heat than the land surface surrounding Omaha.

51. The table below shows the precipitation/potential evapotranspiration ratio (P/E_p ratio) for different types of climates.

Climate Type	P/E_p Ratio
Humid	Greater than 1.2
Subhumid	0.8 to 1.2
Semiarid	0.4 to 0.8
Arid	Less than 0.4

The total annual precipitation (P) for a city in California is 420 millimeters. The total annual potential evapotranspiration (E_p) is 840 millimeters. What type of climate does this city have?
(1) humid   (3) semiarid
(2) subhumid   (4) arid

52. Bodies of water have a moderating effect on climate primarily because
   (1) water gains heat more rapidly than land does
   (2) water surfaces are flatter than land surfaces
   (3) water temperatures are always lower than land temperatures
   (4) water temperatures change more slowly than land temperatures do

53. Adjacent water and landmasses are of equal temperature at sunrise. They are heated by the morning sun on a clear, calm day. After a few hours, a surface wind develops. Which diagram best represents this wind's direction?

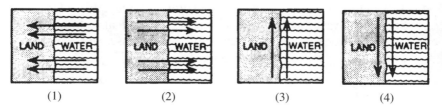

   (1)   (2)   (3)   (4)

54. The seasonal temperature changes in the climate of New York State are influenced mostly by the
   (1) rotation of the Earth on its axis
   (2) changing distance of the Earth from the Sun
   (3) changing angle at which the Sun's rays strike the Earth's surface
   (4) changing speed at which the Earth travels in its orbit around the Sun

55. As a lake's water temperature decreases on a cloudy night, what occurs at the interface between the lake's surface and the air above the lake?
   (1) Energy given up by the lake is lost directly to outer space.
   (2) More energy is gained by the lake than is gained by the air.
   (3) The temperature of the air remains constant.
   (4) Energy is gained by the air from the lake.

56. Water budget graphs for four cities, A, B, C, and D are shown below.

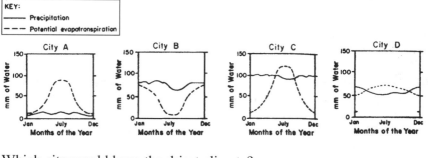

   Which city would have the driest climate?
   (1) A   (3) C
   (2) B   (4) D

409

57. The map below indicates four locations, *A*, *B*, *C*, and *D*, which have the same elevation and latitude. Which location would most likely experience the smallest range of annual temperature?

(1) *A*   (3) *C*
(2) *B*   (4) *D*

58. The diagram below shows the positions of the cities of Seattle and Spokane, Washington. Both cities are located at approximately 48° North latitude, and they are separated by the Cascade Mountains.

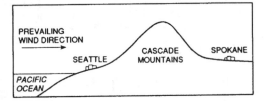

How does the climate of Seattle compare with the climate of Spokane?
(1) Seattle—hot and dry, Spokane—cool and humid
(2) Seattle—hot and humid, Spokane—cool and dry
(3) Seattle—cool and humid, Spokane—warm and dry
(4) Seattle—cool and dry, Spokane—warm and humid

59. The potential evapotranspiration of an area is dependent upon the amount of evaporation surface and the
(1) surface runoff   (3) stream discharge
(2) soil storage   (4) average temperature

60. For a given region, the amount of streamflow (discharge) during drought (dry) periods is controlled mainly by the
(1) shape of the stream channels
(2) available ground water supply
(3) slope of the landscape
(4) elevation of the region

61. During which period will potential evapotranspiration usually be the greatest in New York State?
(1) December 21 to March 21
(2) March 21 to May 21
(3) June 21 to August 21
(4) September 21 to November 21

62. The main source of moisture for the local water budget is
(1) potential evapotranspiration
(2) actual evapotranspiration
(3) groundwater storage
(4) precipitation

63. The map below shows the general path of ocean currents in a portion of the Northern Hemisphere. Locations *A*, *B*, *C*, and *D* are at the shoreline.

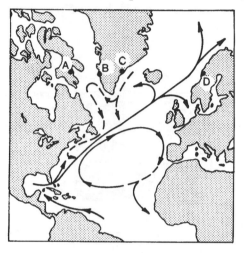

Key
- - -▶ COLD OCEAN CURRENT
———▶ WARM OCEAN CURRENT

NORTH ↑

Which location most likely has the warmest climate?
(1) *A* (3) *C*
(2) *B* (4) *D*

64. When does a moisture deficit occur in a local water budget?
(1) when precipitation is less than potential evapotranspiration and the soil storage is zero
(2) when precipitation plus surplus is greater than potential evapotranspiration
(3) when precipitation is greater than potential evapotranspiration and the soil is saturated
(4) when precipitation is less than potential evapotranspiration

65. In order for soil moisture to decrease, the amount of precipitation must be
(1) greater than the actual evapotranspiration
(2) greater than the amount of recharging taking place
(3) less than the amount of soil-moisture storage
(4) less than the potential evapotranspiration

Base your answers to questions 66 through 70 on your knowledge of earth science, the *Earth Science Reference Tables*, and the information and diagrams below, which describe an investigation with soils.

Three similar tubes, each containing a specific soil of uniform particle size and shape, were used to study the effect that different particle size has on porosity, capillarity, and permeability. A fourth tube containing soil which was a mixture of the same sizes found in the other tubes was also studied and its data are recorded in the table. [Assume that the soils were perfectly dry between each part of the investigation.]

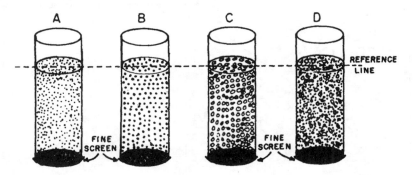

Tube	Particle Size (diameter in cm)	Porosity (%)	Capillarity (mm)	Permeability (s)
A	Fine (0.025)	40	20	14
B	Medium (0.1)	40	15	8
C	Coarse (0.3)	40	7	6
D	Mixed (0.025 to 0.3)	20	12	20

66. When water was poured into the top of each tube at the same time, which tube allowed the water to pass through most quickly?
    (1) *A* (3) *C*
    (2) *B* (4) *D*

67. According to the *Earth Science Reference Tables*, the soil in tube *C* would be classified as
    (1) sand (3) silt
    (2) cobbles (4) pebbles

68. Each tube was placed in a shallow pan of water. In which tube did the water rise the highest?
    (1) *A* (3) *C*
    (2) *B* (4) *D*

69. The bottom of each tube was closed and water was slowly poured into each tube until the water level reached the dotted reference line. Which statement best describes the amount of water held by the tubes?
(1) Tube *D* held more water than any other tube and tube *A* the least.
(2) Tube *C* held more water than any other tube and tube *D* the least.
(3) Tubes *A* and *D* held the same amount of water and twice as much as tubes *B* and *C*.
(4) Tubes *A*, *B*, and *C* held the same amount of water and tube *D* half as much.

70. A handful of material from tube *D* was dropped into a fifth tube filled with water only. In which order would the particle sizes of this soil probably settle in the tube from the bottom of the tube upward?
(1) fine on the bottom, then medium, then coarse
(2) fine on the bottom, then coarse, then medium
(3) coarse on the bottom, then medium, then fine
(4) coarse on the bottom, then fine, then medium

Base your answers to questions 71 through 75 on your knowledge of earth science, the *Earth Science Reference Tables*, and the diagram below. The diagram represents a closed energy system consisting of air and equal masses of copper, granite, and water in a perfectly insulated container. The temperatures were taken at the time the materials were placed inside the closed system.

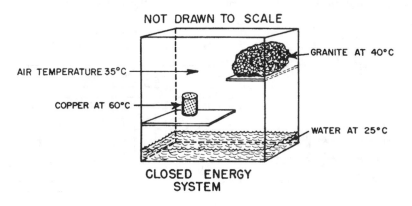

71. In this system, which material is a heat sink for another material?
(1) The water is a heat sink for the air.
(2) The copper is a heat sink for the granite.
(3) The granite is a heat sink for the water.
(4) The copper is a heat sink for the air.

72. Which material in the energy system has the highest specific heat?
(1) copper   (3) dry air
(2) granite   (4) water

413

73. The mass of the granite is 2,000 grams. How much heat would have to be added to raise its temperature 20C°?
(1) 76 cal
(3) 5,400 cal
(2) 4,000 cal
(4) 7,600 cal

Note that questions 74 and 75 have only three choices.

74. In the first day after the materials were placed in the system, the temperature of the water would probably
(1) decrease
(2) increase
(3) remain the same

75. As time passes, the total energy in the system will
(1) decrease
(2) increase
(3) remain the same

Base your answers to questions 76 through 79 on your knowledge of earth science, the *Earth Science Reference Tables*, and the diagram and graph below. In the diagram, equal masses of water and soil are located at identical distances from the lamp. Both were heated for 10 minutes and then the lamp was removed. The water and soil were then allowed to cool for 10 minutes. The graph shows the temperature data obtained during the investigation.

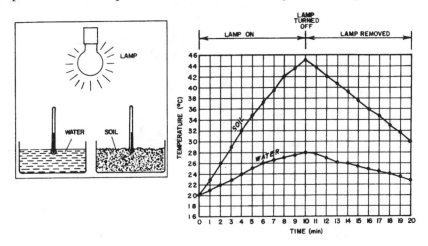

76. What were the temperature readings of the water and soil at the time the lamp was turned off?
(1) The water was 20°C and the soil was 20°C.
(2) The water was 23°C and the soil was 30°C.
(3) The water was 28°C and the soil was 45°C.
(4) The water was 45°C and the soil was 28°C.

414

77. By which process was most of the energy transferred between the lamp and the water during the first 10 minutes of the investigation?
(1) conduction   (3) reflection
(2) convection   (4) radiation

78. What was the rate at which the soil temperature changed during the first 10 minutes of the investigation?
(1) 0.8 C°/min   (3) 8 C°/min
(2) 2.5 C°/min   (4) 25 C°/min

79. Compared to the water, the soil became warmer during the heating period because the soil
(1) has a lower specific heat (3) reradiated less heat
(2) was closer to the lamp (4) has a lower density

Base your answers to questions 80 through 84 on the graph below and your knowledge of earth science. The graph illustrates the relationship between average air temperature, average soil temperature at a depth of 1 meter, and the duration of insolation for a New York State location.

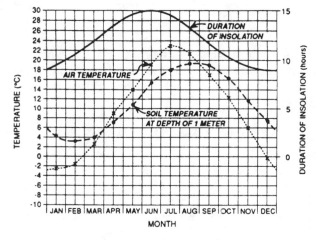

80. During which month was the difference between the soil temperature and the air temperature greatest?
(1) March   (3) October
(2) July   (4) December

Note that question 81 has only three choices.

81. During the month when the highest air temperature occurred, the duration of insolation was
(1) decreasing
(2) increasing
(3) remaining constant

82. What was the average rate of soil temperature change for the period of time between mid-March and mid-July?
(1) 1.2 C°/month   (3) 3.5 C°/month
(2) 5.0 C°/month   (4) 14 C°/month

83. The air temperatures at this New York State location are lower in winter than in summer because in winter this location has
(1) less duration of insolation and is farther from the Sun
(2) less duration of insolation and receives less intense insolation
(3) greater duration of insolation and is farther from the Sun
(4) greater duration of insolation and receives less intense insolation

84. Which graph best represents duration of insolation data collected 20° north of this location?

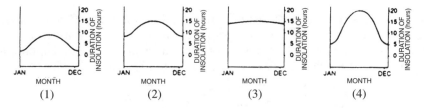

(1)   (2)   (3)   (4)

Base your answers to questions 85 through 89 on your knowledge of earth science and the graph below. This graph shows the varying amounts of insolation received at Brockport, New York, on three different dates under clear or partly cloudy skies.

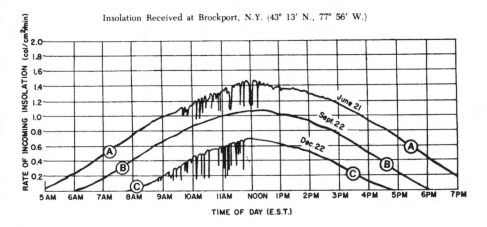

85. The duration of insolation at Brockport on September 22 was approximately
(1) 6 hr   (3) 12 hr
(2) 9 hr   (4) 15 hr

86. Which chart most nearly represents the maximum rate of insolation on the three dates plotted above?

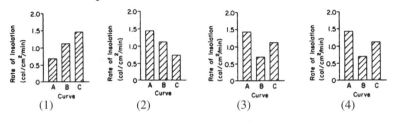

(1) (2) (3) (4)

87. Why is there a difference in the time of sunrise and sunset for each of the three curves?
(1) The duration of insolation varied for each of the 3 days.
(2) The amount of cloud cover varied for each of the 3 days.
(3) The daily temperature varied for each of the 3 days.
(4) The total energy output of the Sun varied for each of the 3 days.

88. What most likely caused the irregular changes in insolation shown on curves *A* and *C* between 9 A.M. and noon?
(1) temperature changes within the atmosphere
(2) variations in the Sun's energy output
(3) instrument error
(4) changes in the amount of cloud cover

89. How would insolation curves for the same three dates in Rome, N.Y. (43°13′N., 75°27′W) compare to the curves for Brockport, N.Y. (43°13′N., 77°56′W), assuming the same type of sky and cloud conditions?
(1) The Rome curves would appear essentially the same.
(2) The Rome curves would be higher and longer.
(3) The Rome curves would he lower and shorter.
(4) The Rome curves *A* and *B* would be higher and longer, but curve *C* would be lower and shorter.

Base your answers to questions 90 through 94 on your knowledge of earth science and the water and climate classification chart below. The water budget data are for a location in the south central United States. The values shown are in millimeters of water.

WATER BUDGET DATA

Month	J	F	M	A	M	J	J	A	S	O	N	D	Yearly Totals
Preciptation (P)	125	102	118	128	127	94	85	84	80	72	104	104	1223
Potential Evapo-transpiration (E_P)	8	10	31	65	112	151	176	160	114	64	23	10	924
$P - E_P$	117	92	87	53	15	−57	−91	−76	−34	8	81	94	–
Changes in Storage (ΔSt)	0	0	0	0	0	−57	−43	0	0	8	81	11	–
Storage (St)	100	100	100	100	100	43	0	0	0	8	89	100	–
Actual Evapo-transpiration (E_A)	8	10	31	65	112	151	128	84	80	64	23	10	766
Deficit (D)	0	0	0	0	0	0	48	76	34	0	0	0	158
Surplus (S)	117	92	87	63	15	0	0	0	0	0	0	83	457

CLIMATE CLASSIFICATION CHART

Climate Type	Total Yearly P/E_P Ratio
Humid	Greater than 1.2
Subhumid	0.8 to 1.2
Semiarid	0.4 to 0.8
Arid	Less than 0.4

90. During which month is soil-moisture recharge taking place?
(1) February　　　　　　(3) September
(2) June　　　　　　　　(4) November

91. A deficit occurs during the summer because the soil-moisture storage is
(1) 100 mm and P is greater than E_P
(2) 100 mm and E_P is greater than P
(3) 0 mm and P is greater than E_P
(4) 0 mm and E_P is greater than P

92. Which graph best represents the actual evapotranspiration (E_A) for this location?

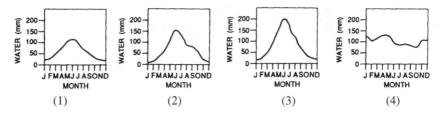

(1)　　　　　　　　(2)　　　　　　　　(3)　　　　　　　　(4)

93. The potential evapotranspiration (E_P) of this area depends primarily on the
(1) amount of precipitation　(3) storage capacity of the soil
(2) average air temperature　(4) amount of actual evapotranspiration

94. According to the climate classification chart, the climate of this location is
(1) humid   (3) semiarid
(2) subhumid   (4) arid

Base your answers to questions 95 through 98 on your knowledge of earth science and the *Earth Science Reference Tables*.

95. What is the dew point temperature when the air temperature is 20°C and the wetbulb temperature is 13°C?
(1) 25°C   (3) 10°C
(2) 13°C   (4) 7°C

96. A student's measurement of the mass of a rock is 30 grams. If the accepted value for the mass of rock is 33 grams, what is the percent deviation (percent of error) of the student's measurement?
(1) 9%   (3) 30%
(2) 11%   (4) 91%

97. At which temperature could water vapor in the atmosphere change directly into solid ice crystals?
(1) 20°F   (3) 10°C
(2) 40°F   (4) 100°C

98. An air pressure of 1,023 millibars is equal to how many inches of mercury?
(1) 30.10   (3) 30.19
(2) 30.15   (4) 30.21

Base your answers to questions 99 through 103 on your knowledge of earth science and on the water budget data shown in the graph below, collected at a location near the center of a continent.

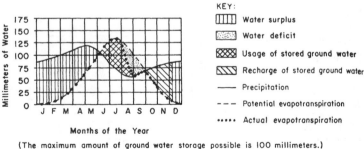

KEY:
⬚ Water surplus
⬚ Water deficit
⬚ Usage of stored ground water
⬚ Recharge of stored ground water
— Precipitation
- - - Potential evapotranspiration
••••• Actual evapotranspiration

Months of the Year
(The maximum amount of ground water storage possible is 100 millimeters.)

99. The greatest water deficit occurs during the month of
(1) March   (3) August
(2) July   (4) November

100. Streams in this area probably carry the greatest volume of water during the month of
(1) April   (3) August
(2) June   (4) November

101. Compared to the winter climate, the summer climate of this location is
(1) cooler and wetter   (3) warmer and wetter
(2) cooler and drier   (4) warmer and drier

102. Data are collected at a new location of the same latitude but on the windward side of a mountain range along an ocean coast. Compared to the original water budget, the water budget of the new location would probably
(1) have less precipitation and less potential evapotranspiration
(2) have greater total precipitation and approximately the same total potential evapotranspiration
(3) have less total precipitation and greater total potential evapotranspiration
(4) have approximately the same total precipitation and approximately the same total potential evapotranspiration

Note that question 103 has only three choices.

103. For a location in the Southern Hemisphere approximately the same distance from the Equator, the amount of potential evapotranspiration for the month of July would normally be
(1) lower
(2) higher
(3) the same

Base your answers to questions 104 through 108 on your knowledge of earth science and the adjacent diagram, which represents an imaginary continent on the Earth with surface locations *A* through *E*. An ocean surrounds the continent and two mountain ranges are located as shown.

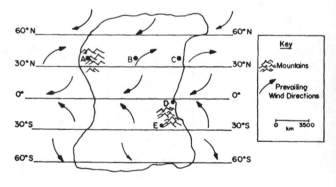

104. At which location would the total yearly potential evapotranspiration be the greatest?
(1) *A*   (3) *C*
(2) *B*   (4) *D*

105. What type of climate would most likely be found at location *E*?
  (1) arid   (3) humid
  (2) subhumid   (4) tropical rain forest

106. Location *A* will probably have a greater
  (1) range in yearly temperatures than location *B*
  (2) total yearly rainfall than location *B*
  (3) altitude of the Sun in the summer than location *B*
  (4) variation in the length of daylight hours than location *B*

107. The maximum insolation will occur at location *E* during the month of
  (1) March   (3) September
  (2) June   (4) December

108. Which graph best represents the relationship between temperature and time of year for location *B*?

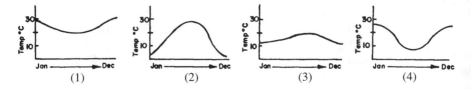

Base your answers to questions 109 through 113 on your knowledge of earth science, the *Earth Science Reference Tables*, and the diagram and data below. The diagram represents a closed glass greenhouse. The data table shows the air temperatures inside and outside the greenhouse from 6 A.M. to 6 P.M. on a particular day.

AIR TEMPERATURE

Time	Average Outside Temperature	Average Inside Temperature
6 A.M.	10°C	13°C
8 A.M.	11°C	14°C
10 A.M.	12°C	16°C
12 noon	15°C	20°C
2 P.M.	19°C	25°C
4 P.M.	17°C	24°C
6 P.M.	15°C	23°C

GREENHOUSE

109. The highest temperature was recorded at
  (1) 12 noon outside the greenhouse
  (2) 2 P.M. outside the greenhouse
  (3) 12 noon inside the greenhouse
  (4) 2 P.M. inside the greenhouse

110. By which process does air circulate inside the greenhouse due to differences in air temperature and air density?
(1) absorption
(3) convection
(2) radiation
(4) conduction

111. Several objects made of the same material, but with different surface characteristics, are tested in the greenhouse to determine which object will absorb the most sunlight. The object that absorbs the most sunlight most likely has a surface that is
(1) dark-colored and smooth
(2) dark-colored and rough
(3) light-colored and smooth
(4) light-colored and rough

112. Which statement best explains what happens to the insolation reaching the greenhouse?
(1) Most of the insolation is absorbed by the glass.
(2) All of the insolation is reflected by the glass.
(3) Insolation absorbed inside the greenhouse is reradiated at longer wavelengths.
(4) Insolation absorbed inside the greenhouse is reradiated at shorter wavelengths.

113. At approximately what rate did the temperature rise inside the greenhouse between 8 A.M. and 10 A.M.?
(1) 1.0 C°/hr
(3) 0.5 C°/hr
(2) 2.0 C°/hr
(4) 12.0 C°/hr

Base your answers to questions 114 through 118 on your knowledge of earth science and on the accompanying graph, which shows the insolation received during a cloudless 24-hour period by a square kilometer of the Earth's surface located in the middle latitudes. The amount of
ground radiation sent out by this same area on the same day is also shown. The dash line is a reference line for a specific amount of heat energy.

114. At what time did the minimum ground radiation occur?
(1) 12 midnight
(3) 3 P.M.
(2) 5 A.M.
(4) 8 P.M.

115. At what time did the maximum air temperature probably occur?
    (1) 4 A.M.   (3) 3 P.M.
    (2) 12 noon   (4) 8 P.M.

116. At about what time did the heat gained by the area equal the heat lost?
    (1) 5 A.M.   (3) 5 P.M.
    (2) 12 noon   (4) 12 midnight

Note that question 117 has only three choices.

117. Which graph best represents the amount of insolation that would be received by this same area on the same day if thick clouds were present in the atmosphere?

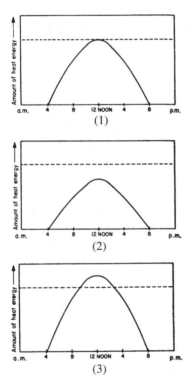

118. If the data had been recorded 6 months later, the insolation curve would have had a
    (1) lower peak and a narrower time range
    (2) lower peak and a greater time range
    (3) higher peak and a narrower time range
    (4) higher peak and a greater time range

UNIT EIGHT _____

The Earth in Space

KEY IDEAS Observations of the universe as seen from the Earth can be explained by a number of models. Although it violates common sense, the heliocentric model most accurately represents the actual behavior of the Earth.

Newton's laws of motion make it possible to explain the elliptical orbits of planets, moons, and comets; this tides; Earth's equatorial bulge; and the motion of falling objects. The rotation, revolution, and spatial orientation of the Earth relative to the Sun affect seasons, weather, climate, tides, and other terrestrial processes. The Moon orbits the Earth, causing tides, eclipses, and phases of the Moon.

Earth is a relatively small, rocky planet, the third in a system of nine planets orbiting the Sun. The Sun is one of billions of stars orbiting the center of the Milky Way galaxy, which in turn is one of the very large number of galaxies in the universe.

KEY OBJECTIVES

Upon completion of this unit, you should be able to:

- Describe the apparent motions of the Sun, Moon, stars, and planets through the sky.
- Compare and contrast the geocentric and heliocentric models in terms of explaining the observed motions of celestial objects.
- Explain how the Earth's tilted axis of rotation and its revolution around the Sun cause the seasons.
- Interpret diagrams of the Earth's orbit to determine how the Sun's path through the sky and the relative position of the noon Sun vary with the seasons, and how the length of daylight varies throughout the year at different locations.

- Explain how the Moon's revolving around the Earth accounts for the phases of the Moon, tides, and eclipses.
- Describe the Earth's place in the universe.

A. HOW CAN WE ACCOUNT FOR OUR OBSERVATIONS OF CELESTIAL OBJECTS?

People were much more aware of the sky in earlier times. They used the changing pattern of stars in the nighttime sky to plan the planting of crops, to schedule religious festivals, and to navigate. The patterns of stars and the "wandering" movements of the planets, the Sun, and the Moon became a part of the myths, philosophies, and religions of different cultures. The scientific effort to understand the universe is part of humanity's attempt to find its place in the cosmic scheme of things. Our modern view of the universe grew out of earlier attempts to explain the universe in terms of what can be seen from our vantage point on Earth.

Celestial objects are objects that can be seen in the sky but that are beyond the Earth's atmosphere. The Sun, the Moon, the stars, the planets, and comets are all examples of celestial objects. Clouds, rainbows, halos, and other phenomena seen in the sky that are part of, or occur in, the Earth's atmosphere are *not* considered celestial objects.

A-1. Apparent Daily Motion of All Celestial Objects

Celestial objects all appear to move across the sky from east to west along a path that is an arc, or part of a circle. All celestial objects move along their circular paths at a constant rate of 15° per hour, or one complete circle every day (24 hours/day × 15°/hour = 360°/day). This is called *apparent daily motion*. In the Northern Hemisphere, all of the circles formed by completing the arcs along which celestial objects move are centered very near the star Polaris (see Figure 8.1). The apparent circular motion of celestial objects causes them to come into view from below the eastern horizon and to sink from view beneath the western horizon (to rise in the east and to set in the west).

We have used the word *apparent* or *appear* when referring to motions so far because the motion of an object is always judged with respect to some other object or point. The idea of absolute motion or rest is misleading because there are several possible

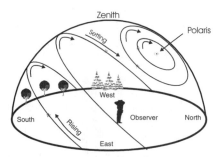

Figure 8.1 Circular Paths Around Polaris. In the Northern Hemisphere, celestial objects appear to move in circles roughly centered on the star Polaris.

reasons why an object may appear to be moving to an observer. One possibility is that the observer is standing still and the object *is* moving. Another possibility is that the object is standing still and the *observer* is moving. When you are in a car speeding down a highway and you look out of the car window, trees along the side of the road seem to whiz by. Of course, your brain tells you that the trees are rooted in the ground and that they only seem to move because you are riding past them at high speed. To your eyes alone, however, you are not moving and the image of the trees appear to be moving from one side of your window to the other.

A-2. The Geocentric and Heliocentric Models: A Historical Perspective

The problem of determining which is moving, the object or the observer, is not always easy to solve. If cues that tell the body it is moving are removed, an observer may not realize that he or she is moving. Then, any object seen to change position will be interpreted as a moving object by the observer.

Early observers of the sky reasoned that, when they looked at the sky, they were standing still because they had no sensory cues that they were moving. They interpreted the changing positions of celestial objects to mean that these objects were moving. They visualized all celestial objects as revolving around a stationary Earth. This Earth-centered or *geocentric model* of the universe was used successfully for thousands of years to explain most observations of celestial objects.

Early observers noted that the positions of celestial objects change in a daily and yearly cyclic pattern. Since these positions of change with time and location, such changes can be used to determine time, find one's position on the Earth, or navigate. Since the distribution of stars is random, early observers invented *constellations*, imaginary patterns of stars, to help them keep track of the changing positions of celestial objects.

In the second century A.D., an Egyptian astronomer named Ptolemy devised a mathematical model of the universe in which all celestial objects moved in circles around the Earth, or in circles around points moving in a circle. He adjusted all of the sizes of the circles and rates of motion until they agreed with what was observed. With this model he was able to predict the motions of the Sun, Moon, stars, and planets. When, over time, the observed positions of celestial objects began to differ from the positions predicted, Ptolemy's model could be adjusted again by changing the sizes of circles or the rates of motion in the model. Because of this flexibility the model continued to be used for more than 1,400 years. However, the constant need to change the model, as well as the increasing complexity of the changes needed, signaled that something was inherently wrong with the model.

In the sixteenth century Nicolaus Copernicus, a Polish astronomer, devised a mathematical model that explained the same celestial motions in terms of the Earth's rotating once a day and revolving around the Sun once

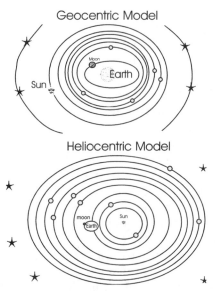

Geocentric Model

Heliocentric Model

Figure 8.2 Geocentric Versus Heliocentric Model. In the geocentric model, the Earth is at the center of the universe and all celestial objects revolve around the Earth or around points that revolve around the Earth. In the heliocentric model, the Earth rotates on its axis and revolves around the Sun.

a year. This Sun-centered or *heliocentric model* was rejected by almost everyone because it defied common sense and was contrary to the belief that the Earth was the center of the universe. Then, Johannes Kepler, a German astronomer, showed that the heliocentric model worked well if uniform circular motion was replaced by uneven motion along off-center ellipses. Using the newly invented telescope to view the sky, Galileo observed many things that supported Copernicus' idea and brought the issue to the attention of the educated people of the time. Isaac Newton showed that the uneven, elliptical motion of planetary orbits proposed by Kepler was a natural consequence of the laws of motion. Although no simpler than Ptolemy's model, the heliocentric model did a better job of predicting the positions of celestial objects and was eventually accepted. Our modern view of planetary motions in the solar system is based on the heliocentric model.

Figure 8.2 contrasts the geocentric and heliocentric models.

A-3. The Earth's Motions

Rotation

In the modern heliocentric model, the Earth rotates at a rate of 15° per hour, or one complete rotation every 24 hours. *Rotation* is a motion in which every part of an object is moving in a circular path around a central line called the *axis of rotation*, as an ice skater spins around a line through his or her body. The Earth's axis of rotation is a line passing through the North Pole, the center of the Earth, and the South Pole. It is almost directly aligned with the star Polaris and is tilted at an angle of 23½° from a perpendicular to the plane passing through the centers of the Earth and Sun (see Figure 8.3).

Revolution

As the Earth rotates, it revolves around the Sun once every 365¼ days. *Revolution* is the motion of one body around another body in a path called

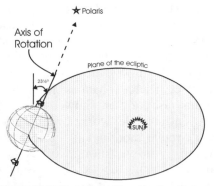

Figure 8.3 Rotation. The Earth rotates from west to east once every 24 hours around an axis that runs through the poles.

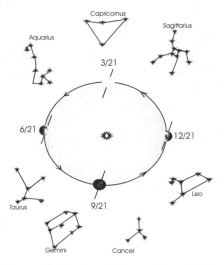

Figure 8.4 Revolution. The Earth revolves around the Sun once every 365.25 days. During each season, different parts of the universe are visible at night.

an *orbit*. The revolving body is called a *satellite* of the body it orbits, and the body a satellite orbits is its *primary*. Thus, the Earth is a satellite of the Sun. The plane of the Earth's orbit is called the *ecliptic* because eclipses occur when the Earth, Moon, and Sun align in this plane.

The Earth's axis of rotation is tilted 23½° from a perpendicular to the plane of its orbit. Spinning like a top, the Earth holds its axis fixed in space as it moves around the Sun. Another way to describe this is to say that the Earth's axis of rotation at any two points in Earth's orbit is parallel. As a result, the Northern Hemisphere is tipped toward the Sun in June and away from the Sun in December.

As the Earth revolves around the Sun, the side of the Earth facing the Sun experiences day and the side facing away experiences night. Since the stars are visible only at night, the portion of the universe whose stars can be seen by an observer on Earth varies cyclically as Earth revolves around the Sun (see Figure 8.4).

Precession

Precession is the very slow change in the direction of the Earth's axis of rotation. The Earth's axis sweeps around in a cone-shaped path like the wobbling of a spinning top. Each sweep takes almost 26,000 years and causes the North Pole to move slowly with respect to Polaris. Precession is caused by the gravitational pull of the Sun and Moon.

Since Earth's precession takes place over such a long period of time, it has little effect during a human lifetime. During your lifetime, the Earth's axis will point slightly closer to Polaris. It will point closest to Polaris around the year 2100 and then will begin to move away. In 13,000 years, Earth's axis will point at the star Vega instead of Polaris.

B. HOW DOES A REVOLVING SPHERE WITH A TILTED SPIN AXIS MODEL YEARLY CELESTIAL OBSERVATIONS?

The value of any model lies in its ability to explain past and present observations and predict future ones. A revolving and rotating planet with its spin axis tilted at 23½° to a line perpendicular to its orbital plane fits all terrestrial and celestial observations.

B-1. The Sun's Apparent Path

The Sun's apparent path through the sky from sunrise to sunset is an arc, like the paths of all other celestial objects, and is the result of the Earth's rotating on its axis. As the Earth (and any observer on the Earth's surface) rotates from west to east, the Sun appears to move from east to west. Look at Figure 8.5, frames *A–E*.

In *A*, an observer at sunrise would see the Sun to the east and low in the sky, at the horizon. In *B*, the Earth has rotated from west to east 45° and an observer would see the Sun to the southeast and higher up above the horizon. In *C*, the Earth has rotated another 45° from west to east and an observer would see the Sun to the south and at its highest point above the horizon—noon. In *D*, the

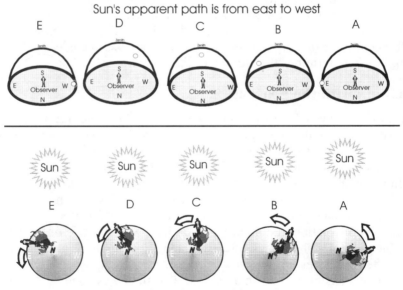

Figure 8.5 The Sun's Apparent Motion Is Due to the Earth's Actual Motion. The bottom portion of this diagram shows an observer's position in relation to the Sun as the Earth rotates on its axis. The top portion shows where the Sun would appear to be in the sky to the same observer.

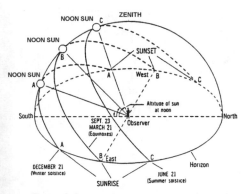

Figure 8.6 The Sun's Apparent Path. An observer at 42° N latitude would see the sun move through different paths across the sky on different dates.
Source: *Earth Science: A Study of a Changing Planet;* Daley, Higham and Matthias, Prentice-Hall.

Earth has rotated a further 45° and an observer would see the Sun to the southwest and lower above the horizon than at noon. In *E*, the Earth has rotated another 45° and now an observer would see the Sun set to the west, low in the sky and at the horizon.

Note that, as the earth rotated through 180° from west to east, an observer saw the Sun move from east to southeast to south to southwest to west. The Sun's path through the sky began at sunrise at the eastern horizon in frame *A*, arced upward, reaching a high point at noon, and then arced downward, ending in sunset at the western horizon in frame *E*. What was seen by the observer as Earth rotated can be summarized in a single drawing, as shown in Figure 8.6.

B-2. Changes in the Sun's Path

The length and the position of the Sun's path vary with the seasons and the latitude. The length of the path determines the length of the daylight period. The longer the path, the more hours of daylight. The position of the Sun's path determines the angle of insolation and thus the intensity of the insolation. It also determines the altitude of the Sun at noon.

Changes with the Seasons

On different days of the year, an observer would see the Sun follow different paths through the sky. This is so because the Sun's position relative to an observer changes as Earth orbits the Sun. The points of sunrise and sunset vary, as does the altitude of the Sun at noon (see Figure 8.7).

In December, the position of sunrise to an observer in the United States is south of east and the position of sunset is south of west. The length of the Sun's path is shortest, so the daylight period is shortest. Because of Earth's tilted axis of rotation, the noon Sun is at its lowest altitude of the year. The combination of a short daylight period and a low angle of insolation results in low temperatures.

In June, the sun rises north of east, sets north of west, and rises to a higher altitude. The Sun's path is at its longest, so the daylight period is longest. The high altitude of the Sun results in a more direct angle of insolation. The combination of a long daylight period and a more direct angle of insolation results in high temperatures.

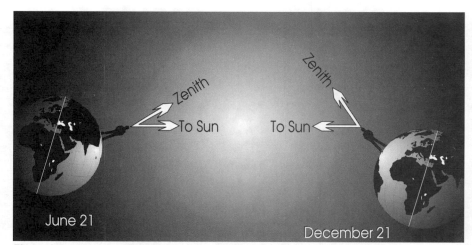

Figure 8.7 Altitude of the Noon Sun on Different Dates. An observer would see the noon Sun at different altitudes on different dates.

In March and September, the Sun rises due east and sets due west. The length of the Sun's path results in exactly equal periods of daylight and darkness. The Sun's altitude at noon is halfway between its highest and lowest points. The length of day and the angle of insolation are intermediate between the two extremes; therefore the temperatures are moderate.

From December to June, the Sun's path gets longer each day and the altitude of the Sun at noon increases. On June 21, the altitude of the Sun at noon stops increasing. Therefore, this date is called the *summer solstice*, meaning summer "Sun stop." From June to December, the Sun's path gets shorter each day and the altitude of the Sun at noon decreases. On December 21, the altitude of the sun at noon stops decreasing. Therefore, this date is called the *winter solstice*, meaning winter "Sun stop."

March 21 and September 21, when the daylight and darkness periods are equal, are called, respectively, the *spring equinox* and *fall equinox*, meaning spring and fall "equal night." This cyclic pattern of change repeats in an annual cycle.

Changes with Latitude

Now consider what two observers at different latitudes would see on a given day as the Earth rotated. Figure 8.8 shows that an observer near the Equator would see the Sun at a higher altitude at noon than an observer in New York State on the same day.

The altitude of the noon Sun at any location can be determined quite easily if you know the latitude of the location and the latitude at which the Sun is directly overhead on that day. To find the altitude of the Sun at noon, we find the difference between the latitude of the location and the latitude at which the Sun is directly overhead, and then subtract this number of degrees from 90°.

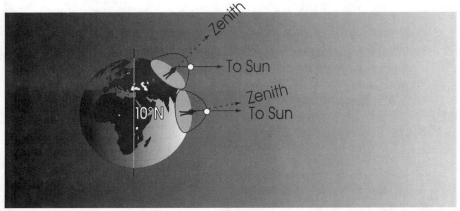

Figure 8.8 Altitude of the Noon Sun at Different Latitudes. Observers at different latitudes would see the noon Sun at different altitudes on the same date.

Example

On March 21, the Sun is directly overhead at the Equator—0° latitude. On that day, what is the altitude of the Sun at noon in New York City?

Latitude of New York City – Latitude where Sun is directly overhead = Difference in latitude

41° N – 0° = 41°

90° – Difference in latitude = Altitude of Sun at noon in New York City

90° – 41° = 49°

The altitude of the Sun in New York City on March 21 is 49°.

Since the farthest the Earth tips toward the Sun is 23½°, the Sun is never directly overhead north of 23½° N (Tropic of Cancer) or south of 23½° S (Tropic of Capricorn). There is always a difference between the latitude of a location in the continental United States and the latitude at which the Sun is directly overhead; therefore, the altitude of the Sun at noon is always less than 90° in the United States.

Because of New York State's latitude, an observer there will never see the noon Sun at zenith, its highest altitude on June 21, at about 72° above the horizon, and at its lowest altitude on December 21, at only about 26° above the horizon.

C. WHAT FORCE KEEPS SATELLITES IN THEIR ORBITS?

C-1. Gravity

Working with the discoveries of Galileo and Kepler, Isaac Newton derived three laws of motion and the law of universal gravitation. Newton's laws of motion made it possible to predict how an object would move if the forces acting on it were known. Newton considered the forces that would be needed to keep a satellite moving in orbit around another object. In thinking about the Moon, Newton realized that the Moon would circle the Earth only if some force was pulling the Moon toward the Earth's center; otherwise the Moon would continue moving in a straight line off into space.

Newton's genius was to realize that the force that kept the Moon in orbit around the Earth was the same force—gravity—that causes objects close to the Earth to fall to the ground, as apples fall from a tree. Newton realized that the force of gravity is universal, that all objects are attracted to one another with a force that depends on the masses of the objects and their distances from one another. He expressed this relationship in a simple mathematical formula:

$$F = -G\frac{m_1 m_2}{r^2}$$

In this equation, F is the force of gravity acting between two masses, m_1 and m_2; G is the gravitational constant; and r is the distance between the masses. Gravity is the force that keeps all satellites moving in a curved orbit.

C-2. Orbital Motion

How does gravity keep satellites moving in a curved orbit? Imagine that a cannonball is shot out of a cannon aimed horizontally. If there were no gravity, the cannonball would fly in a straight line parallel to the Earth's surface until some force stopped it. With gravity pulling the cannonball downward toward the Earth as the cannonball is flying horizontally, however, the path of the cannonball curves downward and eventually the missile strikes the earth. If a more powerful charge is used in the cannon, the cannonball will travel farther horizontally before it strikes the Earth. With a sufficiently powerful charge, the cannonball would travel far enough horizontally that, as its path curved downward because of gravity, the Earth's surface would curve away because of its spherical shape and the ball would never strike the Earth. Instead, gravity would cause the cannonball to fall downward at the same rate that the Earth's surface curves away from it, and the

Figure 8.9 Orbital Velocity. As a fired cannonball travels through the air, it is drawn downward by gravity in a curved path until it strikes the Earth's surface. If it is fired with more energy, it will travel farther in a curved path before crashing into the Earth's surface. If it is fired with enough energy, it's path will curve downward at the same rate that the Earth curves away from its path, and it will go into orbit.

433

missile would fall unendingly in a circular path around the Earth—it would be in orbit (see Figure 8.9).

For an orbit just above the Earth's surface, the cannonball would have to be shot out of the cannon at 7.9×10^3 meters per second, or about 18,000 miles per hour. If an object is sent off at a higher speed, it will travel outward farther before being pulled back by gravity, and the orbit will be elliptical instead of circular. If the object is sent off at a velocity equal to or greater than 11.2×10^3 meters per second, or about 25,000

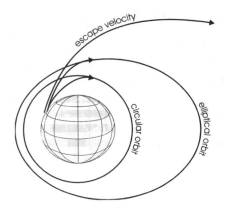

Figure 8.10 Circular Orbit, Elliptical Orbit, and Escape Velocity

miles per hour, the object will be able to escape the Earth's gravity and to fly out of orbit (see Figure 8.10).

C-3. Orbital Motion in the Solar System

In our solar system, all of the planets are satellites of the Sun. Some planets, including Earth, also have satellites, called *moons*. Tycho Brahe, an astronomer, and his student Johannes Kepler made precise observations of celestial objects over a long period of time. Kepler spent 6 years trying to work out the orbit of the planet Mars, using Ptolemy's system of the planet moving in a small circle that moved in a larger circle around the Sun. No matter how hard he tried, however, he could not get the theoretical orbit to match its observed orbit. Finally Kepler came to realize that the orbit of Mars was elliptical, or oval, and that Mars moved at a speed that varied with its distance from the Sun. After years of studying observations of celestial objects, Kepler discovered three simple laws that planets obey as they revolve around the Sun:

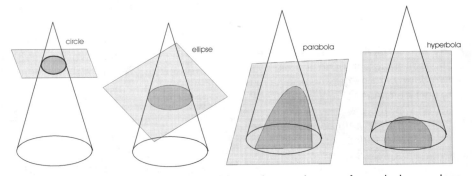

Figure 8.11 Conic Sections. An ellipse is one of several curves formed when a plane intersects a cone.

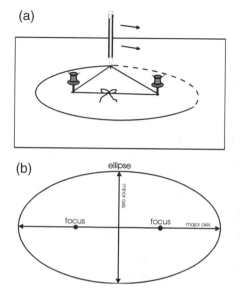

1. **Each planet revolves around the Sun in an elliptical orbit with the Sun at one focus.** An ellipse is a closed figure obtained by cutting a circular cone with a plane. Ellipses are part of a family of curved figures, called *conic sections*, that include circles, parabolas, and hyperbolas (see Figure 8.11). The more the plane is tilted, the more elongated the ellipse.

An ellipse has a major axis, and a minor axis which are lines connecting the two points farthest apart and the two points closest together on the ellipse. It also contains two special points along the major axis, each called a *focus*

Figure 8.12 The Way to Draw an Ellipse (a) and the Main Parts of an Ellipse (b).

(plural, foci). The fact that the distance from one focus to any point on the ellipse and back to the other focus is always the same makes it very easy to draw an ellipse using two tacks and a loop of string. Press the tacks into a board, loop the string around them, and place a pencil in the loop as shown in Figure 8.12a. If you keep the string taut as you move the pencil, it will trace out an ellipse (Figure 8.12b).

The closer together the foci, the more circular the ellipse; the farther apart the foci, the flatter the ellipse. The flatness of an ellipse is called its *eccentricity*. Eccentricity is expressed as the ratio of the distance between the foci and the length of the major axis:

$$e = \frac{d}{l}$$

where e = eccentricity, d = distance between the foci, and l = length of the major axis.

A perfect circle has an eccentricity of 0; a straight line, an eccentricity of 1.

Since the orbit of each planet is an ellipse, the distance from a planet to the Sun varies during its orbit. For example, Earth's distance to the Sun varies from 147×10^6 kilometers on January 3, at its closest (perihelion), to 152×10^6 kilometers on July 6, at its farthest (aphelion). The difference between these distances is the distance between the foci of the Earth's elliptical orbit—5×10^6 kilometers. This distance is very small compared to the 299×10^6 = kilometer length of the major axis, so the eccentricity of the Earth's orbit is very small (0.17). Thus, the Earth's orbit is very nearly a circle. Although many illustrations show the Earth's

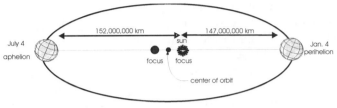

distances are approximate

Figure 8.13 View of the Earth's Elliptical Orbit with the Sun at One Focus. The Earth is closest to the Sun at perihelion and farthest away at aphelion. Distances are approximate.

orbit around the Sun in a perspective view that exaggerates its eccentricity, if viewed from directly overhead the orbit would appear very nearly circular (see Figure 8.13). (In this connection, it is interesting to note that Ptolemy's system of circular orbits *almost* worked because the Earth's orbit is *almost* a circle. However, the slight difference from a perfect circle was enough to throw Ptolemy's system off over time.)

As the distance between the Earth and the Sun changes, the apparent diameter of the Sun changes in a cyclic manner. When the Earth is closest to the Sun, on January 3, the Sun has its greatest apparent diameter. On July 6, when the Sun is farthest away, it has its smallest apparent diameter.

2. **Each planet moves so that a line between the Sun and the planet sweeps out equal areas of the planet's elliptical orbit in equal times.** In Figure 8.14, the elliptical shape of a planet's orbit is exaggerated to show this idea more clearly. The planet sweeps out equal areas of its orbit in equal time periods. The times required for the planet to move from 1 to 2, from 3 to 4, and from 9 to 10 are all equal. Notice, however, that the distance from 1 to 2 is less than the distance from 9 to 10 even though they are covered in the same time. This difference means that the planet is moving fastest when it is closest to the Sun and slowest when it is farthest from the Sun. The cyclically changing velocities of the planets as they move around the Sun are due to changes in the gravitational force between the planet and the Sun as distance changes. As a planet approaches the Sun (distance decreases), gravitational force increases, and since it is acting in the same direction in which the planet

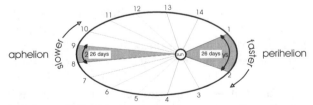

Figure 8.14 Kepler's Law of Equal Areas. A planet sweeps out equal areas of its elliptical orbit in equal time. Since the planet travels less distance in the 26 days it takes to move from 8 to 9 on the ellipse above, it is traveling slower than when it moves from 1 to 2.

436

is moving, it causes the planet to move faster. As a planet moves away from the Sun (distance increases), gravitational force decreases, and gravity, which is now acting opposite to the direction in which the planet is moving, causes the planet to slow down.

3. **The ratio of the square of the time it takes a planet to make one revolution and its average distance from the Sun is the same for all the planets (a constant).** The time required for a planet to make one revolution is called its *period of revolution*. If we use Earth as an example, this relationship can be simplified. Let Earth's period of revolution equal 1 year, and Earth's average distance from the Sun equal 1 unit of distance, called an *astronomical unit* (AU). Then Kepler's third law simplifies as follows:

$$\frac{T^2}{r^3} = K \text{ (a constant)}$$

For Earth, T is 1 earth-year, r is 1 astronomical unit, and the constant is

$$\frac{1^2}{1^2} = \frac{1}{1} = 1$$

Thus, if the period of any planet is measured in earth-years and its distance from the Sun in astronomical units, then:

$$\frac{T^2}{r^3} = 1 \text{ and } T^2 = r^3$$

Example

Mars is 1.52 AU from the Sun and completes one revolution in 1.88 earth-years.

$$0.88^2 = 3.53$$
$$0.52^3 = 3.51$$
$$\frac{3.53}{3.51} = 1 \text{ (rounded off)}$$

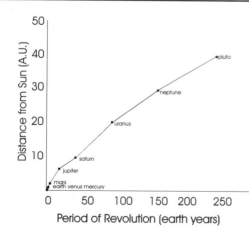

Figure 8.15 Relationship between Period of Revolution and Distance from the Sun-

437

This relationship for the planets in our solar system can be plotted as a curve on a graph (see Figure 8.15), using the values for the planets given in Table 8.1.

TABLE 8.1 VALUES OF *T* AND *R* FOR THE PLANETS

Planet	*T* (Period of Revolution (earth-years)	r (Distance from Sun) (AU)
Mercury	0.24	0.39
Venus	0.62	0.72
Earth	1.0	1.0
Mars	1.88	1.52
Jupiter	11.86	5.23
Saturn	29.46	9.54
Uranus	84.0	19.18
Neptune	164.8	30.05
Pluto	247.7	39.44

C-4. The Earth-Moon-Sun System

Few of the events that can be observed in the sky are as striking as those involving the Earth-Moon-Sun system. Phases of the Moon are probably the most familiar phenomena, but others involving this system include tides and eclipses of the Sun and Moon.

As the Earth revolves around the Sun, Earth's Moon revolves around the Earth in an elliptical orbit with a period of 27.32 days. The Moon's orbit is tilted at an angle of about 5° from the plane of the Earth's orbit around the Sun. The Moon moves rapidly in its orbit, covering 13° every day. As a result, its position against the backdrop of stars changes daily by 13°, or about 26 times its apparent diameter. The Moon also rotates on its axis once every 27.32 days. Thus, the same side of the Moon always faces the Earth (see Figure 8.16).

Phases of the Moon

The Moon does not produce visible light of its own; it is visible only by light from the Sun that is reflected from its surface. An observer on Earth can see only that portion of the Moon that is illuminated by the Sun.

One revolution of the moon

One rotation of the moon

Figure 8.16 Revolution and Rotation of the Moon. The Moon completes one rotation in the same time it makes one revolution. Therefore, the same side of the Moon is always facing the Earth.

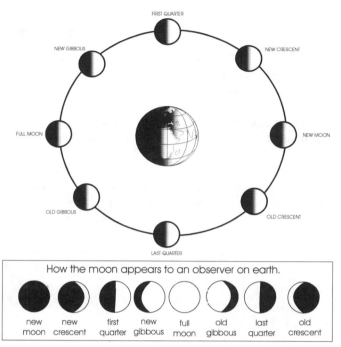

Figure 8.17 Phases of the Moon

As the moon Moves around the Earth, different parts of the side of the Moon facing the Earth are illuminated by sunlight and the Moon passes through a cycle of phases (see Figure 8.17).

Although the Moon makes 1 revolution in 27.32 days, it takes 29.5 days to go through a complete cycle of phases. Why the extra 2 days? At the same time that the Moon is revolving around the Earth, the Earth is revolving around the Sun at a rate of about 1° per day. When the Moon has completed 1 revolution, the Sun is no longer where it was when the Moon started its revolution; in 27 days the Sun has moved 27°. Moving at about 13° per day, the Moon takes about 2 days to catch up to the Sun and align with it in a new moon phase. The word *month* has its origin in "moon-th," which referred to this 29.5-day cycle of phases.

Eclipses of the Sun and Moon

A *solar eclipse* occurs when the Moon passes directly between the Earth and the Sun, casting a shadow on the Earth and blocking our view of the Sun. Both the Moon and the Earth are illuminated by the Sun and cast shadows in space. As you can see in Figure 8.18, the umbras (dark regions) of the shadows cast by the Earth and the Moon are extremely long and narrow.

The umbra of the Moon's shadow barely reaches the Earth, and the small, circular shadow the Moon casts on the Earth's surface is never more than 269 kilometers in diameter. The 5° tilt of the Moon's orbit together with the small

439

Figure 8.18 Moon's Shadow. The 5° tilt of the Moon's orbit and the small size of the Moon's shadow makes it easy for the shadow to miss the Earth.

size of the shadow that can reach the Earth, makes it easy for the shadows to miss the Earth at full and new moon. Thus, total eclipses of the Sun, are rare. For an eclipse to occur, the plane of the Moon's orbit must intersect the shadows being cast by the Earth and the Moon (see Figure 8.19).

Whether an observer sees a total or a partial eclipse depends on what part of the Moon's shadow passes over the observer. The Moon's shadow consists of 2 parts, an umbra and a penumbra. The *umbra* is a region in which all of the light has been blocked. In the *penumbra*, only part of the light is blocked, so the light is dimmed but not totally absent. An observer in the Moon's umbra would see a total eclipse. The path of the Moon's umbra over the

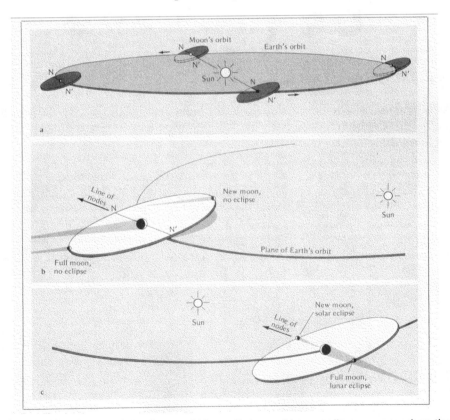

Figure 8.19 Eclipses of the Sun and Moon. Solar and lunar eclipses occur when the Moon intersects the Earth's orbit in new-moon or full-moon position.

From *Horizons: Exploring the Universe,* by Michael A. Seeds, Wadsworth Publishing Co., 1987.

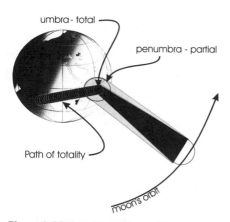

umbra- total

penumbra - partial

Path of totality

moon's orbit

Figure 8.20 A Solar Eclipse. Observers in the umbra experience a total eclipse; those in the penumbra, a partial eclipse.

Earth's surface is therefore called the *path of totality.* An observer in the penumbra would see a partial eclipse. An observer outside the Moon's shadow would see no eclipse. Total and partial solar eclipses, showing umbra and penumbra, are diagramed in Figure 8.20.

The Moon's orbit is elliptical, so the Moon's distance from the Earth varies. If the Moon lines up directly with the Sun at a point in its orbit when it is farthest away from the Earth, the umbra of the Moon's shadow does not reach the Earth's surface. In that case, there is no total eclipse. However, a type of partial eclipse, called an *annular eclipse,* occurs under these conditions. In an annular eclipse the Moon is too far from the Earth to completely block our view of the sun. During the eclipse a narrow ring, or *annulus,* of light is visible around the edges of the Moon (see Figure 8.21).

Viewed from space

Viewed from Earth

Figure 8.21 An Annular Eclipse and the Way It Appears When Viewed from Space and from Earth.

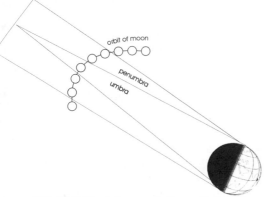

orbit of moon

penumbra

umbra

Figure 8.22 A Lunar Eclipse. If the Moon passes only through the Earth's penumbra, a partial eclipse of the Moon is observed. If the Moon passes through the Earth's umbra, there is a total eclipse of the Moon.

A *lunar eclipse* (see Figure 8.22) occurs when the Moon moves through the shadow of the Earth at full moon. If the Moon moves into the Earth's umbra, a total lunar eclipse is seen. If the Moon moves into the Earth's penumbra, the result is a partial lunar eclipse. When the Moon is totally in the Earth's umbra, it does not completely disappear from view. While no direct sunlight reaches the Moon, some light that is bent as it passes through the Earth's atmosphere arrives at the Moon. Since only the long waves of red light are bent far enough to reach the Moon, the Moon glows with a dull red color during a total lunar eclipse. During a partial lunar eclipse, the Moon is only partially dimmed as some of the Sun's light is blocked by the Earth. Partial lunar eclipses are not very impressive.

Tides

Every day you feel the mutual attraction of gravity between your body and the Earth, pulling you downward with a force called your *weight*. But the Earth's gravity is not the only gravity that acts on you. Although the Moon is farther away from you than the Earth's center and has less mass than the Earth, it still exerts a measurable force on you and everything else on Earth.

The surface of the Earth that faces the Moon is about 6,000 kilometers closer to the Moon than to the center of the Earth. Therefore, the force of gravity the Moon exerts on the Earth's surface is stronger than the gravity exerted on the Earth's center. Although the Earth's surface is solid, it is not absolutely rigid, and the Moon's gravity causes it to flex outward, forming a bulge several inches high. As the Moon moves around the Earth, the bulge moves across the Earth's surface as it remains beneath the Moon.

An inches-high bulge in the bedrock spread over half the Earth's surface is barely noticeable. But water is a fluid, and can move much more readily than rock in response to the Moon's gravity. Water in the oceans, attracted by the Moon's gravity, flows into a bulge on the side of the Earth facing the Moon. A bulge of water forms also on the far side of the Earth. Because the Moon pulls more strongly on the Earth's center than on the Earth's far side, the Earth is pulled away from the oceans on the far side and the water flows into this space, creating a bulge. The water flows into these bulges from the area between them, creating a deep region and a shallow region in the ocean waters.

As the Earth rotates on its axis, the positions of the tidal bulges remain fixed in line with the Moon. As the rotating Earth carries a location into a tidal bulge, the water deepens and the tide rises on the beach. As the location rotates out of the tidal bulge, the water becomes shallower and the tide falls. Since there are two bulges, on opposite sides of the Earth, the tide rises and falls twice a day.

The Sun also produces tidal bulges in the Earth's surface and oceans. At new moon and full moon, the Sun's tidal bulges and the Moon's tidal bulges align with one another and combine. The result is very high and very low tides, called *spring tides* because they "spring so high," not because they occur during the spring season. Spring tides occur at every new and full moon whatever the season. When the Sun's tidal bulges and Moon's tidal bulges are at right angles to each other during the first- and third-quarter moons, they nearly cancel each other. Then *neap tides*, in which there is very little difference between high and low tides, occur. Spring tides and neap tides are diagramed in Figure 8.23.

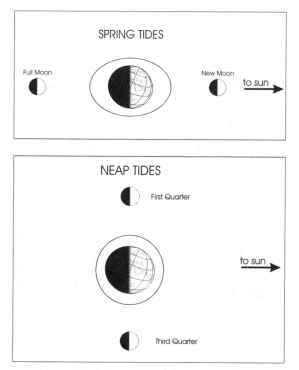

Figure 8.23 Spring and Neap Tides. These are due to changes in the relative directions of the gravities exerted by the Moon and Sun.

D. WHAT IS THE EARTH'S PLACE IN THE UNIVERSE?

Earth is one of several planets that orbit a star—the Sun. The Sun is millions of times closer to Earth than is any other star. Light travels at a finite speed of 300,000 kilometers per second, and light from the Sun reaches Earth in less than 10 *minutes*. Light from the next-closest star takes several *years* to get to Earth. The light from very distant stars takes several *billion years* to reach Earth. When we look at distant stars, we see them as they were when the light that we see left them—in other words, we look back in time.

The universe is so large that the distance light travels in a year, or a light-year, is used to measure its distances. A light-year is a distance of about 9.5 trillion kilometers. With our fastest rockets, it would take thousands of years to reach the nearest star beyond the Sun. Compared with the vast distances between stars, the distances between the Sun and its planets are small. Most astronomers estimate the universe to be about 15 billion light-years in radius, and it is therefore thought to be about 15 billion years old.

Stars are not scattered evenly throughout the universe. Gravity has drawn them together in huge clumps, or galaxies. A *galaxy* is a system of hundreds of billions of stars. The universe contains many billions of galaxies, each, in turn, containing billions of stars. Our solar system is located near the edge of a disk-shaped galaxy of stars called the Milky Way galaxy (see Figure 8.24), which gets its name from the faint white band of its stars that can be seen from Earth on a clear, dark night. The Milky Way galaxy contains more than 100 billion stars revolving in huge orbits around the center of the galaxy.

Figure 8.24 The Milky Way Galaxy. The Sun is one of billions of stars in this galaxy. In this diagram its position is marked with "X".

REVIEW QUESTIONS FOR UNIT 8

1. Why do stars appear to move through the night sky at the rate of 15° per hour?
   (1) The Earth actually moves around the Sun at a rate of 15° per hour.
   (2) The stars actually move around the center of the galaxy at a rate of 15° per hour.
   (3) The Earth actually rotates at a rate of 15° per hour.
   (4) The stars actually revolve around the Earth at a rate of 15° per hour.

2. Which real motion causes the Sun to appear to rise in the east and set in the west?
   (1) the Sun's revolution   (3) the Earth's revolution
   (2) the Sun's rotation   (4) the Earth's rotation

3. An observer took a time-exposure photograph of Polaris and five nearby stars. How many hours were required to form these star paths?
   (1) 6
   (2) 2
   (3) 8
   (4) 4

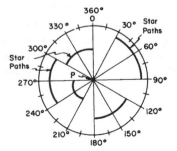

4. In New York State, a recorded length of the shadow cast by a tree at noon each day from January to May will most probably indicate that the length of the shadow
(1) continuously decreases
(2) continuously increases
(3) remains the same
(4) increases and then decreases

5. How would a 3-hour time-exposure photograph of stars in the northern sky appear if the Earth did *not* rotate?

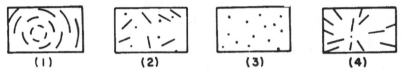

6. Which planetary model allows a scientist to predict the exact positions of the planets in the night sky over many years?
(1) The planets' orbits are circles in a geocentric model.
(2) The planets' orbits are ellipses in a geocentric model.
(3) The planets' orbits are circles in a heliocentric model.
(4) The planets' orbits are ellipses in a heliocentric model.

7. Which diagram best represents the motions of celestial objects in a heliocentric model?

Key: P = Planet, M = Moon, S = Sun

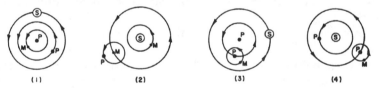

8. What is the total number of degrees that the Earth rotates on its axis during a 12-hour period?
(1) 1°   (3) 180°
(2) 15°   (4) 360°

9. The time required for 1 Earth rotation is about
(1) 1 hour   (3) 1 month
(2) 1 day   (4) 1 year

10. What is the exact shape of the Earth's orbit around the Sun?
(1) The orbit is a slightly eccentric ellipse.
(2) The orbit is a very eccentric ellipse.
(3) The orbit is an oblate spheroid.
(4) The orbit is a perfect circle.

Note that question 11 has only three choices.

11. As the Earth revolves in orbit from its July position to its position in January, the Earth's distance from the sun will
    (1) decrease
    (2) increase
    (3) remain the same

12. In the Northern Hemisphere, during which season does the Earth reach its greatest distance from the Sun?
    (1) winter   (3) summer
    (2) spring   (4) fall

13. Some constellations (star patterns) observed in the summer skies in New York State are different from those observed in the winter skies. The best explanation for this observation is that
    (1) the Earth revolves around the Sun
    (2) the Earth rotates on its axis
    (3) constellations are moving away from the Earth
    (4) constellations revolve around the Earth

14. The Earth's axis of rotation is tilted $23\frac{1}{2}°$ from a line perpendicular to the plane of its orbit. What would be the result if the tilt was increased to $33\frac{1}{2}°$?
    (1) an increase in the amount of solar radiation received by the Earth
    (2) colder winters and warmer summers in New York State
    (3) less difference between winter and summer temperatures in New York State
    (4) shorter days and longer nights at the Equator

15. The diagrams below represent flat horizontal surfaces at four different locations on the Earth. The arrows represent the Sun's rays striking each location at noon on March 21. Which location is farthest from the Equator?

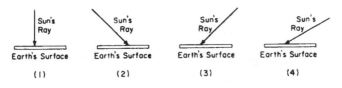

16. In New York State, how do the points of sunrise and sunset change during the course of 1 year?
    (1) They vary with each season in a cyclic manner.
    (2) They move toward the north in the autumn months.
    (3) They move toward the south in the spring months.
    (4) They remain the same during the four seasons.

Base your answers to questions 17 and 18 on the diagram below which represents the apparent daily path of the Sun across the sky in the Northern Hemisphere on the dates indicated.

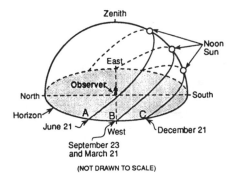

(NOT DRAWN TO SCALE)

17. At noon on which date would the observer cast the longest shadow?
   (1) June 21   (3) March 21
   (2) September 23   (4) December 21

18. Which observation about the Sun's apparent path at this location on June 21 is best supported by the diagram?
   (1) The Sun appears to move across the sky at a rate of 1° per hour.
   (2) The Sun's total daytime path is shortest on this date.
   (3) Sunrise occurs north of east.
   (4) Sunset occurs south of west.

19. On March 21, two observers, one at 45° north latitude and the other at 45° south latitude, watch the "rising" Sun. In which direction(s) must they look?
   (1) Both observers must look westward.
   (2) Both observers must look eastward.
   (3) The observer at 45° N must look westward while the other must look eastward.
   (4) The observer at 45° S must look westward while the other must look eastward.

20. In New York State at 3 P.M. on September 21, the vertical pole shown in the diagram casts a shadow. Which line best approximates the position of that shadow?
   (1) *OA*
   (2) *OB*
   (3) *OC*
   (4) *OD*

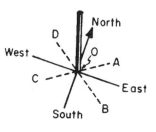

21. The diagram below shows the rotating Earth as it would appear from a satellite over the North Pole.

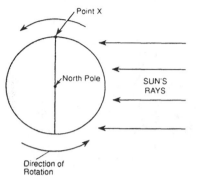

The time at point *X* is closest to
(1) 6 A.M.   (3) 6 P.M.
(2) 12 noon   (4) 12 midnight

22. On June 21, the altitude of Polaris is observed from New York City and found to be 41°. If the altitude is observed again on December 21, it will be
(1) 23½°   (3) 49°
(2) 41°   (4) 64½°°

23. During a period of 1 year, what would be the greatest altitude of the Sun at the North Pole?
(1) 90°   (3) 23½°
(2) 66½°   (4) 0°

24. If the distance between the Earth and the Sun were increased, which change would occur?
(1) The apparent diameter of the Sun would decrease.
(2) The amount of insolation received by the Earth would increase.
(3) The time for one Earth rotation (rotation period) would double.
(4) The time for one Earth revolution (orbital period) would decrease.

25. The force of gravity between two objects will be greatest if their masses are
(1) small and they are far apart
(2) small and they are close together
(3) large and they are far apart
(4) large and they are close together

Note that question 26 has only three choices.

26. In order to escape the Earth's gravitational field, an object must have a minimum velocity of 12 kilometers per second. What minimum velocity must an object have to escape the gravitational field of the Moon?
(1) less than 12 km/s
(2) more than 12 km/s
(3) 12 km/s

27. The diagram shows the Earth *E* in orbit about the Sun. If the gravitational force between the Earth and Sun were suddenly eliminated, toward which position would the Earth then move?
(1) 1
(2) 2
(3) 3
(4) 4

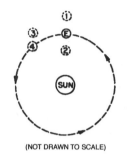

(NOT DRAWN TO SCALE)

28. If the distance from Earth to the Sun were doubled, the gravitational attraction between the Sun and Earth would become
(1) one-fourth as great
(2) one-third as great
(3) twice as great
(4) four times as great

29. Planet *A* has a greater mean distance from the Sun than Planet *B*. On the basis of this fact, which further comparison can be correctly made between the two planets?
(1) Planet *A* is larger.
(2) Planet *A*'s revolution period is longer.
(3) Planet *A*'s speed of rotation is greater.
(4) Planet *A*'s day is longer.

30. The diagram below represents a planet orbiting a star. Lines are drawn from the star to four positions of the planet. The amount of time required to move between positions is indicated. Area *X* is equal to area *Y*. Why is distance *A–B* greater than distance *C–D*?

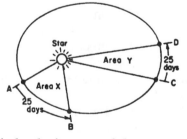

(1) The planet travels at a constant orbital velocity around the star.
(2) The planet travels the same number of degrees per hour.
(3) The length of a planet's day is the same throughout the year.
(4) The planet's orbital velocity is dependent on its distance from the star.

449

31. According to the *Earth Science Reference Tables,* what is the approximate eccentricity of the ellipse shown below?

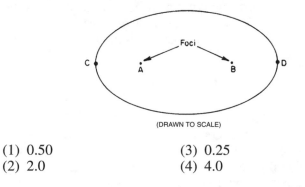

(DRAWN TO SCALE)

(1) 0.50   (3) 0.25
(2) 2.0   (4) 4.0

Base your answers to questions 32 and 33 on the diagram below, which shows the Earth's orbit and the partial orbit of a comet on the same plane around the Sun.

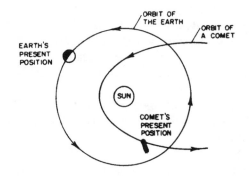

32. Which observation is true for an observer at the Earth's Equator at midnight on a clear night for the positions shown in the diagram?
(1) The comet is directly overhead.
(2) The comet is rising.
(3) The comet is setting.
(4) The comet is *not* visible.

Note that question 33 has only three choices.

33. Compared with the Earth's orbit, the comet's orbit has
(1) less eccentricity
(2) more eccentricity
(3) the same eccentricity

34. The diagram below represents the orbit of a spacecraft around the Sun.

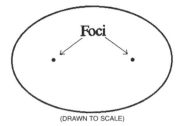

(DRAWN TO SCALE)

According to the *Earth Science Reference Tables,* the eccentricity of the spacecraft's orbit is
(1) more eccentric than Earth's orbit but less eccentric than Mars' orbit
(2) more eccentric than the orbits of planets 300 million km from the Sun but less than the orbits of planets 100 million km from the Sun
(3) more eccentric than the orbit of any planet in the solar system
(4) less eccentric than the orbits of planets with a density less than 5 g/cm^3

35. Which motion causes the Moon to show phases, as seen from the Earth?
(1) the rotation of the Moon on its axis
(2) the rotation of the Earth on its axis
(3) the revolution of the Earth around the Sun
(4) the revolution of the Moon around the Earth

36. The diagram below shows four different positions (*W, X, Y,* and *Z*) of the Moon in its orbit around the Earth. In which position will the full-moon phase be seen from the Earth?
(1) *W*
(2) *X*
(3) *Y*
(4) *Z*

37. Which diagram best represents the full-moon phase of the Moon as seen from the Earth?

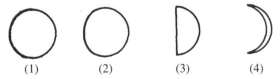

(1)   (2)   (3)   (4)

38. The diagram below shows the relative positions of the Earth, Moon, and Sun for a 1-month period.

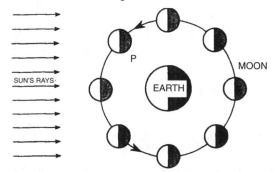

Which diagram best represents the appearance of the Moon at position *P* when viewed from the Earth?

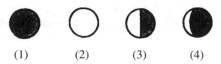

   (1)   (2)   (3)   (4)

39. The diagram below shows the relative positions of the Sun, Earth, and Moon in space. Letters *A, B, C,* and *D* represent locations on the Earth's surface.

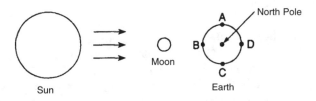

At which location would an observer on the Earth have the best chance of seeing a total solar eclipse?

(1) *A*   (3) *C*
(2) *B*   (4) *D*

40. A total lunar eclipse will occur when the Moon moves into the
   (1) umbra of the Earth   (3) penumbra of the Earth
   (2) umbra of the Moon   (4) penumbra of the Moon

Base your answers to questions 41 through 45 on your knowledge of earth science and the diagram below. The diagram represents a plastic hemisphere upon which lines have been drawn to show the apparent paths of the Sun on 4 days at one location in the Northern Hemisphere. Two of the paths are dated. The protractor is placed over the north-south line. *X* represents the position of a vertical post.

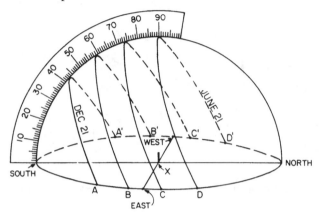

41. For which path is the altitude of the noon Sun 74°?
(1) *A–A′*   (3) *C–C′*
(2) *B–B′*   (4) *D–D′*

42. How many degrees does the altitude of the Sun change from December 21 to June 21?
(1) 43°   (3) 66½°
(2) 47°   (4) 74°

43. Which path of the Sun would result in the longest shadow of the vertical post at solar noon?
(1) *A–A′*   (3) *C–C′*
(2) *B–B′*   (4) *D–D′*

44. Which statement best explains the apparent daily motion of the Sun?
(1) The Earth's orbit is an ellipse.
(2) The Earth's shape is an oblate spheroid.
(3) The Earth is closest to the Sun in winter.
(4) The Earth rotates on its axis.

45. What is the latitude of this location?
(1) 0°   (3) 66½° N
(2) 23½° N   (4) 90° N

Base your answers to questions 46 through 50 on the diagram below and your knowledge of earth science. The diagram represents the Earth at a specific position in its orbit. Arrows indicate radiation from the Sun. Points *A* through *D* are locations on the Earth's surface.

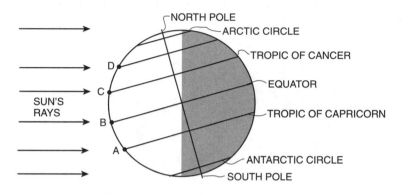

46. During which month could the Earth be in the position shown in the diagram?
(1) February
(2) May
(3) October
(4) December

47. Which location would have the greatest number of daylight hours when the Earth is in this position?
(1) *A*
(2) *B*
(3) *C*
(4) *D*

48. Locations *A*, *B*, *C*, and *D* all have the same
(1) latitude
(2) longitude
(3) intensity of insolation
(4) noontime altitude of the Sun

49. As an observer travels from position *B* to position *D*, the altitude of Polaris in the nighttime sky will
(1) decrease, only
(2) increase, only
(3) increase, then decrease
(4) remain the same

50. Which diagram below best represents the path of the Sun on this date as seen by an observer at location *C*?

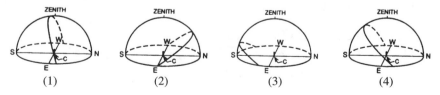

(1)   (2)   (3)   (4)

Base your answers to questions 51 through 55 on your knowledge of earth science, the *Earth Science Reference Tables*, and the diagrams and information below. Diagram I represents the orbit of an Earth satellite, and diagram II shows how to construct an elliptical orbit using two pins and a loop of string. Table 1 shows the eccentricities of the orbits of the planets in the solar system.

DIAGRAM I

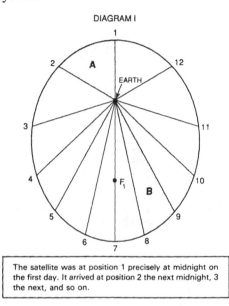

The satellite was at position 1 precisely at midnight on the first day. It arrived at position 2 the next midnight, 3 the next, and so on.

DIAGRAM II

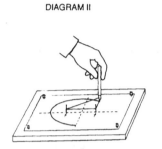

Table 1

Planet	Eccentricity of Orbit
Mercury	0.206
Venus	0.007
Earth	0.017
Mars	0.093
Jupiter	0.048
Saturn	0.056
Uranus	0.047
Neptune	0.008
Pluto	0.250

51. At which position represented in diagram I would the gravitational attraction between the Earth and the satellite be greatest?
(1) 1   (3) 3
(2) 7   (4) 11

52. According to Table 1, which planet's orbit would most closely resemble a circle?
(1) Mercury   (3) Saturn
(2) Venus   (4) Pluto

53. What is the approximate eccentricity of the satellite's orbit?
(1) 0.31   (3) 0.70
(2) 0.40   (4) 2.5

455

Note that questions 54 and 55 have only three choices.

54. The Earth satellite takes 24 hours to move between each two numbered positions on the orbit. How does area *A* (between positions 1 and 2) compare to area *B* (between positions 8 and 9)?
(1) Area *A* is smaller than area *B*.
(2) Area *A* is larger than area *B*.
(3) Area *A* is equal to area *B*.

55. If the pins in diagram II were placed closer together, the eccentricity of the ellipse being constructed would
(1) decrease
(2) increase
(3) remain the same

Base your answers to questions 56 through 60 on your knowledge of earth science, the *Earth Science Reference Tables*, and the diagram below. The diagram represents four planets, *A*, *B*, *C*, and *D*, traveling in elliptical orbits around a star. The center of the star and letter *f* represent the foci for the orbit of Planet *A*. Points 1 through 4 are locations on the orbit of Planet *A*.

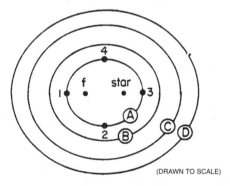

(DRAWN TO SCALE)

56. Which is the order of the planets from shortest period of revolution to longest?
(1) *A, B, C, D*   (3) *C, D, A, B*
(2) *B, A, D, C*   (4) *D, C, B, A*

57. If planets *A, B, C,* and *D* have the same mass and are located at the positions shown in the diagram, the planet that has the greatest gravitational attraction to the star is
(1) *A*   (3) *C*
(2) *B*   (4) *D*

58. Planet *A* travels fastest in its orbit at location
(1) 1   (3) 3
(2) 2   (4) 4

456

59. Using the equation and metric ruler in the *Earth Science Reference Tables*, the eccentricity of Planet A's orbit is found to be approximately

(1) 0.10   (3) 0.50

(2) 0.20   (4) 5.0

60. Which graph best illustrates how the apparent star diameter varies for an observer on Planet *A* as the planet makes one complete orbit around the star?

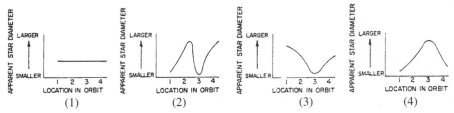

Base your answers to questions 61 through 65 on the *Earth Science Reference Tables,* on your knowledge of earth science, and on the diagram below, which illustrates an imaginary solar system involving two planets, Planet *X* and Planet *Z*.

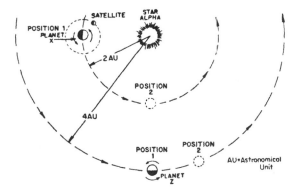

61. Assume that Planet *X* is being orbited by a small natural satellite which is visible from Planet *Z*. To an observer on Planet *Z*, this moon will

(1) always be fully illuminated

(2) always be in complete darkness

(3) be illuminated on one side, only

(4) be in varying amounts of illumination from nearly full to dark

62. An observer on Planet *Z* at midnight will

(1) see only Planet *X* but not its satellite

(2) see only Planet *X*'s satellite but not Planet *X*

(3) see only Star Alpha

(4) not be able to see Star Alpha, Planet *X*, or its satellite

457

Note that questions 63 and 64 have only three choices.

63. As the planets move to their second positions, the angular diameter of Planet Z as observed from Planet X will
(1) decrease
(2) increase
(3) remain the same

64. As the planets move to their second positions, the gravitational attraction between Planet X and Planet Z will
(1) decrease
(2) increase
(3) remain the same

65. If the masses of Planet X and Planet Z are equal, the gravitational attraction between Star Alpha and Planet Z as compared to the gravitational attraction between Star Alpha and Planet X is
(1) ½ as great   (3) ¹⁄₁₆ as great
(2) 2 times as great   (4) ¼ as great

OPTIONAL/EXTENDED
TOPIC F _____

Astronomy Extensions

> **KEY IDEAS** The shift from a geocentric to a heliocentric view of the universe changed the way most people saw themselves in relation to the physical universe. This is an example of the way science works and the ways in which science, mathematics, and technology are interrelated.
>
> The heliocentric view places the Earth as one of nine planets orbiting the Sun—one out of billions of stars in a galaxy. The planets of the solar system vary in composition from small, rocky inner planets to giant, gaseous outer planets.
>
> Most of the evidence collected by observation of celestial objects supports the idea that the universe is expanding as a result of a monumental explosion dubbed the "big bang." Depending on the total mass of the universe, its future may follow one of a number of possible courses including continued expansion, implosion, and oscillation.

KEY OBJECTIVES
Upon completion of this unit, you should be able to:

- Describe the key elements of the geocentric and heliocentric models of the universe.
- Explain how the Foucault pendulum and the Coriolis effect support the heliocentric model.
- Compare and contrast the physical characteristics of the planets in our solar system.

- Describe the evidence supporting the theory that the universe is expanding explosively from an initially dense, hot, chaotic mass.
- Explain why the future of the universe depends on its total mass and describe several possible scenarios for this future.

A. WHAT ARE GEOCENTRIC AND HELIOCENTRIC MODELS OF THE UNIVERSE?

Modern astronomy traces its beginning to the publication in May 1543 of a new heliocentric theory of the universe written by Nicolaus Copernicus.

A-1. Historical Perspective

To appreciate the importance of Copernicus' heliocentric theory, let us consider what two great authorities of early astronomy, Aristotle and Ptolemy, thought of the universe.

Aristotle's Universe

Aristotle was a Greek philosopher who lived from 384 to 322 B.C. He wrote about and taught many subjects, including history, philosophy, drama, poetry, and ethics. His wide-ranging knowledge and insight earned him a prominent place among the great thinkers of antiquity. Aristotle's was a commonsense view of the universe. To the ordinary observer, the sky seems to move. The fixed stars, whose positions relative to one another never change, arc through our view every night. Those near the North Pole never disappear, but circle Polaris, the North Star. Five "stars" do move relative to the fixed stars and are called *planets*, after the Greek word for "wanderer." The Moon and Sun in the sky also move in arcs across our view.

Aristotle explained these observations in terms of a universe consisting of eight crystalline spheres nesting one inside the other like a Chinese-box puzzle, with the Earth at the very center (see Figure F.1). The Sun, Moon, stars, and planets were fixed to the surface of these spheres, which rotated around the unmoving Earth in perfect circles. There were two theories in regard to the motion of the spheres: either they were self-propelled, or, as was thought more likely, their eternal motion was initiated by God. Common sense indicated that the Earth did not move, but there was other evidence as well. If the Earth moved, objects falling in a straight line should fall to the side of points directly beneath them on the Earth. Later astronomers argued against a moving Earth by citing a lack of stellar parallax. *Parallax* is the change in apparent position of an object due to a change in the position of the observer. If the Earth was moving, the apparent positions of the stars should change as the Earth (and the observer) moved.

Figure F.1 Aristotle's Universe. Crystalline spheres were nested one inside the other, with the Earth at the center. The spheres and their attached stars and planets rotated around the Earth.
From *Horizons: Exploring the Universe*, by Michael A. Seeds, Wadsworth, © 1987.

The Earth too was a sphere, the perfect shape, as could be seen when its shadow was seen against the Moon during an eclipse. The natural state of things on Earth was to be at rest, and natural motion on Earth was toward the center of the earth. Unlike the perfect circular motion of the spheres, motion of things on Earth occurred in imperfect straight lines.

All the spheres were perfectly clear and made of ether, a substance that could not be changed or destroyed. There were four other elements: earth, water, air, and fire. All objects were made of mixtures of these four elements and decayed as a result of being forced to move in unnatural directions.

Since Aristotle's universe has the Earth at its center, it is called a *geocentric* or Earth-centered model.

Ptolemy's Universe

There were some problems with Aristotle's view of the universe. The most obvious was visible to the naked eye. There were times when the planets changed course in the sky; for example, Mars would occasionally stop and then move backward in what was called *retrograde motion*. Since the crystalline spheres of the Aristotelian universe could not stop or change direction, this observation was baffling until Claudius Ptolemy proposed an ingenious explanation in the second century (see Figure F.2). Ptolemy's view was that each planet was fixed to a small sphere that was, in turn, fixed to a larger sphere. The smaller sphere and its attached planet turned at the same time that the larger sphere turned. As a result there could be a time when, to an observer on Earth, the planet appeared to be moving backward. Ptolemy called the circular motion of the larger spheres *deferrents* and the motion of the smaller spheres *epicycles*. He also placed the Earth's sphere off the center of its deferent.

With Ptolemy's explanation, Aristotle's universe was able to account for all casual, naked-eye observations of the universe (see Figure F.3). For 1,000 years astronomers studied and preserved Ptolemy's work, but they made no changes in his basic theory. It became part of the accepted thinking of the time and also of church dogma. Thereafter, anyone who questioned it risked being charged with heresy—a crime punishable by death.

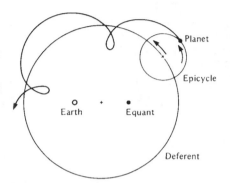

Figure F.2 Ptolemy's Universe. Ptolemy added epicycles to Aristotle's model to explain retrograde motion and changes in apparent diameter.
From *Horizons: Exploring the Universe*, by Michael A. Seeds, Wadsworth, © 1987.

At first, the Ptolemaic system was able to predict the motions of celestial objects with a fair degree of accuracy. However, as the centuries passed, the differences between what the Ptolemaic system predicted and what was actually observed grew so large that they could not be ignored. At first, these differences were blamed on inaccurate observations by earlier astronomers. Arabian and, later, European astronomers corrected the system, recalculated constants, and even added new epicycles. The last great correction was paid for by King Alphonso X of Castile; 10 years of observations and calculations were published as the Alphonsine tables. By the 1500s, however, the Alphonsine tables were inaccurate, often being off by as much as 2°, which is four times the angular diameter of the Moon—an unmistakable error.

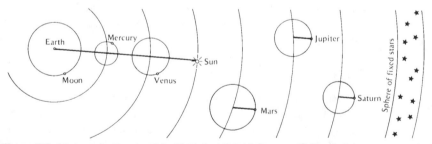

Figure F.3 Ptolemy's Geocentric Model of the Universe. This model was accepted for well over 1,000 years.
From *Horizons: Exploring the Universe*, by Michael A. Seeds, Wadsworth, ©1987.

A-2. Copernicus' Universe

At about the same time that astronomers were struggling with the inaccurate Alphonsine tables, there was a serious need for calendar reform. By the beginning of the 1500s the Julian calendar was off by about 11 days. Easter, a major church holiday, was particularly hard to determine. Both the Hebrew calendar, which was based on the Moon, and the Julian calendar, based on the Sun had to be used to calculate the phase of the Moon upon which Easter depended. A secretary of Pope Sixtus IV asked Nicolaus Copernicus, a priest-mathematician from Poland, to study the problem of calendar reform.

Copernicus recognized that any calendar reform would have to resolve the relationship between the Sun and the Moon. After much study of the problem, Copernicus proposed a mathematically elegant solution in which he suggested a Sun-centered system with a moving Earth. In 1514, he passed around a brief manuscript outlining his ideas, but proceeded discreetly because he recognized the potential dangers of questioning church dogma. Not until his death in 1543 was his full argument in favor of a Sun-centered system published. Even then, he avoided heresy charges by crediting classical Greek sources with the idea, thus implying that it was not the product of his own original thinking.

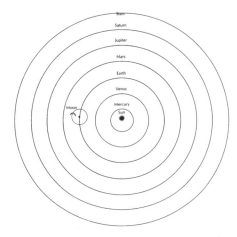

Figure F.4 The Heliocentric Model of the Universe. Copernicus proposed a Sun-centered model in which all planets and stars moved in perfect circles around the Sun.
From *Discovering Astronomy* by Robert D. Chapman. © 1978 by W.H. Freeman, and Company. Used with permission 1978.

Copernicus proposed a *heliocentric*, or Sun-centered universe (see Figure F.4), in which the center of the universe was a point near the Sun. The Earth orbits the Sun and spins once a day on its axis. Retrograde motion occurs because the earth moves faster in its orbit than do planets farther from the Sun. Although the Earth and other planets all move continuously in their orbits around the Sun, the Earth moves toward an outer planet in one part of its orbit, then passes that planet, and moves away from it. Copernicus explained the lack of stellar parallax by stating that stars are so distant that parallax is too small to measure. However, Copernicus avoided the problem of falling objects falling to the side of positions directly below them. Also his system retained epicycles and did no better at predicting the positions of celestial objects than Ptolemy's system.

While Copernicus' system was still wrong, his idea that the universe was heliocentric, or Sun-centered, was correct and gradually gained acceptance. There were probably many reasons why his theory was eventually accepted, but probably the most important were the revolutionary mood of the world at his time and the simple, forthright way in which the theory explained retrograde motion (see Figure F.5).

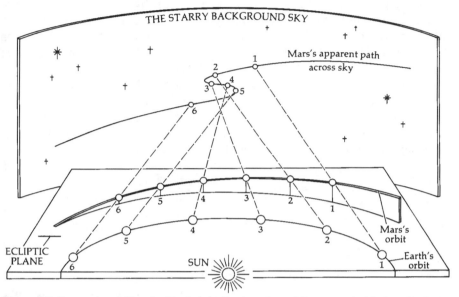

Figure F.5 Copernicus' Simple, Forthright Explanation of Retrograde Motion. Both the Earth and Mars move in a continuous path, but the inner planet (Earth) covers more of its orbit in the same time period, changing its point of view toward the outer planet (Mars).
From *Astronomy: The Cosmic Journey*, by William K. Hartmann, Wadsworth, © 1987.

Contributions of Tycho Brahe and Kepler

Shortly after Copernicus died, a Danish nobleman named Tycho Brahe became interested in astronomy. After observing that the Alphonsine tables were nearly a month off in predicting a conjunction of Jupiter and Saturn, and observing a "new star" produced by a supernova, Tycho also questioned the Ptolemaic system of a perfect, unchanging heaven in a small, widely read book he wrote. With funds from the king of Denmark, Tycho built a world-class astronomical observatory that had no telescopes but included many ingenious devices for precisely measuring celestial motions. When the king of Denmark died, Tycho fell out of favor and accepted a position as court astronomer to the Holy Roman Emperor in Prague. He took with him all of the data from the observatory in Denmark. In Prague, the emperor commissioned him to publish a revision of the Alphonsine tables, and Tycho hired several young mathematicians to help him with his task.

One of these young assistants was Johannes Kepler. Shortly after beginning the project, Tycho died unexpectedly; but before he died, he recommended Kepler to take over his position. Kepler did so and, after analyzing the data from Denmark for 4 years (and making 900 pages of calculations), concluded that the Earth and other planets are moving, not in circular, but in elliptical orbits around the Sun. Kepler published his results in a book and continued his studies. Eventually, he worked out three laws of orbital motion:

1. The orbit of every planet is an ellipse with the Sun at one focus of the ellipse.
2. Planets sweep out equal areas of their elliptical orbits in equal time periods.
3. A planet's period of revolution squared is proportional to its distance from the Sun cubed.

Contributions of Galileo and Newton

Shortly after Kepler's works were published, Galileo Galilei turned his telescope on the heavens and made several discoveries that further undermined the Ptolemaic system. His discovery of imperfections on the Moon's surface, spots on the surface of the Sun, and moons circling Jupiter challenged Aristotle's view of the heavens as perfect and unchanging. He saw Venus go through a full set of phases, an event that was not possible according to the Ptolemaic system of epicycles. Galileo also studied motion and solved Copernicus' falling-object problem. He argued that, if the Earth is in motion, so are all objects on it. Therefore, an object that is dropped moves sideways at the same speed as the Earth and therefore falls on a spot directly beneath its release point. Galileo became an outspoken champion of the heliocentric model.

Eleven months after Galileo died, Isaac Newton was born. During his lifetime, Newton made discoveries in mathematics, mechanics, and optics. He invented calculus, developed his three laws of motion, and proposed an explanation of how gravity works—mutual attraction. His discoveries made it possible to explain the motion of the planets around the Sun as a natural consequence of gravity. With Newton's explanation of the causes of motion, the heliocentric theory began to displace the geocentric theory as the generally accepted model of the universe.

B. WHAT TESTS HAS THE HELIOCENTRIC MODEL MET?

The heliocentric theory requires that the Earth rotate on its axis once every 24 hours. It is one thing to successfully explain away problems like the lack of deflection of a falling object; it is quite another to actually *prove* Earth's rotation. The arguments made in favor of rotation may be quite reasonable, but reasonableness does not constitute scientific proof. In the nineteenth century proof of the Earth's rotation was provided by a French physicist, Jean Foucault, and a French mathematician, Gaspard Coriolis.

B-1. The Foucault Pendulum

Jean Foucault used an ingeniously simple method to prove the rotation of the Earth. Foucault knew that gravity pulls a pendulum only toward the center of the Earth; it does not act laterally to change the plane of the pendulum's swing.

Therefore, if set freely swinging in the absence of any lateral forces, a pendulum should swing in a fixed plane. Foucault further reasoned that the heavier he made the weight of the pendulum, the more force would be required to push the pendulum out of its plane of motion. This fact would make it unlikely that small breezes would deflect the pendulum. Finally, he realized that, if he made the pendulum really long, it would have a long period and would swing for many hours without needing a push to get it going again (a push would introduce a possibility of exerting a lateral force). Foucault argued that, if the Earth was motionless, such a pendulum would swing in a plane whose direction would not change. However, if the Earth rotated, it would rotate beneath the swinging pendulum and change position relative to it. The result would be a pendulum whose plane of swing would appear to rotate relative to the ground in a direction opposite to that of the Earth's rotation.

Foucault's idea is easiest to understand if you visualize a pendulum swinging over the North Pole. As it swings, the Earth moves beneath it. In 24 hours the plane of the swinging pendulum will appear to make one rotation. At the Equator, however, the plane of the pendulum will not appear to change. In between, the period of rotation of the pendulum varies from 24 hours at the poles to infinity at the Equator. At 40° N, the pendulum takes 37 hours make one rotation (see Figure F.6).

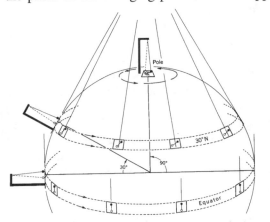

Figure F.6 The Foucault Pendulum It maintains its plane of swing even as the Earth rotates beneath it.
Source: *The Earth Sciences.* © by Arthur N. Stahler, 1971.

In 1851, Foucault suspended a freely swinging, very long (60-meter) pendulum from the inside of the dome of the Pantheon building in Paris. He fastened the heavy weight to one side of the room with a thin cord and sealed all entrances to eliminate any draft that could exert a lateral force on the pendulum. He then burned the cord to set the pendulum swinging smoothly in a fixed plane. Through windows, observers were able to watch the pendulum rotate slowly in a direction opposite to the Earth's rotation, thereby proving that the Earth rotates on its axis.

B-2. The Coriolis Effect

Gaspard Coriolis was a French mathematician who first described the behavior of objects moving in a rotating frame of reference. His work successfully predicted the behavior of objects moving near the rotating Earth.

In a stationary system, fluids such as the atmosphere or the oceans would move directly in a straight line from regions of high pressure to regions of low pressure. However, that behavior is not what occurs on Earth. Large-scale movements of both the atmosphere and the oceans follow curved paths. Study the following diagrams in the *Earth Science Reference Tables*: Surface Ocean Currents (page 4) and Planetary Wind and Moisture Belts in the Troposphere (page 15). Note that ocean currents form large, circular patterns, called *gyres*, that flow clockwise in the Northern Hemisphere and counterclockwise in the Southern Hemisphere. Note too that planetary winds follow paths that curve to the right in the Northern Hemisphere and to the left in the Southern Hemisphere. Such motions should not occur if the Earth was stationary, but are precisely what would be expected if the Earth is rotating. For a full description of why this occurs, refer to the section on the Coriolis effect on page 360.

Since the curving paths of ocean currents and weather systems would not occur on a stationary Earth, their existence is considered proof of the Earth's rotation.

C. HOW CAN THE SOLAR SYSTEM BE DESCRIBED?

C-1. The Structure of the Solar System

The solar system is defined as the Sun, the nine planets that orbit the Sun, the satellites of those planets, and the many small interplanetary bodies such as asteroids and comets (see Figure F.7). The *Sun* is a star, which is composed of gases and emits electromagnetic radiation produced by nuclear reactions in its interior. *Planets* are bodies that are at least partly solid, orbit the Sun, and emit mainly reflected sunlight. *Satellites* are solid bodies that orbit planets.

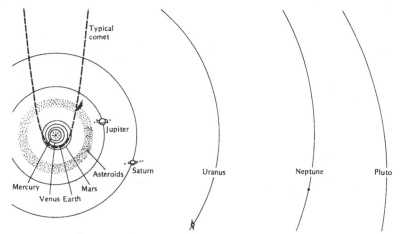

Figure F.7 Our Modern Model of the Solar System.
From *Horizons: Exploring the Universe*, by Michael A. Seeds, Wadsworth, © 1987.

Beginning at the center, the main bodies of the solar system are the Sun, Mercury, Venus, Earth, Mars, Jupiter, Saturn, Uranus, Neptune, and Pluto. A common memory aid for this sequence is *My Very Educated Mother Just Served Us Nine Pickles*. Some basic information about the bodies in the solar system is summarized in Table F.1; an explanation of how these and other solar-system data are obtained is given below the table. Figure F.8 illustrates the use triangulation to determine solar-system distances.

A useful way to grasp the range of sizes in the solar system is to remember that Jupiter is about 10 times the size of Earth and that the Sun is about 10 times the size of Jupiter. Conversely, the smallest planet, Pluto, is about one-fourth the size of Earth. Other useful astronomy measurements are given in Table F.2.

Leaving out Pluto, we can divide the planets into two groups based on their size and composition: the four small, dense *inner planets* and the four large, much less dense *outer planets* of the solar system. This division provides important clues about the origin and early history of the solar system.

TABLE F.1 SOLAR SYSTEM DATA*

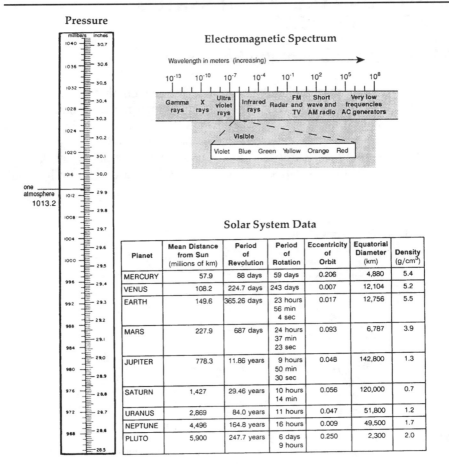

Pressure

Electromagnetic Spectrum

one atmosphere 1013.2

Solar System Data

Planet	Mean Distance from Sun (millions of km)	Period of Revolution	Period of Rotation	Eccentricity of Orbit	Equatorial Diameter (km)	Density (g/cm³)
MERCURY	57.9	88 days	59 days	0.206	4,880	5.4
VENUS	108.2	224.7 days	243 days	0.007	12,104	5.2
EARTH	149.6	365.26 days	23 hours 56 min 4 sec	0.017	12,756	5.5
MARS	227.9	687 days	24 hours 37 min 23 sec	0.093	6,787	3.9
JUPITER	778.3	11.86 years	9 hours 50 min 30 sec	0.048	142,800	1.3
SATURN	1,427	29.46 years	10 hours 14 min	0.056	120,000	0.7
URANUS	2,869	84.0 years	11 hours	0.047	51,800	1.2
NEPTUNE	4,496	164.8 years	16 hours	0.009	49,500	1.7
PLUTO	5,900	247.7 years	6 days 9 hours	0.250	2,300	2.0

*How were these arrived at? The *relative distances* between objects in the solar system were first calculated using Kepler's laws of planetary motion. For example, Kepler's third law states that the ratio of the square of the period of revolution (T) of any two planets, a and b, is equal to the ratio of the cubes (T) of their distance from the Sun (R):

$$\frac{T^2_a}{T^2_b} = \frac{R^3_a}{R^3_b}$$

However, if planet b is Earth, the unit of distance used is the astronomical unit, or AU (1 AU = the average distance between the Earth and the Sun), and time is measured in units of earth-years, then the equation becomes

$$\frac{T^2_a \text{ earth-years}}{1^2_{earth} \text{ earth-years}} = \frac{R^3_a \text{ astronomical units}}{1^3_{earth} \text{ astronomical units}} \quad \text{or}$$

Note that the units cancel out and the proportion reduces to a simple equation. The square of the period of revolution of a planet measured in earth-years equals the cube of its mean distance from the Sun in astronomical units.

According to the Table F.1, the period of revolution of Mars is 687 days. To convert that to earth-years, we divide by the number of days in 1 earth-year, or 365.25 days.

The *period of revolution* of a planet can be measured by directly observing the planet's motions with a telescope. Then this value can be substituted in Kepler's law to solve for mean distance from the Sun. For example:

$$\text{Period of revolution of Mars} = \frac{687 \text{ earth-days}}{365.25 \frac{\text{earth-days}}{\text{earth-year}}} = 1.88 \text{ earth-years}$$

Then we substitute this value for T and solve for R:

$$T^2 = R^3$$
$$(1.88)^2 = R^3$$
$$3.53 = R^3$$
$$R = \sqrt[3]{3.53} = 1.52$$

Thus, Mars is found to be 1.52 AU from the Sun, Venus to be 0.7267 AU from the Sun, and so on. But that still does not tell us exactly how many *kilometers* the Earth, Mars, or Venus is from the Sun.

To convert astronomical units to kilometers, the distance in kilometers between the two bodies of interest must be determined. The best indirect means of measuring interplanetary distances is triangulation. In *triangulation*, a known shift in position of a planet relative to distant stars, as observed from Earth, is used to calculate the distance between planets as shown below. Once this distance is known, the number of kilometers in an astronomical unit can be calculated and then all distances in the solar system can be stated in terms of kilometers.

A direct method of determining distance in the solar system is to bounce radar signals off planets. Radar is electromagnetic radiation and travels at the speed of light, which has been measured very precisely. If the time interval between transmission of the signal and receipt of the reflection that bounced off the surface of the planet is measured precisely, the distance to the planet can be determined with great precision as follows:

$$\text{Speed of light} = \frac{\text{Round-trip distance to planet}}{\text{Radar signal round-trip time}}$$

What Tests Has the Heliocentric Model Met?

For example, suppose 276 seconds elapses between when a radar signal is emitted and when its reflection off Venus is received. To determine the distance to Venus, we substitute as follows:

$$3 \times 10^8 \; \frac{\text{meters}}{\text{second}} = \frac{\text{round-trip distance to Venus}}{276 \text{ seconds}}$$

$$3 \times 10^8 \; \frac{\text{meters}}{\text{second}} \times 276 \text{ seconds} = \text{round-trip distance to Venus}$$

$$828 \times 10^8 \text{ meters} = \text{round-trip distance to Venus}$$

$$414 \times 10^8 \text{ meters} = \text{distance from Earth to Venus}$$

$$41.4 \text{ million kilometers} = \text{distance from Earth to Venus}$$

Since Earth is 1 AU from the Sun and Venus is 0.7267 AU from the Sun, Earth and Venus are 0.2733 AU apart. Using the measurement made above, we find that 0.2733 AU = 41.4 million kilometers and 1 AU = 149.6 million kilometers.

Once the distance to a planet is known, its angular diameter can be used to determine its actual *diameter* in kilometers using triangulation. From this, the *circumference* and *volume* of the planet can be calculated. Then, using Newton's law of universal gravitation, we can determine the mass of the planet, which, together with its volume, yields its *density.*

TABLE F.2 SOME ASTRONOMY MEASUREMENTS

Measurement	Earth	Sun	Moon
Mass (*m*)	5.98×10^{24} kg	1.99×10^{30} kg	7.35×10^{22} kg
Radius (*r*)	6.37×10^3 km	6.96×10^5 km	1.74×10^3 km
Average density (D)	5.52 g/cm^3	1.42 g/cm^3	3.34 g/cm^3

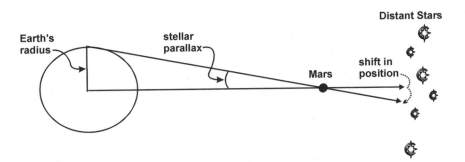

Figure F.8 Use of Triangulation to Determine the Distance in Kilometers between Earth and Mars. The Earth's radius is known, the parallax angle can be measured, and the angle at the center of the Earth is a right angle. When two angles and one side of a triangle are known, trigonometry can be used to calculate the lengths of the unknown sides of the triangle.

470

C-2. The Inner Planets

The four inner planets are Mercury, Venus, Earth, and Mars. The inner planets are terrestrial, resembling the Earth in size and rocky composition. They are also about the same density as Earth.

Mercury

Mercury is one of the five planets that can be seen from Earth with the naked eye. However, Mercury is very difficult to spot because it is never more than 28° from the Sun, whose glare usually hides it. Through a telescope from Earth, Mercury can be seen to exhibit phases like the Moon and Venus, but only dark, fuzzy features are visible on its surface.

Mercury revolves around the Sun once every 88 days and rotates once every 58.65 days. As the closest planet to the Sun, it receives the most intense radiation. Also, the long rotation period causes the side of Mercury facing the Sun to receive nonstop sunlight for a long period of time. At the point where the Sun is directly overhead on Mercury, the surface temperature reaches 700 Kelvin units—hot enough to melt lead. At the same time, the side facing away from the Sun is in darkness for long periods and cools to 100 K.

Like the Moon, Mercury has no atmosphere because its small gravitational field was unable to hold on to any gases. The lack of an atmosphere allows debris from space to strike the surface of Mercury unhindered. A fly-by of Mercury by the Mariner 10 spacecraft showed a cratered surface remarkably like the Moon's.

Venus

Venus is one of the brightest objects seen in the sky; only the Sun and the Moon are brighter. Since its orbit is larger than Mercury's, Venus may be seen as far as 47° from the Sun. This feature, together with its brightness, makes it one of the most obvious of all celestial objects.

Venus is almost identical in size to the Earth. Like Mercury, Venus can be seen to go through a series of phases. However, Venus rotates in a direction opposite to that of Earth and most other planets.

A telescope can detect little about Venus' surface because its dense atmosphere, consisting mostly of carbon dioxide, hides the surface from view. This thick, cloudy atmosphere produces a greenhouse effect that traps heat; several Venera probes soft-landed on Venus by the Soviet Union recorded surface temperatures near 750 K and atmospheric pressures 90 times greater than Earth's. Spectroscopic studies indicate the presence of sulfuric acid, hydrochloric acid, and hydrofluoric acid. In February 1974, the Mariner 10 spacecraft came within 5,800 kilometers of Venus. As it flew by, cameras equipped with special filters and films recorded pictures by ultraviolet light that revealed details of Venus' atmosphere unseen in visible light. Photographs show cloud motions indicating wind speeds of 100 meters per second (200 mph) in the upper atmosphere.

Earth

Earth is the third planet from the Sun and is the largest of the inner planets. Earth's atmosphere is rich in oxygen and nitrogen. From space, many details of the Earth's surface can be seen, including the fact that more than 70 percent is covered with liquid water. Earth has one natural satellite—the Moon.

Mars

Mars is the fourth planet from the Sun. In the sky, it appears as a reddish star. When this planet is viewed through a telescope, its reddish brown, desertlike surface and polar "ice caps" can be seen. Mars's axis is tilted 24° to its orbit (similar to Earth's 23½° tilt), but since its year is nearly twice as long as Earth's, its seasons are longer. The polar "ice caps," which are mostly carbon dioxide with some water ice, melt and refreeze as the seasons pass.

Beginning in 1965, a series of Mariner spacecraft passed Mars and photographed its surface, which is pockmarked with craters, much like the Moon's. The photographs also revealed long, sinuous channels that look just like dry riverbeds. Most analysts think they were cut by liquid water, but have no idea where the water is now. The Viking probes that landed on Mars found a rocky surface with volcanoes (the largest being four times the size of Mt. Everest) and vast, flat plains, but no traces of life. Mars's atmosphere was found to be very thin, with less than 1 percent the surface pressure of Earth. It is composed mainly of carbon dioxide with traces of water vapor, argon, ozone, oxygen, carbon monoxide, and hydrogen.

Mars has two natural satellites—the moons Phobos and Deimos. These moons are tiny—Phobos is only 25 kilometers in diameter, and Deimos is only 15 kilometers. However, they orbit much closer to the surface than our Moon does to Earth and would therefore appear large to an observer on Mars—about one-half the size of our Moon. Phobos revolves around Mars in just under 8 hours–faster than the planet rotates, so its rising and setting are the result of its motion, not the rotation of Mars.

C-3. The Outer Planets

The outer planets—Jupiter, Saturn, Uranus, and Neptune—are all gas giants. Although they have more mass than the inner planets, they are less dense. Gas giants have thick atmospheres (hence their name), composed largely of gaseous hydrogen compounds such as water (H_2O), methane (CH_4), and ammonia (NH_4) surrounding a small rocky or liquid core. Despite their large size, the gas giants rotate very rapidly, causing a distinct equatorial bulge.

Jupiter

Jupiter, the most massive planet in the solar system, contains 70 percent of all the mass outside of the Sun (Jupiter's mass is still only a thousandth of the mass of the Sun). Despite its greater mass, Jupiter is less dense than Earth because of its large size—1,300 times the volume of Earth.

Through even a small telescope, Jupiter can be seen as a disk crossed by narrow, parallel bright and dark bands. Through more powerful telescopes, Jupiter's great red spot can be seen. In 1973 and 1974 Pioneer 10 and 11 flew by Jupiter and took many pictures, which were sent back to Earth. They revealed Jupiter's surface in more detail than had ever before been seen. The visible features of Jupiter's disk are the tops of clouds in the deep atmosphere. A mottled appearance suggests the presence of convective cells. The current thinking is that heated gases from deep within the atmosphere rise and cool, causing clouds to condense and reflect sunlight, forming bright spots. The clouds then spread out to the north and south, sink, and clear up, thus less light is reflected and the regions appear darker. The rapid rotation of Jupiter causes these cloudy and clear regions to swirl together into parallel bands. The great red spot is thought to be the eye of an immense, hurricane-like storm in Jupiter's atmosphere.

Jupiter's interior structure has been inferred from gravity measurements made by Pioneer spacecraft. The core is probably a small, rocky ball similar in composition to Earth. The rest of the planet is mostly hydrogen in two distinct layers: an inner layer of liquid hydrogen that is under such tremendous pressures that it acts like a metal, and an outer layer of liquid hydrogen that acts like hydrogen here on Earth. The outer layer blends gradually from pure hydrogen to hydrogen compounds such as water, ammonia, and ammonium hydrosulfide in the atmosphere.

Jupiter has many natural satellites because of its powerful gravitational field. Four bright moons were first discovered by Galileo. Over the years, ever more powerful telescopes were brought to bear on Jupiter, and by 1975 fourteen moons had been discovered. The four Galilean moons, Io, Europa, Ganymede, and Callisto, are about the same size as Earth's Moon; the rest are much smaller. Voyager 1 discovered a faint ring around Jupiter and showed that one moon, Io, has active volcanoes that cover its surface with red and yellow sulfur compounds. It also showed that Europa has a smooth, ice-covered surface and that Ganymede and Callisto are covered with craters, as is Earth's moon.

Saturn

Saturn resembles Jupiter in composition and structure, but its cloud markings have less contrast. The surface of Saturn is mostly bands of yellowish and tan clouds. The density of Saturn, 0.7 gram per cubic centimeter, is the lowest of any planet and is, in fact, less than that of water. If you could find a large enough body of water, Saturn would float in it!

Saturn is best known for its ring system. When Galileo first saw Saturn through a telescope in 1610, he drew it as a blurry object with other blurry objects on either side and thought it was a triple planet. In 1655, however, Christian Huygens discovered that a ring system surrounds the planet. The rings' dimensions are remarkable; they stretch for 274,000 kilometers from tip to tip but are barely 100 meters thick! In fact, the rings are so thin that observers from Earth lose sight of them when they appear edge-on while Earth passes even with their plane.

Saturn's rings are composed of countless particles ranging in size from that of a golf ball to that of a house. Spectroscopic studies have proved that these particles consist of frozen water or are covered by frozen water. The rings are divided by several large gaps and thousands of finer divisions. The reason the gaps exist is poorly understood, but one theory is that gravitational effects are responsible. Where did the ring particles come from? Some possibilities are that the particles condensed from gas as Saturn formed, that they are fragments of a satellite blown apart by a collision with a comet or an asteroid, or that they are the remains of a comet or asteroid torn apart by tidal forces while passing very close to Saturn.

Like Jupiter, Saturn has an extensive system of natural satellites. It has at least 17 moons, including Titan, which is almost half the size of Earth, and many smaller moons clustered near the rings. Titan is so large that it has an atmosphere of its own, consisting mainly of nitrogen with traces of ethane, acetylene, ethylene, and hydrogen cyanide.

Uranus

Uranus was first discovered in 1781 by William Herschel. It is so far from Earth that it cannot be seen with the naked eye and a telescope cannot resolve any markings on it—only a faint greenish color. In 1986 Voyager flew by Uranus and revealed that its atmosphere has an almost featureless blue haze overlying deeper clouds. Voyager recorded a minimum temperature of 51 K (–368°F) and a composition matching that of Jupiter and Saturn. In 1977 evidence of rings around Uranus was observed. Voyager 2 obtained the first clear pictures of these rings, showing them to be much narrower than those of Jupiter or Saturn.

Like Venus, Uranus' rotation is retrograde, and its axis of rotation, pointing toward the Sun, is almost in line with the plane of its orbit. Uranus has five major moons, and ten smaller moons only recently discovered by Voyager. The larger moons are composed of ice and black soil. All of the moons are heavily cratered, and cracks and canyons can be seen on several.

Neptune

The discovery of Uranus led astronomers to search for other planets. By 1800, observation of Uranus had revealed irregular motions that ran counter to Kepler's laws. Some scientists thought that these motions were caused by a breakdown of gravitation at great distances from the Sun, but others guessed correctly that Uranus was being pulled from its theoretical orbit by an even more distant planet. Using laborious calculations, astronomers set out to predict where the new planet should be located in order to produce the observed effects on Uranus. In 1846, within 12 hours of beginning a search based on the calculations of French mathematician Urbain Leverrier, two German astronomers discovered Neptune.

Neptune's atmosphere has been found to contain hydrogen, helium, and some methane, which gives it its bluish color. It also has faint cloud patterns resembling the bands of Jupiter and Saturn. Like the other gas giants, Neptune rotates rapidly and has a distinct equatorial bulge. The temperature of its atmosphere is 60 K—warmer than expected for a body so far from the Sun and suggesting that Neptune may have an internal source of heat.

Neptune has eight known natural satellites. The largest moon, Triton, revolves around Neptune in a direction opposite that of all other satellites in the solar system. Although its second largest moon, Nereid, revolves in the normal direction, its orbit is highly eccentric.

C-4. Pluto

Pluto, the farthest planet from the Sun, was not discovered until 1930. Pluto is so far from the Sun that the Sun would not appear as a disk from its surface, but would look like a very bright streetlight across the street. In 1992, it was discovered that this tiny planet is made up of 97 percent nitrogen with small amounts of frozen carbon monoxide and methane. Pluto is about 2,300 kilometers in diameter and has one known satellite, Charon, about 1,200 kilometers in diameter. Together, both Pluto and Charon would fit inside the United States.

Pluto's orbit is much more eccentric than that of the other planets and is tilted 17° from the plane of the ecliptic. When it is at perihelion (closest to the Sun), Pluto is closer to the Sun than is Neptune.

Pluto is a special case among the planets, and its origin is still a topic of discussion among astronomers.

C-5. Interplanetary Bodies

In addition to the nine planets, many smaller objects have been found to orbit the Sun, and are therefore also considered part of the solar system. These objects include asteroids, comets, and meteoroids.

Asteroids

Asteroids are solid bodies having no atmosphere. They are tiny "planets" having well-determined orbits. More than 2,000 of these objects have been found in the solar system; most of them orbit the Sun in the gap between Mars and Jupiter. Some asteroids move in very elliptical orbits. The asteroid Icarus comes closer to the Sun than any other asteroid in the solar system.

The composition of asteroids has been studied by examining the wavelength of light that reflects from their surfaces. If asteroids were perfect mirrors, this light would be identical to the sunlight that strikes their surfaces.

Since, however, minerals reflect light of different wavelengths in different ways, the light that reflects from the surfaces of asteroids differs from sunlight. By analyzing this reflected light, scientists can infer the composition of these objects. There are two basic types of asteroids: those with surface reflections characteristic of metals and silicate minerals, and those with reflections characteristic of carbonaceous chondrites (stony with a high carbon content).

Meteoroids

In addition to the asteroids between Mars and Jupiter, chunks of matter circle the Sun in orbits that cross those of the planets. When one of these chunks hits the Earth's atmosphere at very high speed, it vaporizes because of friction with the air. As it streaks through the air and vaporizes, it emits light and, to an observer on the Earth, appears as a streak of light. The chunk of matter is a *meteoroid*; the streak of light seen in the sky is a *meteor*; and any of the matter that survives to strike the ground is called a *meteorite*. Meteoroids vary in size from sand-sized particles to chunks weighing many tons. Meteor showers occur when the Earth crosses the path of a clump or stream of meteoroids (see Figure F.9).

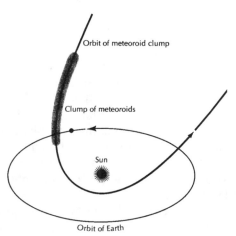

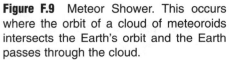

Figure F.9 Meteor Shower. This occurs where the orbit of a cloud of meteoroids intersects the Earth's orbit and the Earth passes through the cloud.

From *Discovering Astronomy* by Robert D. Chapman. © 1978 by W. H. Freeman and Company. Used with permission.

Meteorites tell us a great deal about the solar system. There are three main types of meteorites. *Stony meteorites* are like the Earth's crust, whereas *iron meteorites* resemble its core. *Stony-iron meteorites* are like the iron-rich material of the deep mantle. One explanation for this similarity of meteorites to portions of the Earth is that a process called *gravitational differentiation* took place while the solar system formed. The idea is that the solar system originally had a uniform composition. As gravity drew clumps of the original matter together, however, the material separated into layers because lighter elements do not experience as strong a gravitational attraction as heavier elements. This process resulted in the Earth's layered interior and also layered interiors for all other bodies in the solar system.

The solar system is thought to have originally contained many large asteroids, which collided frequently. With each collision, material was broken off and ejected, some into highly eccentric orbits. If these large asteroids had a

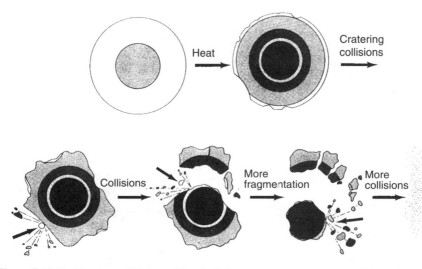

Figure F.10 Production of Meteoroids. Collisions between large asteroids with layered interiors gave rise to the three types of meteorites found on Earth.
From *Discovering Astronomy* by Robert D. Chapman. © 1978 by W. H. Freeman and Company. Used with permission.

layered structure, one would expect them to break into fragments (meteoroids) with different compositions like those of the different layers of the asteroid (see Figure F.10). For this reason, iron meteorites are considered examples of the composition of the Earth's core.

Comets

Comets move in elliptical orbits around the Sun, just as do planets, but their orbital ellipses are highly elongated. Some comets vary from being 1 astronomical unit from the Sun at perihelion to more than 1,000 astronomical units at aphelion. Kepler's law of equal areas tells us that, although such a comet moves very fast while it is near the Sun, it must move very slow when it is so far away. Therefore, some comets take as long as 2 million years to make one orbit of the Sun. Comets are generally classified as long period or short period. Short-period comets orbit the Sun in 200 years or less, long-period comets take more than 200 years to complete an orbit.

Comets are thought to consist of a solid *nucleus*, composed of meteoroid particles embedded in ice. When the nucleus approaches the Sun and heats up, the ice sublimes into a cloud of gas, called a *coma,* around the nucleus. As the comet gets closer to the Sun, the coma increases in size as more gas sublimes, and particles emitted from the Sun collide with the gas molecules and push them out of the coma, forming a *tail.* Under the influence of this solar "wind," the tail always streams away from the Sun. Unlike a contrail from a jet engine, it does *not* stream out along the path of the comet.

The changes that occur in a comet's structure as it orbits the Sun are shown in Figure F.11.

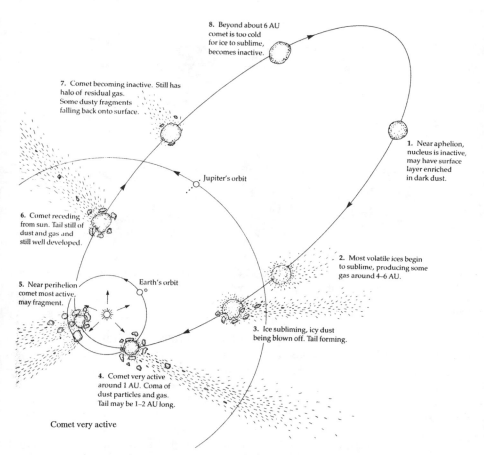

Figure F.11 Changes That Take Place in a Comet as Its Orbit Carries It Past the Sun.
From *Astronomy The Cosmic Journey*, by William K. Hartmann, Wadsworth, © 1987.

C-6. Solar System Exploration

Exploration of the solar system has yielded information about the origin of the Earth and the solar system. Exploration of other planets will help us to understand whether or how they can be of benefit to humans. The asteroids may become a valuable source of metals and mineral resources. In the future, colonization of other planets may be possible or even necessary.

D. HOW DO COSMOLOGISTS EXPLAIN THE ORIGIN, EVOLUTION, AND FUTURE OF THE UNIVERSE?

Cosmology is the scientific inquiry into the origin, evolution, and fate of the universe. By making assumptions that are not at odds with the observable universe, cosmologists create models or theories that attempt to describe the universe, its origin, and its future. The model is then used to make predictions until something is found that contradicts it, whereupon the model is modified or discarded in favor of a new one.

Cosmologists assume that the laws of physics are identical throughout the universe and that the universe has an identical appearance to all observers no matter where they are located. This means that the universe cannot have an edge, or it would appear different to an observer near the edge than to an observer near the center. The geometry of space must be such that all observers see themselves as being at the center. Cosmologists also believe that the only motion that can occur in such a geometry is expansion or contraction of the universe.

D-1. The Expanding Universe

Observations of objects in the universe indicate that the universe is expanding. The evidence of this motion comes from the light reaching us from distant galaxies, which has a lower frequency than expected. The explanation for this shift lies in a phenomenon called the *Doppler effect*.

The Doppler effect is the shift in wavelength of a spectrum line away from its normal wavelength that is caused by motion of the observer toward or away from the source. If the source is approaching the observer, there is a blue shift toward shorter wavelengths of light. If the source is moving away from the observer, there is a red shift toward longer wavelengths.

This phenomenon occurs because light consists of electromagnetic waves. The human eye distinguishes between one wave and another by wavelength, that is, the distance between one crest of a wave and the next. Now, suppose a source and an observer are not moving in relation to each other. The source is emitting waves of a particular length—let's say 500 nanometers (billionths of a meter)—at the speed of light. The observer will perceive a certain number of crests per second, and the light will be "seen" as a particular color, in this case yellow. Now, if the source moves toward the observer at the same time that it is emitting the waves, more crests will reach the eye of the observer per second, as would occur also if the light waves had a shorter length. Therefore, the eye will interpret the light as being a different color, for example, blue even though the source is emitting yellow light. If the source is moving away from the observer, just the opposite happens. Fewer crests per second reach the eye, and yellow light is interpreted as having a longer wavelength, perhaps that of red (see Figure F. 12).

The amount of shift is proportional to the speed at which the source is approaching or receding from the observer:

$$\frac{\text{Shift in wavelength}}{\text{Normal wavelength}} = \frac{\text{Approach or recession speed}}{\text{Speed of light}}$$

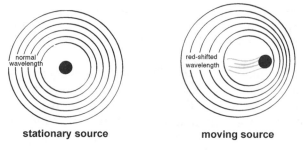

stationary source   moving source

Figure F.12 The Doppler Effect.

When the light from distant galaxies is observed, all are seen to be red-shifted, indicating that they are moving away from us. Furthermore, the farther away the galaxy, the more it is red-shifted. This indicates that distant galaxies are moving away from us faster than nearer galaxies. The only explanation for such an observation is that the universe is expanding in all directions.

D-2. The Big Bang Theory

Since the universe appears to be expanding, it must have been smaller at some time in the past. If the galaxies could be traced back in time, there would be a point at which they were all very close to one another. The observed expansion rate indicates that this close proximity occurred between 10 and 20 billion years ago. Thus, we have a model in which the universe started out with all of its matter in a small volume and then expanded outward in all directions, a motion very similar to an explosion. For this reason, this model of the universe is called the *big bang theory*.

The big bang was, however, quite different from any explosion on Earth. You can observe an explosion on Earth from a distance; it comes from a specific point in space at a particular time. But the big bang did not occur at some preexisting point in space that we can now turn our telescopes to and observe. Instead, the entire universe, including space and time, originated in the big bang. Before the big bang, there was no place from which an observer could stand and watch the explosion and there was no time as we know it. With the big bang, both space and time came into existence simultaneously. What we perceive as movement of galaxies away from us is due to the expansion of space, and what we perceive as the elapsing of time is the expansion of time. Thus, galaxies are not hurtling *through* space; rather, space itself is expanding and carrying the galaxies farther apart in the process.

D-3. The Future of the Universe

The future of the universe depends on the total amount of matter in the universe. Gravity is a strong and pervasive force. If the density of matter in the universe is greater than a certain critical amount, then gravity will be able to slow the expansion of the universe and slowly turn expansion into contraction. The problem is that measurements of the amount of matter that exists in the universe vary so much that none of the possibilities described below can be ruled out. Part of the problem is that the matter we can see, that is, the matter that emits light, represents only a tiny fraction of all the matter in the universe. The "dark" matter, which we cannot see because it doesn't emit light, will determine the fate of the universe.

If the density of matter *falls short of the critical amount*, the gravitational pull needed to stop the expansion cannot be exerted. The universe will continue expanding and cooling. Although stars will continue to form for a while, the amount of matter available for star formation will dwindle until no new stars form. Eventually, all of the mass in the universe will be contained in cold, dead planets and burned-out stars. The universe will end in the form of ever more widely scattered matter in a black void containing no light or heat.

If the universe *has exactly the critical density of matter*, expansion will continue, but will slow down at an ever-decreasing rate. In this case the fate of the universe will be similar to that of an ever-expanding universe, except that at some point the universe will stop moving entirely.

If the density of matter *even slightly exceeds the critical density*, the matter will exert enough gravitational attraction to stop the expansion and start a contraction. Ultimately, all of the matter will collapse back into a small volume—the big crunch. And what then? One theory is that the universe is oscillating and the big crunch would give rise to another big bang. Another theory is that the big crunch would form an enormous black hole that would consume space and time just as the big bang produced it.

Will we ever know which model is correct? Perhaps, but the price of that knowledge will be a continued search for answers.

REVIEW QUESTIONS FOR TOPIC F

1. In the geocentric model (the Earth at the center of the universe), which motion would occur?
   (1) The Earth would revolve around the Sun.
   (2) The Earth would rotate on its axis.
   (3) The Moon would revolve around the Sun.
   (4) The Sun would revolve around the Earth.

2. Which observation can *not* be explained by a geocentric model?
   (1) Stars follow circular paths around Polaris.
   (2) The Sun's path through the sky is an arc.
   (3) A planet's apparent diameter varies.
   (4) A freely swinging pendulum appears to change direction.

3. The actual shape of the Earth's orbit around the Sun is best described as
   (1) a very eccentric ellipse
   (2) a slightly eccentric ellipse
   (3) an oblate spheroid
   (4) a perfect circle

4. Observations of a planet moving among the stars over the course of a year provide evidence that the planet
   (1) rotates on its axis   (3) tilts toward Polaris
   (2) gives off light   (4) revolves around the Sun

5. The diagram below represents a Foucault pendulum in a building in New York State. Points A and A′ are fixed points on the floor.

   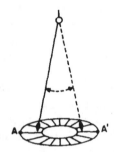

   As the pendulum swings for 6 hours, it will
   (1) appear to change position due to the Earth's rotation
   (2) appear to change position due to the Earth's revolution
   (3) continue to swing between A and A′ due to inertia
   (4) continue to swing between A and A′ due to air pressure

6. The change in the Sun's apparent diameter during a year is supporting evidence that the Earth's orbit is
   (1) a circle   (3) tilted
   (2) an ellipse   (4) parallel

7. In the diagram below, the arrows represent the paths of moving fluids on the surface of the Earth.

   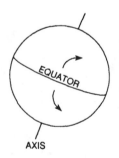

   Which statement best explains why the fluid is deflected?
   (1) The Earth is rotating on its axis.
   (2) The axis of the Earth is tilted.
   (3) The Earth is revolving around the Sun.
   (4) The Earth is moving away from the Sun.

8. The Coriolis effect would be influenced most by a change in the Earth's
   (1) rate of rotation   (3) angle of tilt
   (2) period of revolution   (4) average surface temperature

9. Astronomers have observed a reddish spot on the surface of Jupiter. From observations of this spot, it is possible to estimate the
   (1) period of Jupiter's rotation
   (2) period of Jupiter's revolution
   (3) pressure of Jupiter's atmosphere
   (4) temperature of Jupiter's surface

10. The apparent color of a star is chiefly an indication of its
    (1) composition
    (2) magnitude
    (3) distance from the earth
    (4) surface temperature

11. The bright star Rigel is 650 light-years away. The light from Rigel that is visible in 1995 left Rigel during
    (1) 1995 A.D.
    (2) 1345 A.D.
    (3) the latter part of the ice age
    (4) the Paleozoic era

12. Which of the following statements best describes galaxies?
    (1) They are similar in size to the solar system.
    (2) They contain only one star but hundreds of planets.
    (3) They may contain a few hundred stars in a space slightly larger than the solar system.
    (4) They may contain billions of stars in a space much larger than our solar system.

13. According to the *Earth Science Reference Tables,* approximately how many years ago did the solar system originate?
    (1) 570,000,000
    (2) 1,000,000,000
    (3) 4,500,000,000
    (4) 10,000,000,000

14. A person observes that a bright object streaks across the nighttime sky in a few seconds. What is this object most likely to be?
    (1) a comet
    (2) a meteor
    (3) an aurora
    (4) an orbiting satellite

Base your answers to questions 15 through 19 on the table below, which shows data for the various planets of the solar system, and on your knowledge of earth science.

No.	Planet	Diameter (Earth = 1)	Mass (Earth = 1)	Period of Revolution (earth-years)	Period of Rotation (earth-days)	Maximum Apparent Magnitude
1	Mercury	0.38	0.05	0.24	58	−0.2
2	Venus	0.96	0.81	0.62	247	−4.2
3	Earth	1.0	1.0	1.0	1.0	...
4	Mars	0.53	0.11	1.9	1.0	−2
5	Jupiter	11.2	318.4	11.9	0.41	−2.5
6	Saturn	9.5	95.3	29.5	0.43	−0.7
7	Uranus	3.7	14.5	84	0.45	5.5
8	Neptune	3.5	17.2	164.8	0.65	7.9
9	Pluto	1.0	0.9	248	?	14.9

15. Which of the following planets is smallest ?
(1) Jupiter   (3) Neptune
(2) Uranus   (4) Pluto

16. Which planet has traveled around the Sun more than once in your lifetime ?
(1) Mars   (3) Neptune
(2) Uranus   (4) Pluto

17. Which planet would appear as the brightest object in the sky?
(1) Mercury   (3) Neptune
(2) Venus   (4) Pluto

18. On which of the following planets would the Coriolis effect be greatest?
(1) Mars   (3) Jupiter
(2) Mercury   (4) Neptune

19. If the period of revolution of a planet is directly related to its distance from the Sun, between which two planets is the separation greatest?
(1) Mercury—Venus   (3) Earth—Mars
(2) Venus—Earth   (4) Mars—Jupiter

20. The Foucault pendulum provides evidence that the Earth
(1) rotates on its axis
(2) revolves around the Sun
(3) has an elliptical orbit
(4) has an orbiting Moon

21. Which theory best explains the large-scale movements of ocean currents and wind?
    (1) only the heliocentric theory, in which the Earth rotates
    (2) only the geocentric theory, in which the Earth does not move
    (3) both the heliocentric and geocentric theories
    (4) neither the heliocentric nor the geocentric theory

22. The surface of Venus is much hotter than would be expected, considering its distance from the Sun. Which statement best explains this condition?
    (1) Venus has many active volcanoes.
    (2) Venus has a slow rate of rotation.
    (3) The clouds of Venus are highly reflective.
    (4) The atmosphere of Venus contains a high percentage of carbon dioxide.

23. Which statement best describes how galaxies generally move?
    (1) Galaxies move toward one another.
    (2) Galaxies move away from one another.
    (3) Galaxies move randomly.
    (4) Galaxies do not move.

24. A comparison of the age of the Earth obtained from radioactive dating and the age of the universe based on galactic Doppler shifts suggests that
    (1) the Earth is about the same age as the universe
    (2) the Earth is immeasurably older than the universe
    (3) the Earth was formed after the universe began
    (4) the two dating methods contradict one another

25. The greatest difference in seasons would occur on a planet that has
    (1) a circular orbit
    (2) a slightly elliptical orbit
    (3) its axis of rotation perpendicular to the plane of its orbit around the Sun
    (4) its axis of rotation inclined 4.5° to the plane of its orbit around the Sun

26. On which planet is a large portion of the surface covered by a natural liquid?
    (1) Mercury   (3) Earth
    (2) Venus   (4) Mars

27. The Earth has fewer impact craters than Mercury because of the
    (1) destruction of meteorites in the Earth's upper atmosphere
    (2) more rapid subduction of crustal plates on Mercury
    (3) slower weathering and erosion rates on the Earth
    (4) faster rotational speed of Mercury

485

28. The adjacent diagram represents the orbits of the four inner planets and Earth's moon.

    Which object will be visible from the Earth in the eastern sky after midnight?
    (1) the Moon
    (2) Mars
    (3) Mercury
    (4) Venus

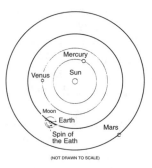

(NOT DRAWN TO SCALE)

29. Which diagram best represents how Venus and the Moon would appear in the sky at sunset to an observer in New York State?

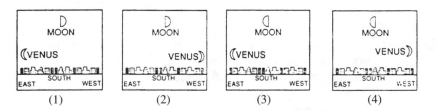

(1)   (2)   (3)   (4)

30. Major ocean and air currents appear to curve to the right in the Northern Hemisphere due to
    (1) Earth's rotation   (3) the Sun's rotation
    (2) Earth's revolution   (4) the Moon's revolution

31. Impact craters on the Moon and Mercury are more obvious than on Earth because
    (1) meteorites have not struck the Earth
    (2) weathering processes on Earth have removed the craters
    (3) Earth is younger than Mercury or the Moon
    (4) all meteorites burn up in Earth's atmosphere

32. Rock samples brought back from the Moon show absolutely no evidence of chemical weathering. This is most likely due to
    (1) the lack of an atmosphere on the Moon
    (2) extremely low surface temperatures on the Moon
    (3) lack of biological activity on the Moon
    (4) large quantities of water in the lunar "seas"

33. The three planets known as gas giants are
    (1) Venus, Neptune, and Jupiter
    (2) Jupiter, Saturn, and Venus
    (3) Jupiter, Saturn, and Uranus
    (4) Venus, Uranus, and Jupiter

34. The average temperature of the planets
(1) increases with greater distance from the Sun
(2) decreases with greater distance from the Sun
(3) has no relationship to the distance from the Sun
(4) depends only on the planets' atmosphere

35. Background radiation detected in space is believed to be evidence that
(1) the universe began with a primeval explosion
(2) the universe is contracting
(3) all matter in the universe is stationary
(4) galaxies are evenly spaced throughout the universe

Note that question 36 has only three choices.

36. Earth and our solar system are
(1) older than the universe
(2) the same age as the universe
(3) younger than the universe

37. Base your answer to the following question on the diagram below and your knowledge of earth science.

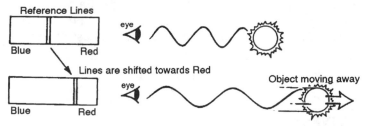

Reference Lines

Blue   Red

Lines are shifted towards Red

Blue   Red

Object moving away

The diagram above indicates that, if the spectral lines produced by the light from a distant galaxy are
(1) blue shifted, the galaxy is moving away from us
(2) blue shifted, the galaxy would be stationary
(3) red shifted, the galaxy is moving away from us
(4) red shifted, the galaxy is moving toward us

38. Which diagram best represents a heliocentric model of a portion of the solar system?

Key: E = Earth, P = Planet, S = Sun

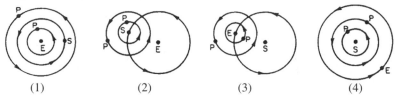

(1)   (2)   (3)   (4)

UNIT NINE _____

Environmental Awareness

> **KEY IDEAS** This unit focuses on the relationships between humans and their geological habitat. It deals with the problems associated with using the Earth and the Earth's response to that use.
>
> If we can understand the dynamic nature of the Earth's physical and biological systems, we may be able to predict the effects of changes and then control or guide the course of these changes.

KEY OBJECTIVES

Upon completion of this unit, you should be able to:

- Describe how the Earth's living and nonliving systems are interdependent.
- Identify problems associated with using the Earth.
- Explain what pollution is and how it relates to population density and industrialization.
- Describe the difference between renewable and nonrenewable resources.
- Give examples of ways in which vital resources can be conserved.
- Discuss the relative merits of different human uses of the Earth and suggest ways to change behaviors so that the maximum benefit is obtained from the Earth while doing it the least harm.

A. HOW CAN WE DESCRIBE THE INTERDEPENDENCE OF EARTH'S LIVING AND NONLIVING SYSTEMS AND THE EFFECTS OF TECHNOLOGY ON THESE SYSTEMS?

A *system* is a group of things or processes that interact to perform some function. A system need not have a specific purpose (e.g., the solar system or a weather system has no "purpose"), but thinking of things or processes as part of a system helps us to understand how they are interrelated.

Although a system may not have a purpose, people define systems with a particular purpose in mind. For example, when thinking about the causes of the phases of the Moon, eclipses, and tides, one might think of a "solar system" consisting only of the Earth, Moon, and Sun. When thinking about retrograde motion, the "solar system" could be expanded to include all of the planets and their moons. And when thinking about a gravitational field, the "solar system" could be expanded even further to include the asteroid belt and comets.

The main purpose of thinking about things in terms of systems is to better understand, predict, or deal with the behavior of the whole system. In a system there is usually input of mass or energy, output of mass or energy, and interactions between the components of the system. If the inputs, outputs, and interactions of a system can be measured, a model can be created to study the theoretical behavior of the system. The model can be used also to identify and study problems or investigate complex phenomena. For example, hydrologists have measured rainfall, runoff, groundwater flow, and channel shapes and sizes to create models of the Mississippi River drainage system. With this model they have been able to predict the timing and extent of flooding.

A-1. Interrelationships of Systems

Any system is usually connected to other systems, both internally and externally. As a result, a system may be thought of both as containing subsystems and as being a subsystem of a larger system. For example, internally, the water cycle consists of many subsystems: ocean water evaporates, condenses to form clouds, and returns to the ocean as rain; rainwater is also absorbed by plants and released as a gas by transpiration; and so on. Externally, the water cycle is part of the erosional-depositional system of streams acting on the Earth's surface.

Living and Nonliving Systems

The Earth consists of a great variety of living and nonliving systems that are related through environmental change. The Earth's nonliving systems are all part of the lithosphere, hydrosphere, or atmosphere that comprise the Earth as a whole. Earth is the only body in the solar system that appears to support life. All of Earth's living things, including humans, are part of and depend on two global interconnected systems called *global food webs*. One global food web consists of the microscopic plants that live in the oceans, or phytoplankton, the animals that eat phytoplankton, and the animals that eat those animals. The other global food web consists of the land plants, the animals that eat plants, and the animals that eat those animals. These living systems continue indefinitely because the once-living things decompose after death to return food material to the environment. These two large living systems consist of many smaller subsystems of living things, such as ecosystems, populations, and biomes.

Effects of Change in One System on Another System

Changing any part of the natural environment can cause dramatic effects on other parts of the environment. For example, climates have sometimes changed abruptly in the past as a result of a change in the Earth's crust, such as a volcanic eruption or the impact of a large meteorite. In 1815, the eruption of the volcano Tamboro in Indonesia spewed vast quantities of ash into the upper atmosphere; the ash blocked enough sunlight to lower summer temperatures worldwide in 1816. Crops in New England failed, there was snow in August, and newspapers dubbed 1816 "the year without a summer."

A-2. Acceleration of Change by Technology

Technology is the application of knowledge to do things. It includes human activities, such as agriculture or manufacturing, as well as the innovative objects or tools people use to accomplish specific purposes (e.g., pencils, microscopes, and television). Technology extends human ability to change the world for better or worse. Using technology, people can shape and form materials, move things from one place to another, and reach farther with their minds and senses than is possible with the human body alone. Thus, technology has greatly increased the scope of change that humans are capable of creating. Since the Industrial Revolution, machines, medicines, and fossil fuels have enabled us to make changes our ancestors could hardly have imagined.

A-3. Good and Bad Effects of Technology

Technologies often have both positive and negative effects. A technology that helps some people or organisms may hurt others. Consider the drawbacks and the benefits associated with pesticides, or atomic energy. Technology may improve our quality of life, but it also causes stress on the natural environment, as the following examples indicate.

Population Increase

The human population of the Earth has increased exponentially in the past several centuries. Advances in technology such as more and better food, sewage disposal, and the development of antibiotics have increased the average life span of humans and thus supported this growth. In 1850 the average life expectancy for a newborn child living in the United States was about 39 years; in 1990 it was 72.7 years. This increased life expectancy here and elsewhere has caused a tremendous increase in population (see Figure 9.1).

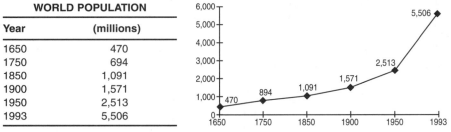

WORLD POPULATION	
Year	(millions)
1650	470
1750	694
1850	1,091
1900	1,571
1950	2,513
1993	5,506

Figure 9.1 World Population from 1650 to 1993

Population growth has caused hostility among humans living in different locations. The fact that natural resources are limited and unequally spread over the Earth has led to competition and disagreements over the sharing of these resources. As the population continues to increase, the intensity of the competition will also increase. At some point, the ability of the environment to support the population will be exceeded, and the population will be forced to decline. Voluntary population control could avert this outcome.

Humans have had a major impact on other species in many ways. As human populations have expanded, they have reduced the amount of the Earth's surface available to other species and have also changed the temperature and chemical composition of these species' habitats. Human activities have interfered with the food sources of other species, and humans have competed with them for food. As human populations have migrated, they have introduced foreign species into the ecosystems in which they settled. In addition, humans have directly altered living things by selective breeding and genetic engineering. The result has been the extinction and endangering of species worldwide. In 1993, a total of 1,087 species of organisms were endangered and 219 were threatened.

Climate Changes

Human activities may possibly change global climate. Naturally occurring greenhouse gases keep the Earth warm enough to be habitable. These greenhouse gases—carbon dioxide, methane, and nitrous oxide—represent only 0.03 percent of the atmosphere, yet they trap 30 percent of the heat being radiated out to space from the Earth. By increasing the concentration of greenhouse gases and adding new ones to the atmosphere, humans are capable of raising the average global temperature. As human population has increased, as agriculture has developed, and as the world has become more industrialized, the concentration of greenhouse gases has increased. Since the Industrial Revolution, the burning of fossil fuels and deforestation have caused a 26 percent increase in the concentration of carbon dioxide in the atmosphere. Chlorofluorocarbons (CFC's) were not part of the atmosphere before their invention in the 1930s; now they are present in measurable quantities.

If global temperatures increase because of the greenhouse effect, other changes in climate can be expected to result. For example, increased temperature would increase evaporation and rainfall, cause ice to melt, and sea levels to rise.

Figure 9.2 shows the concentrations of carbon dioxide, methane, and nitrous oxide from 1750 to the present, and of CFC's from 1950 to the present.

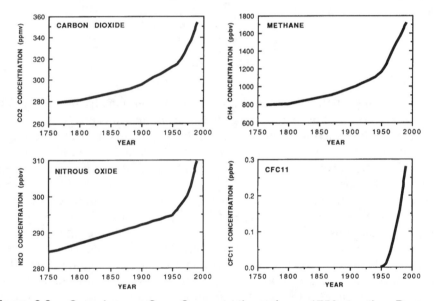

Figure 9.2 Greenhouse Gas Concentrations from 1750 to the Present. Concentrations of carbon dioxide and methane, after remaining relatively constant up to the eighteenth century, have risen sharply since then because of human activities. Concentrations of nitrous oxide have increased since the mid-eighteenth century, especially in the last few decades. CFC's were not present in the atmosphere before the 1930s.

From *Climate Change: The IPCC Scientific Assessment*, edited by J.T. Houghton, G.J. Jenkins, and J.J. Ephraums,Cambridge University Press, 1990.

Ozone Depletion

Chlorofluorocarbons (CFC's) are a group of chemicals used in devices ranging from fire extinguishers to air conditioners and refrigerators to spray cans. The manufacture and widespread use of these compounds have improved our quality of life. They put out fires quickly; keep our homes, food, and cars cool in hot weather; and can be sprayed as everything from deodorants to lubricants wherever we want them. However, they also contain a greenhouse gas and have depleted the atmosphere of much of its ozone, which protects all life on Earth from harmful ultraviolet radiation.

Deforestation

Wood is a valuable natural resource. It is a fuel, a building material, and a raw material for other products such as paper. Consider trying to build or furnish a home without wood. Consider that every book and newspaper you read was once a tree. It is hard to imagine living your life without wood. It is not surprising, then, that the demand for wood is insatiable. Forests represent a valuable commodity to the nations that possess them.

The demand for wood has led to deforestation, which has had far-reaching effects on the environment.

TABLE 9.1 TROPICAL FOREST AREA AND RATE OF DEFORESTATION, 1981–1990
(areas in thousands of hectares)*

Region/ Subregion	Number of Countries Studied	Forest Area, 1980	Forest Area, 1990	Area Deforested Annually, 1981–1990
Latin America	32	923,000	839,000	8,300
Asia	15	310,800	274,900	3,600
Africa	40	650,300	600,100	5,000
Total	87	1,884,100	1,714,800	16,900

From *The Universal Almanac 1994,* © by John W. Wright. Reprinted with permission of Andrews & McMeel. All rights reserved.

Trees absorb carbon dioxide CO_2 and emit oxygen as they carry out photosynthesis. Deforestation decreases the amount of CO_2 removed from the atmosphere by trees, causing a net increase of this gas in the atmosphere. Also, deforested land is susceptible to erosion and loses much of its fertile topsoil during rainstorms. The topsoil ends up in streams and other bodies of water, where it clogs channels and kills aquatic organisms. When land is deforested, many organisms are left without habitats and die. Deforested land also heats up and cools off faster than forested land, causing a change in weather patterns.

Pollution

One of the outcomes of population growth and industrialization has been an increase in pollution. *Pollution* is a concentration of any material or energy form that is harmful to humans. Most pollution is a direct result of the use of technology.

Pollutants are the particular materials or forms of energy that are harmful to humans. Some examples of material pollutants are toxic or radioactive chemicals, petroleum, disease organisms, and human waste. Energy pollutants include heat, harmful radiation, sound, and electromagnetic fields.

Something that contains a pollutant is said to be *contaminated*. We can classify contamination as ground, air, or water pollution.

Ground Pollution

Ground pollution can be harmful in several ways. It can harm humans directly by contact. For example, persons living in homes or attending schools built on land in which toxic heavy metals such as cadmium and manganese were disposed of can come in direct contact with the pollutant by walking, working, or playing on the ground, or by breathing in dust. They can come in contact with the pollutant indirectly by drinking, washing, or bathing with well water that has been contaminated by contact with the ground. Some pollutants are absorbed by plants growing in the ground, or by organisms that live in the ground, and thereby enter the food chain.

A classic example is the pesticide DDT, which was widely sprayed on crops in the 1950s. The spraying of crops contaminated the soil with DDT. Rain seeping through the ground carried the DDT into the groundwater, and runoff washed it into streams. DDT then entered the food chain through small organisms whose bodies absorbed the pesticide. Larger organisms such as fish ate the contaminated small ones and became contaminated themselves. Then birds such as ospreys and eagles that eat fish were contaminated in turn. Because of the DDT in their systems these birds laid eggs so thin shelled and brittle that the weight of the mother bird sitting on them caused them to break. These bird populations plummeted. DDT became so widespread in the food chain that it was even detected in human mothers' milk.

In 1980, Congress enacted the Comprehensive Environmental Response, Cleanup, and Liability Act (CERCLA), also known as the Superfund. The Superfund began as a $1.6-billion, 5-year program to clean up the thousands of hazardous waste sites throughout the country. In order to be listed as a Superfund site, there must be a high degree of risk to the ground, groundwater, surface water, and air. As you can see from Table 9.2, the number of hazardous waste sites has grown dramatically.

TABLE 9.2 SUPERFUND HAZARDOUS WASTE SITES
FOR SELECTED YEARS, 1981–1993*

Year	Number of Sites
1981	115
1982	418
1983	406
1984	538
1986	703
1987	802
1989	981
1990	1187
1992	1183
1993	1202

*From EPA National Priorities List, Supplementary Materials (May 1993).

Air Pollution

Air pollution consists mainly of particles that are small enough to remain suspended in air and gases. Air pollution is one of the most serious threats to humans because it is the most difficult to control once it has occurred. The large-scale patterns of circulation in the atmosphere distribute pollutants throughout the atmosphere very quickly. The unpredictability of the behavior of the atmosphere, coupled with our inability to control its behavior, makes air pollution particularly dangerous.

The Environmental Protection Agency (EPA) monitors the quality of the nation's air at almost 3,000 sites throughout the country. The number of unhealthy days per year is a good indicator of the general levels of air pollution in our country. As you can see from Table 9.3, regulations on auto and factory emissions have had a positive effect on our air quality, which is improving.

TABLE 9.3 NUMBER OF UNHEALTHY DAYS IN
SELECTED METROPOLITAN AREAS, 1980–1991

Metropolitan Area	1980	1985	1988	1989	1990	1991
Atlanta	7	9	15	3	16	5
Boston	8	3	12	2	1	3
Chicago	34	6	18	2	3	8
Dallas	19	15	3	3	5	0
Denver	35	38	18	11	7	7
Detroit	NA	2	17	12	3	7
Houston	10	47	48	32	48	39
Kansas City	13	4	3	2	2	1
Los Angeles	220	196	226	212	164	156
New York	119	60	41	10	12	16
Philadelphia	52	25	34	19	11	24
Pittsburgh	20	5	26	11	11	3
San Francisco	2	5	1	0	1	0
Seattle	33	26	8	4	2	0
Washington, D.C.	38	15	34	7	5	16
Total	610	456	504	330	291	285

Whereas radioactive materials in the ground may be contained, those released into the atmosphere cannot. The Chernobyl nuclear accident led to widespread contamination as wind-borne radioactive particles settled on surrounding land. There was literally nothing that could be done to contain the radioactive particles once they were released into the atmosphere. Entire communities that were downwind of the reactor had to be evacuated, and winds carried the pollution far beyond Russia's borders, causing tension between the Russian government as the source of the pollutant and the other countries that were contaminated.

Water Pollution

Water is an excellent solvent; thus a wide variety of substances may become dissolved in water and contaminate it. Water also has a high specific heat and is therefore widely used in industrial processes as a coolant. As a result, water is widely polluted by heat energy. Although the water cycle separates water from pollutants by evaporation, thereby purifying the water, the pollutants remain in the source from which the water evaporated. Any water entering the polluted source will become contaminated.

Effects of Industry and Population Density on Pollution

Pollution is most intense in areas of industrialization and high population density, where the Earth is used most intensively. All industrial processes and human populations produce waste products and by-products. Where industry and population are concentrated, waste products and by-products will also be concentrated.

Urban areas contain the greatest concentration of industry and human population. Thus, in urban areas the use of the Earth is most intense, society has the most urgent Earth problems, and these Earth problems affect large numbers of people.

A-4. Need to Use Technology to Observe, Monitor, and Understand Earth

The human ability to shape the future comes from a capacity to generate knowledge and develop new technologies. Although technology cannot solve every problem or fulfill every human need, it can be used to help us observe, monitor, and understand the Earth. If we can also understand how our actions will affect the earth, we can modify our behavior to minimize negative outcomes.

B. HOW CAN WE LIVE IN BALANCE WITH OUR NATURAL ENVIRONMENT?

The Earth is a closed system. Whatever material resources it has, such as minerals, soil, clean water, and fresh air, are limited. For example, the Earth is estimated to have 326,000,000 cubic miles of water. That seems like a lot, and it is, but that is *all* there is. Now consider that only one-third of 1 percent is fresh water that is available for human use. When the amount is divided among the current world population of 5 billion, there is only about 85,000 cubic meters per person. That is it, for an entire lifetime, to meet every purpose that water is needed for. And what about the next generation?

B-1. Use of Natural Cycles

Fortunately, many of Earth's resources are part of natural cycles, and this knowledge can be applied to preserve resources. For example:

- Water that is consumed by humans is returned to the environment in their liquid wastes and in the air they exhale. Evaporation returns this water to the atmosphere, and it rejoins the water cycle. As long as stream and lake basins are not polluted, the water cycle will renew the water supply.
- Plants that absorb substances from the soil return these substances to the soil when the plants decompose after dying.
- Forests may be regrown if they are reseeded after trees are cut and cutting is done in such a way that the new growth is able to survive.
- Precipitation, as part of the natural water cycle, removes particulates from the atmosphere. If we do not repollute the atmosphere, this natural process will "cleanse" the air.

B-2. Conservation and Planning

Unfortunately, all resources are not renewed by natural processes; and, as mentioned earlier, pollution of natural resources such as soil and water may not be alleviated by natural cycles. Careful planning, conservation, and regulation however, will help to preserve vital resources. Recycling of materials such as aluminum and paper will alleviate the drain on minerals, trees, and other resources.

Conservation extends the life of existing resources. A limited amount of crude oil exists; estimates range from a 30- to a 100-year supply at present levels of consumption. Decreasing our level of use by conservation could greatly extend the life of this limited resource. Better insulation of homes has cut U.S. residential and commercial use of fuel oil, almost in half, and energy-efficient changes in automobile design, required by law since the late 1980s, have begun to decrease oil consumption for transportation. If we could cut our consumption in half by better design of the devices that consume fuel, we could double the effective lifetime of our oil reserves (see Table 9.4).

TABLE 9.4 U.S. FUEL CONSUMPTION BY TYPE AND END-USE SECTOR, 1950–1991*

	Petroleum (millions of barrels per day)				
Year	Residential and Commercial	Industrial	Transportation	Electric Utilities	Total
1950	1.07	1.82	3.36	0.21	6.46
1955	1.40	2.39	4.46	0.21	8.46
1960	1.71	2.71	5.14	0.24	9.8
1965	1.91	3.25	6.04	0.32	11.52
1970	2.18	3.81	7.78	0.93	14.7
1975	1.95	4.04	8.95	1.39	16.33
1980	1.52	4.84	9.55	1.15	17.06
1985	1.30	4.10	9.85	0.48	15.73
1989	1.32	4.26	11.01	0.74	17.33
1990	1.15	4.32	10.97	0.55	16.99
1991	1.15	4.19	10.78	0.52	16.64

Natural Gas (trillions of cubic feet per year)

Year	Residential and Commercial	Industrial	Transportation	Electric Utilities	Total
1950	1.59	3.43	0.13	0.63	5.78
1955	2.75	4.54	0.25	1.15	8.69
1960	4.12	5.77	0.35	1.72	11.96
1965	5.34	7.11	0.50	2.32	15.27
1970	7.24	9.25	0.72	3.93	21.14
1975	7.43	8.36	0.58	3.16	19.53
1980	7.36	8.20	0.63	3.68	19.87
1985	6.86	6.87	0.50	3.04	17.27
1989	7.50	7.89	0.63	2.79	18.81
1990	7.07	8.21	0.66	2.79	18.73
1991	7.37	8.60	0.80	2.78	19.55

Coal (millions of short tons per year)

Year	Residential and Commercial	Industrial	Transportation	Electric Utilities	Total
1950	114.6	224.6	63.0	91.9	494.1
1955	68.4	217.8	17.0	143.8	447
1960	40.9	177.4	3.0	176.7	398
1965	25.7	200.8	0.7	244.8	472
1970	16.1	186.6	0.3	320.2	523.2
1975	9.4	147.2	<0.05	406.0	562.6
1980	6.5	127.0	<0.05	569.3	702.7
1985	7.8	116.4	<0.05	693.8	818.0
1989	6.2	117.5	<0.05	766.9	890.6
1990	6.7	116.2	<0.05	773.5	896.4
1991	6.3	109.8	<0.05	772.3	888.4

*From *The Universal Almanac,* 1994, Universal Press Syndicate Co.

   Regulations can limit waste of resources and spur efforts to utilize resources more efficiently. For example, regulations requiring mandatory recycling of aluminum, paper, and plastics have greatly increased the percentage of discards recovered. The positive effect has been twofold: both solid wastes and the depletion of natural resources have decreased. Many experts believe the most effective way to decrease waste of resources is by source reduction, that is, by enacting regulations that minimize the volume of products and also the toxic materials they contain.

   Careful planning can stop the unnecessary depletion of natural resources. Consider these few examples. Fishing and hunting regulations that carefully consider the time interval during which animals are harvested can minimize the impact on populations. Harvesting trees individually rather than clear cutting reduces the impact of logging on the habitat. Energy-efficient home design is scarcely more expensive than conventional design and is less expensive when considered over the life of the home. Careful urban planning can minimize the impact of industrial and commercial establishments by siting them wisely.

B-3. Development of Alternatives

As some resources are depleted, we sometimes find other ways to meet our needs. When Europe had denuded its forests to obtain wood for fuel, it turned to coal as an alternative. The switch to coal has removed the pressure on forests, and they have begun to grow back. Eventually, however, fossil fuels will be depleted. Efforts need to be made now, before critical shortages occur, to develop safe, alternative sources of energy. Wind, solar, geothermal, nuclear, and other sources may be the only ones in the future.

Although we have run low on many resources, our quality of life has probably improved. However, that improvement may be short-lived if alternatives are not developed or discovered. And in the case of living things, once a species has been driven to extinction, there is no way to restore it and there is no substitute for it.

B-4. Research and Education

Clearly, certain issues, such as global warming and ozone depletion, which have become causes for concern, warrant more study; current research into matters such as these is inconclusive. The problems associated with the gas ozone are a case in point. There are two main problems with ozone: at ground level there is too much, and in the stratosphere there is too little. The two problems are related. As ozone in the stratosphere is depleted, increased ultraviolet radiation reaching the Earth's surface stimulates the production of ground-level ozone. Does this mean that ozone depletion is self-correcting? We just don't know at this time.

Most of what has been discovered so far indicates that solving or alleviating environmental problems may be costly and complex. But what is the alternative?

Human behavior threatens the very balance of nature. We must carefully study the natural systems that are being affected in order to understand the consequences of this behavior. Then, as we become better informed about environmental issues, we need to change our ways. Research and education will help to guide human activities so as to preserve environmental quality while improving our lives.

In the end, personal responsibility and commitment will make the difference. We need to think globally and act locally; that is, think about the global consequences of our actions and act locally in matters over which we have control.

REVIEW QUESTIONS FOR UNIT 9

1. When the amounts of biologic organisms, sound, and radiation added to the environment reach a level that harms people, these factors are referred to as environmental
   (1) interfaces
   (2) pollutants
   (3) phase changes
   (4) equilibrium exchanges

2. Humans can cause rapid changes in the environment, which may produce catastrophic events. Which statement below is the best example of this concept?
   (1) Mountainside highway construction causes a landslide.
   (2) Lightning causes a forest fire.
   (3) Shifting crustal plates cause an earthquake.
   (4) Changing seasonal winds cause flooding in an area.

3. Recent legislation by federal, state, and local governments dealing with the problem of pollution has come about mainly as a result of
   (1) man's ability to solve this problem with technology
   (2) nature's ability to clean itself, given enough time
   (3) people's belief that natural resources are inexhaustible
   (4) people's action because of their needs, awareness, and attitudes

4. Which graph best represents human population growth?

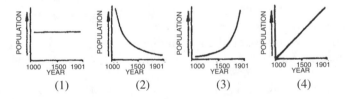

(1)   (2)   (3)   (4)

5. Which pollutant is *not* usually produced or added to the environment by human activities?
   (1) sound
   (2) pollen
   (3) radiation
   (4) smoke

6. Which of the sources of energy listed below is most nearly pollution free?
   (1) nuclear
   (2) solar
   (3) coal
   (4) natural gas

7. Which natural process removes small pollutant particles from the atmosphere?
   (1) the greenhouse effect
   (2) the Coriolis effect
   (3) transpiration
   (4) precipitation

Base your answers to questions 8 and 9 on your knowledge of earth science and on the graph below showing measurements of air pollutants as recorded in a city during a 2-day period.

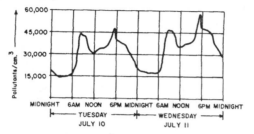

8. What is a probable cause for the increase in pollutants at 8 A.M. and 5 P.M. on the 2 days ?
   (1) change in insolation
   (2) occurrence of precipitation
   (3) high wind velocity
   (4) heavy automobile traffic

9. On the basis of the trends indicated by the graph, at what time on Thursday, July 12, will the greatest amount of pollutants probably be observed ?
   (1) 12 noon   (3) 3 A.M.
   (2) 5 P.M.   (4) 8 A.M.

10. The data table below shows the average dust concentrations in the air over many years for selected cities of different populations.

Population in millions	Dust particles/ meter³
less than 0.7	110
between 0.7 and 1.0	150
greater than 1.0	190

Based on this data table, which graph best represents the general relationship between population and concentration of dust particles?

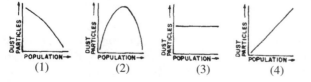

11. Which is the *least* probable source of atmospheric pollution in heavily populated cities?
    (1) human activities   (3) natural processes
    (2) industrial plants   (4) automobile traffic

501

Note that question 12 has only three choices.

12. As the human population of an area increases, the immediate need for land use planning generally will
(1) decrease
(2) increase
(3) remain the same

13. People who live close to major airports are most likely to complain about which form of pollution?
(1) sound
(2) heat
(3) radioactivity
(4) particulates

14. The climates of densely populated industrial areas tend to be warmer than similarly located sparsely populated rural areas. From this observation, what can be inferred about the human influence on local climate?
(1) Local climates are not affected by increases in population density.
(2) The local climate in densely populated areas can be changed by human activities.
(3) In densely populated areas, human activities increase the amount of natural pollutants.
(4) In sparsely populated areas, human activities have stabilized the rate of energy absorption.

15. The addition of pollutants to the atmosphere change the rate of energy absorption and radiation. This could result in
(1) a renewal of volcanic activity
(2) the formation of a new fault zone
(3) a landscape-modifying climate change
(4) a change in the times of high and low tides

16. Some scientists predict that the increase in atmospheric carbon dioxide will cause a worldwide increase in temperature. Which could result from this increase in temperature?
(1) Continental drift will increase.
(2) Isotherms will shift toward the Equator.
(3) Additional landmasses will form.
(4) Icecaps at the Earth's poles will melt.

17. Some scientists believe that high-flying airplanes and the discharge of fluorocarbons from spray cans are affecting the atmosphere. Which characteristic of the atmosphere do they believe is affected?
(1) composition of the ozone layer of the stratosphere
(2) wind velocity of the tropopause
(3) location of continental polar highs
(4) air movement in the doldrums

18. Why were laws passed that made some insecticides, such as DDT, unavailable to the general public?
(1) Some insecticides were not effective against insects.
(2) Some insecticides' colors were causing a change in the environment's physical appearance.
(3) Some insecticides' concentration in the environment became harmful to other animal life as well as to insects.
(4) Some insecticides' effects were weakened when diluted by the atmosphere and hydrosphere.

19. Which statement is best supported by the graph shown below ?

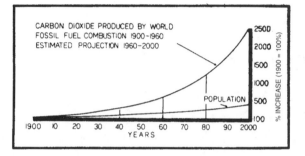

(1) From 1960 to 2000 it is anticipated that there will be a decrease in the use of fossil fuels.
(2) From 1900 to 1960, the average person continuously used a greater quantity of fossil fuels.
(3) By 1980 the world population was approximately 400 million.
(4) From 1970 to 2000 the world population is remaining relatively constant.

20. How will the quality of the environment most likely be affected by the continued dumping of large amounts of industrial waste into a river?
(1) The environment will be unaffected.
(2) Minor, temporary disruptions will occur only near the industrial plant.
(3) Major damage could occur downstream from the industrial plant.
(4) The environment will always adjust favorably to any changes caused by industrial waste.

21. What is the main reason that a high concentration of aerobic bacteria is harmful to a lake?
(1) The bacteria release large amounts of oxygen.
(2) The bacteria use up large amounts of oxygen.
(3) The bacteria cause excessive cooling of the water.
(4) The bacteria provide food for predators.

22. Which graph best represents the typical relationship between population density near a lake and pollution of the lake?

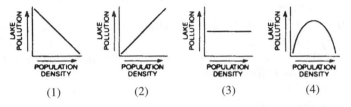

(1)   (2)   (3)   (4)

23. The map below illustrates the distribution of acid rain over the United States on a particular day. The isolines represent acidity measured in pH units.

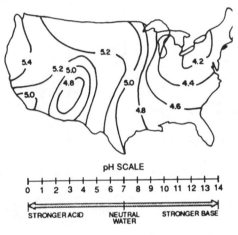

According to the pH scale shown below the map, which region of the United States has the greatest acid rain problem?
(1) northeast   (3) southeast
(2) northwest   (4) southwest

24. Which graph best shows the relationship between atmospheric transparency and the concentration of pollution particles in the air?

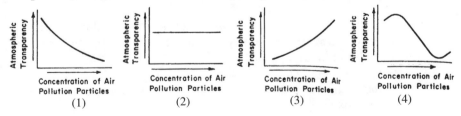

(1)   (2)   (3)   (4)

25. According to the diagram below, at which location would the water probably be most polluted?

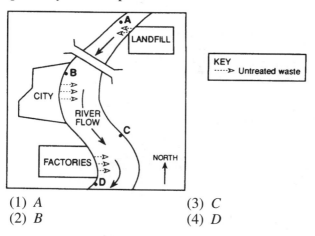

(1) A   (3) C
(2) B   (4) D

26. The graph below shows the amount of noise pollution caused by factory machinery during a 1-week period. Which inference is best supported by the graph?

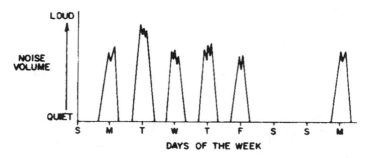

(1) The machinery ran 24 hours a day.
(2) The machinery was turned off on Saturday and Sunday.
(3) The level of pollution remained constant during working hours.
(4) The noise volume reached a peak on Friday.

27. The process of developing and implementing environmental conservation programs is most dependent on
(1) the availability of the most advanced technology
(2) the Earth's ability to restore itself
(3) public awareness and cooperation
(4) stricter environmental laws

Base your answers to questions 28 through 32 on your knowledge of earth science and on the map below, which represents an area of New York State during April.

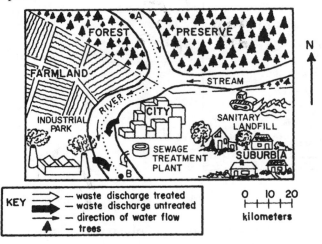

28. Which landscape region has probably been altered *least* by the activities of humans ?
    (1) suburbia   (3) farmland
    (2) sanitary landfill   (4) forest preserve

29. Which graph best represents the probable number of bacteria (anaerobic) along line *A-B* in the river?

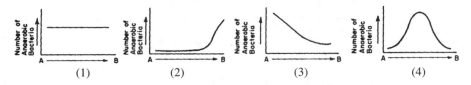

30. If light rains occurred for 1 hour, which area would most likely experience the greatest amount of surface runoff per square kilometer?
    (1) farmland   (3) forest preserve
    (2) suburbia   (4) city

31. Which diagram best illustrates the probable *air pollution field* of this area at an elevation of 100 meters on a windless spring afternoon?

Key: H = High Pollution; L = Low Pollution

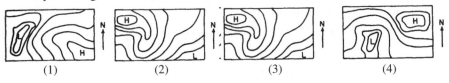

506

32. The people of this area defeated legislation that would have allowed the sale of a large section of the publicly owned forest preserve for the purpose of a second industrial park. They also passed a bond issue providing funds for an additional sewage treatment plant for the city. These actions are an indication that a majority of the voters
(1) are opposed to higher taxes for any reason
(2) feel technology can solve all the problems of the environment
(3) are aware of the delicate balance in nature
(4) feel that nature can take care of itself

Base your answers to questions 33 through 37 on the diagram below, which shows air, water, and noise pollution in a densely populated industrial area.

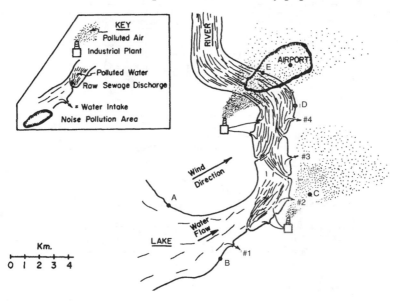

33. Air pollution would probably be greatest at which location?
(1) *A*   (3) *C*
(2) *B*   (4) *D*

34. Water pollution would probably be greatest at which location?
(1) *A*   (3) *C*
(2) *B*   (4) *D*

35. Noise pollution would be greatest at which location?
(1) *B*   (3) *D*
(2) *C*   (4) *E*

36. Which location is subjected to the greatest number of pollution factors?
(1) *A*   (3) *D*
(2) *B*   (4) *E*

507

37. If the water intakes supplied drinking water to the area, which intake would most likely require the most extensive purification procedures?
(1) #1
(3) #3
(2) #2
(4) #4

Base your answers to questions 38 through 42 on your knowledge of earth science, the *Earth Science Reference Tables,* and the diagrams and map below. The graphs in diagram I show the sources of nitrogen and sulfur dioxide emissions in the United States. Diagram II gives information about the acidity of Adirondack lakes. The map shows regions of the United States affected by acid rain.

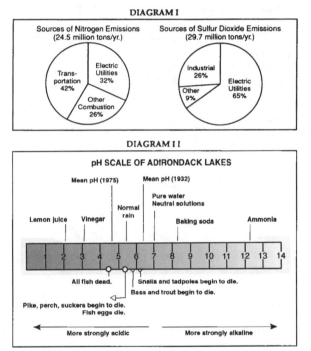

DIAGRAM I

Sources of Nitrogen Emissions (24.5 million tons/yr.)
Transportation 42%
Electric Utilities 32%
Other Combustion 26%

Sources of Sulfur Dioxide Emissions (29.7 million tons/yr.)
Industrial 26%
Other 9%
Electric Utilities 65%

DIAGRAM II

pH SCALE OF ADIRONDACK LAKES

Mean pH (1975)
Mean pH (1932)
Pure water
Neutral solutions
Normal rain
Lemon juice
Vinegar
Baking soda
Ammonia

All fish dead.
Snails and tadpoles begin to die.
Bass and trout begin to die.
Pike, perch, suckers begin to die.
Fish eggs die.

More strongly acidic
More strongly alkaline

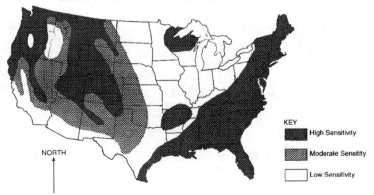

REGIONS OF THE UNITED STATES SENSITIVE TO ACID RAIN

NORTH

KEY
High Sensitivity
Moderate Sensitity
Low Sensitivity

508

38. Which pH level of lake water would *not* support any fish life?
(1) 7.0   (3) 5.0
(2) 6.0   (4) 4.0

39. Which graph best shows the acidity (pH) of Adirondack lakes since 1930?

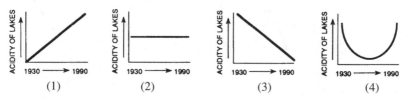

     (1)   (2)   (3)   (4)

40. The primary cause of acid rain is the
(1) weathering and erosion of limestone rocks
(2) decay of plant and animal organisms
(3) burning of fossil fuels by humans
(4) destruction of the ozone layer

41. Acid rain can best be reduced by
(1) increasing the use of high-sulfur coal
(2) controlling pollutants at the source
(3) reducing the cost of petroleum
(4) eliminating all use of nuclear energy

42. In addition to its effects on living organisms, acid rain may cause changes in the landscape by
(1) decreasing chemical weathering due to an increase in destruction of vegetation
(2) decreasing physical weathering due to less frost action
(3) increasing the breakdown of rock material due to an increase in chemical weathering
(4) increasing physical weathering of rock material due to an increase in the circulation of groundwater

Base your answers to questions 43 through 47 on your knowledge of earth science and on the graphs below. Graph I shows the average temperature change on the Earth between the years 1870 and 1955. Graph II shows the amount of carbon dioxide in the atmosphere between the years 1870 and 1962.

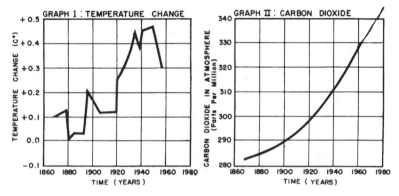

43. Which is the best interpretation that can be made from the graphs for the period between 1870 and 1955?
    (1) The amount of carbon dioxide in the atmosphere increased steadily, and the temperature change on the Earth showed an overall increase.
    (2) The amount of carbon dioxide in the atmosphere and the temperature change on the Earth increased at a steady rate.
    (3) The amount of carbon dioxide in the atmosphere decreased steadily, and the temperature change on the Earth showed an overall decrease.
    (4) The amount of carbon dioxide in the atmosphere decreased at a steady rate, causing a varying change in temperature on the Earth.

Note that question 44 has only three choices.

44. As a result of the changes in temperature and amount of carbon dioxide, what probably happened to the Earth's overall rate of chemical weathering during this time?
    (1) The rate of chemical weathering decreased.
    (2) The rate of chemical weathering increased.
    (3) The rate of chemical weathering remained the same.

45. Which statement best accounts for the relationship between the carbon dioxide and temperature change data shown by the graphs?
    (1) Carbon dioxide is a good absorber of infrared radiation.
    (2) Carbon dioxide is a poor absorber of infrared radiation.
    (3) Temperature decreases usually occur when the carbon dioxide content of the atmosphere increases.
    (4) Temperature changes do not usually occur when the carbon dioxide content of the atmosphere increases.

46. If the trend shown in graph II continued into 1980, the amount of carbon dioxide in the atmosphere in 1980 was probably
  (1) less than 30 parts per million
  (2) between 100 and 280 parts per million
  (3) between 300 and 320 parts per million
  (4) greater than 340 parts per million

47. What was the approximate overall change in the carbon dioxide content between 1900 and 1962?
  (1) 330 parts per million   (3) 40 parts per million
  (2) 290 parts per million   (4) 0.4 part per million

APPENDIX A _____

Answers to Review Questions

UNIT ONE

1.	**1**	10.	**4**	19.	**3**	28.	**4**	37.	**3**	46.	**4**	55.	**3**	64.	**3**
2.	**1**	11.	**4**	20.	**1**	29.	**1**	38.	**4**	47.	**4**	56.	**2**	65.	**3**
3.	**2**	12.	**2**	21.	**3**	30.	**4**	39.	**2**	48.	**2**	57.	**2**	66.	**2**
4.	**4**	13.	**4**	22.	**4**	31.	**3**	40.	**4**	49.	**4**	58.	**4**	67.	**2**
5.	**2**	14.	**1**	23.	**1**	32.	**1**	41.	**4**	50.	**2**	59.	**1**	68.	**3**
6.	**2**	15.	**1**	24.	**1**	33.	**2**	42.	**1**	51.	**2**	60.	**3**	69.	**4**
7.	**3**	16.	**4**	25.	**1**	34.	**1**	43.	**3**	52.	**2**	61.	**1**	70.	**1**
8.	**4**	17.	**3**	26.	**1**	35.	**1**	44.	**4**	53.	**4**	62.	**4**	71.	**2**
9.	**1**	18.	**4**	27.	**3**	36.	**4**	45.	**3**	54.	**1**	63.	**3**		

UNIT TWO

1.	**1**	9.	**4**	17.	**2**	25.	**1**	33.	**2**	41.	**4**	49.	**4**	57.	**2**
2.	**4**	10.	**1**	18.	**4**	26.	**4**	34.	**3**	42.	**3**	50.	**2**	58.	**3**
3.	**2**	11.	**1**	19.	**2**	27.	**1**	35.	**2**	43.	**3**	51.	**4**	59.	**2**
4.	**3**	12.	**4**	20.	**4**	28.	**4**	36.	**2**	44.	**2**	52.	**1**	60.	**4**
5.	**4**	13.	**1**	21.	**3**	29.	**4**	37.	**4**	45.	**3**	53.	**2**	61.	**4**
6.	**1**	14.	**2**	22.	**4**	30.	**2**	38.	**2**	46.	**4**	54.	**4**		
7.	**1**	15.	**2**	23.	**4**	31.	**4**	39.	**1**	47.	**3**	55.	**3**		
8.	**3**	16.	**4**	24.	**2**	32.	**4**	40.	**2**	48.	**1**	56.	**4**		

TOPIC A

1.	**1**	10.	**2**	19.	**4**	28.	**4**	37.	**2**	46.	**1**	55.	**4**	64.	**2**
2.	**2**	11.	**3**	20.	**4**	29.	**4**	38.	**4**	47.	**4**	56.	**2**	65.	**2**
3.	**1**	12.	**1**	21.	**1**	30.	**3**	39.	**1**	48.	**1**	57.	**2**		
4.	**2**	13.	**2**	22.	**1**	31.	**2**	40.	**2**	49.	**3**	58.	**3**		
5.	**1**	14.	**1**	23.	**3**	32.	**4**	41.	**2**	50.	**4**	59.	**4**		
6.	**3**	15.	**3**	24.	**4**	33.	**1**	42.	**4**	51.	**1**	60.	**3**		
7.	**1**	16.	**4**	25.	**2**	34.	**2**	43.	**3**	52.	**2**	61.	**3**		
8.	**2**	17.	**3**	26.	**3**	35.	**3**	44.	**1**	53.	**3**	62.	**1**		
9.	**3**	18.	**4**	27.	**1**	36.	**2**	45.	**2**	54.	**2**	63.	**1**		

UNIT 3

1. 1	6. 3	11. 4	16. 3	21. 2	26. 4	31. 4	36. 4
2. 3	7. 4	12. 4	17. 1	22. 4	27. 4	32. 1	37. 4
3. 4	8. 3	13. 1	18. 4	23. 4	28. 3	33. 1	
4. 2	9. 3	14. 2	19. 1	24. 4	29. 1	34. 3	
5. 1	10. 3	15. 1	20. 4	25. 4	30. 2	35. 2	

TOPIC B

1. 2	13. 3	25. 1	37. 3	49. 1	61. 2	73. 4	85. 4
2. 2	14. 4	26. 2	38. 2	50. 3	62. 2	74. 2	86. 4
3. 2	15. 1	27. 4	39. 4	51. 1	63. 3	75. 2	87. 1
4. 4	16. 2	28. 1	40. 2	52. 1	64. 1	76. 3	88. 4
5. 4	17. 4	29. 3	41. 2	53. 1	65. 4	77. 2	89. 1
6. 3	18. 4	30. 1	42. 1	54. 1	66. 2	78. 2	90. 4
7. 4	19. 3	31. 1	43. 1	55. 1	67. 4	79. 4	91. 3
8. 3	20. 3	32. 4	44. 1	56. 4	68. 4	80. 2	
9. 2	21. 1	33. 2	45. 4	57. 2	69. 3	81. 3	
10. 3	22. 4	34. 2	46. 2	58. 3	70. 4	82. 1	
11. 3	23. 3	35. 2	47. 2	59. 2	71. 1	83. 3	
12. 2	24. 1	36. 3	48. 4	60. 3	72. 1	84. 2	

UNIT 4

1. 2	13. 3	25. 3	37. 2	49. 3	61. 3	73. 3	85. 4
2. 2	14. 2	26. 1	38. 2	50. 4	62. 2	74. 1	86. 1
3. 1	15. 4	27. 1	39. 1	51. 4	63. 4	75. 1	87. 1
4. 2	16. 2	28. 2	40. 2	52. 1	64. 2	76. 2	88. 1
5. 3	17. 3	29. 4	41. 2	53. 3	65. 2	77. 2	89. 4
6. 4	18. 2	30. 2	42. 3	54. 2	66. 4	78. 1	90. 3
7. 4	19. 4	31. 4	43. 2	55. 1	67. 3	79. 4	91. 2
8. 4	20. 2	32. 1	44. 4	56. 4	68. 1	80. 2	92. 4
9. 3	21. 3	33. 1	45. 4	57. 1	69. 4	81. 3	93. 1
10. 4	22. 3	34. 4	46. 1	58. 2	70. 4	82. 1	94. 2
11. 3	23. 3	35. 3	47. 3	59. 3	71. 4	83. 2	95. 4
12. 4	24. 4	36. 1	48. 4	60. 2	72. 4	84. 3	96. 3

TOPIC C

1.	**1**	5.	**4**	9.	**1**	13.	**2**	17.	**2**	21.	**1**	25.	**1**	29.	**1**
2.	**3**	6.	**3**	10.	**1**	14.	**3**	18.	**3**	22.	**2**	26.	**1**	30.	**1**
3.	**3**	7.	**3**	11.	**1**	15.	**1**	19.	**4**	23.	**4**	27.	**4**		
4.	**1**	8.	**2**	12.	**4**	16.	**2**	20.	**3**	24.	**4**	28.	**1**		

TOPIC D

1.	**2**	7.	**2**	13.	**1**	19.	**3**	25.	**4**	31.	**3**	37.	**4**	43.	**2**
2.	**4**	8.	**3**	14.	**2**	20.	**2**	26.	**1**	32.	**2**	38.	**2**	44.	**1**
3.	**4**	9.	**4**	15.	**3**	21.	**1**	27.	**3**	33.	**4**	39.	**2**	45.	**3**
4.	**3**	10.	**4**	16.	**4**	22.	**4**	28.	**2**	34.	**1**	40.	**4**	46.	**3**
5.	**1**	11.	**3**	17.	**1**	23.	**1**	29.	**2**	35.	**3**	41.	**3**	47.	**3**
6.	**1**	12.	**4**	18.	**2**	24.	**1**	30.	**1**	36.	**2**	42.	**3**	48.	**3**

UNIT 5

1.	**2**	12.	**4**	23.	**4**	34.	**3**	45.	**1**	56.	**2**	67.	**3**	78.	**1**
2.	**4**	13.	**3**	24.	**4**	35.	**1**	46.	**3**	57.	**3**	68.	**1**	79.	**3**
3.	**3**	14.	**4**	25.	**4**	36.	**4**	47.	**4**	58.	**4**	69.	**4**	80.	**2**
4.	**4**	15.	**2**	26.	**1**	37.	**1**	48.	**1**	59.	**4**	70.	**3**	81.	**2**
5.	**3**	16.	**4**	27.	**2**	38.	**2**	49.	**4**	60.	**2**	71.	**1**	82.	**1**
6.	**2**	17.	**3**	28.	**4**	39.	**4**	50.	**3**	61.	**2**	72.	**3**	83.	**3**
7.	**4**	18.	**2**	29.	**2**	40.	**2**	51.	**1**	62.	**4**	73.	**3**	84.	**3**
8.	**3**	19.	**3**	30.	**4**	41.	**3**	52.	**4**	63.	**1**	74.	**4**	85.	**1**
9.	**1**	20.	**4**	31.	**3**	42.	**1**	53.	**3**	64.	**4**	75.	**4**		
10.	**1**	21.	**4**	32.	**1**	43.	**1**	54.	**4**	65.	**1**	76.	**4**		
11.	**1**	22.	**3**	33.	**3**	44.	**4**	55.	**2**	66.	**2**	77.	**2**		

UNIT 6

1.	2	14.	4	27.	1	40.	1	53.	3	66.	3	79.	3	92.	3
2.	2	15.	1	28.	1	41.	1	54.	2	67.	2	80.	1	93.	1
3.	1	16.	2	29.	1	42.	4	55.	3	68.	1	81.	3	94.	1
4.	2	17.	4	30.	4	43.	4	56.	1	69.	4	82.	3	95.	1
5.	4	18.	4	31.	2	44.	3	57.	4	70.	3	83.	2	96.	1
6.	2	19.	3	32.	4	45.	2	58.	2	71.	3	84.	2	97.	4
7.	3	20.	3	33.	2	46.	2	59.	1	72.	3	85.	1	98.	5
8.	1	21.	4	34.	2	47.	4	60.	1	73.	2	86.	4	99.	4
9.	1	22.	3	35.	1	48.	3	61.	1	74.	2	87.	1	100.	2
10.	4	23.	4	36.	2	49.	1	62.	4	75.	1	88.	2	101.	1
11.	4	24.	3	37.	1	50.	2	63.	2	76.	3	89.	3	102.	5
12.	1	25.	4	38.	3	51.	3	64.	4	77.	4	90.	2		
13.	2	26.	3	39.	3	52.	1	65.	3	78.	4	91.	1		

TOPIC E

1.	3	9.	1	17.	3	25.	4	33.	2	41.	3	49.	1	57.	3
2.	2	10.	3	18.	1	26.	1	34.	2	42.	1	50.	3		
3.	3	11.	1	19.	3	27.	3	35.	3	43.	4	51.	3		
4.	4	12.	2	20.	1	28.	4	36.	2	44.	2	52.	1		
5.	4	13.	3	21.	2	29.	2	37.	1	45.	4	53.	2		
6.	2	14.	2	22.	1	30.	2	38.	1	46.	1	54.	4		
7.	3	15.	2	23.	2	31.	3	39.	2	47.	3	55.	1		
8.	4	16.	2	24.	2	32.	1	40.	4	48.	2	56.	1		

UNIT 7

1.	3	16.	1	31.	4	46.	1	61.	3	76.	3	91.	4	106.	2
2.	4	17.	2	32.	2	47.	1	62.	4	77.	4	92.	2	107.	4
3.	4	18.	1	33.	1	48.	1	63.	4	78.	2	93.	2	108.	2
4.	3	19.	4	34.	3	49.	3	64.	1	79.	1	94.	1	109.	4
5.	4	20.	1	35.	3	50.	4	65.	4	80.	4	95.	4	110.	3
6.	4	21.	4	36.	3	51.	3	66.	3	81.	1	96.	1	111.	2
7.	4	22.	3	37.	3	52.	4	67.	4	82.	3	97.	1	112.	3
8.	4	23.	3	38.	2	53.	1	68.	1	83.	2	98.	4	113.	1
9.	2	24.	1	39.	3	54.	3	69.	4	84.	4	99.	3	114.	2
10.	3	25.	3	40.	3	55.	4	70.	3	85.	3	100.	1	115.	3
11.	1	26.	4	41.	1	56.	1	71.	1	86.	2	101.	4	116.	1
12.	1	27.	3	42.	4	57.	4	72.	4	87.	1	102.	2	117.	2
13.	3	28.	2	43.	1	58.	3	73.	4	88.	4	103.	1	118.	1
14.	3	29.	4	44.	4	59.	4	74.	2	89.	1	104.	4		
15.	4	30.	2	45.	4	60.	2	75.	3	90.	4	105.	1		

UNIT 8

1. **3**	10. **1**	19. **2**	28. **1**	37. **1**	46. **2**	55. **1**	64. **2**
2. **4**	11. **1**	20. **1**	29. **2**	38. **4**	47. **3**	56. **1**	65. **4**
3. **4**	12. **3**	21. **3**	30. **4**	39. **2**	48. **2**	57. **1**	
4. **1**	13. **1**	22. **2**	31. **1**	40. **1**	49. **2**	58. **3**	
5. **3**	14. **2**	23. **3**	32. **4**	41. **3**	50. **1**	59. **3**	
6. **4**	15. **4**	24. **1**	33. **2**	42. **2**	51. **1**	60. **4**	
7. **4**	16. **1**	25. **4**	34. **3**	43. **1**	52. **2**	61. **4**	
8. **3**	17. **4**	26. **1**	35. **4**	44. **4**	53. **2**	62. **4**	
9. **2**	18. **3**	27. **3**	36. **3**	45. **2**	54. **3**	63. **2**	

TOPIC F

1. **4**	6. **2**	11. **2**	16. **1**	21. **1**	26. **3**	31. **2**	36. **3**
2. **4**	7. **1**	12. **4**	17. **2**	22. **4**	27. **1**	32. **1**	37. **3**
3. **2**	8. **1**	13. **3**	18. **3**	23. **2**	28. **2**	33. **3**	38. **4**
4. **4**	9. **1**	14. **2**	19. **4**	24. **3**	29. **2**	34. **2**	
5. **1**	10. **4**	15. **4**	20. **1**	25. **4**	30. **1**	35. **1**	

UNIT 9

1. **2**	7. **4**	13. **1**	19. **2**	25. **4**	31. **2**	37. **4**	43. **1**
2. **1**	8. **4**	14. **2**	20. **3**	26. **2**	32. **3**	38. **4**	44. **2**
3. **4**	9. **2**	15. **3**	21. **2**	27. **3**	33. **3**	39. **1**	45. **1**
4. **3**	10. **4**	16. **4**	22. **2**	28. **4**	34. **4**	40. **3**	46. **4**
5. **2**	11. **3**	17. **1**	23. **1**	29. **2**	35. **4**	41. **2**	47. **3**
6. **2**	12. **2**	18. **3**	24. **1**	30. **4**	36. **4**	42. **3**	

APPENDIX B _____

Earth Science Reference Tables

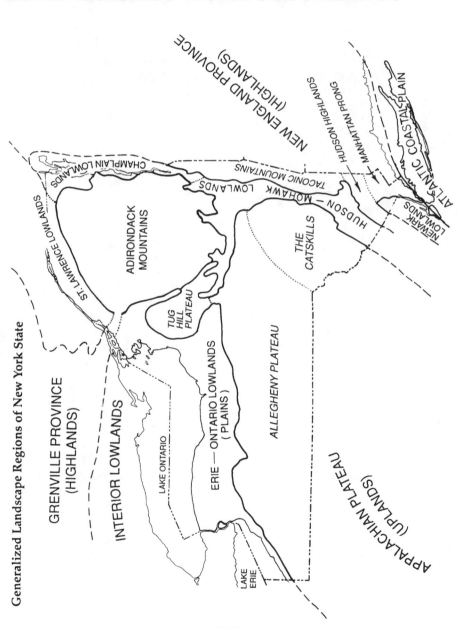

Generalized Landscape Regions of New York State

Generalized Bedrock Geology of New York State

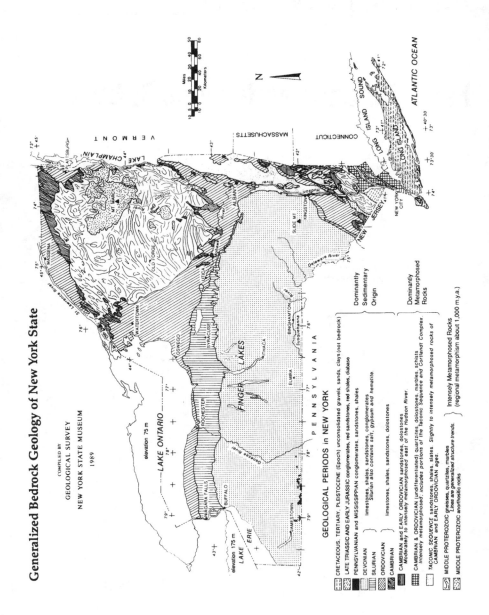

COMPILED BY

GEOLOGICAL SURVEY

NEW YORK STATE MUSEUM

1989

GEOLOGICAL PERIODS in NEW YORK

CRETACEOUS, TERTIARY PLEISTOCENE (Epoch) unconsolidated gravels, sands, clays (not bedrock)

LATE TRIASSIC AND EARLY JURASSIC conglomerates, red sandstones, red shales, diabase

PENNSYLVANIAN and MISSISSIPPIAN conglomerates, sandstones, shales

DEVONIAN limestones, shales, sandstones, conglomerates

SILURIAN limestones, shales, sandstones, shales Silurian also contains salt, gypsum and hematite

ORDOVICIAN limestones, shales, sandstones, dolostones

CAMBRIAN

CAMBRIAN and EARLY ORDOVICIAN sandstones, dolostones Moderately to intensely metamorphosed east of the Hudson River.

CAMBRIAN & ORDOVICIAN (undifferentiated) quartzites, dolostones, marbles, schists Intensely metamorphosed; includes portions of the Taconic Sequence and Cortland Complex

TACONIC SEQUENCE sandstones, shales, slates Slightly to intensely metamorphosed rocks of CAMBRIAN and EARLY ORDOVICIAN ages

MIDDLE PROTEROZOIC gneisses, quartzites, marbles Lines are generalized structure trends.

MIDDLE PROTEROZOIC anorthositic rocks

Intensely Metamorphosed Rocks (regional metamorphism about 1,000 m.y.a.)

Dominantly Sedimentary Origin

Dominantly Metamorphosed Rocks

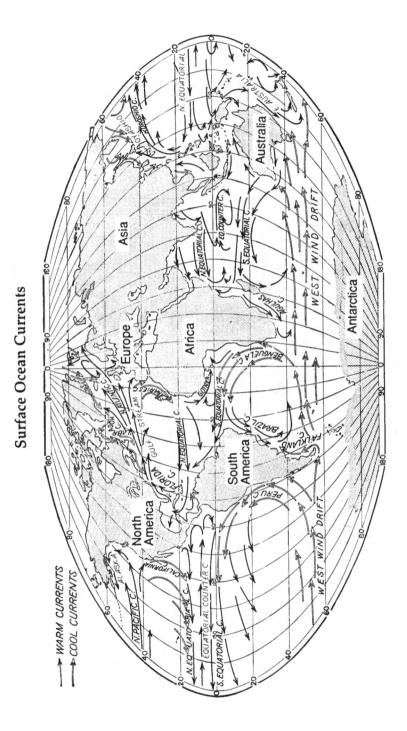

Surface Ocean Currents

Tectonic Plates

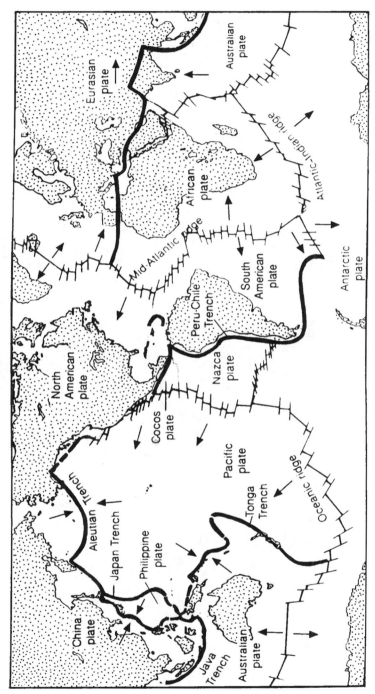

Rock Cycle in Earth's Crust

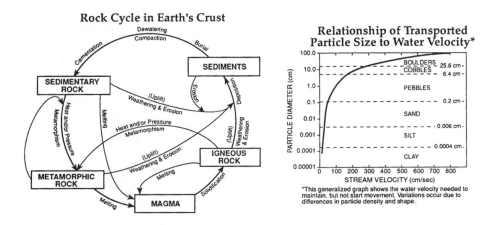

Relationship of Transported Particle Size to Water Velocity*

*This generalized graph shows the water velocity needed to maintain, but not start movement. Variations occur due to differences in particle density and shape.

Scheme for Igneous Rock Identification

ENVIRONMENT OF FORMATION							GRAIN SIZE	TEXTURE
INTRUSIVE (Plutonic)	Granite	Diorite	Gabbro	Perid-otite	Dunite	1 mm or larger	Coarse
EXTRUSIVE (Volcanic)	Rhyolite	Andesite	Basalt Scoria	Rare	Rare	less than 1 mm	Fine
	Pumice *Obsidian	*Obsidian	Basalt Glass	Rare	Rare	Non-crystalline	Glassy

* Obsidian may appear black.

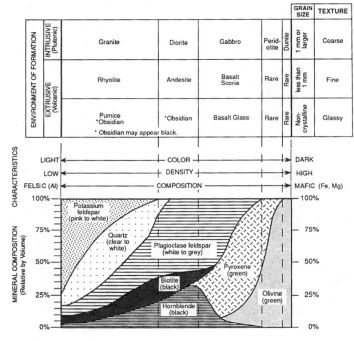

CHARACTERISTICS			
LIGHT	← COLOR →		DARK
LOW	← DENSITY →		HIGH
FELSIC (Al)	← COMPOSITION →		MAFIC (Fe, Mg)

Note: The intrusive rocks can also occur as exceptionally coarse-grained rock, Pegmatite.

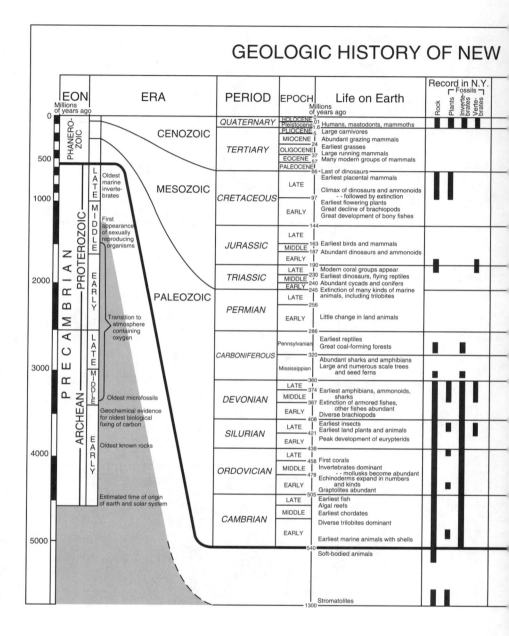

GEOLOGIC HISTORY OF NEW

YORK STATE AT A GLANCE

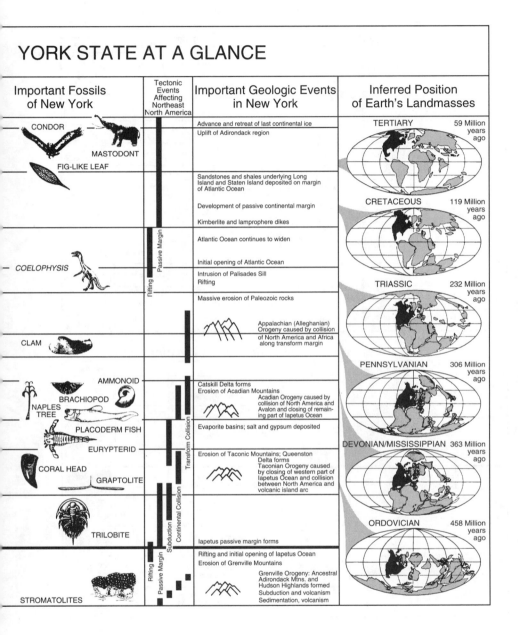

Important Fossils of New York	Tectonic Events Affecting Northeast North America	Important Geologic Events in New York	Inferred Position of Earth's Landmasses

Important Fossils of New York:
- CONDOR
- MASTODONT
- FIG-LIKE LEAF
- COELOPHYSIS
- CLAM
- AMMONOID
- BRACHIOPOD
- NAPLES TREE
- PLACODERM FISH
- EURYPTERID
- CORAL HEAD
- GRAPTOLITE
- TRILOBITE
- STROMATOLITES

Tectonic Events Affecting Northeast North America:
- Passive Margin
- Rifting
- Transform Collision
- Continental Collision
- Subduction
- Rifting
- Passive Margin

Important Geologic Events in New York:
- Advance and retreat of last continental ice
- Uplift of Adirondack region
- Sandstones and shales underlying Long Island and Staten Island deposited on margin of Atlantic Ocean
- Development of passive continental margin
- Kimberlite and lamprophere dikes
- Atlantic Ocean continues to widen
- Initial opening of Atlantic Ocean
- Intrusion of Palisades Sill
- Rifting
- Massive erosion of Paleozoic rocks
- Appalachian (Alleghanian) Orogeny caused by collision of North America and Africa along transform margin
- Catskill Delta forms
- Erosion of Acadian Mountains
- Acadian Orogeny caused by collision of North America and Avalon and closing of remaining part of Iapetus Ocean
- Evaporite basins; salt and gypsum deposited
- Erosion of Taconic Mountains; Queenston Delta forms
- Taconian Orogeny caused by closing of western part of Iapetus Ocean and collision between North America and volcanic island arc
- Iapetus passive margin forms
- Rifting and initial opening of Iapetus Ocean
- Erosion of Grenville Mountains
- Grenville Orogeny: Ancestral Adirondack Mtns. and Hudson Highlands formed
- Subduction and volcanism
- Sedimentation, volcanism

Inferred Position of Earth's Landmasses:
- TERTIARY — 59 Million years ago
- CRETACEOUS — 119 Million years ago
- TRIASSIC — 232 Million years ago
- PENNSYLVANIAN — 306 Million years ago
- DEVONIAN/MISSISSIPPIAN — 363 Million years ago
- ORDOVICIAN — 458 Million years ago

Scheme for Sedimentary Rock Identification

INORGANIC LAND-DERIVED SEDIMENTARY ROCKS					
TEXTURE	GRAIN SIZE	COMPOSITION	COMMENTS	ROCK NAME	MAP SYMBOL
Clastic (fragmental)	Mixed, silt to boulders (larger than 0.001 cm)	Mostly quartz, feldspar, and clay minerals; May contain fragments of other rocks and minerals	Rounded fragments	Conglomerate	
			Angular fragments	Breccia	
	Sand (0.006 to 0.2 cm)		Fine to coarse	Sandstone	
	Silt (0.0004 to 0.006 cm)		Very fine grain	Siltstone	
	Clay (less than 0.0006 cm)		Compact; may split easily	Shale	

CHEMICALLY AND/OR ORGANICALLY FORMED SEDIMENTARY ROCKS					
TEXTURE	GRAIN SIZE	COMPOSITION	COMMENTS	ROCK NAME	MAP SYMBOL
Nonclastic	Coarse to fine	Calcite	Crystals from chemical precipitates and evaporites	Chemical Limestone	
	Varied	Halite		Rock Salt	
	Varied	Gypsum		Rock Gypsum	
	Varied	Dolomite		Dolostone	
	Microscopic to coarse	Calcite	Cemented shells, shell fragments, and skeletal remains	Fossil Limestone	
	Varied	Carbon	Black and nonporous	Bituminous Coal	

Scheme for Metamorphic Rock Identification

TEXTURE		GRAIN SIZE	COMPOSITION	TYPE OF METAMORPHISM	COMMENTS	ROCK NAME	MAP SYMBOL
FOLIATED	Slaty	Fine	CHLORITE MICA QUARTZ FELDSPAR AMPHIBOLE GARNET PYROXENE	Regional	Low-grade metamorphism of shale	Slate	
	Schistose	Medium to coarse			Medium-grade metamorphism; Mica crystals visible from metamorphism of feldspars and clay minerals	Schist	
	Gneissic	Coarse		(Heat and pressure increase with depth, folding, and faulting)	High-grade metamorphism; Mica has changed to feldspar	Gneiss	
NONFOLIATED		Fine	Carbonaceous		Metamorphism of plant remains and bituminous coal	Anthracite Coal	
		Coarse	Depends on conglomerate composition		Pebbles may be distorted or stretched; Often breaks through pebbles	Meta-conglomerate	
		Fine to coarse	Quartz	Thermal (including contact) or Regional	Metamorphism of sandstone	Quartzite	
			Calcite, Dolomite		Metamorphism of limestone or dolostone	Marble	
		Fine	Quartz, Plagioclase	Contact	Metamorphism of various rocks by contact with magma or lava	Hornfels	

Inferred Properties of Earth's Interior

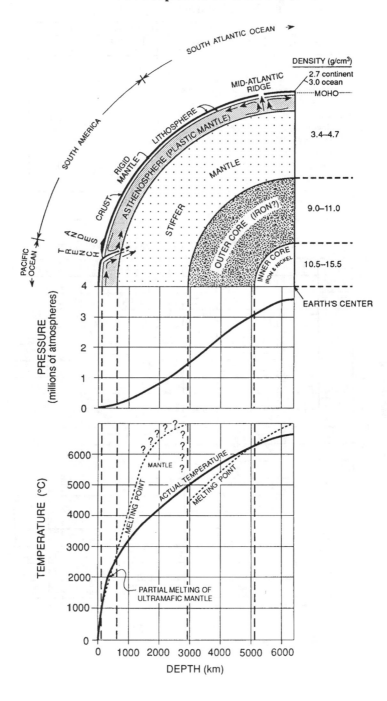

Average Chemical Composition
of Earth's Crust, Hydrosphere, and Troposphere

ELEMENT (symbol)	CRUST		HYDROSPHERE	TROPOSPHERE
	Percent by Mass	Percent by Volume	Percent by Volume	Percent by Volume
Oxygen (O)	46.40	94.04	33	21
Silicon (Si)	28.15	0.88		
Aluminum (Al)	8.23	0.48		
Iron (Fe)	5.63	0.49		
Calcium (Ca)	4.15	1.18		
Sodium (Na)	2.36	1.11		
Magnesium (Mg)	2.33	0.33		
Potassium (K)	2.09	1.42		
Nitrogen (N)				78
Hydrogen (H)			66	

Earthquake P-wave and S-wave Travel Time

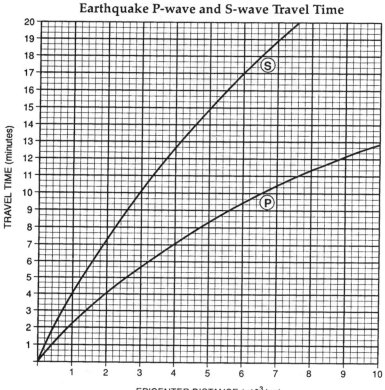

TRAVEL TIME (minutes)

EPICENTER DISTANCE (x10³ km)

526

Dewpoint Temperatures

Dry-Bulb Temperature (°C)	Difference Between Wet-Bulb and Dry-Bulb Temperatures (C°)														
	1	2	3	4	5	6	7	8	9	10	11	12	13	14	15
-20	-33														
-18	-28														
-16	-24														
-14	-21	-36													
-12	-18	-28													
-10	-14	-22													
-8	-12	-18	-29												
-6	-10	-14	-22												
-4	-7	-12	-17	-29											
-2	-5	-8	-13	-20											
0	-3	-6	-9	-15	-24										
2	-1	-3	-6	-11	-17										
4	1	-1	-4	-7	-11	-19									
6	4	1	-1	-4	-7	-13	-21								
8	6	3	1	-2	-5	-9	-14								
10	8	6	4	1	-2	-5	-9	-14	-28						
12	10	8	6	4	1	-2	-5	-9	-16						
14	12	11	9	6	4	1	-2	-5	-10	-17					
16	14	13	11	9	7	4	1	-1	-6	-10	-17				
18	16	15	13	11	9	7	4	2	-2	-5	-10	-19			
20	19	17	15	14	12	10	7	4	2	-2	-5	-10	-19		
22	21	19	17	16	14	12	10	8	5	3	-1	-5	-10	-19	
24	23	21	20	18	16	14	12	10	8	6	2	-1	-5	-10	-18
26	25	23	22	20	18	17	15	13	11	9	6	3	0	-4	-9
28	27	25	24	22	21	19	17	16	14	11	9	7	4	1	-3
30	29	27	26	24	23	21	19	18	16	14	12	10	8	5	1

Relative Humidity (%)

Dry-Bulb Temperature (°C)	Difference Between Wet-Bulb and Dry-Bulb Temperatures (C°)														
	1	2	3	4	5	6	7	8	9	10	11	12	13	14	15
-20	28														
-18	40														
-16	48	0													
-14	55	11													
-12	61	23													
-10	66	33	0												
-8	71	41	13												
-6	73	48	20	0											
-4	77	54	32	11											
-2	79	58	37	20	1										
0	81	63	45	28	11										
2	83	67	51	36	20	6									
4	85	70	56	42	27	14									
6	86	72	59	46	35	22	10	0							
8	87	74	62	51	39	28	17	6							
10	88	76	65	54	43	33	24	13	4						
12	88	78	67	57	48	38	28	19	10	2					
14	89	79	69	60	50	41	33	25	16	8	1				
16	90	80	71	62	54	45	37	29	21	14	7	1			
18	91	81	72	64	56	48	40	33	26	19	12	6	0		
20	91	82	74	66	58	51	44	36	30	23	17	11	5	0	
22	92	83	75	68	60	53	46	40	33	27	21	15	10	4	
24	92	84	76	69	62	55	49	42	36	30	25	20	14	9	4
26	92	85	77	70	64	57	51	45	39	34	28	23	18	13	9
28	93	86	78	71	65	59	53	47	42	36	31	26	21	17	12
30	93	86	79	72	66	61	55	49	44	39	34	29	25	20	16

Lapse Rate

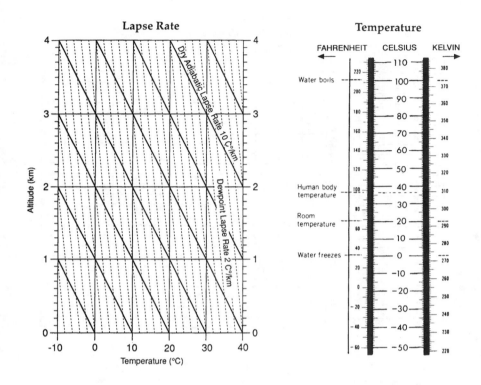

Altitude (km)

Dry Adiabatic Lapse Rate 10 C°/km

Dewpoint Lapse Rate 2 C°/km

Temperature (°C)

Temperature

FAHRENHEIT CELSIUS KELVIN

Water boils

Human body
temperature

Room
temperature

Water freezes

Weather Map Information

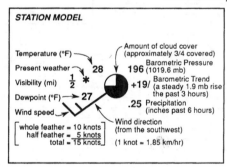

STATION MODEL

Temperature (°F)
Present weather — 28
Visibility (mi) ½ ✳
Dewpoint (°F) — 27
Wind speed

Amount of cloud cover
(approximately 3/4 covered)
Barometric Pressure
196 (1019.6 mb)
+19/ Barometric Trend
(a steady 1.9 mb rise
the past 3 hours)
.25 Precipitation
(inches past 6 hours)
Wind direction
(from the southwest)
(1 knot = 1.85 km/hr)

whole feather = 10 knots
half feather = 5 knots
total = 15 knots

PRESENT WEATHER SYMBOLS

Drizzle	Rain	Showers	Hail	Thunder-storms
Snow	Sleet	Freezing Rain	Fog	Haze

AIRMASSES

cP Continental polar
cT Continental tropical
mT Maritime tropical
mP Maritime polar

FRONT SYMBOLS

Cold
Warm
Stationary
Occluded

Electromagnetic Spectrum

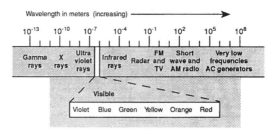

Pressure

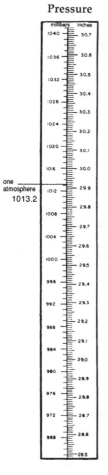

Solar System Data

Planet	Mean Distance from Sun (millions of km)	Period of Revolution	Period of Rotation	Eccentricity of Orbit	Equatorial Diameter (km)	Density (g/cm³)
MERCURY	57.9	88 days	59 days	0.206	4,880	5.4
VENUS	108.2	224.7 days	243 days	0.007	12,104	5.2
EARTH	149.6	365.26 days	23 hours 56 min 4 sec	0.017	12,756	5.5
MARS	227.9	687 days	24 hours 37 min 23 sec	0.093	6,787	3.9
JUPITER	778.3	11.86 years	9 hours 50 min 30 sec	0.048	142,800	1.3
SATURN	1,427	29.46 years	10 hours 14 min	0.056	120,000	0.7
URANUS	2,869	84.0 years	11 hours	0.047	51,800	1.2
NEPTUNE	4,496	164.8 years	16 hours	0.009	49,500	1.7
PLUTO	5,900	247.7 years	6 days 9 hours	0.250	2,300	2.0

Selected Properties of Earth's Atmosphere

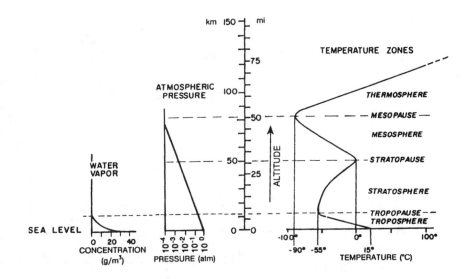

Planetary Wind and Moisture Belts in the Troposphere

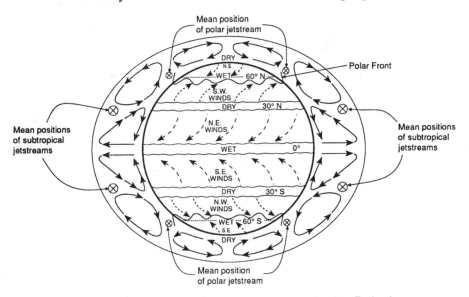

The drawing shows the locations of the belts near the time of an equinox. The locations shift somewhat with the changing latitude of the Sun's vertical ray. In the Northern Hemisphere the belts shift northward in summer and southward in winter.

Equations and Proportions

Equations

Percent deviation from accepted value	deviation (%) = $\dfrac{\text{difference from accepted value}}{\text{accepted value}} \times 100$	
Eccentricity of an ellipse	eccentricity = $\dfrac{\text{distance between foci}}{\text{length of major axis}}$	
Gradient	gradient = $\dfrac{\text{change in field value}}{\text{change in distance}}$	
Rate of change	rate of change = $\dfrac{\text{change in field value}}{\text{change in time}}$	
Circumference of a circle	$C = 2\pi r$	
Eratosthenes' method to determine Earth's circumference	$\dfrac{\angle a}{360°} = \dfrac{s}{C}$	
Volume of a rectangular solid	$V = \ell wh$	
Density of a substance	$D = \dfrac{m}{V}$	
Latent heat	$\begin{cases} \text{solid} \longleftrightarrow \text{liquid} \quad Q = mH_f \\ \text{liquid} \longleftrightarrow \text{gas} \quad Q = mH_v \end{cases}$	
Heat energy lost or gained	$Q = m\,\Delta T C_p$	

C_p = specific heat
C = circumference
d = distance
D = density
F = force
h = height
H_f = heat of fusion
H_v = heat of vaporization
$\angle a$ = shadow angle
ℓ = length
s = distance on surface
m = mass
Q = amount of heat
r = radius
ΔT = change in temperature
V = volume
w = width

Note: $\pi \approx 3.14$

Proportions

Kepler's harmonic law of planetary motion	(period of revolution)$^2 \propto$ (mean radius of orbit)^3
Universal law of gravitation	force $\propto \dfrac{\text{mass}_1 \times \text{mass}_2}{(\text{distance between their centers})^2}$ $\left(F \propto \dfrac{m_1\, m_2}{d^2} \right)$

EURYPTERID
New York State Fossil

Physical Constants

Properties of Water

Latent heat of fusion (H_f)	80 cal/g
Latent heat of vaporization (H_v)	540 cal/g
Density (D) at 3.98°C	1.00 g/mL

Specific Heats of Common Materials

MATERIAL		SPECIFIC HEAT (C_p) (cal/g·C°)
Water	solid	0.5
	liquid	1.0
	gas	0.5
Dry air		0.24
Basalt		0.20
Granite		0.19
Iron		0.11
Copper		0.09
Lead		0.03

Radioactive Decay Data

RADIOACTIVE ISOTOPE	DISINTEGRATION	HALF-LIFE (years)
Carbon-14	$C^{14} \rightarrow N^{14}$	5.7×10^3
Potassium-40	$K^{40} \begin{smallmatrix} \nearrow Ar^{40} \\ \searrow Ca^{40} \end{smallmatrix}$	1.3×10^9
Uranium-238	$U^{238} \rightarrow Pb^{206}$	4.5×10^9
Rubidium-87	$Rb^{87} \rightarrow Sr^{87}$	4.9×10^{10}

Astronomy Measurements

MEASUREMENT	EARTH	SUN	MOON
Mass (m)	5.98×10^{24} kg	1.99×10^{30} kg	7.35×10^{22} kg
Radius (r)	6.37×10^3 km	6.96×10^5 km	1.74×10^3 km
Average density (D)	5.52 g/cm^3	1.42 g/cm^3	3.34 g/cm^3

Glossary of Earth Science Terms

ablation The melting and evaporation of glacial ice.

ablation debris Material on the surface of a glacier that was suspended in the ice but is now exposed due to ablation.

abrasion The breaking down of rocks by rubbing together.

absolute humidity The number of grams of water vapor in a cubic meter of air.

abyssal plain Wide, flat areas adjacent to the mid-ocean ridges.

acid rain Rain that is acidic due to substances that have become dissolved in it.

adiabatic A change that does not involve the addition or removal of heat.

adiabatic lapse rate The rate at which the temperature of air changes due to a decrease in air pressure rather than the addition or removal of heat.

agent of erosion A medium set in motion by gravity that can exert a force on sediments causing them to be moved.

air current Air that is moving in a vertical direction.

air mass A large region of air with uniform characteristics at any given level.

air pressure The force exerted on a unit of area by the air.

alluvial fan A flat, fan-shaped mound of sediment deposited where a stream slows down because its gradient suddenly becomes less steep (i.e., where a mountain valley empties onto a plain).

altitude 1. The angle between the horizon and an object seen in the sky with the observer at the vertex. 2. Distance above the earth's surface.

anemometer An instrument that measures wind speed.

angle of repose The greatest angle at which a pile of sediments is stable.

angular unconformity A place where sedimentary layers at one angle are in contact with sedimentary layers at a different angle.

annular eclipse A solar eclipse in which the moon's shadow does not actually reach the surface of the Earth; to an observer on Earth, the sun's disk is obscured by the Moon except for a narrow ring around the perimeter.

anticline An upward, arched fold.

anticyclone A flow of air that is clockwise and outward from a high pressure center.

apparent daily motion The apparent motion of all celestial objects along a circular path at a constant rate of 15° per hour, or one complete circle every day.

aquiclude A rock or sediment layer that is impermeable to water.

aquifer A rock or sediment layer in which the pore spaces are saturated with water.

arête A narrow, knife-edged ridge between two adjacent cirques.

arrival time The precise time at which a seismic wave reaches a seismograph and is recorded.

artesian well A well in which water rises to the surface unassisted.

asteroid Solid body having no atmosphere that orbits the sun.

asthenosphere A region of the upper mantle between 100 and 350 kilometers in depth that behaves like a fluid.

astronomical unit The average distance from the Earth to the Sun; 149,598,000 kilometers.

atmosphere The thin shell of gases bound to the Earth by gravity; it consists of gases, water, ice, dust, and other particles.

atmospheric variables The characteristics of the atmosphere that change.

axis of rotation The central line around which all parts of an object move during rotation.

banding A structure in a metamorphic rock of nearly parallel bands of different textures or minerals; i.e., the nearly parallel bands of light and dark minerals seen in gneiss.

barrier bar   A sandbar running parallel to a shoreline that shelters the water behind it from waves.

batholith   A large, irregularly shaped pluton covering an area greater than 75 square kilometers.

beach   A strip of loose materials along the shore of a body of water formed and washed by waves.

bed   A layer of sediment with a particular physical or structural character.

bedrock   A general term for the solid rock underlying the soil or loose material covering the Earth's surface.

bench mark   A permanent metal marker set into the ground that gives the exact location and precisely measured elevation of that point.

big bang theory   The idea that the universe started out with all of its matter in a small volume and then expanded outward in all directions.

body waves   Seismic waves that travel through the body of the earth; P-waves and S-waves are body waves.

calcareous ooze   Sediment rich in the carbonate shells of marine organisms.

capillary fringe   A narrow region just above the water table in which narrow pore spaces are filled with water that is drawn up by capillary action.

carbonaceous chondrite   A stony meteorite containing hydrated claylike minerals and a great variety of organic compounds.

carbonation   A chemical reaction in which carbon dioxide combines with a substance.

celestial objects   Objects that can be seen in the sky, but that are beyond the Earth's atmosphere.

cementation   The binding together of sediment particles by a cementing material.

chaos theory   The idea that small, unpredictable changes can cause large-scale changes in a system over time.

chemical sediments   Particles (usually crystals) that crystallized out of solutions at or near the Earth's surface.

chemical weathering   Processes that break down rock by changing its chemical composition.

cinder cone volcano   A narrow, steep-sided volcano formed from the accumulation of tephra around the vent.

cirque   The open, half-bowl-shaped hollow high on a mountainside at the head of a valley glacier.

clastic sediments   Rock or mineral fragments formed by the breakdown of rock due to weathering.

cleavage   The tendency of a mineral to break along flat surfaces parallel to atomic planes in its crystalline structure.

climate   The weather in a region averaged over a long period of time.

cloud base   The elevation of the base of a cloud; the elevation at which condensation begins.

cloud ceiling   The elevation of the base of a layer of cloud.

cloud cover   The fraction of the sky obscured by clouds.

coastal zone   The part of a continent that is adjacent to the ocean; it includes features such as the coast, shoreline, beaches, marshes, estuaries, lagoons, and deltas.

comet   Bodies composed of meteoroids embedded in ice that revolve around the Sun in highly elliptical orbits.

comet coma   The cloud of gas surrounding a comet's nucleus when ice in the nucleus sublimes as the comet approaches the Sun.

comet nucleus   The solid center of a comet thought to consist of meteoroids embedded in ice.

comet tail   The elongation of the cloud of gas surrounding a comet caused by collisions with particles emitted by the Sun.

compaction   The reduction in volume or thickness of a sediment layer due to overlying weight or pressure.

composite cone volcano   A large symmetrical volcano consisting of alternating layers of lava flows and tephra.

compound   A substance that consists of two or more elements joined together in a definite proportion and have properties different from the elements that compose it.

compression Forces acting toward each other along the same line of action causing rock to be squeezed.

conchoidal fracture A tendency to break along a curved, concave surface that looks somewhat like a mussel shell.

condensation The process by which a gas changes into a liquid.

condensation nuclei Particles suspended in the air that provide a surface upon which condensation can occur.

conic section A family of figures obtained by cutting a circular cone with a plane.

conservation The act of decreasing the consumption of a natural resource, usually by more efficient use.

contact metamorphism Changes in rock that result from the extreme heat produced by contact with magma or lava.

continental drift The theory that the continents drift slowly across the Earth's surface, sometimes colliding and breaking into pieces.

continental glacier Immense masses of ice that form at high latitudes where temperatures are always cold and that spread out over the entire land surface rather than being confined to valleys.

continental rise A gently sloping feature between the continental slope and the abyssal plain.

continental shelf The shallow part of the ocean floor adjacent to the coastal zone.

continental slope The portion of the ocean floor that slopes steeply from the continental shelf toward the abyssal plain.

contour interval The difference in elevation between two adjacent contour lines.

convection cell A circular pattern of motion in which warm, moist air rises, expands outward, cools, and sinks back to the surface.

convection current A fluid current set in motion by differences in density within the fluid.

convergent boundary Places where edges of adjacent plates are colliding.

coordinates Two numbers that describe the position of any point on a map in terms of the intersection of two lines.

core The innermost zone of the Earth's interior.

Coriolis effect A deflection of motion from a straight-line path due to the rotation of the Earth.

correlation The process of determining that rock layers in different areas are the same age.

cosmology The scientific inquiry into the origin, evolution, and fate of the universe.

crevasse A deep, nearly vertical fissure in glacial ice.

cross-bedding Layers of sediment lying at an angle to the plane in which sediment layers are normally deposited.

crystal A solid whose internal structural pattern is expressed as plane faces that can be seen with the unaided eye.

crystalline Having a definite internal structural pattern.

crystallization The process of forming crystals.

cutoff A new channel formed when a stream cuts through a narrow strip of land between adjacent meanders.

cyclone A flow of air in which motion is counterclockwise and toward a low pressure center.

decarbonation A chemical reaction in which carbon dioxide is given off.

decay product The new element formed when a radioisotope decays.

deferent In Ptolemy's geocentric model of the universe, a circle with the Earth near its center. The center of a planet's epicycle moves around the circumference of a deferent.

deficit A condition in which potential evapotranspiration exceeds the supply of moisture in the environment.

deflation   The pick up and removal of fine sediments from the Earth's surface by wind.

deflation hollow   A shallow depression formed by the removal of sediments by wind.

degassing   The process by which water is released from the molecules of minerals when heated to high temperatures.

dehydration   A chemical reaction in which water is given off.

delta   A flat, fan-shaped mound of sediment deposited where a stream enters a body of standing water and slows down.

density   The amount of matter in a given amount of space; usually expressed as the mass per unit of volume of a substance.

desert pavement   A continuous layer of sediments too heavy to be moved by wind that remains after finer sediments have been removed by wind, and shields underlying materials from further deflation.

dew   Droplets of water that condense on ground surfaces due to radiative cooling of the ground overnight.

dew point   The temperature at which the water vapor in air will begin to condense.

dike   A pluton that cuts across existing rock layers.

disconformity   An irregular erosional surface between parallel layers of rock.

divergent boundary   Places where edges of adjacent plates are spreading apart.

Doppler effect   The shift in wavelength of a spectrum line from its normal position due to relative motion between the source and the observer.

drift   A general term for all sediments transported and deposited by a glacier.

drizzle   Very fine droplets of water falling slowly and close together.

drumlin   Low mounds of till having an elongated teardrop shape formed when a glacier rides over a previously deposited pile of sediment.

dry bulb temperature   The temperature recorded by a thermometer whose bulb is kept wet.

ductile deformation   A permanent change in rock that occurs when stress exceeds the strength of a rock's internal bonds.

dune   A mound of sand deposited by wind.

earthflow   The slow, downhill movement of a water-saturated layer of soil and vegetation.

earthquake   A sudden trembling of the ground.

eccentricity   The flatness of an ellipse, expressed as the ratio of the distance between the foci of the ellipse to the length of its major axis.

elastic deformation   A temporary change in which rock strains due to stress but returns to its original condition after the stress is removed.

electromagnetic spectrum   A continuum in which electromagnetic waves are arranged in order of wavelength.

electromagnetic wave   An oscillation in the electromagnetic field surrounding a particle.

element   A substance that cannot be broken down into simpler substances by ordinary chemical means.

ellipse   A closed figure obtained by cutting a circular cone with a plane; its appearance is that of a flattened circle.

end moraine   A pile of till that forms at the melting edge of a glacier.

eon   The largest unit of geologic time.

epicenter   The point on the Earth's surface directly above the focus of an earthquake.

epicycle   In Ptolemy's geocentric model of the universe, the circular orbit of a planet. The center of an epicycle travels around the circumference of another circle called the deferent.

epoch   The unit of geologic time into which periods are divided.

equator   A circle on which all points are midway between the Earth's poles.

equatorial diameter   Diameter of the Earth measured from a point on the Equator through the center of the Earth to the Equator (12,756 kilometers).

equinox   Either of the two times during a year when the Sun crosses the celestial Equator and when the length of day and night are approximately equal.

era   The unit of geologic time into which eons are divided.

erosion   Any process that transports sediments.

erratic   An isolated boulder deposited by a glacier.

escape velocity   The smallest velocity a body must attain in order to escape from the gravitational attraction of another body.

esker   An elongated pile of till deposited in the pipelike tunnels of sub-glacial meltwater streams.

evaporite   Sediments, rocks, or minerals formed when water containing dissolved minerals evaporates.

evapotranspiration   The combined processes of evaporation and transpiration, the two mechanisms by which water is returned to the atmosphere.

evolution   The idea that current life-forms have developed from earlier, different life-forms.

exfoliation   The scaling off, or peeling, of successive layers from the surface of rocks.

extrusive igneous rock   A rock formed by the rapid cooling of lava at the surface of the Earth.

faulting   A sudden movement of rock along planes of weakness in the Earth's crust called faults.

felsic   Rocks rich in the minerals feldspar and silica (quartz).

field   A region of space in which a measurable quantity is present at every point.

field map   Represents any quantity that varies in a region of space.

firn   Snow that has been compacted, melted, and refrozen into a denser material that is transitional between snow and glacial ice.

floodplain   A broad, flat strip of land adjacent to a stream consisting of sediments deposited when the stream is in flood.

focus   The point where rock first breaks or moves in an earthquake.

focus of an ellipse   One of two points along the major axis of an ellipse with the characteristic that the total distance from one focus to any point on the ellipse and back to the other focus is always the same.

fog   A cloud whose base is at ground level.

fold   A bend or warp in layered rock.

fossil   Any remains, trace, or imprint of a living thing that has been preserved in the Earth's crust.

fossil fuel   A fuel derived from the remains of once-living things; e.g., coal, petroleum, and natural gas.

Foucault pendulum   A freely swinging pendulum whose path appears to change direction relative to the Earth's surface in a predictable manner due to the Earth's rotation.

fracture   The tendency not to break along any particular direction.

fragmental sedimentary rock   A rock composed of rock fragments cemented or compacted together in a solid mass.

frequency   The number of waves that passes a given point in a unit of time.

friction cracks   Crescent-shaped cracks in bedrock due to friction between the bedrock and a glacier moving over it.

front   The boundary between two adjacent air masses.

frost   Ice crystals that form on ground surfaces by sublimation due to radiative cooling of the ground overnight.

frost action   The breaking down of rock by the force exerted by water expanding as it freezes into ice.

galaxy   A system consisting of hundreds of billions of stars.

gas giant Planets that have thick atmospheres surrounding a small, rocky core.

geocentric model A model of the universe in which the Earth is at the center and all other objects revolve around the Earth.

geologic column The correlation of rocks worldwide into a single sequence showing their relative ages.

geologic time scale The division of the Earth's past into a sequence of time units.

glacial advance The forward and downslope movement of a glacier that results when accumulation exceeds ablation.

glacial basin Depressions formed when fast-moving glaciers remove large quantities of rock material plucked from the bedrock surface.

glacial polish A bedrock surface smoothed by abrasion as a glacier flows over it.

glacial retreat A decrease in the length of a glacier that results from ablation exceeding accumulation.

glacial striations Long, narrow grooves inscribed in bedrock by rock fragments embedded in a glacier as the glacier moves over the bedrock.

glacier A solid mass of ice formed by the compaction of accumulated snow that flows downslope and outward under the influence of gravity.

glaze Rain that forms a layer of ice as soon as it comes into contact with below-freezing surfaces.

global wind belt A region of prevailing winds caused by a difference in air pressure between adjacent regions on a global scale.

graded bedding A layer of sediments in which the particle size changes gradually from large on the bottom to small on the top.

gradient The rate at which a field changes; usually measured in terms of change in field value per unit of distance.

granular snow Snow crystals that have become small and rounded due to evaporation.

gravitational differentiation The process by which the material from which the solar system formed became separated into layers according to density, because atoms of lighter elements do not exert as strong a gravitational attraction as atoms of heavier elements.

gravity The force of attraction that exists between all matter.

great circle A circle that bisects the Earth.

greenhouse effect The process by which solar radiation that is re-radiated at a longer wavelength is absorbed by carbon dioxide, water vapor, and other gases in the atmosphere rather than escaping into space.

ground moraine A thin, widespread layer of till deposited beneath a glacier.

groundmass A mass of fine mineral crystals surrounding the large, isolated phenocrysts in a porphyry.

gully A long, narrow trough eroded in soil by running water.

hail Balls of ice with an internal structure of concentric layers of ice and snow formed when ice pellets fall through turbulent air and are alternately carried through layers of air that are above and below freezing.

hanging valley A valley of a tributary glacier whose floor is above, or hanging over, the floor of the main glacier that it feeds.

hardness A mineral's ability to resist being scratched.

heliocentric model A model of the universe in which the Sun is at the center of our solar system and the planets revolve around the Sun.

horizontal sorting The gradual change in the size of sediments deposited horizontally along the bed of a stream as the water slows down.

horn A pyramidlike peak that forms when a peak is surrounded by three or four cirques.

humidity   The amount of water vapor in the air.

hurricane   Large, cyclonic storms in which the winds exceed 119 kilometers per hour.

hydration   A chemical reaction in which water combines with a substance.

hydrolysis   A reaction in which water separates into hydrogen ions and hydroxide ions that replace ions in a mineral.

hydrosphere   All of the water that rests on the lithosphere; it includes all oceans, lakes, streams, groundwater, and ice.

ice age   A period of time during which conditions exist that cause ice sheets more than 1 million square kilometers in area and thousands of meters thick to develop on nonpolar continents.

iceberg   A mass of floating ice that detached from a glacier into a body of water.

igneous intrusion   A mass of magma that penetrates an opening in an existing rock and solidifies.

igneous rock   Rock formed by the cooling and crystallization of molten minerals (magma or lava).

immature soil   A soil in which boundaries between forming horizons are indistinct.

in solution   Sediments that are transported as a consequence of being dissolved in the water flowing in a stream.

index fossil   A fossil of an organism that had distinctive body features and was abundant over a wide geographical area, but only existed for a short period of time.

infiltrate   The process by which water seeps into the ground and through interconnected pore spaces.

inorganic   Non-living and not derived from a living thing.

insolation   Solar radiation that reaches the surface of the Earth; short for incoming solar radiation.

interglacial   Time periods during an ice age when glaciers retreat or even disappear.

international date line   The 180° meridian; it marks the transition from one date to another.

intrusive igneous rock   A rock formed by the slow cooling of magma within the earth.

ion   An atom or molecule in which the electrical charges are unbalanced resulting in a net positive or negative electrical charge.

isobar   A line connecting points of equal air pressure.

isoline   Lines connecting points of equal field value.

isostasy   The idea that the Earth's crust floats on hot fluid rock in the mantle.

isosurface   Surfaces on which every point has the same field value.

isotherm   A line connecting points of equal temperature on a field map.

isotope   Varieties of the same element that differ slightly in mass.

joint   A fracture in a rock along which movement has not occurred.

kame   Isolated, conical hills of till formed when a hole in the ice penetrates to the bed and is filled with debris by meltwater streams.

kame terrace   A wide, flat bench of till deposited along the sides of a glacial valley by meltwater streams flowing over the lateral moraine of a glacier.

kettle   A depression formed when a block of ice buried in till melts.

key bed   Well-defined, easily identifiable rock layers that have distinctive characteristics or fossil content that allow(s) them to be used in correlation.

laccolith   A pluton formed when magma intrudes between rock layers and pushes upward and solidifies creating a rock structure that has a flat bottom and an arched top.

lagoon   The sheltered body of water between a barrier bar and the mainland.

land breeze   A local wind blowing from land toward an adjacent body of water because nighttime cooling increases the air pressure over the land.

latent heat of fusion   The amount of heat energy given off or absorbed in causing a change between the liquid and solid states that does not result in a change in temperature.

latent heat of vaporization   The amount of heat energy given off or absorbed in causing a change between the liquid and gaseous states that does not result in a change in temperature.

lateral moraine   A mound of till deposited along the sides of a glacier.

latitude   The angular distance north or south of the equator.

Laurentide ice sheet   The glacier that advanced several times across the northern United States during the Pleistocene epoch.

lava   Magma (molten rock) that has emerged onto the Earth's surface.

lava plateau   A flat sheet of igneous rock formed when a very fluid lava erupts from a long fissure instead of from a central vent.

leeward   The side of a feature facing away from the wind.

levee   A low, thick, ridgelike deposit along the banks of a stream consisting of coarse sediments deposited when a stream overflows its banks.

light year   The distance light travels in one year; roughly 9.5 trillion kilometers, or 6 trillion miles.

lithification   The process of converting sediments into coherent, solid rock.

lithosphere   Earth's solid outer layer of soil and solid brittle rock extending from the surface to a depth of about 100 kilometers; it includes the crust and the upper mantle.

lodgement till   Till deposited beneath a glacier consisting of larger particles aligned in the direction of motion of the ice and smaller particles compacted into deposits with a streamlined shape.

loess   A fine, buff-colored sediment deposited by wind.

longitude   The angular distance east or west of the prime meridian.

longitudinal waves   Waves in which the medium oscillates parallel to the direction of wave motion.

longshore current   An ocean current flowing parallel to a shore caused by the approach of waves at an angle to the coast.

lunar eclipse   An event in which the Moon passes into the Earth's shadow during the full moon phase.

luster   The way light reflects from the surface of a mineral.

mafic   Rocks rich in minerals containing magnesium and fe (iron).

magma   Molten rock beneath the Earth's surface.

magma chamber   A pocket of molten magma in the crust or upper mantle.

mantle   The zone of the Earth's interior between the crust and the core.

map   A model of the Earth's surface.

map scale   The ratio between distance on a map and actual distance on the ground; usually expressed as a ratio.

mass   The amount of matter in an object.

mass wasting   The downhill movement of sediments under the direct influence of gravity.

mature soil   A soil in which distinct horizons have formed.

meander   A curving loop in a stream channel.

medial moraine   Elongated piles of till along the margin between two glaciers that flowed together.

meridian   Semicircles on which all points have the same longitude.

mesopause   The boundary between the mesosphere and the thermosphere.

mesosphere   The layer of the atmosphere ranging from 30 to 50 kilometers; temperatures in this layer decrease with altitude.

metamorphic rock   Rocks that form as a result of physical and chemical changes in existing rocks.

metamorphism   The processes that change the physical and chemical properties of existing rocks.

meteor   The streak of light seen in the sky as a meteoroid is vaporized by frictional heating as it falls through the atmosphere.

meteorite   The portion of a meteoroid that survives to strike the ground.

meteoroid   A chunk of matter moving through space.

mid-ocean ridge   A chain of undersea mountains running through the center of an ocean.

Milky Way galaxy   The galaxy in which our solar system is located.

mineral   A naturally occurring, inorganic, crystalline solid with a composition that can be expressed by a chemical symbol or a formula.

mineral composition   The minerals present in a rock.

mixture   Consisting of one or more substances that retain their characteristic properties and can be separated by relatively simple means.

modified Mercalli scale   A system that measures the strength of an earthquake based upon perception of motion by human observers and damage to structures built by humans.

Moho   The boundary between the dense rock of the mantle and the less dense rock of the crust. It was named for the seismologist Andrija Mohorovicic, who recognized its existence after analyzing seismic wave behavior inside the Earth.

Mohorovicic discontinuity   The boundary between the Earth's crust and mantle. It marks the interface between rocks within the Earth of two different average densities.

monocline   A fold in which only one side of the fold is inclined.

moraine   A deposit of unsorted sediments deposited directly from glacial ice.

mudflow   The rapid downhill flow of a mixture of rock fragments, soil, and water.

natural selection   The process by which organisms that are better suited to their environment survive and reproduce while poorly suited organisms do not.

neap tide   A tidal cycle in which there is very little difference between the water level at high and low tides; caused by the canceling out effect of the Moon's gravity acting at right angles to the Sun's.

nodule   A small, irregularly shaped mass of mineral deposited from seawater directly onto a rock or other small object.

nonconformity   A place where sedimentary rock lies directly on another type of rock, such as igneous or metamorphic rock.

numerical forecasting   Weather forecasting based upon mathematical models of the physical behavior of the atmosphere.

oasis   A place where deflation has lowered the surface in a desert to the water table allowing plants to take root and grow.

oblate spheroid   A slightly flattened sphere; the Earth's shape.

ocean-floor spreading   The idea that the ocean floors were spreading sideways away from the mid-ocean ridges.

orbit   The path along which a satellite revolves around a primary.

orbital velocity   The smallest velocity a body must attain in order to enter a stable orbit around another body.

ordinary well   A hole dug in the ground so that it penetrates the water table in which water accumulates and can be pumped to the surface.

organic sediments   Particles produced by the life activities of plants or animals.

origin time   The time at which an earthquake occurred.

original horizontality, law of   The observation that sediments deposited in water form flat level layers.

orographic effect The net increase in temperature of a parcel of air that is forced over a mountain range due to adiabatic cooling on the windward side followed by adiabatic heating on the leeward side.

outwash plain A flat plain next to the leading edge of a glacier formed by the deposition of sediments from meltwater streams.

overland flow The downhill flow of water in a thin sheet.

oxbow lake A crescent-shaped body of water formed when a cutoff isolates the water in a meander from the main channel of the stream.

oxidation A chemical reaction in which a substance combines with oxygen.

ozone A gas whose molecules consist of three atoms of oxygen joined together.

P-waves Longitudinal seismic waves produced by an earthquake that travel the fastest and are therefore the first to be recorded by a seismograph.

Pacific ring of fire A narrow zone of active volcanoes surrounding the Pacific Ocean.

paleomagnetism A record of the orientation of the Earth's magnetic field preserved in the crystalline structure of igneous rock containing magnetic minerals.

parallel Circles on which all points have the same latitude north or south of the Equator.

parting The tendency to break along flat surfaces that follow structural weaknesses.

path of totality The path of the Moon's umbra over the surface of the Earth during a solar eclipse.

pedalfer Soils rich in aluminum and iron compounds.

pedocal Soils rich in calcium compounds.

pelagic Sediments consisting of the skeletal remains of marine organisms or particles precipitated from seawater.

penumbra The portion of a shadow in which only part of the light has been blocked, so the light is dimmed but not totally absent.

period The unit of geologic time into which eras are divided.

period of revolution The length of time it takes for a satellite to complete one revolution around its primary.

permeability The rate at which water can infiltrate a material.

phases of matter The form in which matter may exist; solid, liquid, and gas.

phases of the Moon The changing shape of the illuminated portion of the Moon that can be seen by an observer on Earth.

phenocryst Isolated large crystals in a rock.

photosynthesis The process by which cells convert water and carbon dioxide into carbohydrates and oxygen using light as an energy source.

physical weathering Processes that break down rock without changing their chemical composition.

phytoplankton Plankton that carry out photosynthesis.

planet A body that is at least partly solid, orbits the Sun, and emits mainly reflected sunlight.

plant action The breaking down of rock by the force exerted by the growth of plant roots and stems or by chemicals produced as a by-product of plant growth.

plate tectonics The theory that the Earth's crust consists of large rigid slabs of rock, or plates, resting on a fluidlike layer of denser rock and that these plates slide around causing their edges to collide, separate, and slide past each other.

Pleistocene The time period, which began 1.6 million years ago, during which climates grew colder worldwide.

plucking The tearing loose of rock frozen to the base of a glacier as the glacier moves along.

pluton A rock structure formed by the solidification of an igneous intrusion.

polar diameter Diameter of the Earth measured from a pole through the center of the Earth to the opposite pole (12,714 kilometers).

polar front A boundary between polar and tropical air that forms in the prevailing westerly global wind belts.

Polaris A star whose position is almost directly over the Earth's north geographic pole.

pollutant The particular materials or forms of energy that are harmful to humans.

pollution The concentration of any material or energy form that is harmful to humans.

porosity The percentage of empty space in a material.

porphyry A rock consisting of large crystals called phenocrysts embedded in a mass of fine mineral grains.

potential evapotranspiration The maximum amount of water that could be returned to the atmosphere by evapotranspiration under a given set of conditions.

precession The very slow change in the orientation of a rotating body's axis of rotation.

precipitate A solid that crystallizes out of a solution and settles to the bottom of the solution.

precipitation Condensed moisture that falls to the ground.

pressure unloading The release of stress on rock as the stress of the weight of overlying rock is removed as the latter is worn away, or the removal of the stress as the weight of the ice in a glacier is removed as the glacier melts away.

primary The body a satellite orbits.

prime meridian The zero meridian, or starting point, for measuring the angular distance east or west of any other meridian.

profile What a cross section of the land between two points would look like viewed from the side.

radiative balance The condition in which the Earth's average level of heat energy remains constant over a long period of time.

radioactive decay The process by which unstable isotopes break apart spontaneously giving off energy, subatomic particles, or both.

radioisotope Isotopes that are unstable and undergo radioactive decay.

recessional moraine A series of end moraines marking the edge of a retreating glacier that has undergone brief episodes of advance and retreat.

recharge Water that seeps into the ground and replaces soil moisture that was lost during a period of drought.

regional metamorphism Changes in rock over an extensive area that occur due to the pressure and high temperatures associated with either deep burial or movements of the Earth's crust.

relative age The age of an object or event relative to another object or event.

relative humidity The ratio of water vapor in the air to the maximum amount of water vapor the air can hold at that temperature.

relief The differences in the height of landforms in an area.

renewable resource A material that can be renewed, or restored to its original form after being used, by means of a natural process; i.e., pure water that is polluted by human use is purified by the natural processes of the water cycle.

replacement reaction A chemical reaction in which one element replaces another element in a molecule.

residual soil A soil formed by the weathering of bedrock in place.

respiration The oxidation of food within living cells by which the chemical energy of food molecules is released in a series of metabolic steps involving the consumption of oxygen and the liberation of carbon dioxide and water.

retrograde motion An apparent motion of a planet in a direction opposite to its normal motion.

revolution The motion of one body around another body.

Richter scale A system that measures the strength of an earthquake by the amount of motion of the Earth's crust that occurs during the earthquake.

rill Tiny grooves eroded in soil by a trickling stream of water flowing downhill.

roche moutonée A knob of bedrock sculpted by a glacier having a smooth, elongated upstream side and a plucked downstream side that is steep and rough.

rock The naturally formed, solid material that makes up the Earth's crust.

rock cycle A sequence of events that shows how rocks are formed, changed, destroyed, and reformed from the same parent material.

rock-forming minerals The one hundred or so common minerals that make up more than 95 percent of the rock in the Earth's crust.

rockfall Mass wasting process in which rock fragments broken loose by weathering fall from cliffs or bounce by leaps down steep slopes.

rockslide Mass wasting process in which rock fragments slide downhill.

rotation Motion in which every part of an object is moving in a circular path around a central line called the axis of rotation.

runoff Runoff is precipitation that does not evaporate or sink into the ground, but runs downhill along the Earth's surface.

S-waves Transverse seismic waves produced by an earthquake that travel slower than P-waves and are therefore the second to be recorded by a seismograph.

salinity The total amount of dissolved matter in seawater expressed in parts per thousand.

saltation A type of motion in which sediments transported in a stream move along the stream bed in a series of bounces, hops, or leaps.

sandbar A long, narrow pile of sand deposited in open water.

satellite A solid body that orbits another body; i.e., the Moon is a satellite of the Earth.

sea breeze A local wind blowing from a body of water toward land that develops because daytime heating lowers the air pressure over the land.

sea cliff A steep cliff formed by the undercutting of steep slopes by waves and subsequent collapse of the overhanging material.

sea stack An isolated pillar of resistant rock that was detached from a headland by wave erosion.

sea terrace A flat platform of sediments deposited by wave action on the ocean floor adjacent to a steep coastline.

sea waves Wind-generated sea waves that are directly affected by the wind.

sediment Solid particles that have been transported and then deposited by agents of erosion.

sedimentary rock Rock formed by the lithification of sediments.

seismic wave A vibration of the Earth's crust produced by faulting.

seismogram A line recorded on paper by a seismograph that represents the motion during an earthquake.

seismograph A device that detects the motion of the Earth's crust during an earthquake.

seismology The study of earthquakes and their effects.

shadow zone A region in which no seismic waves from an earthquake are received by seismograph stations because the waves have been refracted or blocked as they traveled through the Earth.

shear Forces acting along different lines of action causing rock to twist or tear.

sheet erosion Erosion caused by overland flow.

shield volcano A wide, gently sloping volcano formed from successive layers of solidified lava flows.

silicaceous ooze Sediment rich with the silica skeletons of marine organisms.

silicates   Minerals composed of compounds of silicon and oxygen together with other elements; the most abundant mineral group on Earth.

silicon tetrahedron   An ion of silicon joined with four ions of oxygen in the shape of a tiny tetrahedron (pyramid-like shape).

sill   A pluton that forms between, and parallel to, existing rock layers.

sleet   Clear pellets of ice that form when raindrops freeze as they fall through layers of air at below-freezing temperatures.

slump   Mass wasting process in which a mass of bedrock or soil slides downward along a curved plane of weakness.

snow   Hexagonal crystals of ice or needlelike ice crystals formed by the sublimation of water vapor in the atmosphere.

soil   The accumulation of loose, weathered material that covers much of the land surface of the Earth.

soil creep   The invisibly slow, downhill movement of soil.

soil horizon   Recognizable layers in soil with characteristic properties.

soil moisture storage   Water that is retained in the soil after water seeps into the ground.

soil moisture zone   A moist layer of soil in which water clinging to soil particles supplies plant roots with water.

soil profile   A cross section of soil from surface to bedrock.

solar eclipse   An event in which the Moon passes directly between the Earth and the Sun, casting a shadow on the Earth and partly or completely blocking an observer's view of the Sun.

solstice   The two times of the year when the Sun's altitude at noon reaches a maximum or a minimum. The summer solstice in the Northern Hemisphere occurs about June 21, when the Sun is in the zenith at the Tropic of Cancer; the winter solstice occurs about December 21, when the Sun is over the Tropic of Capricorn. The summer solstice is the longest day of the year and the winter solstice is the shortest.

source region   The geographical region in which an air mass originated.

specific gravity   The density of a mineral compared to the density of water.

specific heat capacity   The number of degrees the temperature of one gram of a material will increase if one calorie of heat is added to it.

spit   A sandbar attached at one end to the mainland.

splash erosion   A process in which the impact of falling raindrops lifts some soil and drops it in a new position.

spring   A place where groundwater seeps out of the ground.

spring tide   A tidal cycle in which there are very high and very low tides; caused by the additive effect of the Sun and Moon's gravity acting together along the same line of action.

station model   A shorthand symbolic representation of the weather conditions at a particular location.

statistical forecasting   Weather forecasting based upon the statistical analysis of historical weather data.

stellar parallax   The change in the position of a near object relative to distant objects as an observer's position changes.

stock   An irregularly shaped pluton covering an area less than 75 square kilometers.

strain   The change in size or shape of a rock due to stress.

stratopause   The boundary between the stratosphere and the mesosphere.

stratosphere   The layer of the atmosphere ranging from about 6 to 30 kilometers above the Earth's surface; it contains ozone that absorbs ultraviolet light; temperatures in this layer rise with altitude.

streak   The color of a powdered mineral; created by rubbing a mineral against an unglazed piece of porcelain.

stream A body of water that moves under the influence of gravity to progressively lower levels in a narrow, but clearly defined channel.

stream banks The sides of the channel containing the water of a stream.

stream bed The bottom of the channel containing the water of a stream.

stream channel Clearly defined path along which the water in a stream flows.

stream gradient The angle between the stream bed and the horizontal; the slope of the stream.

stream load The sediments transported by the water flowing in a stream.

stream mouth The site at which a stream flows into a standing body of water.

stream source The site at which the water in a stream originates.

stress Forces acting on a rock.

subduction The sliding of a denser ocean plate beneath a less dense continental plate resulting in the melting of the ocean plate as it plunges into the hot mantle.

sublimation The process by which a solid changes directly into a gas without passing through a liquid stage.

superposition The idea that the bottom layer of a sedimentary series is the oldest, unless it was overturned or had older rock thrust over it, because the bottom layer was deposited first.

surf Waves that are breaking along a shoreline.

surplus A condition in which moisture entering the environment exceeds the amount that can be absorbed by the soil or lost by potential evapotranspiration.

suspended load The sediments carried in suspension in the water flowing in a stream.

swells Long wavelength sea waves that settle into a uniform pattern once they leave the area of turbulent winds that formed them.

syncline A downward, valleylike fold.

synoptic forecasting Weather forecasting based upon analysis of a sequence of synoptic weather maps and charts.

synoptic weather map A map that summarizes weather data from many locations using station models.

system A group of things or processes that interact to perform some function.

talus A mound of rock fragments that accumulates at the base of a rock cliff that is weathering.

technology The application of knowledge to do things.

temperature A specific degree of hotness or coldness as indicated on or referred to a standard scale.

tension Forces acting away from each other along the same line of action causing rock to stretch.

tephra A collective name for particles formed when globs of lava ejected during an eruption cool and solidify as they fall to the ground.

terminal moraine A mound of till deposited along the leading edge of a glacier.

terrigenous Sediments derived from the land or a continent.

texture The size, shape, and arrangement of mineral crystals or grains in a rock.

thermocline A layer of seawater in which temperature drops sharply with depth.

thermosphere The uppermost layer of the atmosphere ranging from about 50 to 75 kilometers in altitude where it grades into space; this layer contains the ionosphere, which consists of charged particles or ions that deflect harmful radiation and reflect radio waves.

thunderstorm A vigorous, turbulent convection cell in the atmosphere resulting in a transient, violent storm of thunder and lightning, often accompanied by rain and sometimes hail.

tombolo A sandbar that connects an island to the mainland.

topographic map A field map in which the field value is elevation above sea level; it shows the three-dimensional shape of the Earth's surface in two dimensions.

tornado   A rotating column of air whirling at speeds of up to 800 kilometers per hour, usually accompanied by a funnel-shaped downward extension of a cumulonimbus cloud.

traction   The process by which particles that are not frozen into the ice are dragged along slipping, sliding, and rolling beneath the moving ice.

transform boundary   Places where edges of plates are sliding laterally past each other.

transpiration   The process by which plants release water into the atmosphere through their leaves.

transported soil   A soil that formed from material that was deposited in a region after being transported from another place.

transverse waves   Waves in which the medium oscillates perpendicular to the direction of wave motion.

travel time   The time it takes for a seismic wave to travel from the epicenter to a seismograph.

trench   A deep crevice in the ocean floor produced by downward warping of the ocean floor where plates collide.

tropopause   The boundary between the troposphere and the stratosphere.

troposphere   The layer of the atmosphere ranging from the Earth's surface to about 6 kilometers in altitude; it contains most of the air in the atmosphere; temperatures in this layer decrease with altitude.

tsunami   Waves caused by undersea movements of the Earth's crust such as slumping, earthquakes, or volcanic eruptions.

umbra   The portion of a shadow in which all light has been blocked.

unconformity   A break or gap in the sequence of a series of rock layers.

uniformitarianism   The idea that the Earth's features have formed gradually by natural processes that are still occurring, not by instantaneous creation or by catastrophic events.

usage   Moisture lost from the soil during a period of drought.

valley glacier   A glacier that forms between mountain slopes at high elevations.

velocity   The speed and direction in which something is moving.

vent   The central opening of a volcano out of which lava erupts.

ventrifact   A rock particle that has developed a flat side facing the prevailing winds due to abrasion of its exposed surface.

vertical sorting   The separation and deposition of sediments in successive layers caused by differences in the settling rate of the sediment particles; particles that settle fastest form the bottom-most layer, whereas those that settle most slowly form the top layer.

visibility   The horizontal distance through which the eye can distinguish objects.

volcanic glass   A rock formed by cooling that was so rapid that no crystalline structure was able to form; such rocks have a glassy appearance.

volcanism   The processes by which magma rises into the crust is extruded onto the Earth's surface.

volcano   The opening in the Earth's crust through which lava emerges and the moundlike structure formed by the accumulation of lava.

walking the outcrop   Physically following layers of rock from one place to another in order to correlate them.

water cycle   The process by which the Earth's water cycles in and out of the atmosphere.

water table   The top of the subsurface zone in which the pore spaces in a soil are saturated with water.

wave cyclone   Warm and cold fronts that move in a counterclockwise direction around a low pressure center.

wavelength   The distance between two successive wave crests or troughs.

weather   The present condition of the atmosphere at any location.

weathering   The breakdown of rocks into smaller particles by natural processes.

weight   The gravitational force exerted on an object by the Earth.

Wentworth Scale   A scheme for classifying and naming sediment particles according to their size.

wet-bulb temperature   The temperature recorded by a thermometer whose bulb is kept wet; it is usually lower than dry-bulb temperature due to evaporative cooling.

whalebacks   A smooth, elongated bedrock feature formed by glacial abrasion of the bedrock.

wind   Air that is moving horizontally.

windward   The side of a feature facing the wind.

zenith   A point directly overhead to an observer.

zone of aeration   The subsurface zone in which the pore spaces in a soil are filled mainly with air.

zone of saturation   The subsurface zone in which the pore spaces in a soil are saturated with water.

Examination
June 1999
Earth Science

PART I

Answer all 55 questions in this part. [55]

Directions (1–55): For *each* statement or question, select the word or expression that, of those given, best completes the statement or answers the question. Record your answers in the spaces provided.

1 The diagram below is a three-dimensional model of a landscape region.

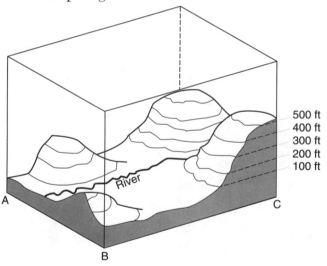

Which map view best represents the topography of this region?

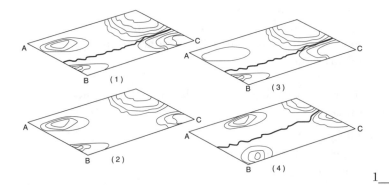

1_____

2 The drawing below shows the effects of an earthquake on a small part of Earth's surface. Letters *A* and *B* indicate land on opposite sides of a cliff that formed along a fault during the earthquake.

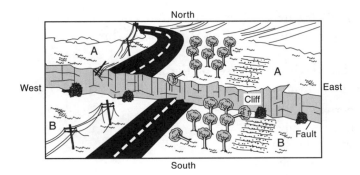

In relation to the position of side *B*, the movement of side *A* was

1 eastward and downward
2 eastward and upward
3 westward and downward
4 westward and upward

2_____

3 Science investigators initially use classification systems to

   1 extend their powers of observation
   2 make more accurate inferences
   3 organize their observations in a meaningful way
   4 make direct comparisons with standard units of measurement   3_____

4 Which process requires the most absorption of energy by water?

   1 melting 1 gram of ice
   2 condensing 1 gram of water vapor
   3 vaporizing 1 gram of liquid water
   4 freezing 1 gram of liquid water   4_____

5 The circumference of Earth is approximately 40,000 kilometers at the Equator. What is Earth's approximate rate of rotation, in kilometers per hour, at the Equator?

   (1) 1,667 km/hr   (3) 16,667 km/hr
   (2) 9,600 km/hr   (4) 66,000 km/hr   5_____

6 As measured by an observer on Earth over the course of a year, the apparent diameter of the Moon will

   1 decrease, only
   2 increase, only
   3 remain the same
   4 vary in a cyclic manner   6_____

7 Which diagram best shows the altitude of Polaris observed near Buffalo, New York?

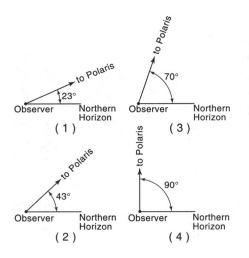

7_____

8 An object's weight at sea level at 90° North latitude is slightly more than the weight of the same object at sea level at 0° latitude. Which statement about Earth can best be inferred from this evidence?

1 Earth's orbit is slightly elliptical.
2 Earth's axis is tilted $23\frac{1}{2}°$ to the plane of its orbit.
3 Earth's shape is slightly bulged at the Equator.
4 Earth rotates counterclockwise as viewed from above the Equator.

8_____

9 The shaded portion of the map below indicates areas
of night and the unshaded portion indicates areas of
daylight.

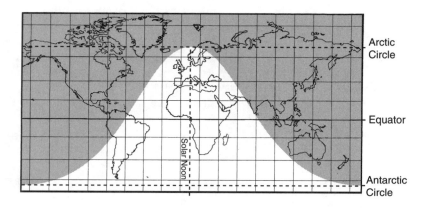

What day of the year is best represented by the map?

1 March 21   3 September 21

2 June 21   4 December 21   9 _____

10 The diagram below shows information needed to use Eratosthenes' method to find the circumference of a planet. At noon, a vertical stick at the planet's equator casts no shadow. At a location farther north, another vertical stick casts a shadow that makes an angle of 25° with the rays of the Sun. The distance between the two sticks is 2,000 kilometers.

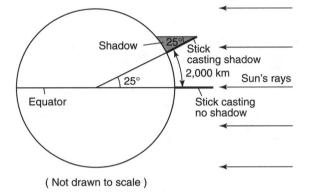

(Not drawn to scale)

What is the approximate circumference of this planet?

(1) 18,800 km   (3) 36,200 km
(2) 28,800 km   (4) 50,000 km   10_____

11 During how many days of a calendar year is the Sun directly overhead at noon in New York State?

(1) only 1 day   (3) 365 days
(2) only 2 days   (4) 0 days   11_____

12 A barefoot student steps on a hot concrete surface. Most of the heat transferred to the student's skin by this contact is by the process of

1 convection   3 vaporization
2 conduction   4 radiation   12_____

13 In the visible spectrum, which color has the longest wavelength?

1 red   3 orange
2 green   4 violet   13_____

14 How do the rates of warming and cooling of land surfaces compare to the rates of warming and cooling of ocean surfaces?

1 Land surfaces warm faster and cool more slowly.
2 Land surfaces warm more slowly and cool faster.
3 Land surfaces warm faster and cool faster.
4 Land surfaces warm more slowly and cool more slowly.   14_____

15 The diagram below shows the apparent path of a projectile fired from Earth's North Pole toward a target located at 45° North latitude.

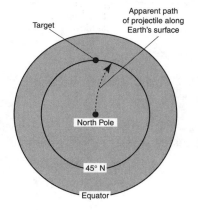

The apparent curving of the projectile's path, as seen by observers on Earth, is caused mainly by

1 prevailing winds   3 Earth's rotation
2 convection currents   4 gravitational forces   15_____

16 Which diagram best represents the wavelength of most of the sunlight energy absorbed and the wavelength of infrared energy reradiated by the roof of a building at 2 p.m. on a clear summer day?

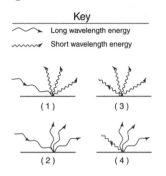

16_____

17 A student set up the activity shown in the diagram below to demonstrate how convection cells in Earth's mantle could cause crustal plates to converge.

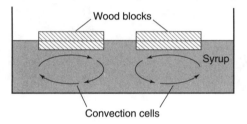

Which diagram shows the best placement of heat sources to cause the blocks to converge?

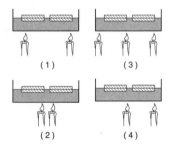

17_____

18 The diagram below represents Earth.

Which diagram best represents Mars, drawn to the same scale?

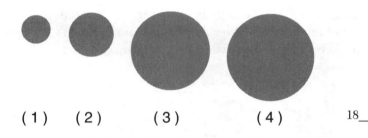

(1)   (2)   (3)   (4)   18_____

Base your answers to questions 19 and 20 on the map below, which shows North American air-mass source regions, the resulting air-mass names, and typical air-mass tracks.

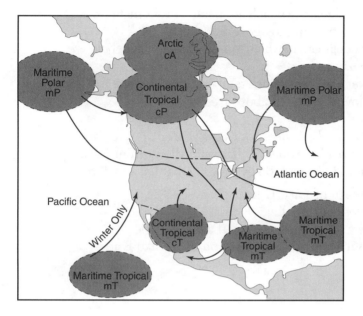

19 A maritime polar air mass approaching New York State would most likely bring

1 cool, moist air from the north
2 warm, moist air from the south
3 cool, dry air from the southeast
4 warm, dry air from the southwest   19_____

20 Polar air-mass characteristics differ from tropical air-mass characteristics primarily because the source regions differ in their

1 nearness to land   3 latitude
2 nearness to water   4 longitude   20_____

21 Which type of climate has the greatest amount of rock weathering caused by frost action?

　1　a dry climate in which temperatures remain below freezing

　2　a dry climate in which temperatures alternate from below freezing to above freezing

　3　a wet climate in which temperatures remain below freezing

　4　a wet climate in which temperatures alternate from below freezing to above freezing　　　　　　21_____

22 The *Generalized Bedrock Geology Map of New York State* provides evidence that water flows from Lake Erie into Lake Ontario by showing that Lake Ontario

　1　is north of Lake Erie

　2　is deeper than Lake Erie

　3　has a larger surface area than Lake Erie

　4　has a lower surface elevation than Lake Erie　　　22_____

23 The cartoon below shows a humorous view of a weather phenomenon.

BLOOM COUNTY / By Berke Breathed

Rain fell from the cloud because

1 too few condensation nuclei were in the cloud
2 the cloud's water droplets combined and became large enough to fall
3 evaporation in the upper portion of the cloud caused the cloud to become smaller and "squeezed" water out
4 the temperature in the cloud rose above the dewpoint

23_____

24 A mercury barometer that is used to measure air pressure is shown below.

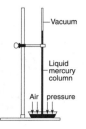

A decrease in the height of the mercury column usually indicates the approach of a

1 low-pressure system and stormy weather
2 low-pressure system and clear weather
3 high-pressure system and stormy weather
4 high-pressure system and clear weather   24_____

25 The graph below shows the air pressure recorded at the same time at several locations between Niagara Falls and Albany.

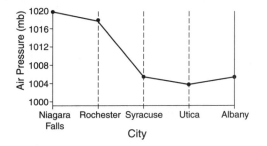

Based on the information in this graph, the wind velocity is probably greatest between which two cities?

1 Niagara Falls and Rochester
2 Rochester and Syracuse
3 Syracuse and Utica
4 Utica and Albany   25_____

26 The map below shows partial weather conditions for weather stations *A* and *B* at 4 p.m. A weather front is located between the two stations.

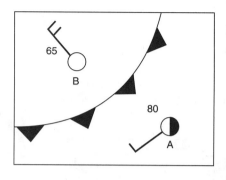

Which graph represents the temperature change that will most likely occur at station *A* as the front passes in the next three hours?

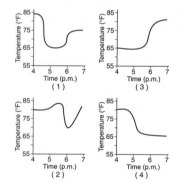

26_____

27 The percentage of open space between grains of soil is called the soil's

1 permeability   3 discharge
2 porosity   4 capillarity

27_____

28 Which current is a cool ocean current that flows completely around Earth?

   1 California Current
   2 Gulf Stream
   3 West Wind Drift
   4 North Equatorial Current

28_____

29 The map below shows the annual amount of surface runoff in part of the Northern Hemisphere.

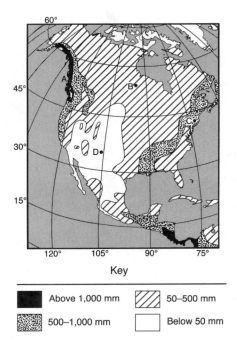

Key

■ Above 1,000 mm		▨ 50–500 mm	
▨ 500–1,000 mm		☐ Below 50 mm	

At which point on the map does the amount of precipitation exceed by the greatest amount both the actual evapotranspiration and the infiltration of water into the ground?

   (1) A   (3) C
   (2) B   (4) D

29_____

30 The cross section below represents a portion of the coastal mountains in the western United States. Location X is near the shore of the Pacific Ocean, and location Y is on the eastern slope of the mountains. Prevailing winds are from west to east.

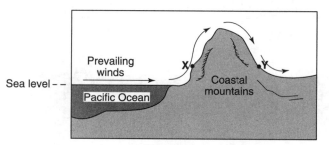

(Not drawn to scale)

Compared to the climate at location X, the climate at location Y is

1 warmer and drier   3 cooler and drier
2 warmer and wetter   4 cooler and wetter   30_____

31 What is the largest particle that can generally be transported by a stream moving at 200 centimeters per second?

1 boulder   3 pebble
2 cobble   4 sand   31_____

32 Quartz particles of varying sizes are dropped at the same time into deep, calm water. Which cross section best represents the settling pattern of these particles?

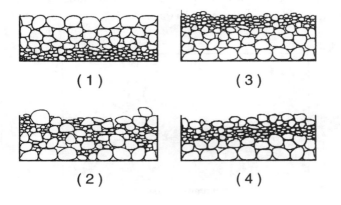

(1)   (3)

(2)   (4)   32_____

33 Which agent of erosion formed the long U-shaped valleys now occupied by the Finger Lakes in central New York State?

1 glacial ice   3 ocean currents
2 wind   4 running water   33_____

34 The cross section below represents bedrock that has been changed by several geologic events.

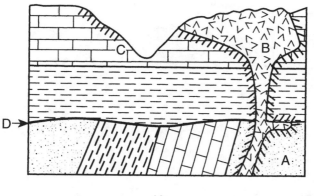

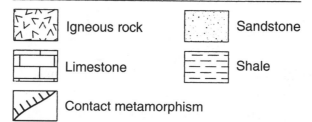

Key

Igneous rock

Sandstone

Limestone

Shale

Contact metamorphism

Which geologic event happened most recently?

1 tilting of rock *A*
2 intrusion of rock *B*
3 formation of the unconformity at *D*
4 erosion of rocks *B* and *C*   34_____

35 The table below provides information about four common silicate minerals.

Which conclusion is best supported by the information in this table?

1 The shape of the tetrahedral unit controls the shape of the broken mineral.
2 The arrangement of the tetrahedral units controls the mineral breakage pattern.
3 The percent of shared oxygen controls the size of the mineral crystal.
4 The type of atoms present controls how much oxygen is shared.   35_____

36 A seismograph records the arrival of a *P*-wave at 11:13 a.m. If the earthquake occurred 4,000 kilometers from the recording station, when did the earthquake occur?

(1) 11:06 a.m.   (3) 11:13 a.m.
(2) 11:11 a.m.   (4) 11:20 a.m.   36_____

37 At which depth below Earth's surface is the density most likely 9.5 grams per cubic centimeter?

(1) 1,500 km   (3) 3,500 km
(2) 2,000 km   (4) 6,000 km   37_____

38 The table below shows the chemical composition of some common minerals found in rocks of the lithosphere.

Mineral	Chemical Composition
Hematite	Fe_2O_3
Calcite	$CaCO_3$
Quartz	SiO_2
Potassium feldspar	$KAlSi_3O_8$

As indicated by its chemical composition, which mineral could have formed when CO_2 (carbon dioxide) combined with another substance?

1 potassium feldspar   3 quartz
2 hematite   4 calcite   38_____

39 Although more than 2,000 minerals have been identified, 90% of Earth's lithosphere is composed of the 12 minerals listed below.

Rock-Forming Minerals	
feldspar	augite
quartz	garnet
mica	magnetite
calcite	olivine
hornblende	pyrite
kaolinite	talc

The best explanation for this fact is that most rocks

1 are monomineralic
2 are composed of recrystallized minerals, only
3 have a number of minerals in common
4 have a 10% nonmineral composition   39_____

40 Deep, parallel grooves and scratches (striations) are found on the surface of some limestone bedrock in New York State. These scratches and grooves suggest that the surface was

1 abraded by windblown sand
2 scraped by rocks in a continental glacier
3 eroded by a meandering stream
4 cracked by the evaporation of a warm, shallow sea   40_____

41 Which rock sample is most likely a foliated meta-morphic rock?

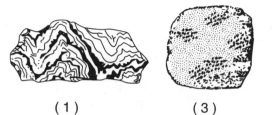

(1)   (3)

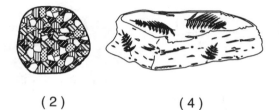

(2)   (4)

41_____

42 Which statement best describes the formation of an intrusive igneous rock?

1 Magma solidifies slowly, resulting in a coarse-grained texture.

2 Magma solidifies slowly, resulting in a fine-grained texture.

3 Magma solidifies rapidly, resulting in a glassy texture.

4 Magma solidifies rapidly, resulting in a clastic texture.

42_____

43 The diagram below shows data received at a seismic
station following an earthquake.

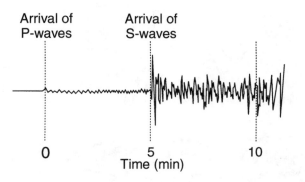

The distance from this seismic station to the epicenter of the earthquake is approximately

(1) 1,300 km   (3) 3,400 km
(2) 2,600 km   (4) 5,000 km   43_____

44 The dots on the map below show the present locations of living coral reefs. Site *X* indicates an area of fossil coral reefs preserved in rocks formed during the Jurassic Period.

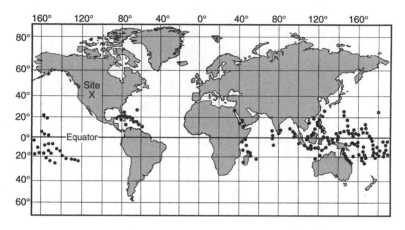

Which inference is best supported by this map?

1 The climate at site *X* during the Jurassic Period was colder than the present climate at site *X*.
2 The coral at site *X* evolved from ocean-dwelling animals into land-dwelling animals after the Jurassic Period.
3 Site *X* has drifted southward since the Jurassic Period.
4 Site *X* was covered by warm ocean water during the Jurassic Period.

44_____

45 The block diagram below represents a portion of the surface of Earth's crust.

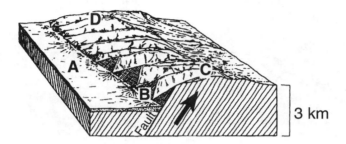

Which letter is located on the boundary between two landscape regions?

(1) *A*   (3) *C*

(2) *B*   (4) *D*   45_____

46 Trilobite fossils were recently discovered in Himalayan Mountain bedrock. During which geologic period could this bedrock have been formed?

1 Tertiary   3 Triassic

2 Cretaceous   4 Cambrian   46_____

47 What is the geologic age of the surface bedrock of most of the Allegheny Plateau landscape region in New York State?

1 Cambrian   3 Silurian

2 Devonian   4 Ordovician   47_____

Base your answers to questions 48 and 49 on the table below. The table shows the characteristics and geological range of the four general fossil types of one kind of echinoderm called crinoids.

Crinoid Fossil Type	Mouth and Food Grooves	Calyx	Calyx Plates	Arms	Geologic Time Range
Camerata	Not exposed	Rigid	Some brachials	Pinnules	Ordovician to Permian
Flexibilia	Exposed	Flexible	Some brachials	No pinnules	Ordovician to Mississippian
Inadunata	Not exposed	Rigid	No brachials	With or without pinnules	Ordovician to Triassic
Articulata	Exposed	Flexible	With or without brachials	Pinnules	Triassic to Recent

48 Which type of crinoid is shown below?

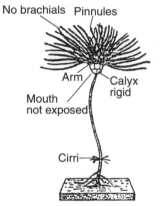

1 Camerata   3 Inadunata

2 Flexibilia   4 Articulata   48_____

49 Evidence shown on the chart suggests that most crinoid types

1 are extinct

2 lacked a mouth

3 had a rigid calyx

4 lived during the Cambrian   49_____

50 What happens to *P*-waves and *S*-waves from a crustal earthquake when the waves reach Earth's outer core?

  (1) *S*-waves are transmitted through the outer core, but *P*-waves are not transmitted.
  (2) *P*-waves are transmitted through the outer core, but *S*-waves are not transmitted.
  (3) Both *P*-waves and *S*-waves are transmitted through the outer core.
  (4) Neither *P*-waves nor *S*-waves are transmitted through the outer core.   50_____

51 Evidence found in igneous rocks suggests that, through geologic time, Earth's magnetic poles have

  1 maintained their present positions
  2 corresponded exactly with Earth's geographic poles
  3 maintained constant strength
  4 reversed their magnetic polarities   51_____

52 A graph of the radioactive decay of carbon-14 is shown below.

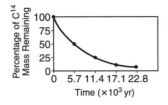

Which graph correctly shows the accumulation of nitrogen-14, the decay product of carbon-14, over the same period?

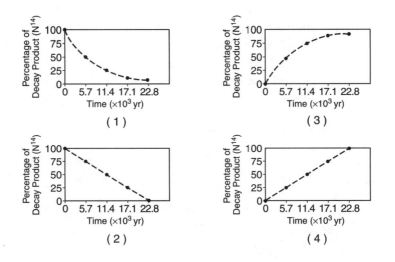

52_____

53 The maps below represent three different stream drainage patterns.

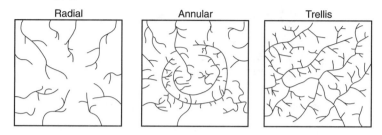

Radial   Annular   Trellis

Which statement is generally true of these three drainage patterns?

1 All are controlled by underlying bedrock structure.
2 All are in the old-age stage of stream development.
3 All are located in semiarid regions.
4 All are located in areas where deposition is greater than erosion.   53_____

54 The table below shows characteristics of three landscape regions, X, Y, and Z.

Landscape Region	Relief	Bedrock
X	Great relief, high peaks, deep valleys	Many types, including igneous and metamorphic rocks; nonhorizontal structure
Y	Moderate to high relief	Flat layers of sedimentary rock or lava flows
Z	Very little relief, low elevations	Many types and structures

Which terms, when substituted for X, Y, and Z, best complete the table?

(1) X = mountains, Y = plains, Z = plateaus
(2) X = plateaus, Y = mountains, Z = plains
(3) X = plains, Y = plateaus, Z = mountains
(4) X = mountains, Y = plateaus, Z = plains   54_____

55 The diagram below shows the relative thickness of soil and major vegetation at various latitudes on Earth.

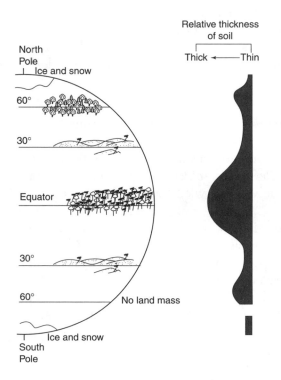

The development of thick soils is most closely associated with areas that have a

1 small amount of vegetation and low yearly precipitation
2 small amount of vegetation and high yearly precipitation
3 large amount of vegetation and low yearly precipitation
4 large amount of vegetation and high yearly precipitation

55_____

PART II

This part consists of ten groups, each containing five questions. Choose seven of these ten groups. Be sure that you answer all five questions in each group chosen. Record the answers to these questions in the spaces provided. [35]

GROUP 1

If you choose this group, be sure to answer questions **56–60**.

Base your answers to questions 56 through 60 on the *Earth Science Reference Tables,* the diagram and data tables below, and your knowledge of Earth science.

The diagram shows a process that is used to separate gypsum from shale on the basis of density. Crushed rock is dropped into a mixture of liquid and small particles, called a slurry. The data tables show the mass and volume of three samples of gypsum and three samples of shale after they were separated by this process.

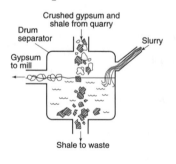

Gypsum		
Sample	Mass (g)	Volume (cm³)
1	2.0	1
2	8.0	4
3	16.0	8

Shale		
Sample	Mass (g)	Volume (cm³)
1	3.0	1
2	12.0	4
3	18.0	6

56 What is the average density of the three samples of gypsum?

(1) 0.5 g/cm³
(2) 2.0 g/cm³
(3) 3.0 g/cm³
(4) 8.7 g/cm³

56_____

578

57 Since gypsum floats in the slurry and shale sinks in the slurry, which sequence shows the substances arranged in order from the least dense to the most dense?

1 shale, gypsum, slurry
2 gypsum, slurry, shale
3 shale, slurry, gypsum
4 slurry, gypsum, shale   57_____

58 In addition to the difference in density, how else are shale and gypsum different?

1 Shale is an evaporite and gypsum is composed of land-derived sediments.
2 Shale is a mineral and gypsum is a rock.
3 Shale is nonsedimentary and gypsum is sedimentary.
4 Shale is clastic and gypsum is nonclastic.   58_____

59 Which graph best represents the mass and volume of the gypsum and shale samples?

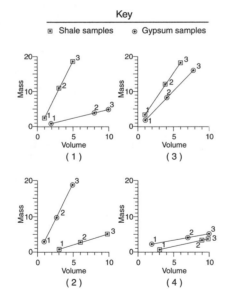

59_____

60 A student weighs sample 3 of the shale and incorrectly determines the mass to be 16 grams. Which equation should be used to calculate the student's percentage of error (percent deviation)?

1 percentage of error = $\frac{18 - 16}{18} \times 100$

2 percentage of error = $\frac{18 - 16}{16} \times 100$

3 percentage of error = $\frac{18}{16} \times 100$

4 percentage of error = $\frac{16}{18} \times 100$ 60_____

GROUP 2

If you choose this group, be sure to answer questions **61–65**.

Base your answers to questions 61 through 65 on the *Earth Science Reference Tables,* the map below, and your knowledge of Earth science. The map shows a portion of the eastern United States with New York State shaded. The isolines on the map indicate the average yearly total snowfall, in inches, recorded over a 20-year period. Points *A* through *D* are locations on Earth's surface. Latitude and longitude coordinates are shown along the border of the map.

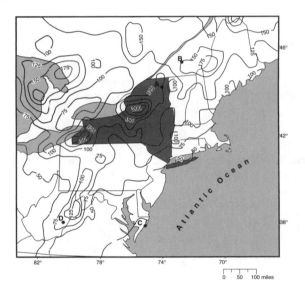

61 The latitude and longitude coordinates indicate that this map covers an area that is located

1 south of the Equator and west of the Prime Meridian
2 south of the Equator and east of the Prime Meridian
3 north of the Equator and west of the Prime Meridian
4 north of the Equator and east of the Prime Meridian   61_____

62 The greatest portion of which New York State landscape region averages more than 175 inches of snowfall each year?

1 Atlantic Coastal Plain
2 Adirondack Mountains
3 Allegheny Plateau
4 Tug Hill Plateau

62_____

63 The diagram below shows the location of five cities in New York State.

Which graph best represents the total average annual snowfall for each of the five cities?

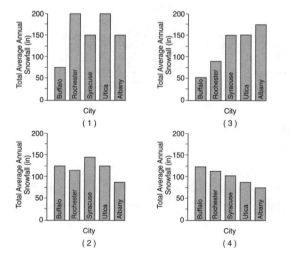

63_____

64 Location *C* has a lower average yearly snowfall than location *D* primarily because location *C* has a

1 coastal location
2 higher longitude
3 higher elevation
4 different prevailing wind direction   64_____

65 What is the approximate average yearly total snowfall gradient between locations *A* and *B*?

(1) 0.25 in/mi   (3) 0.40 in/mi
(2) 2.50 in/mi   (4) 4.00 in/mi   65_____

GROUP 3

If you choose this group, be sure to answer questions **66–70**.

Base your answers to questions 66 through 70 on the *Earth Science Reference Tables*, the weather map below, and your knowledge of Earth science. The map shows weather systems over the central and eastern United States and weather data for several cities.

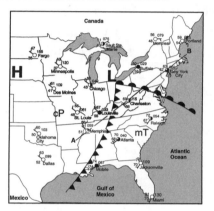

66 Which city has the greatest wind velocity?

   1 Louisville, Kentucky   3 Dallas, Texas

   2 Chicago, Illinois   4 Atlanta, Georgia   66_____

67 Which city has the highest relative humidity?

   1 Mobile, Alabama

   2 Minneapolis, Minnesota

   3 Miami, Florida

   4 Memphis, Tennessee   67_____

68 If the low-pressure center shown on the map follows a typical storm track, the system will move toward the

   1 northeast   3 southeast

   2 northwest   4 southwest   68_____

69 Which map correctly shows the movement of surface air associated with the high-pressure and low-pressure systems?

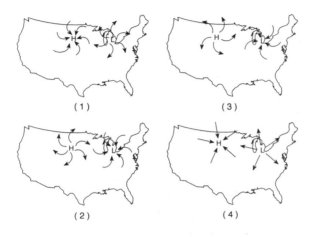

(1)   (3)

(2)   (4)   69_____

70 Which cross-sectional diagram of the lower atmosphere along line AB best represents the fronts and the movement of air masses?

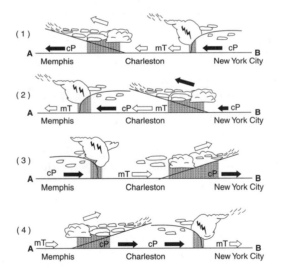

70_____

GROUP 4

If you choose this group, be sure to answer questions **71–75**.

Base your answers to questions 71 through 75 on the *Earth Science Reference Tables,* the diagrams below, and your knowledge of Earth science. The diagrams show two views of the same river flowing from a lake to an ocean bay.

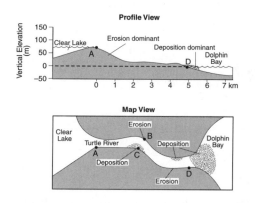

71 The deposition in Dolphin Bay near location *D* is caused mainly by

  1 the increased wave action within Dolphin Bay

  2 the presence of a meander in Turtle River just above Dolphin Bay

  3 the difference in salt content in the waters of Clear Lake and Dolphin Bay

  4 a decrease in the current velocity of Turtle River as it enters Dolphin Bay   71_____

72 How do the size and the density of the particles deposited in Dolphin Bay generally change as distance from the shoreline increases?

  1 Both size and density decrease.

  2 Both size and density increase.

  3 Size decreases and density increases.

  4 Size increases and density decreases.   72_____

73 Which diagram best represents the profile of the river bottom between points C and B?

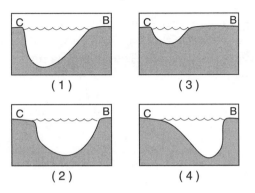

74 Which change will occur if the volume (discharge) of Turtle River increases?

1 Downcutting will decrease and riverbank erosion will increase.
2 Downcutting will increase and riverbank erosion will decrease.
3 Both downcutting and riverbank erosion will decrease.
4 Both downcutting and riverbank erosion will increase.

74_____

75 If no further uplift occurs within the mapped area over the next 200 years, the average gradient of the riverbed between points A and D will

1 decrease, only
2 increase, only
3 decrease, then increase
4 remain the same

75_____

GROUP 5

If you choose this group, be sure to answer questions **76–80**.

Base your answers to questions 76 through 80 on the *Earth Science Reference Tables,* the diagram below, and your knowledge of Earth science. The diagram represents the orbits of four planets and Halley's Comet. The period of revolution is shown for the comet. The orbital positions of Halley's Comet are shown for the years 1910, 1948, and 1986.

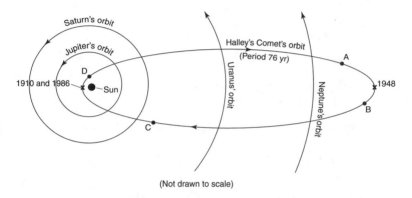

(Not drawn to scale)

76 The solar system model shown in the diagram is best classified as

1 geocentric   3 topographic

2 heliocentric   4 lunar   76_____

77 At which of these points in its orbit does Halley's Comet have the greatest orbital velocity?

(1) *A*   (3) *C*

(2) *B*   (4) *D*   77_____

78 The period of time each of these planets takes to orbit the Sun depends on its average

  1 distance from the Sun
  2 equatorial diameter
  3 density
  4 rate of rotation   78_____

79 Halley's Comet should be nearest the Sun again during the year

  (1) 2024   (3) 2062
  (2) 2040   (4) 2100   79_____

80 Which graph best represents the gravitational attraction between the Sun and Halley's Comet from 1910 to 1986

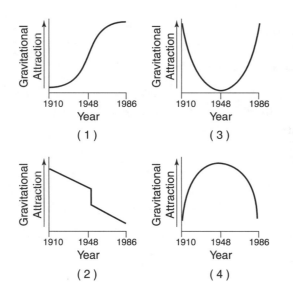

                                       80_____

GROUP 6

If you choose this group, be sure to answer questions **81–85**.

Base your answers to questions 81 through 85 on the *Earth Science Reference Tables,* the table below, and your knowledge of Earth science. The table is a summary of soil water budget information for four United States cities, represented by letters *W, X, Y,* and *Z*. Symbols in the table indicate water budget activity during each month.

City	Jan.	Feb.	Mar.	Apr.	May	June	July	Aug.	Sept.	Oct.	Nov.	Dec.
W	R	R	U	U/D	D	D	D	D	D	D	D	R
X	S	S	S	S	S	U	U/D	D	D	R	R	R/S
Y	S	S	S	S	U	U	U	U/D	D	R	R	R/S
Z	S	S	S	S	U	U	R	R/S	S	S	S	S

KEY:
U = water usage
U/D = usage ends and deficit begins
D = water deficit
R = water recharge
R/S = recharge ends and surplus begins
S = water surplus

81 For which city did ground-water storage become full in August?

(1) W   (3) Y

(2) X   (4) Z   81_____

82 The water budgets of cities X and Y have

    1 surplus, but no deficit
    2 deficit, but no surplus
    3 surplus and deficit, but no recharge and usage
    4 usage, deficit, recharge, and surplus   82_____

83 Which conditions caused the moisture deficits in these water budgets?

    1 Precipitation was more than potential evapotranspiration, and ground storage was empty.
    2 Precipitation was more than potential evapotranspiration, and ground storage was full.
    3 Precipitation was less than potential evapotranspiration, and ground storage was empty.
    4 Precipitation was less than potential evapotranspiration, and ground storage was full.   83_____

84 What is the most likely moisture source for these soil water budgets?

    1 nearby lakes   3 ground water
    2 precipitation   4 runoff   84_____

85 The four cities have their highest potential evapotranspiration values in July. Which statement best explains this observation?

    1 The lowest temperatures occur in July.
    2 The highest temperatures occur in July.
    3 The ground-water storage capacity is lowest in July.
    4 The ground-water storage capacity is highest in July.   85_____

GROUP 7

If you choose this group, be sure to answer questions **86–90**.

Base your answers to questions 86 through 90 on the *Earth Science Reference Tables,* the map below, and your knowledge of Earth science. The map shows the island of Hawaii (approximately 20° N 157° W) and the locations of recent volcanic eruptions and earthquake epicenters. The volcanic eruptions are inferred to be caused by the movement of the lithospheric plate over a hot spot in the mantle below.

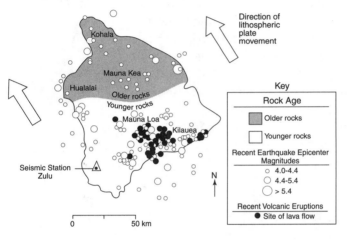

86 Where are earthquakes on and around Hawaii located?

1 along shorelines, only
2 in the ocean, only
3 in older rocks, only
4 scattered across the area   86_____

87 Rock collected from the side of Kilauea Volcano has the following mineral composition: 5% plagioclase feldspar, 68% pyroxene, 25% olivine, and 2% hornblende. What type of rock is this?

1 andesite   3 rhyolite

2 scoria   4 peridotite   87_____

88 Which inference concerning the distribution of earthquakes and recent volcanic eruptions on this island is most accurate?

1 Earthquakes and recent volcanic eruptions are located mainly along the boundary between the older rocks and the younger rocks.

2 Recent volcanic eruptions are more common than earthquakes.

3 Many earthquakes occur near recent volcanic eruptions.

4 Earthquakes and recent volcanic eruptions rarely occur in the same areas.   88_____

89 A scientist wants to film underwater volcanic activity that is forming a new island in the Hawaiian Island chain. In which direction from Hawaii should she concentrate her efforts?

1 northeast   3 southeast

2 northwest   4 southwest   89_____

90 Hawaii is located near the middle of which tectonic plate?

1 Philippine plate

2 Nazca plate

3 North American plate

4 Pacific plate   90_____

GROUP 8

If you choose this group, be sure to answer questions **91–95**.

Base your answers to questions 91 through 95 on the *Earth Science Reference Tables,* the diagram below, and your knowledge of Earth science. The diagram is a geologic cross section of the rocks that underlie parts of southern New York State and northern New Jersey. Landscape regions are indicated across the top of the diagram.

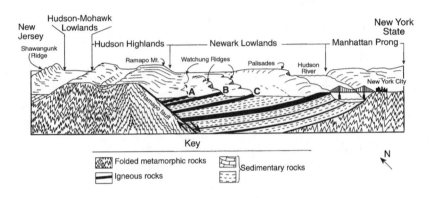

91 The portion of New York City shown in the geologic cross section is located in which landscape region?

    1 Hudson Highlands

    2 Manhattan Prong

    3 Newark Lowlands

    4 Atlantic Coastal Plain   91_____

92 Which inference explains the existence of the ridges and valleys of the Newark Lowlands?

    1 The igneous rocks are more resistant to weathering and erosion than the sedimentary rocks are.

    2 The sedimentary rocks are more resistant to weathering and erosion than the igneous rocks are.

    3 Both the sedimentary and the igneous rocks are highly resistant to weathering and erosion.

    4 Neither the sedimentary nor the igneous rocks are resistant to weathering and erosion.   92_____

93 The boundaries between the landscape regions represented in the cross section are marked by

    1 a fault, only

    2 a change in rock type, only

    3 a fault, or a change in rock type and rock structure, only

    4 a fault, or a change in rock type, rock structure, and surface elevation   93_____

94 Which important New York State fossil is most likely to be found in the Triassic Age rocks in the Newark Lowlands?

    1 eurypterid   3 *Coelophysis*

    2 mastodont   4 Naples tree   94_____

95 A more detailed view of the Ramapo fault region is shown below.

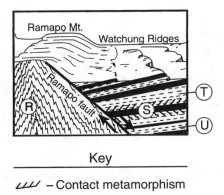

The geologic evidence shown in the diagram indicates that

1 rock *S* is older than rock *T*
2 rock *T* is older than rock *U*
3 rock *S* is older than rock *R*
4 rock *R* is older than the Ramapo fault

95____

GROUP 9

If you choose this group, be sure to answer questions **96–100**.

Base your answers to questions 96 through 100 on the *Earth Science Reference Tables*, the diagram below, and your knowledge of Earth science. The diagram represents a scheme for classifying rocks. The letters *A, B, C* and *X, Y, Z* represent missing labels.

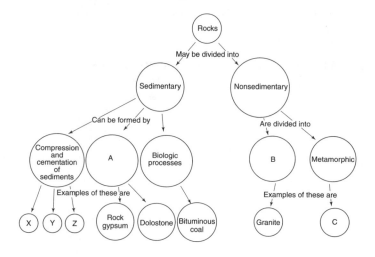

96 Which processes would form the type of rock that is represented by circle *B*?

1 deposition and compaction
2 weathering and erosion
3 melting and solidification
4 faulting and folding   96_____

97 If the rock in circle *C* formed from limestone, it would be called

1 schist   3 marble
2 anthracite coal   4 quartzite   97_____

98 Dolostone and granite are similar because both are

1 monomineralic   3 foliated

2 clastic   4 crystalline   98_____

99 Which rocks could be represented by circles X, Y, and Z?

1 sandstone, conglomerate, and siltstone

2 bituminous coal, slate, and schist

3 anthracite coal, metaconglomerate, and rock salt

4 breccia, gneiss, and rhyolite   99_____

100 The classification of rocks into sedimentary or nonsedimentary groups is based primarily on the rocks'

1 origin   3 color

2 density   4 age   100_____

GROUP 10

If you choose this group, be sure to answer questions **101–105**.

Base your answers to questions 101 through 105 on the *Earth Science Reference Tables* and on your knowledge of Earth science.

101 The diagram below represents the construction of a model of an elliptical orbit of a planet traveling around a star. The focal point and the center of the star represent the foci of the orbit.

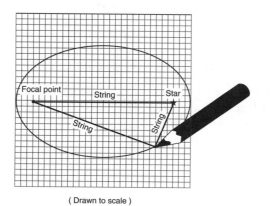

(Drawn to scale)

The eccentricity of this orbit is approximately

(1) 1.3   (3) 0.5
(2) 0.8   (4) 0.3   101_____

102 A parcel of air has a dry-bulb temperature of 18°C and a wet-bulb temperature of 10°C. What are the dewpoint and the relative humidity?

(1) 5°C and 19%   (3) 2°C and 33%
(2) –5°C and 19%   (4) –2°C and 33%   102_____

103 Equal masses of granite, iron, basalt, and liquid water are at room temperature. Each substance then absorbs the same amount of additional heat energy. Assuming no phase change, which substance would change *least* in temperature?

1 granite   3 basalt
2 iron   4 liquid water   103_____

104 A station model is shown below.

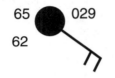

65   029
62

What is the air pressure at this location?

(1) 902.9 mb   (3) 1029.0 mb
(2) 1002.9 mb   (4) 9029.0 mb   104_____

105 The isoline map below shows the variations in relative strength of Earth's magnetic field from 1 (strong) to 11 (weak).

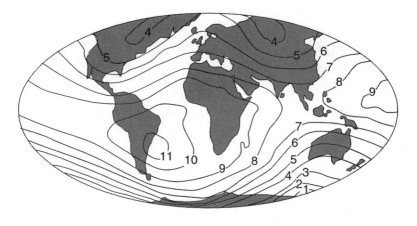

Which of Earth's tectonic plates has the *weakest* magnetic field strength?

1 South American plate   3 North American plate
2 African plate   4 Pacific plate   105_____

Examination
June 1999
Earth Science —
Program Modification Edition

PART I

Answer all 40 questions in this part.

Directions (1–40): For *each* statement or question, select the word or expression that, of those given, best completes the statement or answers the question. Record your answers in the spaces provided. Some questions may require the use of the *Earth Science Reference Tables*. [40]

1 The diagram below is a three-dimensional model of a landscape region.

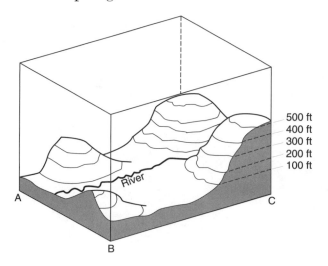

Which map view best represents the topography of this region?

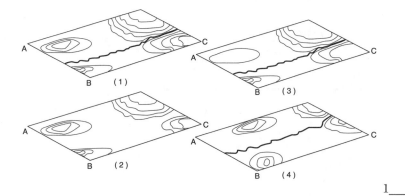

(1) (3) (2) (4)

1_____

2 The map below shows the location and diameter, in kilometers, of four meteorite impact craters, *A, B, C,* and *D,* found in the United States.

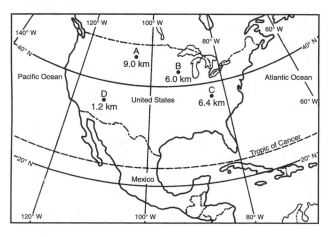

What is the approximate latitude and longitude of the largest crater?

(1) 35° N 111° W   (3) 44° N 90° W

(2) 39° N 83° W   (4) 47° N 104° W   2_____

3 Which diagram best shows the altitude of Polaris observed near Buffalo, New York?

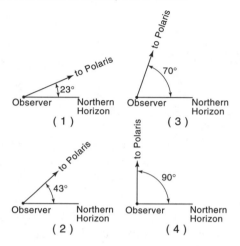

3____

4 Although more than 2,000 minerals have been identified, 90% of Earth's lithosphere is composed of the 12 minerals listed below.

Rock-Forming Minerals	
feldspar	augite
quartz	garnet
mica	magnetite
calcite	olivine
hornblende	pyrite
kaolinite	talc

The best explanation for this fact is that most rocks

1 are monomineralic
2 are composed only of recrystallized minerals
3 have a number of minerals in common
4 have a 10% nonmineral composition

4____

Base your answers to questions 5 through 8 on the diagram below, which represents a scheme for classifying rocks. The letters *A, B, C* and *X, Y, Z* represent missing labels.

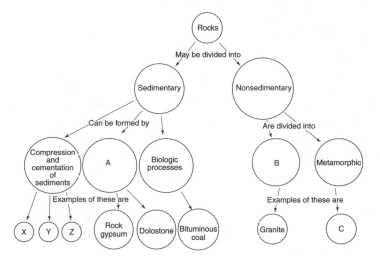

5 The classification of rocks into sedimentary or nonsedimentary groups is based primarily on the rocks'

1 origin   3 color
2 density   4 age   5_____

6 Which processes would form the type of rock that is represented by circle *B*?

1 deposition and compaction
2 weathering and erosion
3 melting and solidification
4 faulting and folding   6_____

7 If the rock in circle *C* formed from limestone, it would be called

1 schist   3 marble
2 anthracite coal   4 slate   7_____

8 Which rocks could be represented by circles *X*, *Y*, and *Z*?

   1 shale, slate, and schist
   2 sandstone, shale, and siltstone
   3 anthracite coal, metaconglomerate, and rock salt
   4 breccia, gneiss, and rhyolite   8_____

---

9 What is the largest particle that can generally be transported by a stream that is moving at 200 centimeters per second?

   1 sand   3 cobble
   2 pebble   4 boulder   9_____

10 Which agent of erosion formed the long U-shaped valleys now occupied by the Finger Lakes in central New York State?

   1 running water   3 wind
   2 ocean currents   4 glacial ice   10_____

11 The diagram below shows two landscapes, *A* and *B*.

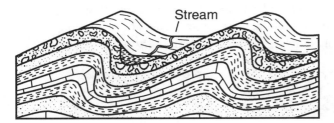

A   B

The difference in appearance of these two landscapes was caused mainly by a difference in the

1 climate   3 rock type
2 amount of uplift   4 rock structure   11_____

12 The diagram below shows a cross section of a portion of Earth's crust.

Stream

The hills of this area were formed primarily by

1 bedrock folding   3 stream erosion
2 bedrock faulting   4 volcanic activity   12_____

13 Which type of climate has the greatest amount of rock weathering caused by frost action?

    1 a wet climate in which temperatures remain below freezing

    2 a wet climate in which temperatures alternate from below freezing to above freezing

    3 a dry climate in which temperatures remain below freezing

    4 a dry climate in which temperatures alternate from below freezing to above freezing   13_____

14 The *Generalized Bedrock Geology Map of New York State* provides evidence that water flows from Lake Erie into Lake Ontario by showing that Lake Ontario

    1 is north of Lake Erie

    2 is deeper than Lake Erie

    3 has a larger surface area than Lake Erie

    4 has a lower surface elevation than Lake Erie   14_____

15 Quartz particles of varying sizes are dropped at the same time into deep, calm water. Which cross section best represents the settling pattern of these particles?

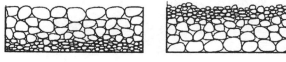
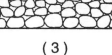

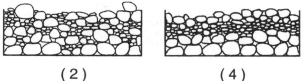

      (1)   (3)

      (2)   (4)   15_____

16 The diagrams below show the stages, *A* through *D*, in the formation of an oxbow lake over a period of time. [The arrows indicate the direction of streamflow.]

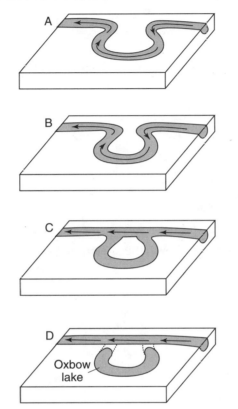

Oxbow lakes are generally formed by

1 erosion, resulting in a sudden increase in the stream's gradient
2 deposition, resulting in a sudden increase in the stream's gradient
3 erosion along the outside banks of the curve in a meandering stream
4 deposition along the outside banks of the curve in a meandering stream

16_____

17 Trilobite fossils were recently discovered in Himalayan Mountain bedrock. During which geologic period could these organisms have lived?

1 Tertiary   3 Triassic

2 Cretaceous   4 Cambrian   17_____

18 A graph of the radioactive decay of carbon-14 is shown below.

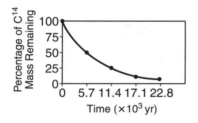

Which graph correctly shows the accumulation of nitrogen-14, the decay product of carbon-14, over the same period?

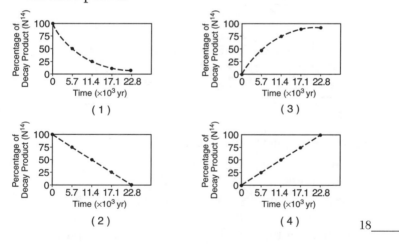

18_____

19 What is the geologic age of the surface bedrock of most of the Allegheny Plateau landscape region in New York State?

1 Cambrian   3 Silurian

2 Devonian   4 Ordovician   19_____

20 The dots on the map below show the present locations of living coral reefs. Site *X* indicates an area of fossil coral reefs preserved in rocks formed during the Jurassic Period.

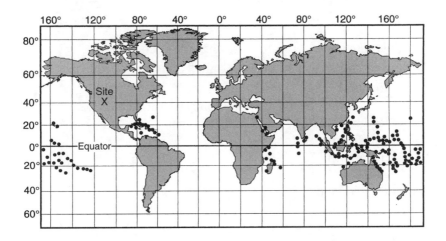

Which inference is best supported by this map?

1 The climate at site *X* during the Jurassic Period was colder than the present climate at site *X*.

2 Site *X* was covered by warm ocean water during the Jurassic Period.

3 Site *X* has drifted southward since the Jurassic Period.

4 The coral at site *X* evolved from ocean-dwelling animals into land-dwelling animals after the Jurassic Period.   20_____

21 The diagrams below show the sequence of events that formed sedimentary rock layers *A*, *B*, *C*, and *D*.

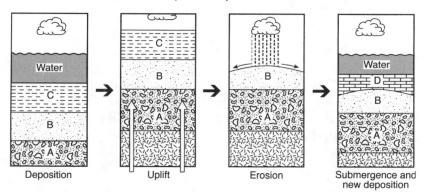

Deposition   Uplift   Erosion   Submergence and new deposition

This sequence of events best illustrates the

1 formation of a buried erosional surface (unconformity)
2 movement of rock layers along a fault between layers *B* and *D*
3 overturning of rock layers
4 metamorphism of sandstone (layer *B*) into quartzite   21____

22 The table below shows characteristics of three landscape regions, *X*, *Y*, and *Z*.

Landscape Region	Relief	Bedrock
X	Great relief, high peaks, deep valleys	Many types, including igneous and metamorphic rocks, nonhorizontal structure
Y	Moderate to high relief	Flat layers of sedimentary rock or lava flows
Z	Very little relief, low elevations	Many types and structures

Which terms, when substituted for X, Y, and Z, best complete the table?

(1) X = mountains, Y = plains, Z = plateaus
(2) X = plateaus, Y = mountains, Z = plains
(3) X = plains, Y = plateaus, Z = mountains
(4) X = mountains, Y = plateaus, Z = plains   22_____

Base your answers to questions 23 through 25 on the weather map below, which shows weather systems over the central and eastern United States and weather data for several cities.

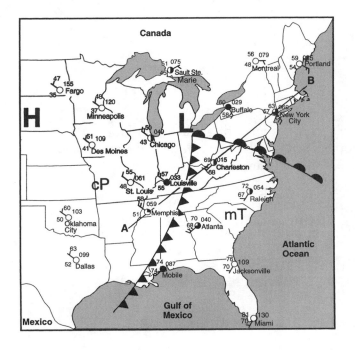

23 The cP air mass shown on the map most likely developed over

1 central Canada   3 the Gulf of Mexico
2 central Mexico   4 the North Atlantic   23_____

24 Which map correctly shows the movement of surface air associated with the high-pressure and low-pressure systems?

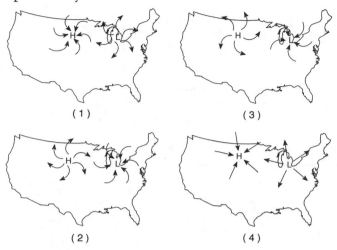

(1) (3)

(2) (4)

24 _____

25 Which cross-sectional diagram of the lower atmosphere along line *AB* best represents the movement of the fronts and air masses?

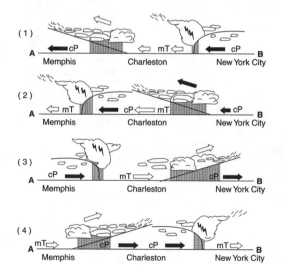

25 _____

Base your answers to questions 26 through 28 on the map below. The map shows a portion of the eastern United States with New York State shaded. The isolines on the map indicate the average yearly total snowfall, in inches, recorded over a 20-year period. Points *A*, *B*, and *C* are locations on Earth's surface. Latitude and longitude coordinates are shown along the border of the map.

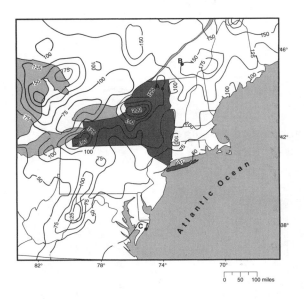

26 Location *C* has a lower average yearly snowfall than location *A* primarily because location *C* has a

   1 lower latitude
   2 higher longitude
   3 higher elevation
   4 different prevailing wind direction   26_____

27 What is the approximate average yearly total snowfall gradient between locations *A* and *B*?

   (1) 0.25 in/mi   (3) 0.40 in/mi
   (2) 2.50 in/mi   (4) 4.00 in/mi   27_____

28 The diagram below shows the location of five cities in New York State.

Which graph best represents the total average annual snowfall for each of the five cities?

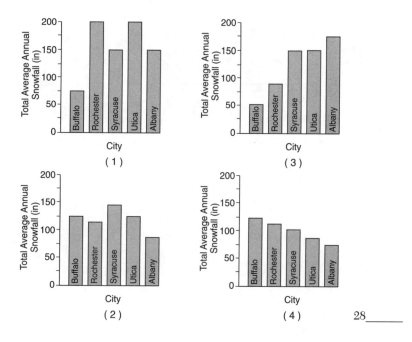

28_____

29 The shaded portion of the map below indicates areas of night and the unshaded portion indicates areas of daylight.

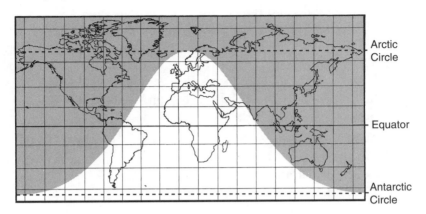

What day of the year is best represented by the map?

1 March 21   3 September 21

2 June 21   4 December 21   29_____

30 The graph below shows the air pressure recorded at the same time at several locations between Niagara Falls and Albany.

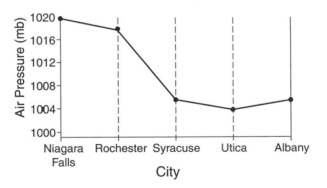

Based on the information in this graph, the wind velocity is probably greatest between which two cities?

1 Niagara Falls and Rochester
2 Rochester and Syracuse
3 Syracuse and Utica
4 Utica and Albany   30_____

31 Which current is a cool ocean current that flows completely around Earth?

1 West Wind Drift
2 Gulf Stream
3 North Equatorial Current
4 California Current   31_____

32 During how many days of a calendar year is the Sun directly overhead at noon in New York State?

(1) only 1 day   (3) 365 days
(2) only 2 days   (4) 0 days   32_____

33 Ozone is important to life on Earth because ozone

1 cools refrigerators and air-conditioners
2 absorbs energy that is reradiated by Earth
3 absorbs harmful ultraviolet radiation
4 destroys excess atmospheric carbon dioxide   33_____

Base your answers to questions 34 and 35 on the diagram below, which shows the Moon in four different positions, *A*, *B*, *C*, and *D*, as it orbits Earth.

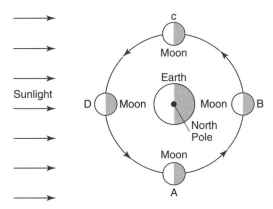

34 How does the Moon appear to an observer in New York State when the Moon is located at position *A*?

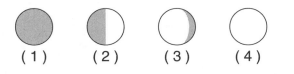

(1)   (2)   (3)   (4)   34_____

35 The cartoon below shows a comical view of an eclipse as viewed from Earth.

B.C.

The type of eclipse represented in the cartoon might occur when the Moon is located at position

(1) *A*   (3) *C*
(2) *B*   (4) *D*   35_____

36 The diagram below represents the construction of a model of an elliptical orbit of a planet traveling around a star. The focal point and the center of the star represent the foci of the orbit.

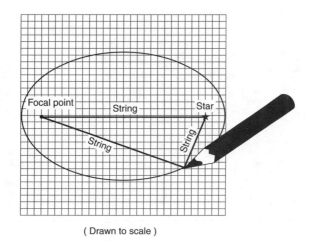

(Drawn to scale)

The eccentricity of this orbit is approximately

(1) 1.3   (3) 0.5

(2) 0.8   (4) 0.3   36_____

620

Base your answers to questions 37 and 38 on the diagram below. The diagram shows twelve constellations that are visible in the night sky to an observer in New York State, over the course of a year. Different positions of Earth are represented by letters *A* through *D*. The arrows represent the direction of Earth's motion around the Sun.

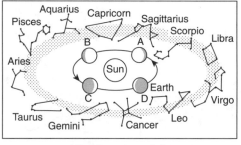

(Not drawn to scale)

37 Which constellations are both visible at midnight to an observer in New York State when Earth is located at position *D*?

  1 Aries and Taurus
  2 Pisces and Libra
  3 Leo and Virgo
  4 Aquarius and Scorpio   37_____

38 The constellations observed from New York State when Earth is at position *A* are different from the constellations observed when Earth is at position *C* because

  1 Earth moves in its orbit
  2 Earth is tilted on its axis
  3 the lengths of day and night are different
  4 the stars move around Earth as shown by star trails   38_____

39 The diagram below represents Earth.

Which diagram best represents Mars, drawn to the same scale?

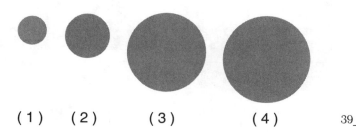

(1)   (2)   (3)   (4)   39_____

40 Which graph shows the most probable effect of environmental pollution on the chances of human survival?

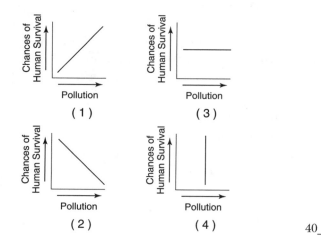

40_____

PART II

This part consists of six groups, each containing five questions. Choose any *two* of these six groups. Be sure that you answer all five questions in each of the two groups chosen. Record the answers to these questions in the spaces provided. Some questions may require the use of the *Earth Science Reference Tables.* [10]

GROUP A—Rocks and Minerals
If you choose this group, be sure to answer questions 41–45.

Base your answers to questions 41 and 42 on the rock sample shown below.

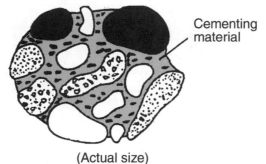

Cementing material

(Actual size)

41 The rounded pebbles of this rock have been cemented together to form

1 granite, an igneous rock
2 conglomerate, a sedimentary rock
3 siltstone, a sedimentary rock
4 gneiss, a metamorphic rock   41_____

42 The average size of the pebbles in the sample is approximately

(1) 1.2 cm   (3) 6.4 cm
(2) 0.2 cm   (4) 13.2 cm   42_____

623

43 Different arrangements of tetrahedra in the silicate group of minerals result in differences in the minerals'

    1 age, density, and smoothness
    2 cleavage, color, and abundance
    3 hardness, cleavage, and crystal shape
    4 chemical composition, size, and origin   43_____

44 Some Moon rock samples have coarse intergrown crystals composed of plagioclase feldspar, hornblende, and olivine. These Moon rock samples are most similar to Earth rock samples of

    1 gabbro   3 breccia
    2 marble   4 pumice   44_____

45 Which common mineral fizzes when dilute hydrochloric acid (HCl) is placed on it?

    1 calcite   3 quartz
    2 feldspar   4 talc   45_____

GROUP B—**Plate Tectonics**

*If you choose this group, be sure to answer questions **46–50**.*

Base your answers to questions 46 and 47 on the map below. The map shows epicenters of some of the earthquakes that occurred in North America during a 2-week period. Five epicenters are labeled *A* through *E*. Denver and New York City are also indicated.

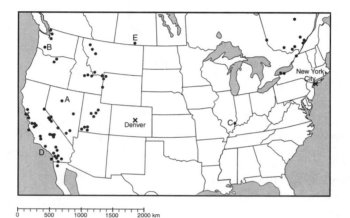

46 A seismograph station at Denver recorded the arrival of *P*-waves at 8:00 a.m. and the arrival of *S*-waves at 8:02 a.m. Which epicenter is located above the source of this earthquake?

(1) *A*   (3) *C*

(2) *B*   (4) *D*   46_____

47 The distance from epicenter *E* to New York City is 3,000 kilometers. What was the approximate travel time for the *P*-waves from this epicenter to New York City?

(1) 1 min 20 sec   (3) 7 min 30 sec

(2) 5 min 40 sec   (4) 10 min 00 sec   47_____

48 The cutaway diagram below shows the paths of earthquake waves generated at point X.

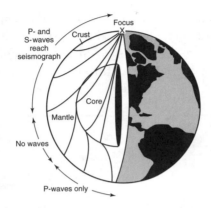

Only *P*-waves reach the side of Earth that is opposite the focus because *P*-waves

1 are stronger than *S*-waves
2 travel faster than *S*-waves
3 bend more than *S*-waves
4 can travel through liquids and *S*-waves cannot   48____

49 Magnetic readings taken across mid-ocean ridges provide evidence that

1 the seafloor is spreading
2 the ocean basins are older than the continents
3 the mid-ocean ridges are higher than the nearby plains
4 Earth's rate of rotation has changed   49____

50 Compared to Earth's oceanic crust, Earth's continental crust is

1 thinner and composed of granite
2 thinner and composed of basalt
3 thicker and composed of granite
4 thicker and composed of basalt   50____

GROUP C—**Oceanography**

If you choose this group, be sure to answer questions **51–55**.

51 The cross section below shows the ocean floor between two continents. Points *A* through *D* represent locations on the ocean floor where samples of oceanic crust were collected.

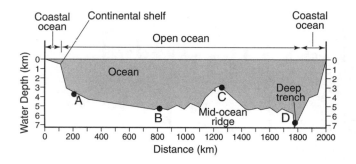

The youngest rock sample most likely was collected from location

(1) *A*   (3) *C*

(2) *B*   (4) *D*   51_____

52 Waste produced by people in New York State has been dumped into the Atlantic Ocean, where it is distributed by surface ocean currents. Which coastal area is most likely to become polluted by this waste?

1 western coast of Europe

2 southern coast of South America

3 western coast of Mexico

4 eastern coast of Africa   52_____

53 Tsunamis are caused by

    1 Earth's rotation   3 hurricane winds

    2 dynamic equilibrium   4 earthquakes   53_____

Base your answers to questions 54 and 55 on the map below, which shows Rockaway Peninsula, part of Long Island's south shore.

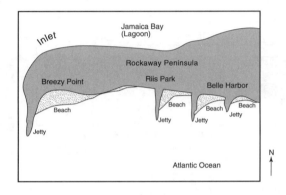

54 Students compared recent photographs of the beaches with photographs taken three years ago and discovered that parts of the shoreline have changed. Which characteristic of the shoreline probably has changed most?

    1 composition of the beach sand

    2 size of the beaches

    3 positions of the jetties

    4 length of the peninsula   54_____

55 Toward which direction is sand being transported along the shoreline within the zone of breaking waves?

    1 northeast   3 southeast

    2 south   4 west   55_____

GROUP D—Glacial Processes
*If you choose this group, be sure to answer questions **56–60**.*

Base your answers to questions 56 and 57 on the three maps below, which show the ice movement and changes at the ice front of an alpine glacier from the years 1874 to 1882. Points *A, B, C, D,* and *E* represent the positions of large markers placed on the glacial ice and left there for a period of eight years.

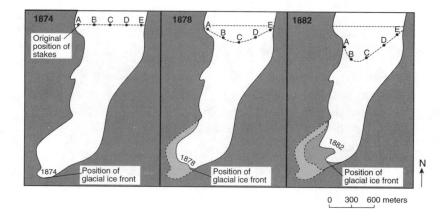

56 The changing positions of markers *A, B, C, D,* and *E* show that the glacial ice is

    1 slowly becoming thicker
    2 forming smaller crystals
    3 gradually shifting northward
    4 moving fastest near the middle   56 _____

57 Which statement best describes the changes happening to this glacier between 1874 and 1882?

  1 The ice front was advancing, and the ice within the glacier was advancing.

  2 The ice front was advancing, and the ice within the glacier was retreating.

  3 The ice front was retreating, and the ice within the glacier was advancing.

  4 The ice front was retreating, and the ice within the glacier was retreating.   57_____

---

58 As a result of glaciation, New York State has

  1 few lakes

  2 many V-shaped valleys

  3 many sand and gravel deposits

  4 thick soils formed "in place" from underlying bedrock   58_____

Base your answers to questions 59 and 60 on the chart below, which shows the changing climatic conditions that led to alternating glacial and interglacial periods.

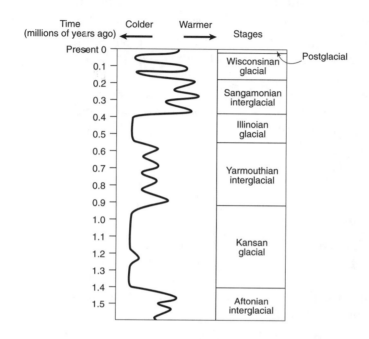

59 The interglacial stages were most likely caused by

1 a drop in worldwide sea levels
2 an increase in average worldwide temperature
3 crustal plate movement
4 a large increase in the amount of snowfall   59_____

60 The chart represents climatic conditions that occurred mostly during which geological time period?

1 Triassic Period   3 Quaternary Period
2 Ordovician Period   4 Cretaceous Period   60_____

GROUP E—**Atmospheric Energy**

If you choose this group, be sure to answer questions **61–65**.

61 At which latitudes do currents of dry, sinking air cause the dry conditions of Earth's major deserts?

(1) 0° and 30° N   (3) 30° N and 30° S

(2) 60° N and 60° S   (4) 60° S and 90° S   61_____

62 At what approximate altitude would clouds begin to form when the surface air temperature is 30°C and the dewpoint is 14°C?

(1) 1.1 km   (3) 2.5 km

(2) 2.0 km   (4) 3.7 km   62_____

63 A large amount of latent heat is absorbed by water during

1 evaporation   3 condensation

2 freezing   4 precipitation   63_____

64 When would the water in a New York State pond evaporate fastest?

1 in January, when the pond is frozen

2 in March, when the pond ice is melting

3 in May on a calm, sunny day

4 in July on a hot, windy day   64_____

Note that question 65 has only three choices.

65 Compared to the accuracy of the 24-hour weather forecasts of the 1930's, the 24-hour weather forecasts of the 1990's are usually

1 less accurate

2 more accurate

3 equally accurate   65_____

GROUP F—**Astronomy**

If you choose this group, be sure to answer questions **66–70**.

66 Which diagram represents a geocentric model?
[Key: *E* = Earth, *P* = Planet, *S* = Sun]

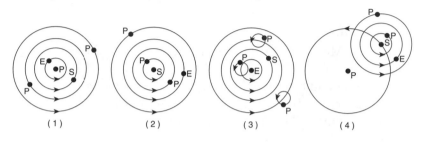

(1)　　　　　　(2)　　　　　　(3)　　　　　　(4)

66_____

67 According to the big bang theory, the universe began as an explosion and is still expanding. This theory is supported by observations that the stellar spectra of distant galaxies show a

1 concentration in the yellow portion of the spectrum
2 concentration in the green portion of the spectrum
3 shift toward the blue end of the spectrum
4 shift toward the red end of the spectrum

67_____

68 Major ocean and air currents appear to curve to the right in the Northern Hemisphere because

1 Earth has seasons
2 Earth's axis is tilted
3 Earth rotates on its axis
4 Earth revolves around the Sun

68_____

69 Why are impact structures more obvious on the Moon than on Earth?

   1 The Moon's gravity is stronger than Earth's gravity.
   2 The Moon has little or no atmosphere.
   3 The rocks on the Moon are weaker than those on Earth.
   4 The Moon rotates at a slower rate than Earth does.   69_____

70 If the amount of greenhouse gases were to increase for a very long time, the atmosphere on Earth would become most like the atmosphere of

   1 Mercury   3 Jupiter
   2 Venus   4 Saturn   70_____

PART III

This part consists of questions 71 through 88. Be sure that you answer *all* questions in this part. Record your answers in the spaces provided. Some questions may require the use of the *Earth Science Reference Tables.* [25]

Base your answers to questions 71 and 72 on the data table and profile below. The data table gives the average annual precipitation for locations *A* and *B*. The profile represents a mountain in the western United States. Points *A* and *B* are locations on different sides of the mountain.

Data Table

Location	Average Annual Precipitation (cm)
A	120
B	35

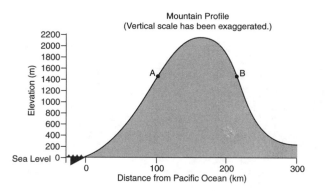

Mountain Profile
(Vertical scale has been exaggerated.)

71 State the elevation of location *A*. [1]

_____

_____

_____

72 State one probable reason for the difference in average annual precipitation between location *A* and location *B*. [1]

_____

_____

_____

_____

73 State one way in which a hurricane differs from a tornado. [1]

_____

_____

_____

Base your answers to questions 74 through 76 on the geologic cross section below. The cross section shows an outcrop in which the layers have not been overturned. Rock units are labeled *A* through *E*.

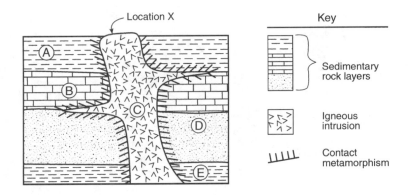

74 Using letters *A* through *E*, list the rock units in order from oldest to youngest. [2]

_____

75 State the name of the sediment that was compacted to form rock unit *A*.   [1]

_____

76 State one observation about the crystals at location *X* that would provide evidence that igneous rock unit *C* was formed by very slow cooling of magma.   [1]

_____

_____

_____

77 A parcel of air has a dry-bulb temperature of 18°C and a wet-bulb temperature of 10°C. State the relative humidity of this parcel of air.   [1]

_____

_____

_____

Base your answers to questions 78 through 80 on the weather map below. The map shows temperature readings at weather stations in the continental United States.

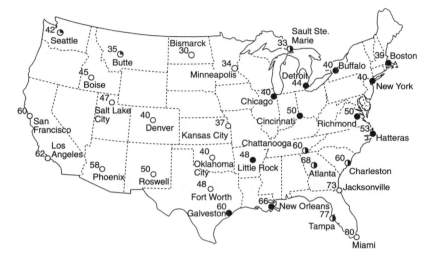

78 On the weather map provided, draw three isotherms: the 40°F isotherm, the 50°F isotherm, and the 60°F isotherm.   [2]

79 In Richmond, Virginia, the wind direction is from the east at a speed of 20 knots. On the station model provided, draw the correct symbols for wind direction and windspeed.   [2]

50

Richmond

80 In addition to temperature, one other weather variable for each weather station is shown on the map. State the other weather variable.   [1]

_____

_____

Base your answers to questions 81 and 82 on the information and map below.

The eruption of Mt. St. Helens in 1980 resulted in the movement of volcanic ash across the northwestern United States. The movement of the ash at 1.5 kilometers above sea level is shown as a shaded path on the map. The times marked on the path indicate the length of time the leading edge of the ash cloud took to travel from Mt. St. Helens to each location.

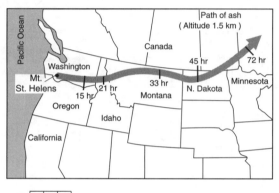

81 Calculate the average *rate* of movement of the volcanic ash for the first 15 hours, following the directions below.

a Write the equation used to determine the average rate of the volcanic ash movement. [1]

_____

b Substitute values into the equation. [1]

_____

c Solve the equation and label the answer with the correct units. [2]

_____

_____

82 The movement of the ash occurred at an altitude of 1.5 kilometers. State the name of the layer of Earth's atmosphere in which the ash cloud traveled. [1]

_____

_____

Base your answers to questions 83 through 85 on the cross section of a portion of Earth's interior below. The cross section shows the focal depth of some earthquakes that occurred west of the Tonga Trench. Data were collected along the 22° S parallel of latitude.

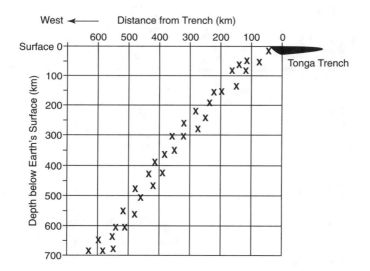

83 State the relationship between the depth of an earthquake's focus and the earthquake's distance from the Tonga Trench. [1]

_____

_____

_____

84 The Tonga Trench is the crustal surface boundary between two tectonic plates. State the names of the two plates.   [1]

_____

_____

85 The focal depth pattern shown on the cross section represents the location of the subsurface boundary between the two tectonic plates. Describe the relative motion of the plates along this boundary.   [1]

_____

_____

_____

Base your answers to questions 86 through 88 on the tables below. Table 1 shows the average distance from the Sun in astronomical units (AU) and the average orbital speed in kilometers per second (km/s) of the nine planets in our solar system. Table 2 lists five large asteroids and their average distances from the Sun.

Table 1

Planet	Average Distance from Sun (AU)	Average Orbital Speed (km/s)
Mercury	0.4	48.0
Venus	0.7	35.0
Earth	1.0	30.0
Mars	1.5	24.0
Jupiter	5.2	13.0
Saturn	9.6	10.0
Uranus	19.0	7.0
Neptune	30.0	5.1
Pluto	39.0	4.7

Table 2

Asteroid	Average Distance from Sun (AU)
Ceres	2.8
Pallas	2.8
Vesta	2.4
Hygiea	3.2
Juno	2.7

86 On the grid provided, plot the average distance from the Sun and the average orbital speed for each of the nine planets listed in table 1. Connect the nine points with a line. [2]

Planets' Average Orbital Speed vs. Average Distance from Sun

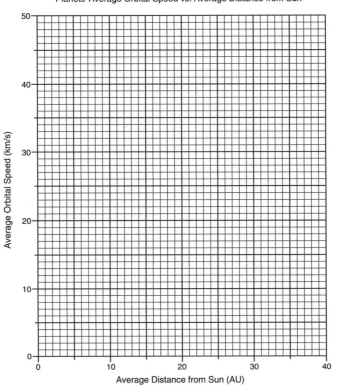

87 State the relationship between a planet's average distance from the Sun and the planet's average orbital speed. [1]

_____

_____

_____

88 The orbits of the asteroids listed in table 2 are located between two adjacent planetary orbits. State the names of the two planets. [1]

_____

_____

Answers to the Regents Examinations

JUNE 1999

Part I

1. 1	12. 2	23. 2	34. 4	45. 2
2. 4	13. 1	24. 1	35. 2	46. 4
3. 3	14. 3	25. 2	36. 1	47. 2
4. 3	15. 3	26. 4	37. 3	48. 3
5. 1	16. 4	27. 2	38. 4	49. 1
6. 4	17. 1	28. 3	39. 3	50. 2
7. 2	18. 2	29. 1	40. 2	51. 4
8. 3	19. 1	30. 1	41. 1	52. 3
9. 4	20. 3	31. 2	42. 1	53. 1
10. 2	21. 4	32. 3	43. 3	54. 4
11. 4	22. 4	33. 1	44. 4	55. 4

Part II

56. 2	66. 4	76. 2	86. 4	96. 3
57. 2	67. 1	77. 4	87. 2	97. 3
58. 4	68. 1	78. 1	88. 3	98. 4
59. 3	69. 2	79. 3	89. 3	99. 1
60. 1	70. 3	80. 3	90. 4	100. 1
61. 3	71. 4	81. 4	91. 2	101. 2
62. 4	72. 1	82. 4	92. 1	102. 3
63. 2	73. 4	83. 3	93. 4	103. 4
64. 1	74. 4	84. 2	94. 3	104. 2
65. 1	75. 1	85. 2	95. 4	105. 1

JUNE 1999 PROGRAM MODIFICATION EXAMINATION

Part I

1. 1	9. 3	17. 4	25. 3	33. 3
2. 4	10. 4	18. 3	26. 1	34. 2
3. 2	11. 1	19. 2	27. 1	35. 4
4. 3	12. 1	20. 2	28. 2	36. 2
5. 1	13. 2	21. 1	29. 4	37. 3
6. 3	14. 4	22. 4	30. 2	38. 1
7. 3	15. 3	23. 1	31. 1	39. 2
8. 2	16. 3	24. 2	32. 4	40. 2

Part II

Group A	Group B	Group C	Group D	Group E	Group F
41. 2	46. 1	51. 3	56. 4	61. 3	66. 3
42. 1	47. 2	52. 1	57. 3	62. 2	67. 4
43. 3	48. 4	53. 4	58. 3	63. 1	68. 3
44. 1	49. 1	54. 2	59. 2	64. 4	69. 2
45. 1	50. 3	55. 4	60. 3	65. 2	70. 2

Part III

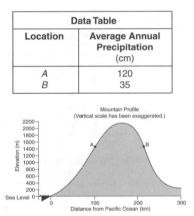

Data Table	
Location	Average Annual Precipitation (cm)
A	120
B	35

Mountain Profile
(Vertical scale has been exaggerated.)

71 To determine the elevation of location A, trace left from location A on the profile to the elevation scale. Location A has an elevation of 1,500 meters.

One credit is allowed for the answer **1,500 meters** (± 50 meters).

72 The question states that the mountain is located in the western United States, and the profile indicates that it lies along the Pacific Ocean. Find the Planetary Wind and Moisture Belts in the Troposphere diagram in the *Earth Science Reference Tables*. Note that the prevailing winds in this region are from the southwest. Thus, winds blow moisture-laden air from over the Pacific Ocean inland and up the windward side of the mountain. As the air rises along the mountainside past location A, it undergoes adiabatic expansion, cools to its dewpoint, and forms clouds; precipitation will occur. Then, as the air, having lost moisture, drops down on the leeward side of the mountain, past location B, it is warmed by compression; and evaporation, rather than condensation, is likely to occur. Thus, precipitation is less likely to occur at location B.

One credit is allowed for a scientifically correct answer.

Examples of acceptable answers:
Location A is on the windward side of the mountain.
At location B, air is warming by compression.

73 A hurricane is a large, cyclonic storm, often extending over hundreds of kilometers from edge to edge, in which the winds exceed 119 kilometers per hour. Hurricanes can last for weeks and travel over great distances. A tornado is a rotating column of air whirling at speeds up to 800 kilometers per hour, usually accompanied by a funnel-shaped downward extension of a cumulonimbus cloud. Tornadoes seldom last for more than a few hours and are rarely more than a mile wide at their base.

One credit is allowed for a scientifically correct answer.

Examples of acceptable answers:
Tornadoes exist for a shorter period of time.
A hurricane is a larger storm.

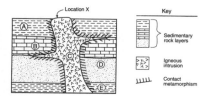

Location X — Key

Sedimentary rock layers

Igneous intrusion

Contact metamorphism

74 The cross section shows five rock units: four sedimentary rock layers, *A*, *B*, *D*, and *E*, and an igneous intrusion, *C*. Note that the igneous intrusion has caused contact metamorphism all along its boundary with all four sedimentary rock units, indicating that those rock units already existed when the igneous intrusion occurred. Thus, rock unit *C* is the youngest. The four sedimentary rock units appear flat and horizontal, so it is logical to infer that they have not been disturbed. The principle of superposition states that, in a series of sedimentary rock layers, the bottom layer is oldest, unless it was overturned or had older rock thrust upon it, because the bottom layer was deposited first. Thus, the ages of the sedimentary rock units, from oldest to youngest is *E*, *D*, *B*, *A*. Therefore, the correct order for all the rock units, from oldest to youngest, is

Oldest ☐ E ☐ D ☐ B ☐ A ☐ C ☐ Youngest

Two credits are allowed for the answer **E, D, B, A, C**.

One credit is allowed for correct placement of the igneous intrusion (**C**) or for the correct order of the sedimentary layers, regardless of the placement of the igneous intrusion.

75 Rock unit *A* is a layer of sedimentary rock. Find the Scheme for Sedimentary Rock Identification in the *Earth Science Reference Tables*. Note that the symbol used in the geologic cross section for rock unit *A* corresponds to the map symbol for shale. Trace to the left, and note that the grains in shale are identified as clay.

One credit is allowed for the answer **clay**.

76 Location *X* is at the surface. When molten rock cools rapidly, the crystals that form are small. When molten rock cools very slowly, the crystals that form can grow very large. If location *X* was at the surface when the igneous intrusion occurred, the rock there would have cooled rapidly and the crystals would be small. The observation of large crystals at location *X* would provide evidence that igneous rock unit *C* was formed by very slow cooling of magma.

One credit is allowed for a scientifically correct answer.

  Examples of acceptable answers:
   large crystals
   coarse texture

77 Find the Relative Humidity chart in the *Earth Science Reference Tables.* Locate, in the column labeled "Dry-Bulb Temperature (°C)," the row cor-

responding to a dry-bulb temperature of 18°C. Since the dry-bulb temperature is 18°C, and the wet-bulb temperature is 10°C, the difference between them is 8°C. Find the column for 8°C along the horizontal scale at the top of the chart labeled "Difference Between Wet-Bulb and Dry-Bulb Temperatures." Trace down the 8°C column to where it intersects the 18°C row and read the relative humidity: 33%.

One credit is allowed for the answer **33%**.

78 Isotherms connect points with the same temperature. Find the Weather Map Information chart in the *Earth Science Reference Tables,* and refer to the Station Model section. Note that temperature (°F) is listed to the upper left of a station model. The number on the weather map printed to the upper left of the station model for each city is the temperature reading in °F. For the 40°F isotherm, draw a smooth line connecting cities for which 40 is given. For the 50°F isotherm, draw a smooth line connecting cities for which 50 is given. For the 60°F isotherm, draw a smooth line connecting cities for which 60 is given. The three isotherms are shown on the map below.

Two credits are allowed if all three isotherms are drawn correctly. (If more than the three required isotherms are drawn, all isotherms must be correct for 2 credits.)

One credit is allowed if only one or two isotherms are drawn correctly. (If more than the three required isotherms are drawn, and the three **required** isotherms are drawn correctly, but the additional isotherms are incorrect, 1 credit may be allowed.) Isotherms need not be labeled to receive credit.

79 Find the Weather Map Information chart in the *Earth Science Reference Tables,* and refer to the Station Model section. Note that the wind direction staff is drawn pointing to the direction *from* which the wind is blowing, so here it should be drawn pointing to the east. Also, since each whole feather corresponds to 10 knots, two feathers should be drawn on the staff to indicate a wind-speed of 20 knots. The correct symbols are shown on the station model below.

50

A maximum of 2 credits is allowed: 1 credit for correctly indicating wind direction, *and* 1 credit for correctly indicating windspeed. (Feathers may be placed on either side of the staff.)

80 The only other weather variable shown on the map is the shading in the circles for the weather stations. Find the Weather Map Information chart in the *Earth Science Reference Tables*, and refer to the Station Model section. Note that the shading in a circle indicates the amount of cloud cover.

One credit is allowed for a scientifically correct answer.

Examples of acceptable answers:
  sky conditions
  cloud cover

0 150 300 450 km

81 *a.* Find the Equations and Proportions chart in the *Earth Science Reference Tables*. Locate the equation for rate of change.

One credit is allowed for correctly recording the equation.

Examples of acceptable answers:

$$\text{rate of change} = \frac{\text{change in field value}}{\text{change in time}}$$

$$r = \frac{\Delta d}{\Delta t}$$

b. The change in time is given as 15 hours. Use the map scale to determine the distance between Mt. St. Helens and the 15-hour mark on the map: 300 kilometers. Now substitute values into the equation:

$$\text{rate of change} = \frac{300 \text{ km}}{15 \text{ hr}}$$

One credit is allowed for substituting both acceptable measurements into the equation (±25 km is allowed for distance; time must be 15 hr).

c. Solve the equation, and label your answer with correct units.

A maximum of two credits is allowed: 1 credit for correctly calculating the rate based on your answer in *b*, and 1 credit for recording the proper units.

Examples of acceptable answers:
  rate = 20 kilometers/hour
  r = 20 km/hr

82 Find the Selected Properties of Earth's Atmosphere graphs in the *Earth Science Reference Tables*. Note that an altitude of 1.5 kilometers falls within the layer of the atmosphere called the troposphere.

One credit is allowed for the answer **troposphere**.

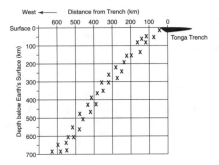

83 The cross section has two scales. The horizontal scale along the top shows the distance from the Tonga Trench. Note that the Tonga Trench is indicated by a black symbol at the upper right-hand corner of the cross section at distance 0. The vertical scale along the left side of the cross section shows depth below the surface. Note that the surface (depth = 0) is at the top of the scale. Each earthquake focus is indicated on the cross section by an **X**, which corresponds to a specific depth below the surface and a specific distance from the Tonga Trench. The example below shows the distances from the trench and depths for two earthquake foci.

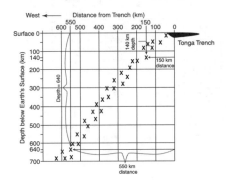

Note that earthquakes that occur farther from the trench have deeper foci than earthquakes that occur near the trench.

647

One credit is allowed for a scientifically correct answer.

Examples of acceptable answers:
As distance increases, depth increases.
The relationship is direct.

84 Find the Tectonic Plates map in the *Earth Science Reference Tables*. Locate the Tonga Trench, marked by a bold black line east (to the right) of Australia. Note that this trench is the boundary between the Australian plate and the Pacific plate.

One credit is allowed for the answer **Australian** and **Pacific**. (Note: The Australian plate is also known as the Indian plate or the Indo-Australian plate.)

85 Find the Tectonic Plates map in the *Earth Science Reference Tables*. Locate the Tonga Trench, marked by a bold black line east of Australia. Note the arrow indicating plate motion pointing at the Tonga Trench. This indicates that the Pacific plate is moving toward, or converging with, the Australian plate. The earthquakes occurring beneath the surface to the west of the trench that are shown in the cross section indicate that the Pacific plate is sliding under the Australian plate.

One credit is allowed for a scientifically correct answer.

Examples of acceptable answers:
The plates are converging.
One plate is sliding under the other.

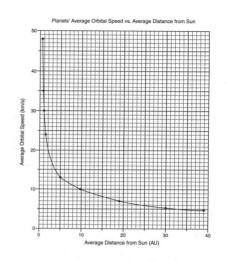

Planets' Average Orbital Speed vs. Average Distance from Sun

A maximum of two credits is allowed: 1 credit if six or more of the points are plotted correctly (±0.5 unit), and 1 credit for connecting all plotted points.

87 As can be seen from Table 1 and the graph plotted for question 86, as distance from the Sun increases, average orbital speed decreases.

One credit is allowed for a scientifically correct answer.

Examples of acceptable answers:
Planets with higher orbital speeds are located closer to the Sun.
Distance from the Sun and orbital speed are inversely related.

88 The average distances from the Sun of the asteroids listed in Table 2 range between 2.4 and 3.2 AU. According to Table 1, these distances would place the asteroids between Mars at 1.5 AU and Jupiter at 5.2 AU.

One credit is allowed for the answer **Mars** and **Jupiter**.

Table 1		
Planet	**Average Distance from Sun (AU)**	**Average Orbital Speed (km/s)**
Mercury	0.4	48.0
Venus	0.7	35.0
Earth	1.0	30.0
Mars	1.5	24.0
Jupiter	5.2	13.0
Saturn	9.6	10.0
Uranus	19.0	7.0
Neptune	30.0	5.1
Pluto	39.0	4.7

Table 2	
Asteroid	**Average Distance from Sun (AU)**
Ceres	2.8
Pallas	2.8
Vesta	2.4
Hygiea	3.2
Juno	2.7

86 Plot the nine points using the coordinates in Table 1, and connect the points in a smooth line as shown below:

Index

Divergent boundaries, 106
Doppler effect, 479–80
Drift, 180, 250
Drumlins, 191, 246, 251
Dry adiabatic lapse rate, 356–57
Ductile deformation, 104–5
Dunes, 188–89

E
Earth, 1, 126–27, 472
  circumference of, 6–7
  crust of, 94–107, 125, 134
  history of, 267–83
  mantle and core of, 125–26
  mapping surface features of, 11–20
  motions of, 427–28
  place in universe, 443–44
  shape of, 2–6
  spheres of, 7–11
Earthflow, 171, 183–84
Earthquakes, 102
  causes of, 95
  effects of, 96–97
  measurement of, 96
  seismogram analysis of, 120–24
  wave behavior in, 95–96, 115–20
  zones of activity, 101
Eccentricity, 435
Eclipses, 439–42
Ecliptic, 428
Elastic deformation, 104
Electromagnetic field, 350
Electromagnetic spectrum, 352–53
Electromagnetic waves, 350–53
Elements, 36–37
End moraines, 249–50
Energy
  from electromagnetic waves, 350–53
  from phase changes of moisture and
    adiabatic changes in atmosphere,
    353–59
  solar, 383–89
Eons, 274
Epicycles, 462
Epochs, 275
Equator
  and climate, 389–91

diameter of, 2
and global wind belts, 314–15
Eras, 275
Eratosthenes, 7
Erosion. *See also* Deposition;
  Weathering
  agents of, 168–69
  evidence of, 168
  by glaciers, 179–80, 244–47
  by mass wasting, 169–71
  by moving water, 171–75
  by ocean, 175–76, 230–31
  by wind, 176–79
Erratics, 191, 250
Eskers, 251–52
Evaporation, 378
Evaporites, 74
Evapotranspiration, 395
Evolution, 281
Exfoliation, 158–59
Extrusive rocks, 69

F
Fall equinox, 431
Faults, 95, 105
Felsic group, 71
Fetch, 228
Field, 14, 319
Field maps, 14–15, 319
Firn, 239
Fish, 212
Floodplains, 175, 185
Focus, 435
Fog, 317
Folds, 105
Foliation, 50
Forecasting methods, 326–28, 361
  long-range, 363–64
  numerical, 363
  statistical, 362–63
  synoptic, 362
Fossil fuels, 79
Fossils, 48, 103, 273, 282–83
Foucault pendulum, 465–66
Fracture, 40, 105
Frequency, 117
Friction cracks, 245

NOTES